Advances in Agricultural Technologies

Advances in Agricultural Technologies

Edited by Caroline Walters

MURPHY & MOORE
www.murphy-moorepublishing.com

Published by Murphy & Moore Publishing,
1 Rockefeller Plaza,
New York City, NY 10020, USA
www.murphy-moorepublishing.com

Advances in Agricultural Technologies
Edited by Caroline Walters

International Standard Book Number: 978-1-63987-014-1 (Hardback)

Cataloging-in-Publication Data

Advances in agricultural technologies / edited by Caroline Walters.
p. cm.
Includes bibliographical references and index.
ISBN 978-1-63987-014-1
1. Agricultural innovations. 2. Agricultural industries--Appropriate technology.
3. Agriculture--Technology transfer. 4. Agriculture. I. Walters, Caroline.
S494.5.I5 A28 2022
338.1--dc23

TABLE OF CONTENTS

PREFACE

The main aim of this book is to educate learners and enhance their research focus by presenting diverse topics covering this vast field. This is an advanced book which compiles significant studies by distinguished experts in the area of analysis. This book addresses successive solutions to the challenges arising in the area of application, along with it; the book provides scope for future developments.

The field of agrotechnology or agricultural technology is concerned with the use and application of technology in agriculture, horticulture and aquaculture. The current progress in agricultural yield and efficiency is being driven directly through technological advancement. Whether through the development and application of synthetic fertilizers and pesticides or through machinery such as tractors and agricultural aircraft, agriculture has seen considerable innovation over the centuries. Some of the more recent advancements include genetically modified crops, soilless farming, and improved drip irrigation, among others. Today, satellite photography and sensors, agricultural drones, soil moisture sensors, light and heat control, etc. are increasingly advancing the practices of agriculture. This book traces the progress of agricultural technologies and highlights some of their key applications. It will also provide interesting topics for research which interested readers can take up. Scientists and students actively engaged in this field will find this book full of crucial and unexplored concepts.

It was a great honour to edit this book, though there were challenges, as it involved a lot of communication and networking between me and the editorial team. However, the end result was this all-inclusive book covering diverse themes in the field.

Finally, it is important to acknowledge the efforts of the contributors for their excellent chapters, through which a wide variety of issues have been addressed. I would also like to thank my colleagues for their valuable feedback during the making of this book.

Editor

1

The role of biostimulants and bioeffectors as alleviators of abiotic stress in crop plants

Michael James Van Oosten, Olimpia Pepe, Stefania De Pascale, Silvia Silletti and Albino Maggio*

Abstract

The use of bioeffectors, formally known as plant biostimulants, has become common practice in agriculture and provides a number of benefits in stimulating growth and protecting against stress. A biostimulant is loosely defined as an organic material and/or microorganism that is applied to enhance nutrient uptake, stimulate growth, enhance stress tolerance or crop quality. This review is intended to provide a broad overview of known effects of biostimulants and their ability to improve tolerance to abiotic stresses. Inoculation or application of extracts from algae or other plants have beneficial effects on growth and stress adaptation. Algal extracts, protein hydrolysates, humic and fulvic acids, and other compounded mixtures have properties beyond basic nutrition, often enhancing growth and stress tolerance. Non-pathogenic bacteria capable of colonizing roots and the rhizosphere also have a number of positive effects. These effects include higher yield, enhanced nutrient uptake and utilization, increased photosynthetic activity, and resistance to both biotic and abiotic stresses. While most biostimulants have numerous and diverse effects on plant growth, this review focuses on the bioprotective effects against abiotic stress. Agricultural biostimulants may contribute to make agriculture more sustainable and resilient and offer an alternative to synthetic protectants which have increasingly falling out of favour with consumers. An extensive review of the literature shows a clear role for a diverse number of biostimulants that have protective effects against abiotic stress but also reveals the urgent need to address the underlying mechanisms responsible for these effects.

Keywords: Abiotic stress, Biostimulants, Bioeffectors, Microbial inoculants, Humic acid, Fulvic acid, Protein hydrolysates, Amino acids, Seaweed extracts, Bioprotection

Introduction

Plant biostimulants, sometimes referred to as agricultural biostimulants, are a diverse classification of substances that can be added to the environment around a plant and have positive effects on plant growth and nutrition, but also on abiotic and biotic stress tolerance. Although most plant biostimulants are added to the rhizosphere to facilitate uptake of nutrients, many of these also have protective effects against environmental stress such as water deficit, soil salinization and exposure to sub-optimal growth temperatures [1]. Biostimulants are not nutrients per se; instead they facilitate the uptake of nutrients or beneficially contribute to growth promotion or stress resistance [2]. A newly emerged paradigm emphasizes that plants are not standalone entities within their environments; instead they are host and partner to microorganisms of bacteria and fungi; plants are a host to numerous microbiota and those associations, both outside and within its tissues, allow them to respond and adapt to abiotic and biotic stress [3]. Reasonably, if we functionally optimize these associations, we may strengthen their role in plant stress protection.

The industry definition of biostimulants was originally proposed in 2012 and stated: "Plant biostimulants contain substance(s) and/or microorganisms whose function when applied to plants or the rhizosphere is to stimulate natural processes to enhance/benefit nutrient uptake, nutrient efficiency, tolerance to abiotic stress,

*Correspondence: almaggio@unina.it
Department of Agricultural Science, University of Naples Federico II, Via Università 100, Portici, Italy

and crop quality. Biostimulants have no direct action against pests, and therefore do not fall within the regulatory framework of pesticides". Biostimulants were loosely defined for a long time and often regarded dubiously because of their aggregate nature and the inherent difficulty to determine which specific components were making beneficial contributions. The definition proposed by du Jardin [1] "A plant biostimulant is any substance or microorganism applied to plants with the aim to enhance nutrition efficiency, abiotic stress tolerance and/or crop quality traits, regardless of its nutrients content" represents the clearest and most concise way to define biostimulants.

Our understanding of biostimulants and their potential effects has been expanding at a considerable rate [4]. The role of biostimulants, specifically in regard to growth promotion and nutrient availability, has been reviewed (du Jardin [1, 4–6]). In addition to numerous general reviews, many categories of specific biostimulants have been extensively reviewed such as protein hydrolysates [7], seaweed extracts [8], silicon [9], chitosan [10], humic and fulvic acids [11], the role of phosphite [12], arbuscular mycorrhizal fungi [13], *trichoderma* [14], plant growth-promoting rhizobacteria [15]. These reviews have focused on plant growth promotion and biotic stress but our intent with this review is to comprehensively address what is known about biostimulants ameliorating the effects of abiotic stress (Table 1). The majority of these studies were conducted as greenhouse or field experiments. The literature has mainly focused on crop species with a large representation of cereal crops such as wheat, barley, and corn. Finally, we attempted to map different categories vs. their physiological function in plants.

Algal extracts

Seaweed extracts (SWE) as biostimulants are emerging as commercial formulations for use as plant growth-promoting factors and a method to improve tolerance to salinity, heat, and drought. Algal extracts target a number of pathways to increase tolerance under stress (Fig. 1). Seaweeds are red, green, and brown macroalgae that represent 10% of marine productivity [8]. Macroalgae have been used as organic fertilizers for thousands of years and are still in use [64]. Currently, there are over 47 companies producing and marketing various algal extracts for agricultural use; the majority of the formulations are from the brown algae, *Ascophyllum nodosum* [65].

While the growth-promoting effects of seaweed extracts have been documented in many species [8, 66], very little is actually known about the mechanisms behind these effects. The variable and complex nature of these substances makes it difficult to determine exactly which components are playing a key role. Commercial formulations of SWEs are often proprietary, and the composition is largely dependent on the method of extraction. Indeed, characterization of the actual composition of most common algal-based commercial products would be useful first step to better hypothesize and/or depict a cause–effect relationship of their mechanism of action. Mechanical disruption, pulverization, acid or alkali extractions are some of the more common methods employed [8]. Most commercial products are derived from red (ex *Lithothamnium calcareum*) and brown (ex *Ascophyllum nodosum, Durvillaea potatorum*) macroalgae [67]. The role of SWEs and cold tolerance is now emerging. Very recent work has focused on SWEs and their ability to enhance tolerance to chilling stress. When multiple extracts were tested for their ability to enhance cold tolerance in maize only extracts rich in Zn and Mn were able to enhance tolerance through enhanced ROS responses. In this case, the protective effects likely stem from supplying plants with micronutrients that play a role as co-factors in anti-oxidative enzymes [63]. These results indicate that nutrient deficiency stress induced by cold can be overcome by supplying SWEs rich in micronutrients to improve oxidative stress tolerance. Previous studies with corn seedlings under root chilling stress supplemented with micronutrients demonstrated the utility of nutrient seed priming [68].

Some work has been done in model systems with the goal of determining the physiological and molecular responses induced by SWEs. In order to better understand the active components of *A. nodosum*, Rayirath et al. [55] separated the organic-sub-fractions of extracts and tested them with *Arabidopsis thaliana* and freezing experiments. Plants grown in vitro with sub-fractions added to the substrate or in "Peat pellet freezing assays" irrigated with sub-fractions were tested for freezing tolerance. The authors found that the ethyl acetate extracted fraction, rich in fatty acids and sterols enhanced freezing tolerance over water treated (controls) at temperatures from −2.5 to −5.5 °C. Treated plants maintained faster rates of recovery, greater membrane integrity, and had 70% less chlorophyll damage upon freezing recovery as well as increased expression of key freezing tolerance genes such as RD29A, COR15A, and CBF3 [55]. Priming of key tolerance genes prior to exposure to stress greatly increases tolerance in many cases. The lipophilic components were found to be rich in fatty acids such as butyric acid, palmitic acid, oleic acid, linoleic acid the sterol fucosterol. These extracts increased proline content and total soluble sugars, contributing to freezing tolerance [56]. *A. nodosum* extracts have even been used to reduce cold stress sensitivity in *Kappaphycus alvarezii. Kappaphycus alvarezii* is a red algae and the most important source of carrageenans; which are hydrophilic colloids

Table 1 Summary of species, biostimulant, and stress effect

Type of BE	Crop	Stress and effect	Reference
A. brasilense	*T. aestivum*	Drought tolerance	[16, 17]
A. brasilense	*C. arietinum*	Salt tolerance	[18]
A. brasilense	*V. faba*	Salt tolerance	[18]
A. brasilense	*L. sativa*	Salt tolerance	[19, 20]
A. brasilense	*T. aestivum*	Salt and osmotic stress	[21]
A. brasilense	*L. lycopersicum*	Drought tolerance	[22]
A. brasilense/P. dispersa	*C. annuum*	Salt tolerance	[23]
A. chrococcum	*Z. mays*	Salt tolerance	[24]
A. chrococcum	*T. aestivum*	Salt tolerance	[25]
A. chrococcum	*T. aestivum*	Temperature tolerance	[26, 27]
A. lipoferum	*T. aestivum*	Salt tolerance	[28]
A. nodosum	*Kappaphycus alvarezii*	Cold tolerance	[29]
A. nodosum	*P. dulcis*	Ion homeostasis	[30]
A. nodosum	*C. sinensis*	Drought tolerance	[31]
B. phytofirman,	*Vitis vinifera*	Cold tolerance	[32, 33]
F. glaciei	*Solanum lycopersicum*	Cold tolerance	[34]
Fulvic and humic acids	*F. arundinacea*	Drought tolerance	[35, 36]
Fulvic and humic acids	*A. palustris*	Drought tolerance	[37]
Glycinebetaine	*L. lycopersicum*	Chilling stress	[38]
H. diazotrophicus	*H. vulgare*	Salt tolerance	[39]
Humic acid and phosporous	*C. annuum*	Salt tolerance and ion homeostasis	[40]
Humic acids	*O. sativa*	Oxidative and drought stress	[41]
Humic acids	*P. vulgaris*	Salt tolerance	[42]
Megafol	*L. lycopersicum*	Drought tolerance	[43]
Melatonin	*Z. mays*	Chilling tolerance	[44]
P. frederiksbergensis	*Solanum lycopersicum*	Cold tolerance	[34]
P. putida	*T. aestivum*	Heat tolerance	[45]
P. putida	*S. bicolor*	Heat tolerance	[46]
P. vancouverensis	*Solanum lycopersicum*	Cold tolerance	[34]
P.dispersa	*T. aestivum*	Cold tolerance	[47]
Protein hydrolysates	*H. vulgare*	Ion homeostasis	[48]
Protein hydrolysates	*Z. mays*	Salt tolerance	[49]
Protein hydrolysates	*T. aestivum*	Heavy metal tolerance	[50]
Protein hydrolysates	*L. sativa*	Salt tolerance, cold tolerance	[51, 52]
Protein hydrolysates	*D. kaki/D. lotus*	Salt tolerance	[53]
Protein hydrolysates	*Lolium perenne*	Heat tolerance	[51]
R. leguminosarum	*V. faba*	Salt tolerance	[54]
R. leguminosarum	*P. sativum*	Salt tolerance	[54]
SWE	*A. thaliana*	Cold tolerance	[55, 56]
SWE	*P.* pratensis	Salt tolerance	[57]
SWE	*A. stolonifera*	Heat tolerance	[58]
SWE	*S. oleracea*	Drought tolerance	[59]
SWE	*L. sativa*	Ion homeostasis	[60]
SWE	*V. vinifera*	Drought tolerance and ion homeostasis	[61]
SWE	*S. nipponica*	Drought tolerance	[62]
SWE	*P. eugenioides*	Drought tolerance	[62]
SWE	*Z. mays*	Cold tolerance	[63]

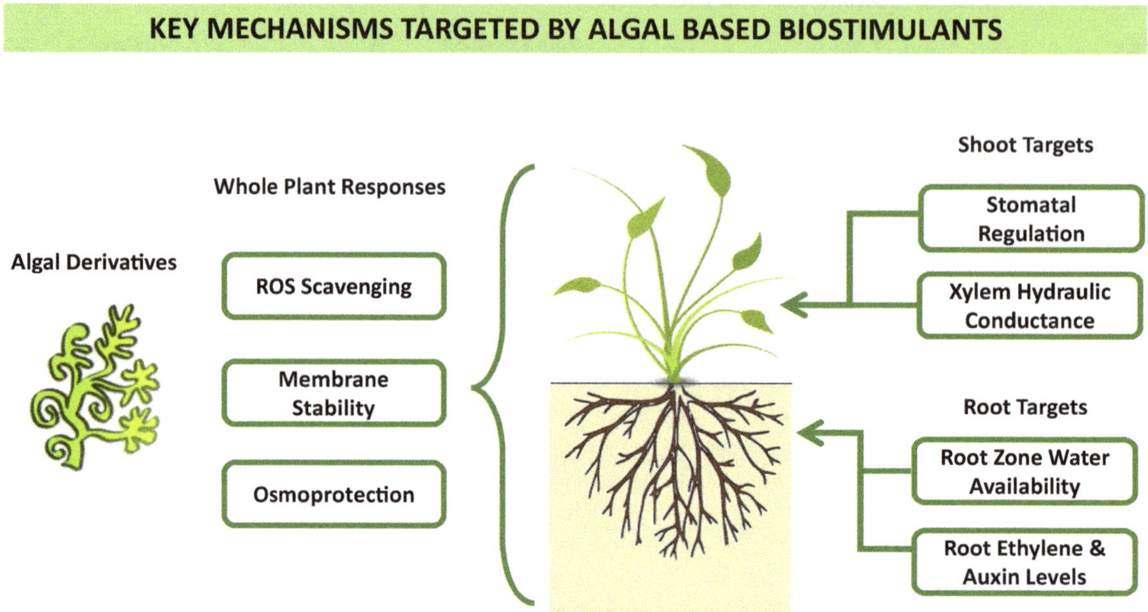

Fig. 1 Summary of main key mechanisms targeted by algal-based biostimulants

largely used in foods and dairy products [29, 69]. Algal extracts have also been used on Kentucky bluegrass (*Poa pratensis* L. cv. Plush) to alleviate salinity stress from saline watering in turfgrass experiments [57]. Similarly SWE-based cytokinins have been used on creeping bentgrass (*Agrostis stolonifera* L.) to improve tolerance to heat stress [58]. SWEs from *A. nodosum* have also been used for ornamental plants, such as *Spiraea nipponica* "Snowmound" and *Pittosporum eugenioides* "Variegatum", to enhance drought tolerance. Treated plants showed higher phenolic, proline, and flavonoid content while demonstrating improved physiology under mild drought stress conditions [62].

In horticultural crops and trees, SWE have been largely used for similar purposes. *A. nodosum* SWE increased RWC, Fresh Weight, and Dry Weight in spinach (*Spinacia oleracea* L.) plants under drought stress with some adverse effects on the nutritional value through reduced ferrous ion chelating ability [59]. SWE applied to seedlings of lettuce (*Lactuca sativa* L.) enhanced cotyledon growth similar to fertilization with potassium [60].

Foliar application of marine bioactive substances (isopropanol extracts from microalgae) to grape plants (*Vitis vinifera* L) increased leaf water potential and stomatal conductance under drought stress [61]. Consistent with an improved stomatal response, it was also observed that K^+ and Ca^{2+} fluxes at the stomatal level were higher in treated plants. Commercial formulations of *A. nodosum* have been tested on almond plants (*Prunus dulcis* [Mill.] D. A. Webb), which demonstrated increased growth and accumulation of K^+. In conditions with ample K^+ both MegaFol and GroZyme (Valagro, Atessa, Chieti, ITALY) increased leaf area and number of leaves greater than controls treated with water or K^+. In K^+-deficiency conditions only MegaFol and a foliar application of K^+ was able to stimulate growth, although at lower levels than observed with adequate K^+ nutrition [30]. Accumulation of K^+ is an essential step in protecting against both ionic and osmotic stress and may contribute to tolerance. Orange trees, *Citrus sinensis* L., subjected to drought stress and treated with commercial extracts of *A. nodosum* had better water relations and increased water use efficiency (WUE) under irrigation at 50% restitution of evapotranspired water [31]. The promise of biostimulants to increase drought tolerance and WUE holds great potential for drought prone regions where horticultural crops and fruit trees are agronomically important but water availability is becoming less reliable due to urbanization and climate change.

As earlier noted, almost all of the above-mentioned experiments with SWE use commercial formulations. This may be of some concern, due to the variable nature of these products and formulation methods. A recent transcriptomic study using *A. thaliana* plants treated with two different commercial *A. nodosum* extracts showed that not all extracts are alike. One commercial product resulted in dysregulation of 4.47% of the transcriptome while the other extract only affected 0.87% [70]. Since transcriptional priming is likely a key component in enhancing abiotic stress tolerance using SWEs, these differences imply significant variability in responses elicited. Compositions of the extracts differed greatly, indicating that choice of commercial product may have a significant effect on plant responses. Commercial formulations are often proprietary and the exact composition and extraction methods, shifting the burden to the research community to analyse and isolate the active components in these products. In order to identify and characterize how these SWEs affect plants, some form of standardization is necessary.

Carbohydrates, proteins, amino acids, and lipids

Protein hydrolysates are mixtures of polypeptides, oligopeptides, and free amino acids derived from partial hydrolysis of agricultural by-products from animals and plants [7]. Carbohydrates, proteins, amino acids, and lipids may increase stress tolerance through different (Fig. 2). The effects of amino acids on ion fluxes across membranes have been clearly established, with most having a positive effect on reducing NaCl-induced potassium efflux [48]. Protein hydrolysates (PH) are often sold as formulations that include plant growth regulators. The bulk of PH products, over 90%, are produced from chemical hydrolysis of animal by-products while enzymatically processed plant-based products are a recent development [7].

Megafol (Valagro, Atessa, Chieti, ITALY) is a commercial biostimulant comprising vitamins, amino acids, proteins, and betaines from plant and algal extracts. Application of Megafol to tomato plants under drought stress enhanced induction of a number of drought responsive genes such as tomato orthologs of *RAB18* and *RD29B*. Treated plants also had higher fresh weight and relative water content under drought stress, indicating a protective effect on water status and stress responsive genes [43, 71]. When hydrolysate-based biostimulants from alfalfa (*Medicago sativa* L.), containing triacontanol (TRIA) and indole-3-acetic acid (IAA), were applied to maize plants under salt stress, the protective effects were amplified. Treated plants had higher flavonoid, proline, and potassium content in salt stress conditions over untreated controls [49]. Extracts that are rich in amino acids may play a role in increasing cold tolerance. When lettuce plants (*Lactuca sativa*) were treated with an amino acid mixture, derived from enzymatic hydrolysis of proteins, (Terra-Sorb) and subjected to cold, treated plants had higher fresh weights and improved stomatal conductance [51]. Use of animal derived amino acid hydrolysates on strawberry plants after transplantation and cold stress did not improve survival though some growth promotions were observed in the absence of stress [72]. Perennial Ryegrass (*Lolium perenne* L.) treated with hydrolyzed amino acids and high temperatures (36 °C) had improved photosynthetic efficiency over control plants [51].

Mutants of *A. thaliana* deficient in production of proline have stress sensitive phenotypes [73]. These plants can have their phenotype rescued with exogenous application of L-proline, a common amino acid available in biostimulant formulations of various amino acids and hydrolysate mixtures [74]. Hydrolysates from wheat germs show strong anti-oxidant and free radical scavenging properties as well as the ability to chelate some metals [50].

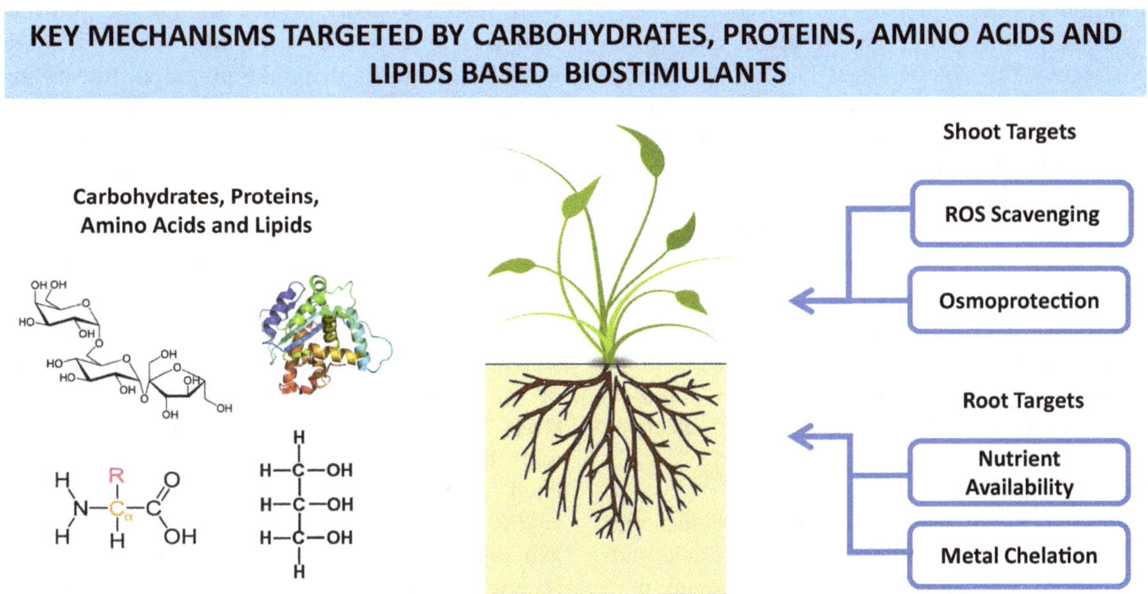

Fig. 2 Summary of main key mechanisms targeted by carbohydrate-, protein-, amino acid-, and lipid-based biostimulants

Lettuce (*Lactuca sativa* L.) is particularly salt sensitive and the addition of plant-derived protein hydrolysates improved fresh yield, dry biomass, and root dry weight as well as increased concentrations of osmoyltes, glucosinolates and the composition of sterols and terpenes [52]. Hydrolysates have applications for trees, which require considerable investment costs and can be vulnerable to drought. Japanese persimmon trees, *Diospyros kaki* L. cv. "Rojo Brillante" grafted on *Diospyros lotus* L., are particularly sensitive to drought stress [53]. Treatment of these trees with calcium protein hydrolysates decreased chloride uptake under saline irrigation, lowered water potentials as well as increased the concentration of compatible solutes [53], all of which would enhance plant growth under saline stress.

Recent reports indicate that melatonin, derived from L-tryptophan via the shikimate pathway, can prime seeds to tolerate adverse environmental conditions at imbibition and germination stages [75]. Corn seeds pre-treated with melatonin show increased tolerance to chilling stress upon germination, indicating a priming effect by melatonin [44]. Melatonin may prove to be an effective biostimulant for improving stress tolerance of seedlings.

Glycinebetaine is a compatible solute accumulated in many plants in response to salt stress [76]. Exogenous application of glycinebetaine has increased tolerance for environmental stresses such as drought, chilling, freezing, salinity, and oxidative stress. Foliar application of glycinebetaine results in rapid uptake by leaves and concentration in meristematic tissues. Rapid uptake and localization of glycinebetaine in these most vulnerable tissues are particularly beneficial in chilling and freezing stress where glycinebetaine can exert a protective effect [77]. Transgenic plants of various species expressing two biosynthetic genes, *codA* and *betA*, produce more glycinebetaine and had an increased tolerance to abiotic stress [38, 78]. Exogenous application of small amounts of compatible solutes such as proline and betaine to barley roots resulted in an immediate reduction of NaCl-induced efflux of K^+, indicating that ion fluxes across the membrane can be affected by relatively low concentrations of compatible solutes [79]. The cause–effect relationship between accumulation of compatible solutes and stress protection still remains to be fully understood [80]. However, a better understanding of the specific mechanisms of action of these molecules is becoming increasingly important if we want to make predictions on which combination of biostimulants can be more effective.

Humic and fulvic acids

Humic and fulvic substances are the major organic components of lignites, soil, and peat. Humic and fulvic acids are produced by the biodegradation of organic matter resulting in a mixture of acids containing phenolate and carboxyl groups. Fulvic acids are humic acids with a higher oxygen content and lower molecular weight [81]. A number of examples exist indicating the potential for these substances to improve abiotic stress tolerance in plants (Fig. 3). Pre-treatment of tall fescue (*Festuca arundinacea* Schreb.) and creeping bentgrass (*Agrostis palustris* Huds. A.) with seaweed extract and humic acid increased leaf hydration under dry soil conditions as well as root growth, shoot growth, and antioxidant capacity [35, 58]. Further studies with bentgrass showed these extracts, high in cytokinins, combined with humic acid increased drought tolerance as well as endogenous cytokinin content [37].

Treatment of bell pepper (*Capsicum annuum* L. cv. Demre) with humic acid and phosphorous resulted in plants with reduced Na content and elevated N, P, K, Ca, Fe, Mg, S, Mn, and Cu ion contents in roots and shoots, which were associated with a general protective effect under mild salinity stress [40]. Application of humic acids to common bean (*Phaseolus vulgaris* L.) under high salinity (120 mM NaCl) increased endogenous proline levels and reduced membrane leakage [42], which are both indicators of better adaptation to saline envirnoments.

Humic acid extracts seem to be beneficial also for field crop monocots. Extracts from vermicompost applied to rice (*Oryza sativa* L.) played a role in activating anti-oxidative enzymatic function and increased ROS scavenging enzymes. These enzymes are required to inactivate toxic-free oxygen radicals produced in plants under drought and saline stress [41]. One possible mode of action for vermicompost may be the differential regulation of proton ATPases located in the vacuolar and plasma membranes. When Micro-Tom tomato plants were treated with vermicompost, plasma membrane proton extrusion was increased by over 40% which facilitated acid growth and nutrient uptake potential. Interestingly, the auxin insensitive mutant *diageotropica* (dgt) showed no increase in proton extrusion, indicating that humic substance may increase root growth through mediating auxin signalling [82].

Microorganisms affecting stress tolerance

While plants are known to establish symbiotic relationships with bacteria, our understanding of those relationships under abiotic stress is rudimentary. However, some of the targets of microorganisms that increase abiotic stress tolerance have been identified (Fig. 4). Bacteria with the potential to act as biostimulants have been isolated from a number of ecosystems with saline, alkaline, acidic, and arid soils. These bacteria belong to several genera such as *Rhizobium*, *Bradyrhizobium*, *Azotobacter*, *Azospirillum*, *Pseudomonas*, and *Bacillus*. Members of these genera have developed strategies to adapt and

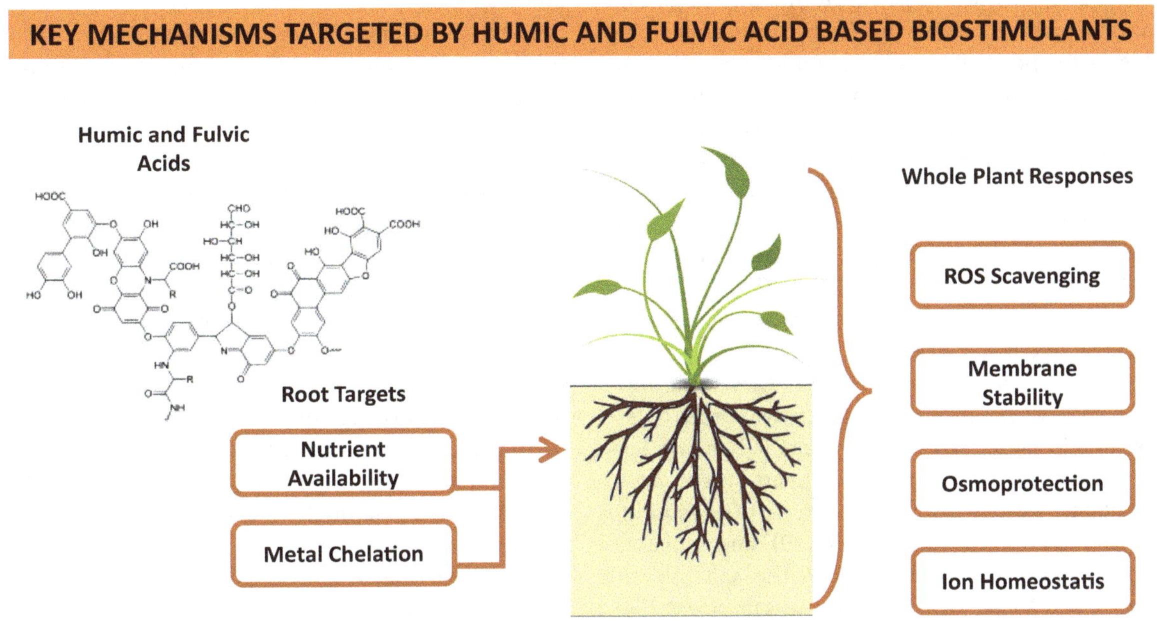

Fig. 3 Summary of main key mechanisms targeted by humic- and fulvic acid-based biostimulants

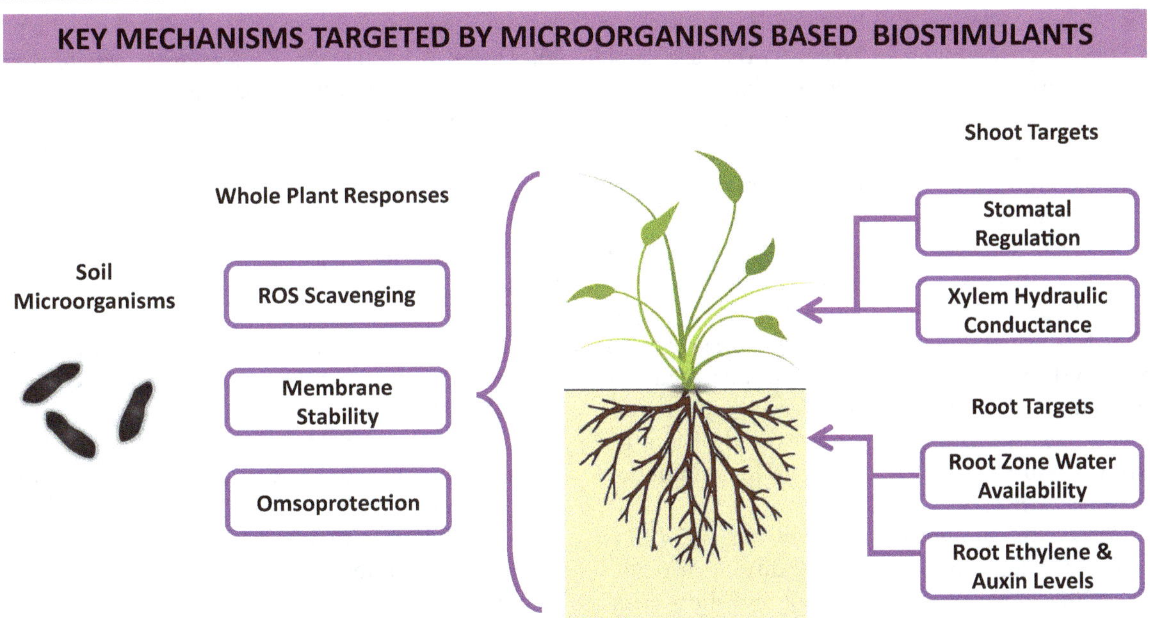

Fig. 4 Summary of main key mechanisms targeted by microorganism-based biostimulants

thrive under adverse conditions [83, 84]. Amongst these adaptations, alterations to the composition of the cell wall and the ability to accumulate high concentrations of soluble solutes are common. These allow for enhanced water retention and increased tolerance to osmotic and ionic stress. Cell wall composition is altered through enrichment for exopolysaccharides (EPS) and lipopolysaccharide–proteins and polysaccharide–lipids which my form a protective biofilm on the root surface [85, 86]. Plant growth-promoting rhizobacteria (PGPR) inoculated soils can ameliorate plant abiotic stress responses. A number of recent reviews have extensively covered the protective effects of *Rhizobium* against abiotic stress in plants [87]. Most documented growth enhancement determined by these bacteria is associated with high level of IAA, which has been proven to alleviate salt stress [88] and EPS production that may help in maintaining a film of hydration around the roots and/or help re-establishing favourable water potential gradients under water limitations. These functions have been proven useful under saline stress [89], extremes of temperature, pH, salinity, and drought [87, 90]. Inoculation of maize with *Azotobacter* strains has been shown to have general positive effects under saline stress by facilitating uptake of K^+ and exclusion of Na^+ as well as increasing phosphorous and nitrogen availability [24]. In wheat, inoculation of salt tolerance *Azobacter* strains increased biomass, nitrogen content, and grain yield under salt stress [25].

Tolerance to salt stress varies within these microorganisms and their tolerance can confer advantages to the host relationship under stress conditions. When two legumes, pea (*Pisum sativum*) and fava bean (*Vicia faba*), were inoculated with two different strains of *Rhizobium leguminosarum*, a salt-tolerant (GRA19) and salt-sensitive (GRL19) plants inoculated with the salt-tolerant strain performed better under moderate salt stress [54]. The authors further found that pea plants had larger nodules and high levels of nitrogen fixation under salt stress when inoculated with GRA19, the salt-tolerant strain of *R. leguminosarum*. Similar results have been observed for non-symbiotic free-living soil bacteria that are capable of fixing nitrogen. *Azospirillum brasilense* is closely associated with the plant rhizosphere and can colonize the surface of roots. When chickpea (*Cicer arietinum* L.) and faba bean were inoculated with *A. brasilense*, they experienced enhanced nodulation by native rhizobia and greater tolerance to salt stress [18]. Another free-living nitrogen-fixing species, *Azotobacter chrococcum* A2 demonstrated salt tolerance. Inoculation with *A. chrococcum* has been shown to increase yields of pea, potato, rice, wheat, and cotton in saline-arid soils. Increased root length and shoot growth was also observed with inoculation [26, 27] with significant positive yield effects for wheat (from 2.8 to 3.5 t ha^{-1} when grown in conjunction with *A. chrococcum*) [26, 27].

In barley, *Hartmannibacter diazotrophicus* E19 (T) is capable of colonizing roots in saline conditions. Inoculation of roots in saline soil increased root and shoot mass significantly, 308 and 189%, respectively. Inoculated roots also had increased relative water content over three and a half times that of control plants [39]. High concentrations of salt can also be inhibitory to rhizobial bacteria. While certain strains of *R. leguminosarum*, such as *viciae* SAAN1, are very salt tolerant and able to withstand up to 0.34 M NaCl, they often show lower rates of nodulation in saline soils. These strains are often less competitive with natural rhizobial populations, however.

The stress protection of bacterial biostimulants to rainfed field crops can be of particular relevance under increasing temperatures foreseen by most climate change prediction models. Wheat inoculated with the thermo tolerant *Pseudomonas putida* strain AKMP7 significantly increased heat tolerance. Inoculated plants had increased biomass, shoot and root length, and seed size. ROS generation under stress treatment was also lessened, with lower levels of expression observed for ROS response genes such as superoxide dismutase, ascorbate peroxidase, and catalase [45]. Similar results have been found with sorghum and other *Pseudomonas putida* AKMP strains [46]. Psychrophilic (cold-adapted) microorganisms are capable of surviving in extreme conditions and their interactions with plants provide potential mechanisms for improving tolerance [91]. While many strains of soil bacteria with growth-promoting properties have been isolated from low-temperature conditions, few have been tested in conjunction with plants subjected to cold stress.

Wheat inoculated with the cold-tolerant plant growth-promoting bacteria *Pantoea dispersa* showed improved growth and nutrient uptake, likely due to the solublization of phosphorous and production of IAA [47].

Inoculation of soil with psychrotolerant (cold tolerant) bacteria can play a role in chilling tolerance. The psychrotolerant soil bacterium, *Burkholderia phytofirman*, is a plant-growth-promoting rhizobacterium (PGPR) that is capable of colonizing multiple plant species. *B. phytofirman* was shown to play a role in enhancing chilling tolerance in *Vitis vinifera* L. by increasing ROS scavenging metabolites and stress-induced genes. Inoculated plants also recovered faster from chilling stress, returning to normal metabolic levels more quickly than controls [33]. *B. phytofirman* inoculation also alters carbohydrate metabolism and accumulation while having a protective effect on net photosynthesis during cold acclimation and stress [32, 92].

Tomato plants (*Solanum lycopersicum* cv Mill) were inoculated with cold-tolerant strains of *Pseudomonas vancouverensis*, and *frederiksbergensis* as well as *Flavobacterium glaciei* that were isolated from agricultural fields during winter. Treated tomato seedlings were subjected to a week of chilling stress at 15C and inoculation three of these strains showed significantly reduced electrolyte leakage and ROS activity [34]. Improved stress tolerance and growth-promoting effects of microorganism treatments have been seen in other species also. Inoculation of lettuce (*Lactuca sativa* L., cv Mantecosa) seeds with *A. brasilense* increased germination in the presence of salt and demonstrated tolerance through higher total fresh and dry weights of plants at harvest [19]. Additional experiments studying these effects have shown increased biomass, chlorophyll, ascorbic acid content, antioxidant content, and post-harvest shelf life after being subjected to salt stress [20]. Sweet pepper (*Capsicum annuum* L.) inoculated with *A. brasilense* and *Pantoea dispersa* was not affected by moderate levels of salinization, up to 80 mM NaCl, while uninoculated control plants demonstrated lower DW starting at 40 mM NaCl [23].

Triticum aestivum cv. Buck Ombú inoculated with *A. brasilense* sp. 245 and subjected to salt stress (320 mM NaCl) and osmotic stress (20 and 30% PEG 6000) had higher FW, DW, and RWC than non-inoculated controls [21]. Analysis of phospholipids and fatty acid composition in inoculated wheat indicated that the distribution profiles of major root phospholipids are altered in inoculated plants, possibly contributing to the increased tolerance [16]. Wheat inoculated with *Azospirillum lipoferum* and irrigated with 80 mM NaCl had significantly higher leaf and root dry weight than uninoculated controls [28].

While the mechanisms by which *A. brasilense* confer tolerance to osmotic stress are not clear, some evidence indicates that inoculation induces wider xylem vessels and greater hydraulic conductance [17]. In inoculated tomato plants subjected to water stress similar changes have been observed, such as larger xylem vessel area, higher stem-specific hydraulic conductivity, thicker stems [22]. Pepper plants co-inoculated with *A. brasilense* and *Pantoea dispersa* accumulated more dry matter under salt stress. Inoculated plants showed higher stomatal conductance and rates of photosynthesis under salt stress. The chlorophyll concentration and efficiency of photosystem II were not affected in inoculated plants under stress conditions [23].

Inhibition of root growth under salt stress conditions is well documented. One of the primary causes of this inhibition is the production and perception of ethylene in the roots [93]. Plants and PGPR both have ACC-deaminases, which possess the ability to lower the concentration of ethylene in the roots and root zone. PGPR-derived ACC-deaminases can reduce ethylene induced inhibition by reducing root zone ethylene [94] and contribute to maintain relatively higher root-to-shoot ration, a trait that would result beneficial under water shortage.

Conclusions

Biostimulant treatments of agricultural crops have the potential to improve plant resilience to environmental perturbations. In order to fine-tune application rates, biostimulant-plant specificities and techniques is identified that may yield highest impact on stress protection; high priority should be given to better understanding of the causal/functional mechanism of biostimulants. Only once a good understanding of these mechanisms has been reached; we will be able to move to the next generation of biostimulants where synergies and com plementary mechanisms can be functionally designed. Comprehension of the specific mechanisms that should be potentiated to overcame a specific stress can be based today on sound/reasonable hypotheses and be more fruitful than the try-and-see approach. A comprehensive and systematic approach has been proposed to discover and characterize novel biostimulants and understand the mode of action for those both known and new using a combined approach utilizing biology, chemistry, and 'omics [95]. Meta-analysis of the effects of biostimulants has been proposed and an extensive meta-level examination of humic substances on plant growth has been conducted. The analysis found that humic substances increased the dry weight of shoots and roots by at least 20% [96]. However, it should be noted that the diverse conditions, compositions, and species tested do not lend to robust meta-level analysis when an excessive number of variables is present. Identification of synergistic/complementary properties of biostimulants can be pivotal to develop specific formulations targeted to enhance plant response to abiotic stress. For example, biostimulants for improving plant resilience in water limiting environments should stimulate root vs. shoot growth which would allow plants to explore deeper soil layer during the drought season and stimulate the synthesis of compatible solutes to re-establish favourable water potential gradients and water uptake at diminishing soil water. Similar positive effects can be given by those microbial biostimulants that create absorption surfaces around the root systems and sequester soil water in favour of the plants.

Authors' contributions

MJVA and AM wrote the first draft of the manuscript and took care of the revisions of the review; MJVA contributed to the section "Algal Extracts" and "Carbohydrates, proteins, amino acids and lipids"; OP contributed to the section "Microorganisms affecting stress tolerance"; SDP contributed to the section "Carbohydrates, proteins, amino acids and lipids"; SS contributed to the section "Humic and Fulvic Acids" and "Microorganisms affecting stress tolerance". All authors read and approved the final manuscript.

Acknowledgements

Not applicable.

Competing interests

All authors declare that they have no competing interests.

Funding

This work was supported by the EU Project BIOFECTOR Plant Growth-Promoting Bio-effectors (#FP7-KBBE-2012-6 Grant Agreement 312117) and MACSUR Modelling European Agriculture with Climate Change for Food Security—a knowledge hub within FACCE-JPI.

References

1. du Jardin P. Plant biostimulants: definition, concept, main categories and regulation. Sci Hortic. 2015;30(196):3–14.
2. Brown P, Saa S. Biostimulants in agriculture. Front Plant Sci [Internet]. 2015;6. http://www.ncbi.nlm.nih.gov/pmc/articles/PMC4550782/. Accessed 13 Oct 2015.
3. Vandenkoornhuyse P, Quaiser A, Duhamel M, Le Van A, Dufresne A. The importance of the microbiome of the plant holobiont. New Phytol. 2015;206(4):1196–206.
4. Colla G, Rouphael Y. Biostimulants in horticulture. Sci Hortic. 2015;30(196):1–2.
5. Du Jardin, P. The science of plant biostimulants—a bibliographic analysis. Ad hoc Study Report to the European Commission DG ENTR; 2012.
6. Calvo P, Nelson L, Kloepper JW. Agricultural uses of plant biostimulants. Plant Soil. 2014;383(1–2):3–41.
7. Colla G, Nardi S, Cardarelli M, Ertani A, Lucini L, Canaguier R, et al. Protein hydrolysates as biostimulants in horticulture. Sci Hortic. 2015;30(196):28–38.
8. Battacharyya D, Babgohari MZ, Rathor P, Prithiviraj B. Seaweed extracts as biostimulants in horticulture. Sci Hortic. 2015;30(196):39–48.
9. Savvas D, Ntatsi G. Biostimulant activity of silicon in horticulture. Sci Hortic. 2015;30(196):66–81.
10. Pichyangkura R, Chadchawan S. Biostimulant activity of chitosan in horticulture. Sci Hortic. 2015;196:49–65.
11. Canellas LP, Olivares FL, Aguiar NO, Jones DL, Nebbioso A, Mazzei P, et al. Humic and fulvic acids as biostimulants in horticulture. Sci Hortic. 2015;30(196):15–27.
12. Gómez-Merino FC, Trejo-Téllez LI. Biostimulant activity of phosphite in horticulture. Sci Hortic. 2015;196:82–90.
13. Rouphael Y, Franken P, Schneider C, Schwarz D, Giovannetti M, Agnolucci M, et al. Arbuscular mycorrhizal fungi act as biostimulants in horticultural crops. Sci Hortic. 2015;30(196):91–108.
14. López-Bucio J, Pelagio-Flores R, Herrera-Estrella A. Trichoderma as biostimulant: exploiting the multilevel properties of a plant beneficial fungus. Sci Hortic. 2015;30(196):109–23.
15. Ruzzi M, Aroca R. Plant growth-promoting rhizobacteria act as biostimulants in horticulture. Sci Hortic. 2015;30(196):124–34.
16. Pereyra MA, Zalazar CA, Barassi CA. Root phospholipids in Azospirillum-inoculated wheat seedlings exposed to water stress. Plant Physiol Biochem. 2006;44(11–12):873–9.
17. Pereyra MA, García P, Colabelli MN, Barassi CA, Creus CM. A better water status in wheat seedlings induced by Azospirillum under osmotic stress is related to morphological changes in xylem vessels of the coleoptile. Appl Soil Ecol. 2012;53(1):94–7.
18. Hamaoui B, Abbadi J, Burdman S, Rashid A, Sarig S, Okon Y. Effects of inoculation with *Azospirillum brasilense* on chickpeas (Cicer arietinum) and faba beans (Vicia faba) under different growth conditions. Agronomie. 2001;21(6–7):553–60.
19. Barassi CA, Ayrault G, Creus CM, Sueldo RJ, Sobrero MT. Seed inoculation with Azospirillum mitigates NaCl effects on lettuce. Sci Hortic. 2006;109(1):8–14.
20. Fasciglione G, Casanovas EM, Quillehauquy V, Yommi AK, Goñi MG, Roura SI, et al. Azospirillum inoculation effects on growth, product quality and storage life of lettuce plants grown under salt stress. Sci Hortic. 2015;195:154–62.
21. Creus CM, Sueldo RJ, Barassi CA. Shoot growth and water status in Azospirillum-inoculated wheat seedlings grown under osmotic and salt stresses. Plant Physiol Biochem. 1997;35(12):939–44.
22. Romero AM, Vega D, Correa OS. Azospirillum brasilense mitigates water stress imposed by a vascular disease by increasing xylem vessel area and stem hydraulic conductivity in tomato. Appl Soil Ecol. 2014;82:38–43.
23. Del Amor FM, Cuadra-Crespo P. Plant growth-promoting bacteria as a tool to improve salinity tolerance in sweet pepper. Funct Plant Biol. 2012;39(1):82–90.
24. Rojas-Tapias D, Moreno-Galván A, Pardo-Díaz S, Obando M, Rivera D, Bonilla R. Effect of inoculation with plant growth-promoting bacteria (PGPB) on amelioration of saline stress in maize (Zea mays). Appl Soil Ecol. 2012;61:264–72.
25. Chaudhary D, Narula N, Sindhu SS, Behl RK. Plant growth stimulation of wheat (*Triticum aestivum* L.) by inoculation of salinity tolerant Azotobacter strains. Physiol Mol Biol Plants. 2013;19(4):515–9.
26. Egamberdiyeva D, Höflich G. Influence of growth-promoting bacteria on the growth of wheat in different soils and temperatures. Soil Biol Biochem. 2003;35(7):973–8.
27. Egamberdiyeva D, Höflich G. Effect of plant growth-promoting bacteria on growth and nutrient uptake of cotton and pea in a semi-arid region of Uzbekistan. J Arid Environ. 2004;56(2):293–301.
28. Bacilio M, Rodriguez H, Moreno M, Hernandez J-P, Bashan Y. Mitigation of salt stress in wheat seedlings by a gfp-tagged Azospirillum lipoferum. Biol Fertil Soils. 2004;40(3):188–93.
29. Loureiro RR, Reis RP, Marroig RG. Effect of the commercial extract of the brown alga Ascophyllum nodosum Mont. on Kappaphycus alvarezii (Doty) Doty ex P.C. Silva in situ submitted to lethal temperatures. J Appl Phycol. 2014;26(1):629–34.
30. Saa S, Olivos-Del Rio A, Castro S, Brown PH. Foliar application of microbial and plant based biostimulants increases growth and potassium uptake in almond (Prunus dulcis [Mill.] D. A. Webb). Front. Plant Sci. 2015;6:87.
31. Spann TM, Little HA. Applications of a commercial extract of the brown seaweed *Ascophyllum nodosum* increases drought tolerance in container-grown "hamlin" sweet orange nursery trees. Hort Science. 2011;46(4):577–82.
32. Fernandez O, Theocharis A, Bordiec S, Feil R, Jacquens L, Clément C, et al. Burkholderia phytofirmans PsJN acclimates grapevine to cold by modulating carbohydrate metabolism. MPMI. 2012;25(4):496–504.
33. Theocharis A, Bordiec S, Fernandez O, Paquis S, Dhondt-Cordelier S, Baillieul F, et al. Burkholderia phytofirmans PsJN Primes *Vitis vinifera* L. and Confers a better tolerance to low nonfreezing temperatures. MPMI. 2012;25(2):241–9.
34. Subramanian P, Kim K, Krishnamoorthy R, Mageswari A, Selvakumar G, Sa T. Cold stress tolerance in psychrotolerant soil bacteria and their conferred chilling resistance in tomato (Solanum lycopersicum Mill.) under low temperatures. PLoS ONE. 2016;11(8):e0161592.
35. Zhang X, Schmidt RE, Ervin EH, Doak S. Creeping bentgrass physiological responses to natural plant growth regulators and iron under two regimes. HortScience. 2002;37(6):898–902.
36. Zhang X, Schmidt RE. Hormone-containing products'; impact on antioxidant status of tall fescue and creeping bentgrass subjected to drought. Crop Sci. 2000;40(5):1344–9.
37. Zhang X, Ervin EH. Cytokinin-containing seaweed and humic acid extracts associated with creeping bentgrass leaf cytokinins and drought resistance. Crop Sci. 2004;44(5):1737.
38. Park E-J, Jeknić Z, Sakamoto A, DeNoma J, Yuwansiri R, Murata N, et al. Genetic engineering of glycinebetaine synthesis in tomato protects seeds, plants, and flowers from chilling damage. Plant J. 2004;40(4):474–87.
39. Suarez C, Cardinale M, Ratering S, Steffens D, Jung S, Montoya AMZ, et al. Plant growth-promoting effects of Hartmannibacter diazotrophicus on summer barley (Hordeum vulgare L.) under salt stress. Appl Soil Ecol. 2015;95:23–30.
40. Çimrin KM, Türkmen Ö, Turan M, Tuncer B. Phosphorus and humic acid application alleviate salinity stress of pepper seedling. Afr J Biotechnol [Internet]. 2010;9(36). http://www.ajol.info/index.php/ajb/article/view/92903. Accessed 16 Nov 2015.
41. García AC, Santos LA, Izquierdo FG, Sperandio MVL, Castro RN, Berbara RLL. Vermicompost humic acids as an ecological pathway to protect rice plant against oxidative stress. Ecol Eng. 2012;47:203–8.
42. Aydin A, Kant C, Turan M. Humic acid application alleviates salinity stress of bean (*Phaseolus vulgaris* L.) plants decreasing membrane leakage. Afr J Agric Res. 2012;7(7):1073–86.
43. Petrozza A, Santaniello A, Summerer S, Di Tommaso G, Di Tommaso D, Paparelli E, et al. Physiological responses to Megafol® treatments in tomato plants under drought stress: a phenomic and molecular approach. Sci Hortic. 2014;22(174):185–92.
44. Kołodziejczyk I, Dzitko K, Szewczyk R, Posmyk MM. Exogenous melatonin improves corn (Zea mays L.) embryo proteome in seeds subjected to chilling stress. J Plant Physiol. 2016;1(193):47–56.
45. Ali SKZ, Sandhya V, Grover M, Linga VR, Bandi V. Effect of inoculation with a thermotolerant plant growth promoting *Pseudomonas putida* strain AKMP7 on growth of wheat (Triticum spp.) under heat stress. J Plant Interact. 2011;6(4):239–46.
46. Ali SKZ, Sandhya V, Grover M, Kishore N, Rao LV, Venkateswarlu B. Pseudomonas sp. strain AKM-P6 enhances tolerance of sorghum seedlings to elevated temperatures. Biol Fertil Soils. 2009;46:45–55.

47. Selvakumar G, Kundu S, Joshi P, Nazim S, Gupta AD, Mishra PK, et al. Characterization of a cold-tolerant plant growth-promoting bacterium Pantoea dispersa 1A isolated from a sub-alpine soil in the North Western Indian Himalayas. World J Microbiol Biotechnol. 2008;24(7):955–60.
48. Cuin TA, Shabala S. Amino acids regulate salinity-induced potassium efflux in barley root epidermis. Planta. 2007;225(3):753–61.
49. Ertani A, Schiavon M, Muscolo A, Nardi S. Alfalfa plant-derived biostimulant stimulate short-term growth of salt stressed Zea mays L. plants. Plant Soil. 2012;364(1–2):145–58.
50. Zhu K, Zhou H, Qian H. Antioxidant and free radical-scavenging activities of wheat germ protein hydrolysates (WGPH) prepared with alcalase. Process Biochem. 2006;41(6):1296–302.
51. Botta A. Enhancing plant tolerance to temperature stress with amino acids: an approach to their mode of action. Acta Horticulturae. 2013. doi:10.17660/ActaHortic.2013.1009.1.
52. Lucini L, Rouphael Y, Cardarelli M, Canaguier R, Kumar P, Colla G. The effect of a plant-derived biostimulant on metabolic profiling and crop performance of lettuce grown under saline conditions. Sci Hortic. 2015;23(182):124–33.
53. Visconti F, De Paz JM, Bonet L, Jordà M, Quiñones A, Intrigliolo DS. Effects of a commercial calcium protein hydrolysate on the salt tolerance of Diospyros kaki L. cv. "Rojo Brillante" grafted on Diospyros lotus L. Sci Hortic. 2015;185:129–38.
54. Del Pilar Cordovilla M, Berrido SI, Ligero F, Lluch C. Rhizobium strain effects on the growth and nitrogen assimilation in *Pisum sativum* and *Vicia faba* plant growth under salt stress. J Plant Physiol. 1999;154(1):127–31.
55. Rayirath P, Benkel B, Mark Hodges D, Allan-Wojtas P, MacKinnon S, Critchley AT, et al. Lipophilic components of the brown seaweed, *Ascophyllum nodosum*, enhance freezing tolerance in *Arabidopsis thaliana*. Planta. 2009;230(1):135–47.
56. Nair P, Kandasamy S, Zhang J, Ji X, Kirby C, Benkel B, et al. Transcriptional and metabolomic analysis of Ascophyllum nodosum mediated freezing tolerance in Arabidopsis thaliana. BMC Genom. 2012;13(1):643.
57. Nabati DA, Schmidt RE, Parrish DJ. Alleviation of salinity stress in Kentucky bluegrass by plant growth regulators and iron. Crop Sci. 1994;34(1):198–202.
58. Zhang X, Ervin EH. Impact of seaweed extract-based cytokinins and zeatin riboside on creeping bentgrass heat tolerance. Crop Sci. 2008;48(1):364–70.
59. Xu C, Leskovar DI. Effects of A. nodosum seaweed extracts on spinach growth, physiology and nutrition value under drought stress. Sci Hortic. 2015;12(183):39–47.
60. Möller M, Smith ML. The significance of the mineral component of seaweed suspensions on lettuce (Lactuca sativa L.) seedling growth. J Plant Physiol. 1998;153(5–6):658–63.
61. Mancuso S, Azzarello E, Mugnai S, Briand X. Marine bioactive substances (IPA extract) improve foliar ion uptake and water stress tolerance in potted Vitis vinifera plants. Adv Hortic Sci. 2006;20(2):156–61.
62. Elansary HO, Skalicka-Woźniak K, King IW. Enhancing stress growth traits as well as phytochemical and antioxidant contents of Spiraea and Pittosporum under seaweed extract treatments. Plant Physiol Biochem. 2016;105:310–20.
63. Bradáčová K, Weber NF, Morad-Talab N, Asim M, Imran M, Weinmann M, et al. Micronutrients (Zn/Mn), seaweed extracts, and plant growth-promoting bacteria as cold-stress protectants in maize. Chem Biol Technol Agric. 2016;3(1):19.
64. Craigie JS. Seaweed extract stimuli in plant science and agriculture. J Appl Phycol. 2011;23:371–93.
65. Sharma HS, Fleming C, Selby C, Rao JR, Martin T. Plant biostimulants: a review on the processing of macroalgae and use of extracts for crop management to reduce abiotic and biotic stresses. J Appl Phycol. 2014;26:465–90.
66. Verkleij FN. Seaweed extracts in agriculture and horticulture: a review. Biol Agric Hortic. 1992;8:309–24.
67. Khan W, Rayirath UP, Subramanian S, et al. Seaweed extracts as biostimulants of plant growth and development. J Plant Growth Regul. 2009;28:386–99. doi:10.1007/s00344-009-9103-x.
68. Imran M, Mahmood A, Römheld V, Neumann G. Nutrient seed priming improves seedling development of maize exposed to low root zone temperatures during early growth. Eur J Agron. 2013;49:141–8.
69. Guiseley K, Stanley N, Whitehouse P. Carrageenan. Handbook of water-soluble gums and resins. 1980. pp. 5–11.
70. Goñi O, Fort A, Quille P, McKeown PC, Spillane C, O'Connell S. Comparative transcriptome analysis of two *Ascophyllum nodosum* extract biostimulants: same seaweed but different. J Agric Food Chem. 2016;64(14):2980–9.
71. de Vasconcelos ACF, Zhang X, Ervin EH, de Kiehl J. C. Enzymatic antioxidant responses to biostimulants in maize and soybean subjected to drought. Scientia Agricola. 2009;66(3):395–402.
72. Marfà O, Cáceres R, Polo J, Ródena J. Animal protein hydrolysate as a biostimulant for transplanted strawberry plants subjected to cold stress. Acta Horticulturae. 2009. doi:10.17660/ActaHortic.2009.842.57.
73. Szabados L, Savouré A. Proline: a multifunctional amino acid. Trends Plant Sci. 2010;15(2):89–97.
74. Nanjo T, Kobayashi M, Yoshiba Y, Sanada Y, Wada K, Tsukaya H, et al. Biological functions of proline in morphogenesis and osmotolerance revealed in antisense transgenic Arabidopsis thaliana. Plant J. 1999;18(2):185–93.
75. Posmyk MM, Szafrańska K. Biostimulators: a new trend towards solving an old problem. Front Plant Sci. 2016;7:48.
76. Peel GJ, Mickelbart MV, Rhodes D. Choline metabolism in glycinebetaine accumulating and non-accumulating near-isogenic lines of Zea mays and Sorghum bicolor. Phytochemistry. 2010;71(4):404–14.
77. Park E-J, Jeknic Z, Chen THH. Exogenous application of glycinebetaine increases chilling tolerance in tomato plants. Plant Cell Physiol. 2006;47(6):706–14.
78. Chen THH, Murata N. Glycinebetaine: an effective protectant against abiotic stress in plants. Trends Plant Sci. 2008;13(9):499–505.
79. Cuin TA, Shabala S. Exogenously supplied compatible solutes rapidly ameliorate NaCl-induced potassium efflux from barley roots. Plant Cell Physiol. 2005;46(12):1924–33.
80. Blum A. Osmotic adjustment is a prime drought stress adaptive engine in support of plant production. Plant Cell Environ. 2016.
81. Bulgari R, Cocetta G, Trivellini A, Vernieri P, Ferrante A. Biostimulants and crop responses: a review. Biol Agric Hortic. 2015;31(1):1–17.
82. Zandonadi DB, Santos MP, Caixeta LS, Marinho EB, Peres LEP, Façanha AR, et al. Plant proton pumps as markers of biostimulant action. Scientia Agricola. 2016;73(1):24–8.
83. Selvakumar G, Joshi P, Mishra PK, Bisht JP, Gupta HS. Mountain aspects influence the genetic clustering of psychrotolerant phosphate solubilizing Pseudomonads in the Uttarkhand Himalayas. Curr Microbiol. 2009;59:432–8.
84. Upadhyay SK, Singh DP. Saikia R Genetic diversity of plant growth promoting rhizobacteria from rhizospheric soil of wheat under saline conditions. Curr Microbiol. 2009;59(5):489–96.
85. Sandhya V, Ali AS, Grover M, Reddy G, Venkateswarlu B. Alleviation of drought stress effects in sunflower seedlings by the exopolysaccharides producing Pseudomonas putida strain GAP-P45. Biol Fertil Soils. 2009;46(1):17–26.
86. Zahran HH. Rhizobium-legume symbiosis and nitrogen fixation under severe conditions and in an arid climate. Microbiol Mol Biol Rev. 1999;63(4):968–89.
87. Gopalakrishnan S, Sathya A, Vijayabharathi R, Varshney RK, Gowda CLL, Krishnamurthy L. Plant growth promoting rhizobia: challenges and opportunities. 3 Biotech. 2015;5(4):355–77.
88. Egamberdiyeva D. Alleviation of salt stress by plant growth regulators and IAA producing bacteria in wheat. Acta Physiol Plant. 2009;31(4):861–4.
89. Paul D, Lade H. Plant-growth-promoting rhizobacteria to improve crop growth in saline soils: a review. Agron Sustain Dev. 2014;34(4):737–52.
90. Kaushal M, Wani SP. Plant-growth-promoting rhizobacteria: drought stress alleviators to ameliorate crop production in drylands. Ann Microbiol. 2015;66(1):35–42.
91. Siddiqui KS, Williams TJ, Wilkins D, et al. Psychrophiles. Ann Rev Earth Planet Sci. 2013;41:87–115. doi:10.1146/annurev-earth-040610-133514.
92. Thao HTB, Yamakawa T. Phosphite (phosphorous acid): fungicide, fertilizer or bio-stimulator? Soil Sci Plant Nutr. 2009;55(2):228–34.
93. Jiang C, Belfield EJ, Cao Y, Smith JAC, Harberd NP. An arabidopsis soil-salinity–tolerance mutation confers ethylene-mediated enhancement of sodium/potassium homeostasis. Plant Cell. 2013;25(9):3535–52.
94. Nadeem SM, Zahir ZA, Naveed M, Ashraf M. Microbial ACC-deaminase: prospects and applications for inducing salt tolerance in plants. Crit Rev Plant Sci. 2010;29(6):360–93.

2

Characterizing phosphorus forms in cropland soils with solution ^{31}P-NMR: past studies and future research needs

Barbara J. Cade-Menun*

Abstract

Understanding the forms and dynamics of soil phosphorus (P) is essential to maintain agricultural productivity while minimizing environmental risks. Since it was first used on soil extracts in 1980, ^{31}P-nuclear magnetic resonance spectroscopy (P-NMR) has emerged as the leading technique to characterize extractable soil organic P forms. However, it is still underutilized in agriculture; of the more than 200 soil P-NMR papers published to date, only 44 have been conducted in non-pasture soils used for the production of annual or perennial crops, and only nine of those have linked identified P forms to agronomic parameters such as yield. This paper reviews these prior studies, suggesting gaps in research with respect to cropping systems and geographical regions. In particular, there have been few recent P-NMR studies that have fully identified P forms in African soils, and few studies of permanent crops such as orchards and vineyards. There is a need to link future P-NMR studies of cropping systems to agronomic parameters, and combine P-NMR with other techniques to fully capture P dynamics in cropping systems.

Keywords: Phosphorus, NMR, Soil, Cropland

Introduction

Phosphorus (P) is an essential element for all organisms, and is needed for energy transfer through adenosine triphosphate (ATP), cell membrane structure (phospholipids and lipoteichoic acids), cell reproduction [deoxyribonucleic acid (DNA)] and gene expression [ribonucleic acid (RNA)]. Phosphorus is classified as a macronutrient for plant growth, and is a limiting nutrient in many soils [1, 2]. Unlike nitrogen (N), P cannot be fixed from atmospheric sources; it can only be obtained by uptake from soil via plant roots as orthophosphate (HPO_4^{2-} or $H_2PO_4^-$ depending on soil pH) [3].

In soil, P originates from the weathering of apatites, which are primary P minerals containing calcium (Ca). The orthophosphate released from weathering will move to the soil solution, where it can precipitate as secondary minerals that are generally associated with Ca, magnesium (Mg), iron (Fe) and aluminum (Al); sorb to soil particles; or be taken up by plants and microbes and converted to organic P forms (immobilization) [3]. Organic P forms can be returned to the soil, where they can sorb to soil particles or be mineralized back to orthophosphate for plant uptake from the soil solution.

Organic P can comprise 35–65% of soil P [4]. Soil organic P forms are grouped into broad compound classes, each of which can contain a number of P forms [4]. Orthophosphate esters are subdivided based on the number of ester linkages. Orthophosphate monoesters have one ester linkage per phosphate, and include sugar phosphates (e.g. glucose 6-phosphate), mononucleotides (e.g. adenosine monophosphate) and inositol phosphates. Inositol phosphates identified in soils include the plant P storage compound phytate (*myo*-inositol hexaphosphate, *myo*-IHP) and several of its stereoisomers (*scyllo*-, D-*chiro*- and *neo*-IHP) [5–7]. Orthophosphate diesters have two esters per phosphate, and include phospholipids, lipoteichoic acids, RNA and DNA. Phosphonates have a carbon (C)–P bond, and include natural compounds such as ciliatine (2-aminoethyl phosphonic

*Correspondence: Barbara.cade-menun@agr.gc.ca
Agriculture & Agri-Food Canada, Swift Current Research and Development Centre, Gate 4 Airport Road, PO BOX 1030, Swift Current, SK S9H 3X2, Canada

acid) and agricultural chemicals such as glyphosate [*N*-(phosphonomethyl) glycine]. Orthophosphate anhydrides are linear chains of orthophosphate, linked with energy-rich phosphoanhydride bonds. These can include inorganic and organic compounds, all of which are generally included with organic P due to similarities in dynamics and analysis. Organic orthophosphate anhydrides include ATP; inorganic compounds include polyphosphates, which are chains of two (pyrophosphate) or more orthophosphates. More details on all of these compounds can be found in Condron et al. [4].

The P that is available for plant uptake from the soil solution can be deficient in many regions, requiring fertilization for optimal crop growth [2, 8]. Animal manure (AM) and other biochemically stabilized organic wastes have been applied as fertilizers for centuries; chemical inorganic P fertilizers (CF) were introduced in the late 1800s, and their use has significantly expanded globally since the 1950s [8]. The widespread application of P fertilizers, both AM and CF, has led to concerns with respect to water quality and eutrophication due to excess P from soil that is transferred to water through erosion and runoff [8, 9]. Additionally, CF is produced from rock phosphate, and the long-term sustainability of this non-renewable resource is uncertain [1, 2, 8]. In light of these concerns, there is a need to improve P use efficiency in agriculture, to minimize the need for P fertilizer application.

Many methods have been proposed to improve P use efficiency in crop production. These include targeted fertilization [8], plant breeding for traits such as rooting form, and production of phosphatase and organic acids [1, 2], and enhanced microbial P cycling in the rhizosphere [1, 2]. Key to all of these is an understanding of soil P dynamics, especially those of organic P forms [9, 10]. Historically, organic P forms have received less attention than inorganic P forms, in part because they are more difficult to study [4]. As such, a detailed understanding of the organic P forms contributed to soils from plants, microbes and organic amendments, and the pathways by which these are recycled to plant-available P forms, is lacking relative to that for inorganic P. Advanced spectroscopic tools are required to characterize P forms, the most widely used of which is solution ^{31}P nuclear magnetic resonance spectroscopy (P-NMR). Since its first use in soil science in 1980 by Newman and Tate [11], more than 200 papers have been published that have used P-NMR to characterize soil P forms. However, relatively few of these have studied cropping systems (defined here as non-pasture systems used for annual or perennial crops), and even fewer of these have linked P forms to management practices or to agronomic indicators such as yield or P balance (the difference between fertilizer inputs and crop removal). This has limited the usefulness of P-NMR studies to guide agricultural practices to improve P use efficiency, including the role of organic P forms in soil fertility. In order to connect P cycling in soil to sustainable management in agriculture, the knowledge gained from past studies needs to be evaluated and used to direct future research. As such, the objectives of this manuscript are (1) to review prior studies using P-NMR to investigate soil P forms and dynamics in field studies of non-pasture cropping systems; and (2) to identify knowledge gaps and suggest directions for future research.

Review

Solution ^{31}P-NMR spectroscopy

It is not my intention to review the principles of NMR spectroscopy in this manuscript, because that information can be obtained elsewhere [e.g. 12, 13]. It is also not my intention to give extensive information related to the use of P-NMR in soil and environmental science, as that has been covered in several other review papers [14–17]. However, a brief overview is needed to help understand the strengths and weaknesses of the studies discussed below.

Solid-state P-NMR is rarely used for soil studies; instead, solution (liquid-state) P-NMR is preferred for the improved spectral resolution [14, 15]. Soil total P concentrations are generally in the mg kg^{-1} range; thus, concentrating soil extracts increases the P concentration per NMR sample, significantly improving the NMR response. In the early years of soil P-NMR work, a number of different extractants were used, with little standardization among research groups [14, 15], making it difficult to compare results. In 1996, Cade-Menun and Preston [18] introduced the use of 0.25 M sodium hydroxide (NaOH) + 0.05 M disodium ethylenediaminetetraacetic acid (Na_2EDTA) as an extractant, based on a method to measure total soil organic P [19]. Once extracted, solutions are lyophilized to concentrate P. This is now the most widely used extraction procedure for soil and other environmental samples [17]. However, as discussed in detail elsewhere [17], other dissimilarities exist among groups with respect to sample extraction (e.g. soil:extractant ratios), dissolution of lyophilized extracts for NMR analysis, NMR acquisition parameters, and methods to identify and quantify specific P forms. In some early studies, research groups concentrated their extracts with dialysis [e.g. 20, 21]. While this improved the resolution of other peaks by removing most of the orthophosphate, it may have also removed other P forms, and definitely altered the relative proportions of all peaks. Even small changes in preparation methods or parameters can affect the final NMR results, so caution is needed when comparing results from different research groups that use different protocols and parameters.

The early soil P-NMR work produced spectra that were often very broad, with little resolution of individual peaks. As such, P forms were identified as compound classes only (e.g. orthophosphate, orthophosphate monoesters) [11, 20–28]. There were few attempts to identify specific P forms, and the identification of some compounds such as choline phosphate [23] has since been shown to be incorrect [29]. Advances over time now allow the majority of peaks to be identified with spiking experiments and compound libraries, although it is still difficult to distinguish some specific forms such as mononucleotides [29]. Figure 1 shows spectra from several recent studies of P in cropping systems [30–32]; more information about the samples can be found in the published studies. The P compounds of interest in soil science generally fall between 30 and −30 ppm, and include (from left to right, Fig. 1) phosphonates, from 30 to ~8 ppm; orthophosphate at ~6 ppm; orthophosphate monoesters, at ~7 to 6.1 ppm and ~5.9 to ~2.5 ppm; orthophosphate diesters from 2.5 to −3.9 ppm; pyrophosphate at about −4.2 ppm; and polyphosphates at −4 (end group) and from −5.3 to −30 ppm (mid-chain). More details of the orthophosphate monoester region are shown in Fig. 2, with spectra from published studies [32, 33], and unpublished data (B. Cade-Menun). The spectra are labeled in the manner used by the Cade-Menun research group. More information about the chemical shifts of these and other P compounds can be found in published compound libraries [29, 34].

One of the biggest improvements in recent years involves the identification of peaks for degradation compounds. These are compounds that were orthophosphate diesters in the soil, but which have degraded to orthophosphate monoesters during extraction and analysis. It is generally now agreed that α- and β-glycerophosphates are degradation products from phospholipids, while most mononucleotides (e.g. adenosine monophosphate) originate from RNA [29, 34–37]. There has long been an interest in quantifying total orthophosphate monoesters and total orthophosphate diesters in soil; diesters are thought to be more labile, while many monoesters are more recalcitrant due to strong sorption to mineral surfaces [4]. As such, the ratio of orthophosphate monoesters to orthophosphate diesters (M:D) has been used to predict organic P cycling in soils [4, 23]. In addition, activities of the soil enzymes acid monoesterase, alkaline monoesterase and diesterase are commonly measured [4], and linked back to total orthophosphate monoester and diester concentrations in models [e.g., 38]. However, if corrections are not made for degradation products, the M:D and models are of questionable value for predicting mineralization of organic P forms [37].

Unless compounds labeled with other P isotopes (^{32}P, ^{33}P) have been used in experiments, all P in soils is ^{31}P. As such, P-NMR is a quantitative method, allowing the concentrations of P forms to be determined. This is can be done by measuring the P concentration in the soil extract if the entire extract is used for the NMR experiment, or by measuring the P in the prepared NMR sample after analysis if only a portion of the lyophilized extract is used. In either case, this is done by digestion and colorimetric analysis or by analysis with inductively coupled plasma optical emission spectroscopy (ICP-OES); the percentage of each peak, determined from the spectrum by integration and/or deconvolution, is then multiplied by the P concentration [14, 15, 17]. Concentrations can also be determined by integration against a peak from a standard with a known P concentration spiked into the sample [17].

Solution ^{31}P-NMR studies of cropland soils

The P-NMR literature for pasture soils was recently reviewed [39]; as such, this section of the manuscript will review studies that have used P-NMR to characterize P in non-pasture soils used to grow annual or perennial crops. Most commercial crop production relies heavily on fertilization with CF or AM to optimize yields, so improvements in P use efficiencies could have the greatest benefits in these systems.

Studies using P-NMR to characterize P in cropping systems are listed in Table 1. These are grouped by country and listed by date of publication. The soil classifications listed are from the original manuscripts; no attempt was made to convert them to a single classification system. Other information given includes the crops, P fertilization history (kg P ha^{-1} year^{-1}), and the treatments the study was designed to investigate. The degree of peak identification is ranked as B, M or D; B is broad compound classes only (e.g. orthophosphate monoesters); M is moderate, with some more specific identifications, usually of *myo*-IHP; and D indicates spectra with detailed peak identifications, generally confirmed with spiking experiments. Studies that have identified degradation peaks and ideally corrected the calculation of total orthophosphate monoesters for diester degradation compounds are indicated with a C. If agronomic data such as soil test P, yield or P balances were included in the study this is indicated in the final column. Blanks in any column for any paper indicate that the information was not included in the original publication. Some studies included other land uses, such as forests, in addition to cropping systems but the information in Table 1 is for soils from cropland only. Please note that I excluded studies of soils that were identified as "cropland" or "arable" if no specific details about crop rotations were given

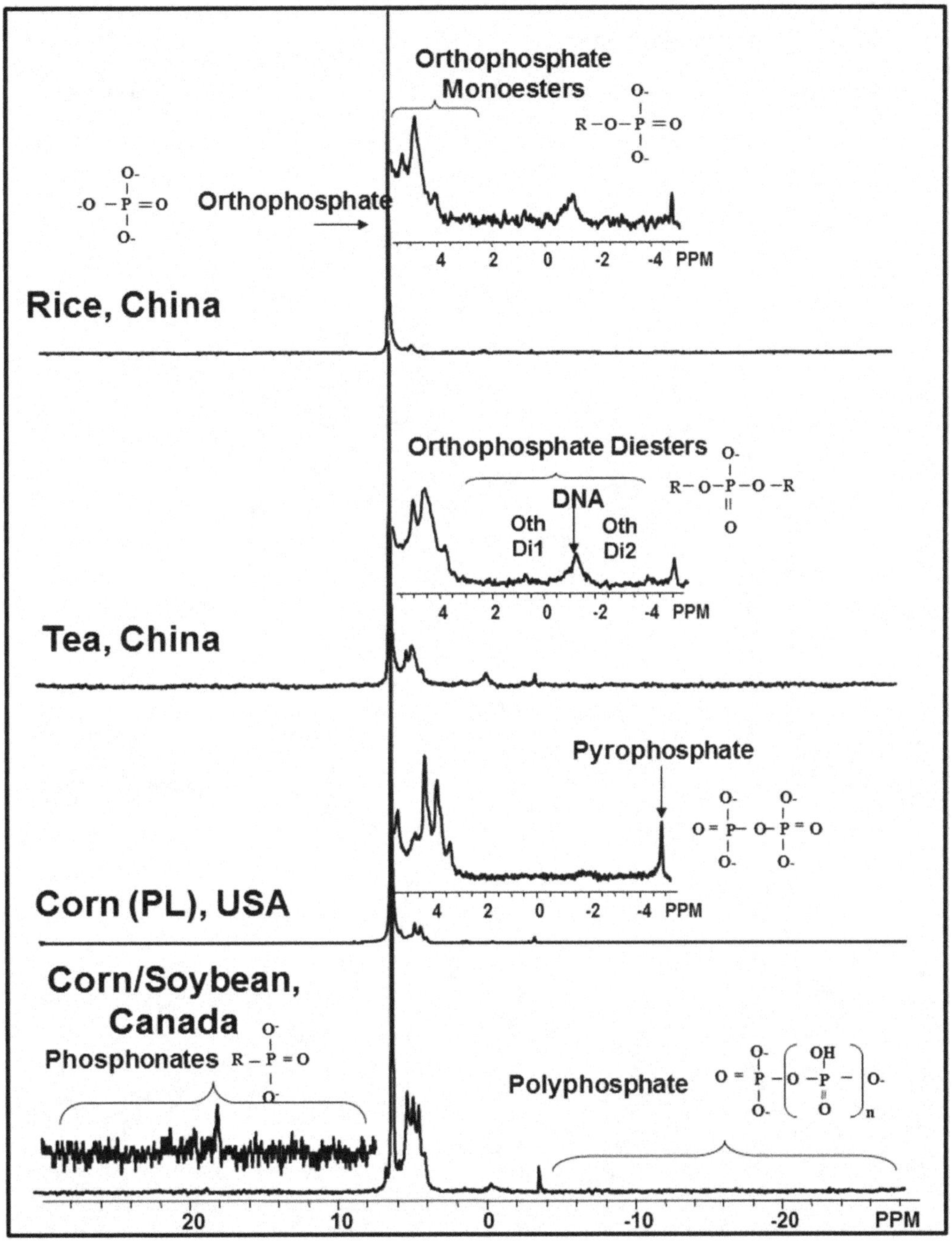

Fig. 1 Example spectra from cropping systems from various locations. The rice and tea spectra are from Liu et al. [30], the corn soil, no till, fertilized with poultry litter, is from Georgia, USA [31], and the corn/soybean sample is from Quebec, Canada [32]. *DNA* deoxyribonucleic acid; *Oth Di1*, *Oth Di 2* orthophosphate diesters other than DNA. Spectra were analyzed without proton decoupling and processed with 7 Hz line-broadening. The full spectra are scaled to the height of the orthophosphate peak; the *insets* were enlarged by different factors, to best show the indicated regions

[e.g. 40–46], as well as incubation studies using soils from crop field [46, 47].

In Africa, there have been two studies from Madagascar and one each from Tanzania, Ethiopia and West Cameroon [24, 26–28, 48]. Four of these studies were published in 2003 or earlier, used NaOH as an extractant [24, 26–28], and identified only broad compound classes. One study used dialysis on sample extracts [26], which likely distorted the relative peak areas. One study from Madagascar [48] used NaOH–EDTA and identified

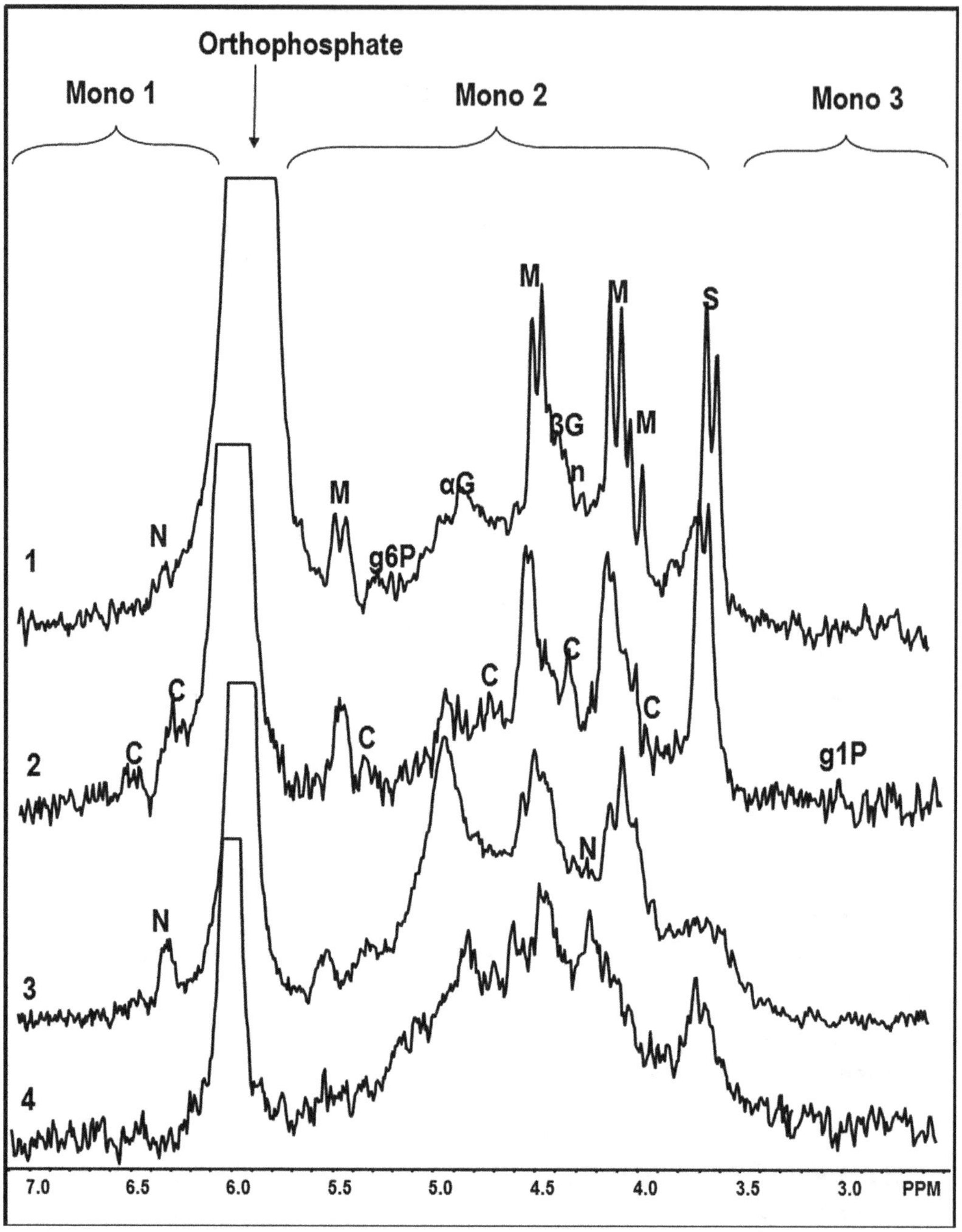

Fig. 2 Example spectra from cropping systems from various locations, showing details of the orthophosphate monoester region. *1* barley/corn, no till, Prince Edward Island, Canada [33]; *2* corn, France; *3* corn/soybean, no till, Quebec, Canada [32]; *4* Rice, China. Samples *2* and *4* are unpublished spectra. *Mono* general orthophosphate monoester regions; *α-G* α-glycerophosphate; *βG* β-glycerophosphate; *C* D-*chiro* inositol hexaphosphate (IHP); *g1P* glucose 1-phosphate; *g6P* glucose 6-phosphate; *M myo*-IHP; *N neo*-IHP; *n* nucleotides; *S scyllo*-IHP. Spectra were analyzed without proton decoupling, processed with 2 Hz line-broadening, and are scaled to the height of the tallest non-orthophosphate peak shown in all spectra

peaks in detail, including degradation compounds. In Asia, there was one study from Northern Thailand [23] and eight studies from various provinces in China [30, 38, 49–54]. One study used NaOH [25], the remainder used NaOH–EDTA [30, 38, 49–54], and none used dialysis. Although the majority of studies were published in 2012 or later, only two included detailed peak identifications and corrections for degradation [30, 54]. Four studies from Australia were included [35, 55–57], two of which sampled from the same fields under the same crops [35,

Table 1 Studies in North America, Europe and South America using ^{31}P-NMR to investigate P in soil under cropping systems, sorted by location

Location	Pub date[a]	Soil[b]	Crop	P fertilization (kg ha^{-1} year^{-1})	Treatment	Peak ID[c]	Agron. data[d]
Africa							
Madagascar [24]	1990	Ferrosol, Andosol	Rice	20	P forms in rice soils	B	
Northern Tanzania [26]	2000	Luvisol	Corn, beans	Various	Forests vs. fertilized (0 CF vs. AM) cropping systems	B	
Ethiopia [27]	2002	Alisol, Nitrisol	Tea, corn, corn/sorghum	Various	Cropping systems; D[e]	B	STP
West Cameroon [28]	2003	Palehumult	Tea plantation, food crop fields	Various	Forest vs. cropping systems (permanent, annual)	B	STP
Madagascar [48]	2006	Oxisols	Rice	Various	Different rice cropping systems; CF, AM, COM	M	STP, Y
Asia							
Northern Thailand [25]	2000	Cambisol	Cabbage	Not given	Cultivated vs. native forests	B	
Shenyan, China [49]	2012	Luvisol	Corn, rice	35 or 47	Land use types, EA	B	STP
Zhejang, China [30]	2013	Ultisol	Rice, tea, vegetables, nursery stock, bamboo	Various	Crops;	D, C	STP
Henan, China [38]	2014	Cambisol	Winter wheat/summer corn	59	Tillage, residue incorporation, aggregates, EA	B	STP
Jilin, China [50]	2014	Phaeozem	Corn–soybean each year	45	Tillage, aggregates	B	STP
Jilin, China [51]	2014	Phaeozem	Continuous corn monoculture	Various	Fertilization, CF and AM, EA	M	STP
Yunnan, China [52]	2015	Ultisols	Vegetables, horticulture	118.5–196.5	Land use, leaching losses; forests vs. open cropland, greenhouses	B	STP
Northeastern China [53]	2015	Gleysol, lithosol	Cropland (winter wheat summer corn), vegetables, orchards		Land use, fertilization, EA, sorption	M	STP
Shandong, China [54]	2016	Gleysol	Greenhouse, double-crop tomatoes	Various	CF, AM; high N fertilizer, leaching	D, C	PB, STP
Australia							
New South Wales [55]	2008	Luvisol	Continuous wheat, wheat–legume rotations	20	Tillage, stubble, rotation, microbial community	B	RP, Y
New South Wales [35, 56]	2009, 2011	Haplorthod, Fragiudult	Vegetables		Pasture vs. conventional and organic vegetables and horticulture	D, C	
New South Wales, Queensland [57]	2014	Vertisol	Grain-fallow-cotton, pulse-fallow-cotton	6 (every 2 years)	Vertisols in Australia	D	STP
Europe, Israel							
Germany, Denmark [20, 21]	1996, 1999	Eutrochept, Hapludalf	Wheat, beet, barley rotation	0–39	Compared fertilized, cultivated fields to native soils; particle size; D	B	STP
Finland [59]	2011	Cambisol	Barley, winter rye, pea, oats rotation	Various	Organic (AM) vs. conventional (CF) rotations	B	STP, PB
Sweden [60]	2013	Cambisol, Phaeozem, Umbrisol	Grain-based rotation	Various	Soil textures, fertilization rates (CF, AM)	M	STP, PB
England [62]	2013	Paleudalf	Winter wheat	0	pH gradient	D	RP
Israel [63]	2014	Chromoxererert	Corn	40 RW, 10 CF	Irrigation, fertilization with CF or RW	D, C	STP

Table 1 continued

Location	Pub date[a]	Soil[b]	Crop	P fertilization (kg ha^{-1} year^{-1})	Treatment	Peak ID[c]	Agron. data[d]
Switzerland [61]	2015	Luvisol	8-year, wheat-based rotation	Various	Fertilization; 0, CF, AM, COM, SS; EH	D, C	PB, RP, Y
Germany [58]	2015	Luvisol	Winter wheat, barley	0	Particle size fractions	B	
United Kingdom [10]	2015	Cambisol, Luvisol, Podzol	Winter wheat, oats, barley	Not given	Compared land uses	B	STP
North America							
Mexico [22]	1985	Vertisol	Tree garden (mango, papaya, avocado); Corn	20 (corn)	Unfertilized gardens, fertilized corn (CF)	B	
Saskatchewan, Canada [23]	1990	Chernozem, Gleysol, Luvisol	Cereal rotation	0	Compared to uncultivated native grasslands	B	
Quebec, Canada [64]	1999	Hapludalf, Humaquet	Corn	0, 44, 132	Fertilization rates	B	STP
New York, USA [66]	2005	Fragiudepts, Dys-trudepts	Corn or corn/alfalfa	Various	Compared short and long-term AM application; D	B	STP
Pennsylvania, USA [67]	2009	Hapludults, Hapludalf, Fragiudults, Dys-trudepts	Various corn-based rotations	29–120	Compared high rates of AM from various sources	M	STP
Delaware, Maryland USA [68]	2009	Hapludult	Corn, soybean		Compared crop soils with AM to poultry feces, composted manure	M	STP
Prince Edward Island, Canada [33]	2010	Podzol	Wheat, barley/soybean, fescue	Various	Tillage	D, C	STP
Quebec, Canada [32]	2014	Humaquept	Corn/soybean rotation	0, 17.5, 35	Tillage, CF fertilization	D, C	STP
Georgia, USA [31]	2015	Kanhapludult	Corn/winter rye, cotton/winter rye	Various	Tillage, CF and AM fertilization	D, C	
Saskatchewan, Canada [65]	2015	Chernozem	Continuous monoculture wheat	0, 10	Long-term with and without CF	D, C	STP, Y
Idaho, USA [69]	2016	Haplocalcid	Wheat–potato–barley	Various	Fertilization: CF, high and low AM	D, C	STP
South America							
Rio Grande do Sul, Brazil [71]	2002	Oxisols, Paleudalt	Various, grain-based	Various	Tillage; various crop rotations	B	STP
Minas Gerais, Brazil [72]	2003	Oxisols	Coffee plantations	5–15	Monoculture coffee vs. agroforestry coffee	B	STP
Rio Grande do Sul, Brazil [73]	2005	Oxisols	Various crops	0, 26, 52	Greenhouse cropping experiment, multiple crop rotation cycles	B	RP
South Chile [70]	2011	Palehumult	Lupin–wheat, oat–wheat rotations	80–90	Rotation, tillage, microbial P, enzymes	B	STP, Y
Santa Catarina, Brazil [74]	2015	Cambisol	Apple orchard	11	Land use; CF, between and in apple tree rows	B	
Rio Grande do Sul and Parana, Brazil [75]	2016	Oxisol, Ultisol	Various summer crops, winter cover crop	9, 27, 70	Grazed and ungrazed winter cover crops, CF	D, C	PB, STB

Within each region, publications are arranged chronologically based on date of publication

AM animal manure, *CF* chemical fertilizer, *COM* compost, *EA* measured phosphomonoesterase and phosphodiesterase activities, *EH* enzyme hydrolysis, *RW* recycled wastewater, *SS* sewage sludge; please see the original publications for more details

[a] Year of publication

[b] As reported in the original publication

[c] Peak identification: B, broad (only compound classes); C, monoesters corrected for diester degradation peaks; D, detailed (most peaks identified); M, moderate (some peaks identified)

[d] Agronomic information was included; PB, P balance(input–removal); RP, resin-extractable P; STP, soil test P (Olsen P, Mehlich P); Y, yield

[e] D: extracted sample dialyzed prior to NMR analysis

56]. Three studies identified peaks in detail, with correction for degradation [35, 56, 57].

In Europe, two studies from Germany and Denmark used the same research plots [20, 21], but both used dialysis and identified only broad peaks. There was a more recent study from Germany [58], one each from Finland, Sweden and Switzerland [59–61]; one specifically from England [62] and one from the United Kingdom in general [10]. The study from Israel is included here too [63]. Three of these studies identified only broad compound classes [10, 58, 59], one identified the majority of peaks [60] and three identified peaks in detail [61–63]. In North America, there was one study from Mexico [22], five from Canada [23, 32, 33, 64, 65] and five from the USA [31, 66–69]. The early studies identified only broad peaks [22, 23, 64, 66]. These used NaOH as the extractant, and one study used dialysis [66]. All remaining studies from North America used NaOH–EDTA as the extractant. The two studies from 2009 identified some of the peaks [67, 68], while the remainder identified peaks in detail. In South America, there was one study from Chile [70] and five from Brazil [71–75]. The majority of studies from South America used NaOH–EDTA as an extractant, although several studies added additional steps such as pre-extraction with resin and a final treatment with Chelex [71–73]. Only one study from South America identified peaks in detail and corrected for degradation [75].

The crops that have been studied include food crops for humans [e.g. wheat (*Triticum aestivum* L.), rice (*Oryza sativa* L.)] and livestock [e.g., corn (*Zea mays* L.), forage oats (*Avena strigose* Screb.)], and cash crops [e.g. tea (*Camellia sinensis* L.), coffee (*Coffea arabica* L.)]. In North America, Europe and Australia, P-NMR has been used most frequently to study annual crops, generally rotations based on wheat or corn. In Africa and Asia, more studies have investigated soils in vegetable and rice production. There have been very few studies of permanent plants, with three studies investigating tea [27, 28, 30], one study of coffee plantations [72], and one in an apple orchard [74].

Cropping systems have often been included in land use studies, and compared to forests [20, 21, 26, 28, 52, 53] or grasslands [10, 23, 75]. With respect to management practices, fertilization was the most studied practice, and included CF and AM, as well as alternate fertilizers such as compost, sewage sludge and waste water [48, 61, 63]. The next most studied management practice is tillage, with studies in China [38, 50] Australia [55], the USA [31], Canada [32, 33], Chile [70] and Brazil [71].

For improved P use efficiency in agriculture, studies need to not only identify P forms, but also relate both concentrations and changes in P forms to agronomic indicators such as yield or P use efficiency. Of the 44 studies listed in Table 1, many do not even include measures of soil test P (STP), and only nine include information on yield or P balances [48, 54, 55, 59–61, 65, 70, 75]. Of these nine, only six identified peaks in enough detail (M or D) to identify trends in P forms other than orthophosphate [48, 54, 60, 61, 65, 75]. I will focus on these studies in more detail, and they are summarized in Table 2.

The studies that included detailed peak identifications and yield information were studies of rice in Madagascar [48], greenhouse-grown tomatoes in China [54], wheat-based crop rotations in Sweden [60] and Switzerland [61], continuous wheat monoculture in Canada [65], and summer rotations [including corn, soybean (*Glycine max* (L.) Merrill), common bean (*Phaseolus vulgaris* L.)] with a grazed winter wheat cover crop in Brazil [75]. One had only CF [65], one had CF plus grazing of winter cover crops [75], two were fertilized with CF and AM [54, 60], one had fertilization with CF, AM, and compost [48], and one studied fertilization with CF, AM, compost and sewage sludge [61]. Information about soil pH, soil total P and soil test P concentrations is included in Table 2. These ranged considerably for each study depending on the location and treatments. All studies extracted soils for P-NMR with NaOH–EDTA. The recoveries of total P in the NaOH–EDTA extracts ranged from 36 to 96% (Table 2), again varying with location and treatment. The P-NMR results are reported as percentages of extracted P in Table 2 to simplify comparisons among studies. All of the soils in the studies from Madagascar, China and Brazil received some P fertilizers, and all contained high percentages of orthophosphate. The study from Canada included samples from plots that were fertilized from 1967 to 2010, and plots that were fertilized from 1967 to 1995, with no P from 1995 to 2010. The studies from Sweden and Switzerland included plots receiving no P for 30 or 60 years, respectively, plus fertilization at various rates with different fertilizers. The percentages of orthophosphate in the studies from Canada, Sweden and Switzerland varied with P fertilization, and were always lowest in soils receiving no P for the length of the study, as were the soil test P concentrations. The percentages of pyrophosphate and DNA were low across all treatments in all studies. The percentage of *myo*-IHP varied with P treatment, and was highest with the lowest fertilization, while *scyllo*-IHP was present in most samples from all studies, also increasing in plots with lowest fertilizer P inputs. The M:D ratio, after correcting for diester degradation, was over 1 for all but the sites with the lowest fertilization rates in Sweden. The ratio was particularly high in the Switzerland samples because no DNA was detected. The P balance was generally positive for all studies except at sites receiving no P fertilizer. There were generally no

Table 2 Summary of P-NMR results from selected studies

Location	Madagascar [48]	China [54]	Sweden [60]	Switzerland [61]	Canada [65]	Brazil [75]
Crop	Rice	Greenhouse, double-crop tomatoes	Grain-based rotation	8-year, wheat-based rotation	Continuous monoculture wheat	Various summer crops, winter cover crop
Fertilizers	Different rice cropping systems; CF, AM, COM	CF, AM; high N fertilizer, leaching	Soil textures, fertilization rates (CF, AM)	Fertilization; 0, CF, AM, COM, SS; EH	Long-term with and without CF	Grazed and ungrazed winter cover crops, CF
Soil pH	4.64–5.82[a]	6.91–7.03	5.8–7.5	4.8–5.9	5.4–6.6	5.2–6.4
Total P (mg kg^{-1})	133–1378	685–2446	288–980	442–1150	440–615	841–2531
Soil test P (mg kg^{-1})[b]	<1–6.9 (O)	20–210 (O) 179–1128 (M)	12–223 (A)	114–58 (R)	3–33 (O) 21–103 (M)	22–71 (M)
P recovery (%)[c]	36–82	49–74	48–98	73–96	36–50	60–88
OrthoP (%)[d]	54–80	90.7–93.4	20.4–75.7	53–80	22.6–44.1	71.6–77.6
Pyro P (%)	0–4	0.3–0.8	0–3.2	0.3–0.5	1.0–1.4	2.0–2.3
Myo-IHP (%)	0–5	0.5–1.7	0–24	3.0–6.6	4.6–8.5	6.5–9.7
Scyllo-IHP (%)	0–4	0.1–0.3	1.4–6.6	1.0–2.7	0.8–2.7	2.9–3.9
Other mono (%)	nd[e]	4.2–8.4	nd	7.3–12.2	29.3–34.2	0.7–1.1
DNA (%)	0–9	0.0–2.0	0–3	0	0.3–1.2	0.3–1.6
Other diesters (%)[f]	nd	0.0–0.7	0–17.5	0.7–1.1	2.8–4.5	6.8–9.8
cM:D[g]	nd	1.5–2.0	0.9–5.6	16–25	1.2–1.5	1.3–1.5

AM animal manure, *CF* chemical fertilizer, *COM* compost, *EA* measured phosphomonoesterase and phosphodiesterase activities, *EH* enzyme hydrolysis, *RW* recycled wastewater, *SS* sewage sludge, please see the original publications for more details

[a] Range for all depths and treatments

[b] Soil test P: A, ammonium lactate; M, Mehlich; O, Olsen; R, Resin

[c] Percentage of total P extracted in NaOH-EDTA

[d] Percentage of spectral area

[e] Could not be determined from information in the publication

[f] Calculated from information for α- and β-glycerophosphate and nucleotides as given in paper, if not already calculated in the paper

[g] Corrected for diester degradation compounds

differences among fertilizer types (CF, AM, etc.) within studies; the main differences resulted from differences in P rates applied to soils from each fertilizer. The use of non-CF fertilizers did not build up organic P; organic P was higher in plots not receiving fertilizer, mainly from the drawdown of inorganic P. In Madagascar, there were no differences in yields with treatments. In Canada, fertilizer cessation had also not affected yields, although orthophosphate and STP had decreased. In Switzerland, yields varied among treatments, but P was not the only limited nutrient. For all studies with a 0-P treatment [60, 61, 65], orthophosphate decreased in those plots, but organic P species remained the same or increased. This suggests that crops are drawing orthophosphate down faster than it is being replenished by mineralizing organic P forms. And the relatively low percentages of DNA and P degradation forms from phospholipids and RNA suggest that the P extracted for NMR in these studies is from other sources than microbes [4].

Information such as that which is shown in Table 2 needs to be obtained from a wider range of studies, with respect to soils, crops and fertilizers. This in turn could be used to enhance P use efficiency, with further testing. For example, does orthophosphate decrease and organic P increase because organic P forms are more tightly sorbed and less available for hydrolysis? If so, then crop breeding or soil microbial inoculation could be used to enhance the production of organic acids or phosphatases to mineralize organic P forms. However, without the detailed information about P forms and their dynamics that can be obtained from the use of advanced studies such as P-NMR, it will be more difficult to develop appropriate strategies to improve P use efficiency.

Future research needs

Solution P-NMR spectroscopy is a powerful technique, producing information not available from other methods. However, the reliability of quantitative data depends on the correct set up of acquisition parameters, e.g. with delay times long enough for full relaxation [17]. This is true for all future P-NMR studies, not just those of cropland soils. Extraction and analytical procedures should also be standardized among research groups, to enhance comparisons of data sets, and all peaks should be identified and confirmed with spiking experiments [29]. Calculations of total orthophosphate monoesters and diesters, and the M:D ratio, should be corrected for diester degradation products [37]. And field replicates need to be analyzed by P-NMR as much as possible, rather than compositing samples, to allow statistical analyses of P-NMR results along with other field data.

For studies of agricultural soils, especially those comparing management practices, published papers need to include as much agronomic data as possible. At the very minimum, some measure of STP needs to be included, as well as yield data and P balances. Ideally, P-NMR studies of management practices should also link changes in P forms to agronomic data related to those practices, which has been poorly done in the past. Although tillage has been shown to alter the distribution of soil P forms in several studies from around the world [31–33, 37, 50, 55, 70, 71], these changes in P forms were not linked to yield in any of these studies, making it difficult to determine if changing P forms makes any difference agronomically. Phosphorus NMR studies of agriculture that do not link the identified P forms to agronomic values become esoteric technical studies rather than practical management studies, and are more limited in their wider use.

The majority of studies to date are static studies, capturing the point in time when samples were collected. Sampling needs to be coupled to active changes in crops during the growing seasons such as seed germination (for annual plants) and flowering, and P-NMR measurements should be paired with other techniques such as enzyme assays or microbial community analysis to relate P forms to P dynamics. Long-term studies of cropping systems are available in many countries [e.g., 60–62, 65, 71, 75]; these should be coupled with shorter greenhouse studies to capture P dynamics at different time scales [e.g. 73]. And if orthophosphate is indeed the P form most affected by management, then there is a need to pair P-NMR studies with complementary techniques that better characterize orthophosphate forms, such as sequential fractionations and P-XANES [e.g., 30, 65, 69].

There is a need to understand the inputs into soil from various plant materials (roots, shoots, etc.), and changes in these inputs from changing soil P status [76–78]. And while the P forms in manures have been fairly well-characterized [16] and their dynamics in soil generally well-studied, there is a need for more information on alternate fertilizers, particularly recycled wastes [e.g., 60, 79–81] and their behavior in soil [61, 80]. The range of cropping systems investigated with P-NMR has been limited to date, as have the countries in which it has been used to characterize soil P. There is a particular need to characterize P forms and their dynamics in tropical soils, which sorb P more strongly, and where there are greater concerns with respect to food security [28].

Conclusions

Since its first use in soil science in 1980, P-NMR has been a valuable tool to characterize P forms. In agriculture, there is recognition of the need to improve P use efficiency, arising from concerns about eutrophication from soil P loss and the long-term sustainability of rock phosphate stores from which chemical fertilizers are made. The prior publications reviewed here show that P-NMR can be a valuable tool in studies to improve agricultural P use efficiency, when conducted properly to maximize quantitative information. However, P-NMR studies to date have been limited with respect to cropping systems and geographic locations. Therefore, further research using P-NMR is needed globally on a wider range of crops to fully understand the dynamics in soil of this valuable crop nutrient.

Acknowledgements

I acknowledge and thank the other participants at the Organic Phosphorus meeting and the 8th International Phosphorus Workshop, held in England and Germany, respectively, in September 2016. Discussions at these meetings revealed the need for this review topic, and influenced its focus. I gratefully acknowledge my many research collaborators on past and future projects.

Competing interests

The author declares that she has no competing interests.

References

1. Richardson AE, Lynch JP, Ryan PR, Delhaize E, Smith FA, Smith SE, Harvey PR, Ryan MH, Veneklaas EJ, Lambers H, Oberson A, Culvenor RA, Simpson RJ. Plant and microbial strategies to improve the phosphorus efficiency of agriculture. Plant Soil. 2011;349:121–56.
2. Faucon MP, Houben D, Reynoird JP, Mercadal-Dulaurent AM, Armand R, Lambers H. Advances and perspectives to improve the phosphorus availability in cropping systems for agroecological phosphorus management. Adv Agron. 2015;134:51–79.
3. Pierzynski GM, McDowell RW, Sims JT. Chemistry, cycling and potential movement of inorganic phosphorus in soils. In: Sims JT, Sharpley AN, editors. Phosphorus, agriculture and the environment. Monograph no 46. Madison: Soil Science Society of America; 2005. p. 53–86.
4. Condron LM, Turner BL, Cade-Menun BJ. Chemistry and dynamics of soil organic phosphorus. In: Sims JT, Sharpley AN, editors. Phosphorus, agriculture and the environment. Monograph no 46. Madison: Soil Science Society of America; 2005. p. 87–121.
5. Turner BL, Papházy MJ, Haygarth PM, McKelvie ID. Inositol phosphates in the environment. Phil Trans R Soc Lond. 2002;357:449–69.
6. Turner BL, Cheesman AW, Godage HY, Riley AM, Potter BVL. Determination of *neo*- and D-*chiro*-inositol hexakisphosphate in soils by solution ^{31}P NMR spectroscopy. Environ Sci Technol. 2012;46:4994–5002.
7. Giles CD, Cade-Menun BJ. Phytate in animal manure and soils: Abundance, cycling and bioavailability. Chapter 9. In: He Z, Zhang H, editors. Applied manure and nutrient chemistry for sustainable agriculture and environment. New York: Springer; 2014. p. 163–90.

8. Withers PJA, Sylvester-Bradley R, Jones DL, Healey JR, Talboys PJ. Feed the crop not the soil: rethinking phosphorus management in the food chain. Environ Sci Technol. 2014;48:6523–30.
9. Dodd RJ, Sharpley AN. Recognizing the role of soil organic phosphorus in soil fertility and water quality. Resour Conserv Recycl. 2015;105B:282–93.
10. Stutter MI, Shand CA, George TS, Blackwell MSA, Dixon L, Bol R, MacKay RL, Richardson AE, Condron LM, Haygarth PM. Land use and soil factors affecting accumulation of phosphorus species in temperate soils. Geoderma. 2015;257–258:29–39.
11. Newman RH, Tate KR. Soil phosphorus characterization by ^{31}P nuclear magnetic resonance. Commun Soil Sci Plant Anal. 1980;11:835–42.
12. Keeler J, Understanding NMR. Spectroscopy. 2nd ed. Hoboken: Wiley; 2011.
13. Claridge TDW. High-resolution NMR techniques in organic chemistry. 3rd ed. Amsterdam: Elsevier; 2016.
14. Cade-Menun BJ. Using phosphorus-31 nuclear magnetic resonance spectroscopy to characterize phosphorus in environmental samples. In: Turner BL, Frossard E, Baldwin D, editors. Organic phosphorus in the environment. Wallingford: CABI Publishing; 2005. p. 21–44.
15. Cade-Menun BJ. Characterizing phosphorus in environmental and agricultural samples by 31P nuclear magnetic resonance spectroscopy. Talanta. 2005;66:359–71.
16. Cade-Menun BJ. Characterizing phosphorus in animal waste with solution 31P NMR spectroscopy. In: He Z, editor. Environmental chemistry of animal manure. New York: Nova Science Publishers, Inc; 2011. p. 275–99.
17. Cade-Menun BJ, Liu CW. Solution 31P-NMR spectroscopy of soils from 2005 to 2013: a review of sample preparation and experimental parameters. Soil Sci Soc Am J. 2014;78:19–37.
18. Cade-Menun BJ, Preston CM. A comparison of soil extraction procedures for ^{31}P NMR spectroscopy. Soil Sci. 1996;161:770–85.
19. Bowman RA, Moir JO. Basic EDTA as an extractant for soil organic phosphorus. Soil Sci Soc Am J. 1993;57:1516–8.
20. Guggenberger G, Christensen BT, Rubæk G, Zech W. Land-use and fertilization effects on P forms in two European soils: resin extraction and ^{31}P-NMR analysis. Eur J Soil Sci. 1996;47:605–14.
21. Rubæk GH, Guggenberger G, Zech W, Christensen BT. Organic phosphorus in soil size separates characterized by phosphorus-31 nuclear magnetic resonance and resin extraction. Soil Sci Soc Am J. 1999;63:1123–32.
22. Zech W, Alt HG, Zucker A, Kögel I. ^{31}P-NMR-spectroscopic investigations of NaOH-extracts from soil with different land use in Yucatan (Mexico). Z Pflanzenernaehr Bodenk. 1985;148:626–32.
23. Condron LM, Frossard E, Tiessen H, Newman RH, Stewart JWB. Chemical nature of organic phosphorus in cultivated and uncultivated soils under different environmental conditions. J Soil Sci. 1990;41:41–50.
24. Zech W, Forster JC, Rakotoasimbola S, Fardeau JC. Caractérisation, par spectrométrie RMN du 31P, de constituants phosphatés extraites de sols de rizières malgaches. Cah. ORSTOM sér Pédol. 1990;XXV:243–51.
25. Möller A, Kaiser K, Amelung W, Niamskul C, Udomsri S, Puthawong M, Haumaier L, Zech W. Forms of organic C and P extracted from tropical soils as assessed by liquid-state ^{13}C-and ^{31}P-NMR spectroscopy. Aust J Soil Res. 2000;38:1017–35.
26. Solomon D, Lehmann J. Loss of phosphorus from soil in semi-arid northern Tanzania as a result of cropping: evidence from sequential extraction and ^{31}P-NMR spectroscopy. Eur J Soil Sci. 2000;51:699–708.
27. Solomon D, Lehmann J, Mamo T, Fritzsche F, Zech W. Phosphorus forms and dynamics as influenced by land use changes in the sub-humid Ethiopian highlands. Geoderma. 2002;105:21–48.
28. Tchiendoua Zech W. Chemical and spectral characterization of soil phosphorus under three land uses from an Andic Palehumult in West Cameroon. Agric Ecosyst Environ. 2003;100:193–200.
29. Cade-Menun BJ. Improved peak identification in ^{31}P-NMR spectra of environmental samples with a standardized method and peak library. Geoderma. 2015;257–258:102–14.
30. Liu J, Yang J, Cade-Menun BJ, Liang X, Hu Y, Liu CW, Zhao Y, Li L, Shi J. Complementary phosphorus speciation in agricultural soils by sequential fractionation, solution ^{31}P NMR, and P K-edge XANES spectroscopy. J Environ Qual. 2013;42:1763–70.
31. Cade-Menun BJ, He Z, Zhang H, Endale DM, Schomberg HH, Liu CW. Stratification of phosphorus forms from long-term conservation tillage and poultry litter application. Soil Sci Soc Am J. 2015;79:504–16.
32. Abdi D, Cade-Menun BJ, Ziadi N, Parent LE. Long-term impact of tillage practices and P fertilization on soil P forms as determined by ^{31}P-NMR spectroscopy. J Environ Qual. 2014;43:1431–41.
33. Cade-Menun BJ, Carter MR, James DC, Liu CW. Phosphorus forms and chemistry in the soil profile under long-term conservation tillage: a phosphorus-31 nuclear magnetic resonance study. J Environ Qual. 2010;39:1647–56.
34. Turner BL, Mahieu N, Condron LM. Phosphorus-31 nuclear magnetic resonance spectral assignments of phosphorus compounds in soil NaOH–EDTA extracts. Soil Sci Soc Am J. 2003;67:497–510.
35. Doolette AL, Smernik RJ, Dougherty WJ. Spiking improved solution phosphorus-31 nuclear magnetic resonance identification of soil phosphorus compounds. Soil Sci Soc Am J. 2009;73:919–27.
36. Smernik RJ, Doolette AL, Noack SR. Identification of RNA hydrolysis products in NaOH–EDTA extracts using ^{31}P NMR spectroscopy. Commun Soil Sci Plant Anal. 2015;46:2746–56.
37. Schneider KD, Cade-Menun BJ, Lynch DH, Voroney RP. Soil phosphorus forms from organic and conventional forage fields. Soil Sci Soc Am J. 2016;80:328–40.
38. Wei K, Chen ZH, Zhu A, Zhang JB, Chen LJ. Application of ^{31}P NMR spectroscopy in determining phosphatase activities and P composition in soil aggregates influenced by tillage and residue management practices. Soil Tillage Res. 2014;138:35–43.
39. Nash DM, Haygarth PM, Turner BL, Condron LM, McDowell RW, Richardson AE, Watkins M, Heaven MW. Using organic phosphorus to sustain pasture productivity: a perspective. Geoderma. 2014;221–222:11–9.
40. Turner BL, Cade-Menun BJ, Westermann DT. Organic phosphorus composition and potential bioavailability in semi-arid arable soils of the Western United States. Soil Sci Soc Am J. 2003;67:1168–79.
41. Hansen JC, Cade-Menun BJ, Strawn DG. Phosphorus speciation in manure-amended alkaline soils. J Environ Qual. 2004;33:1521–7.
42. Briceño M, Escudey M, Galindo G, Borchardt D, Chang A. Comparison of extraction procedures used in determination of phosphorus species by ^{31}P-NMR in Chilean volcanic soils. Commun Soil Sci Plant Anal. 2006;37:1553–69.
43. Ohno T, Hiradate S, He Z. Phosphorus solubility of agricultural soils: a surface charge and phosphorus-31 NMR speciation study. Soil Sci Soc Am J. 2011;75:1704–11.
44. Jarosch KA, Doolette AL, Smernik RJ, Tamburini F, Frossard E, Bünemann EK. Characterization of soil organic phosphorus in NaOH–EDTA extracts: a comparison of ^{31}P NMR spectroscopy and enzyme addition assays. Soil Biol Biochem. 2015;91:298–309.
45. Moata MRS, Doolette AL, Smernik RJ, McNeil AM, McDonald LM. Organic phosphorus speciation in Australian Red Chromosols: stoichiometric control. Soil Res. 2016;54:11–9.
46. Leytem AB, Smith DR, Applegate TJ, Thacker PA. The influence of manure phytic acid on phosphorus solubility in calcareous soils. Soil Sci Soc Am J. 2006;70:1629–38.
47. Doolette AL, Smernik RJ, Dougherty WJ. Rapid decomposition of phytate applied to a calcareous soil demonstrated by a solution ^{31}P NMR study. Eur J Soil Sci. 2010;61:563–75.
48. Turner BL. Organic phosphorus in Madagascan rice soils. Geoderma. 2006;136:279–88.
49. Zhang A, Chen Z, Zhang G, Chen L, Wu Z. Soil phosphorus composition determined by ^{31}P NMR spectroscopy and relative phosphatase activities influenced by land use. Eur J Soil Biol. 2012;52:73–7.
50. Wei K, Chen ZH, Zhang XP, Liang WJ, Chen LJ. Tillage effects on phosphorus composition and phosphatase activities in soil aggregates. Geoderma. 2014;217–218:37–44.
51. Zhang G, Chen Z, Zhang A, Chen L, Wu Z, Ma X. Phosphorus composition and phosphatase activities in soils affected by long-term application of pig manure and inorganic fertilizers. Commun Soil Sci Plant Anal. 2014;45:1866–76.
52. Li M, Hu Z, Zhu X, Zhou G. Risk of phosphorus leaching from phosphorus-enriched soils in the Dianchi catchment, Southwestern China. Environ Sci Pollut Res. 2015;22:8460–70.
53. Zhou B, Vogt RD, Lu X, Yang X, Lu C, Mohr CW, Zhu L. Land use as an explanatory factor for potential phosphorus loss risk, assessed by P indices and their governing parameters. Environ Sci Processes Impacts. 2015;17:1443–54.
54. Yan Z, Chen S, Li J, Alva A, Chen Q. Manure and nitrogen application enhances soil phosphorus mobility in calcareous soils in greenhouses. J Environ Manage. 2016;181:26–35.

55. Bünemann EK, Marschner P, Smernik RJ, Conyers M, McNeill AM. Soil organic phosphorus and microbial community composition as affected by 26 years of different management strategies. Biol Fertil Soils. 2008;44:717–26.
56. Doolette AL, Smernik RJ, Dougherty WJ. A quantitative assessment of phosphorus forms in some Australian soils. Soil Res. 2011;49:152–65.
57. McLaren TI, Smernik RJ, Guppy CN, Bell MJ, Tighe MK. The organic composition of vertisols as determined by ^{31}P NMR spectroscopy. Soil Sci Soc Am J. 2014;78:1893–902.
58. Jiang X, Bol R, Willbold S, Vereecken H, Klumpp E. Speciation and distribution of P associated with Fe and Al oxides in aggregate-sized fraction of an arable soil. Biogeosciences. 2015;12:6443–52.
59. Soinne H, Uusitalo R, Sarvi M, Turtola E, Hartikainen H. Characterization of soil phosphorus in differently managed clay soil by chemical extraction methods and 31P NMR spectroscopy. Commun Soil Sci Plant Anal. 2011;42:1995–2011.
60. Ahlgren J, Djodjic F, Börjesson G, Mattsson L. Identification and quantification of organic phosphorus forms in soils from fertility experiments. Soil Use Manage. 2013;29:24–55.
61. Annaheim KE, Doolette AL, Smernik RJ, Mayer J, Oberson A, Frossard E, Bünemann EK. Long-term addition of organic fertilizer has little effect on soil organic phosphorus as characterized by ^{31}P NMR spectroscopy and enzyme additions. Geoderma. 2015;257–258:67–77.
62. Turner BL, Blackwell MSA. Isolating the influence of pH on the amounts and forms of soil organic phosphorus. Eur J Soil Sci. 2013;64:249–59.
63. Zohar I, Cade-Menun B, Paytan A, Shaviv A. Phosphorus transformations from reclaimed wastewater to irrigated soil: a ^{31}P NMR study. Soil Sci Soc Am J. 2014;78:1884–92.
64. Zhang TQ, Mackenzie AF, Sauriol F. Nature of soil organic phosphorus as affected by long-term fertilization under continuous corn (*Zea mays* L.): a ^{31}P NMR study. Soil Sci. 1999;164:662–70.
65. Liu J, Hu YF, Yang JJ, Abdi D, Cade-Menun BJ. Investigation of soil legacy phosphorus transformations in long-term agricultural fields using sequential fractionation, P K-edge XANES and solution P-NMR spectroscopy. Environ Sci Technol. 2015;49:168–76.
66. Lehmann J, Lan Z, Hyland C, Sato S, Solomon D, Ketterings QM. Long-term dynamics of phosphorus forms and retention in manure-amended soils. Environ Sci Technol. 2005;39:6672–80.
67. Dou Z, Ramberg CF, Toth JD, Wang Y, Sharpley AN, Boyd SE, Chen CR, Williams D, Xu ZH. Phosphorus speciation and sorption-desorption characteristics in heavily manured soils. Soil Sci Soc Am J. 2009;73:93–101.
68. Hill JE, Cade-Menun BJ. Phosphorus-31 nuclear magnetic resonance spectroscopy transect study of poultry operations on the Delmarva Peninsula. J Environ Qual. 2009;37:1–9.
69. Weyers E, Strawn DG, Peak D, Moore AD, Baker LL, Cade-Menun B. Phosphorus speciation in calcareous soils following annual dairy manure amendments. Soil Sci Soc Am J. 2016;80:1531–42.
70. Redel YD, Escudey M, Alvear M, Conrad J, Borie F. Effects of tillage and crop rotation on chemical phosphorus forms and some related biological activities in a Chilean Ultisol. Soil Use Manage. 2011;27:221–8.
71. Rheinheimer DS, Anghinoni I, Flores AF. Organic and inorganic phosphorus as characterized by phosphorus-31 nuclear magnetic resonance in subtropical soils under management systems. Commun Soil Sci Plant Anal. 2002;33:1853–71.
72. Cardosa IM, Van der Meer P, Oenema O, Janssen BH, Kuyper TW. Analysis of phosphorus by ^{31}PNMR in Oxisols under agroforestry and conventional coffee systems in Brazil. Geoderma. 2003;112:51–70.
73. Gatiboni LC, dos Santos Rheinheimer D, Claro Flores AF, Anghinoni I, Kaminski J, de Lima MS. Phosphorus forms and availability assessed by 31P-NMR in successive cropped soil. Commun Soil Sci Plant Anal. 2005;36:2625–40.
74. De Oliveira CM, Erich MS, Gatiboni LC, Ohno T. Phosphorus fractions and organic matter chemistry under different land use on Humic Cambisols in Southern Brazil. Geoderma Reg. 2015;5:140–9.
75. Deiss L, de Moraes A, Dieckow J, Franzluebbers AJ, Gatiboni LC, Sassaki GI, Carvalho PCF. Soil phosphorus compounds in integrated crop-livestock systems of subtropical Brazil. Geoderma. 2016;4:88–96.
76. Noack SR, McLaughlin MJ, Smernik RJ, McBeath TM, Armstrong RD. Phosphorus speciation in manure wheat and canola plants as affected by phosphorus supply. Plant Soil. 2014;378:125–37.
77. Noack SR, Smernik RJ, McBeath TM, Armstrong RD, McLaughlin MJ. Assessing crop residue phosphorus speciation using chemical fractionation and solution ^{31}P nuclear magnetic resonance spectroscopy. Talanta. 2014;126:122–9.
78. Doolette AL, Smernik RJ. Phosphorus speciation of dormant grapevine (*Vitis vinifera* L.) canes in the Barossa Valley, South Australia. Aust J Grape Wine Res. 2016;22:462–8.
79. Preston CM, Cade-Menun BJ, Sayer BG. Characterization of Canadian backyard composts: chemical and spectroscopic analyses. Compost Sci Util. 1998;6:53–66.
80. Smith MTE, Cade-Menun BJ, Tibbett M. Soil phosphorus dynamics from sewage sludge at different stages in a treatment stream. Biol Fertil Soils. 2006;42:186–97.
81. Stutter MI. The composition, leaching and sorption behavior of some alternative sources of phosphorus for soils. Ambio. 2015;44:S207–16.

3

Effect of arbuscular mycorrhizal fungi and water stress on ultrastructural change of glandular hairs and essential oil compositions in *Ocimum gratissimum*

Zakaria Hazzoumi*, Youssef Moustakime and Khalid Amrani Joutei

Abstract

Background: The objective of this work is to study the influence of arbuscal mycorrhizal fungi (*Glomus intraradices*) and water stress on the ultrastructural change of basil glandular hair and also on the essential oil synthesis, especially on the isomerization of the two main compounds methyl chavicol and *trans*-anethole.

Results: The results of this study show that mycorrhization increases the yield of essential oils, with a maximum recorded in mycorrhizal stressed plants (0.33%) and a minimum among non-mycorrhizal unstressed plants (0.22%). Oxygenated monoterpenes represent the main family of essential oils. Furthermore, the chemical composition of these essential oils changed under the condition in which the plant is. However, the main compound methyl chavicol was found to be at maximum concentration (93%) in mycorrhizal stressed plants and minimum in non-mycorrhizal stressed plants (84%). Contrariwise, its isomer *trans*-anethole reached the maximum (8%) among non-mycorrhizal non-stressed plants and minimum (2%) in non-mycorrhizal stressed plants.

Conclusion: *Ocimum gratissimum* L. leaves showed the presence of two types of peltate glands: the first one with 4 secreting cells and the second with 8 secreting cells. The diameter of these glands changed with the hydric state of plants; in the absence of mycorrhization, stress causes a decrease in the extracellular cavity diameter of glandular trichomes and their deflation which would cause a concentration of essential oil in the glands.

Keywords: AMF, *Glomus intraradices*, m-Chavicol, *t*-Anethole, Glandular trichomes, Water stress, Extracellular cavities

Background

Water stress causes significant changes in essential oil yield and composition in several aromatic plants [26]. There has been a reduction of yield in the rosemary (*Rosmarinus officinalis* L.) and anise (*Pimpinella anisum* L.) [29, 36] following a water deficit, whereas in pepper (*Capsicum annuum* L.), the water deficit led to an increase in the essential oil yield.

In *Ocimum basilicum* plant, water stress increases the essential oil yield and the levels of major compounds (linalool and methyl chavicol) [14]. Hendawy and Khalid [12] confirmed these results in *Salvia officinalis*. Similarly, Solinas et al. [30] confirmed that water stress strongly affected the secondary metabolism of plants in a general way and essential oils in a specific way.

In our previous work performed on *Ocimum gratissimum* L. plants, Hazzoumi et al. [10] have confirmed that water stress may slightly increase the essential oil yield to an average of 10%.

Furthermore, mycorrhizae strongly affect the plant metabolism, leading to the induction of chemical defense [24]. This change affects the synthesis of several families of molecules, terpenoids [25], essential oils (Copetta et al. [2, 3]), and glucosinolates [33].Gupta et al. [8] reported that, in *Mentha arvensis* L., mycorrhizal inoculation

*Correspondence: Zakaria.hazzoumi@yahoo.fr; Zakaria.hazzoumi@usmb.ac.ma
Laboratory of Bioactive Molecules, Faculty of Science and Technology, Sidi Mohamed Ben Abdellah University, Road of Imouzzer, B. P. 2202, Fez, Morocco

significantly increased the essential oil content. Freitas et al. [4] have also observed that the inoculation of *M. arvensis* with mycorrhizal fungi leads to an increase in the essential oil yield by 89% in the same way as the major compound (menthol). In *O. gratissimum* plants inoculated with *Glomus intraradices*, Hazzoumi et al. [10] have shown that the levels of essential oils increased significantly and related this increase to the abundance of secreting glands in leaves.

The accumulation of essential oils in plants is usually limited to specialized secretory structures which are present in several different types, namely glandular trichomes. They are multicellular epidermal hairs found in some families like the Lamiaceaes, Asteraceaes, and Solanaceae, which secrete terpenes in the extracellular cavity at the apex of the trichome [34, 35].

The storage of terpenoid in these structures can be used to limit the toxicity risk to the plant itself, because it is proven that many terpenoids are potentially toxic to plant tissues when monoterpenes are released to the surrounding cells [20]. Injuries were also found when some sesquiterpenes were artificially deposited on sheets during tests on their ability to deter herbivores [23]. Therefore, sequestration of terpenoids in specific compartments by sensitive metabolic processes can be essential to avoid harmful effects. However, the morphology of these structures varies with the irrigation conditions and also depends on the toxicity of intracuticular content.

According to Gershenzon et al. [6], the glandular trichome consists of several cells providing different functions. The essential oils secreted by secretory cells are passed through a space situated at the apex of subcuticular trichome and get accumulated; however, the basal cells provide attachment to the skin structure (Fig. 1).

The question of gland maturity and the number of secreting cells per glands remains a subject of debate between scientists [34, 35]. It is found that basil plants have 4 or more secreting cells per gland, whose maturity is reflected by a slight folding in the cuticle. However, Gang et al. [5] have limited the number of secreting cells to 4 and showed that folded glands do not present

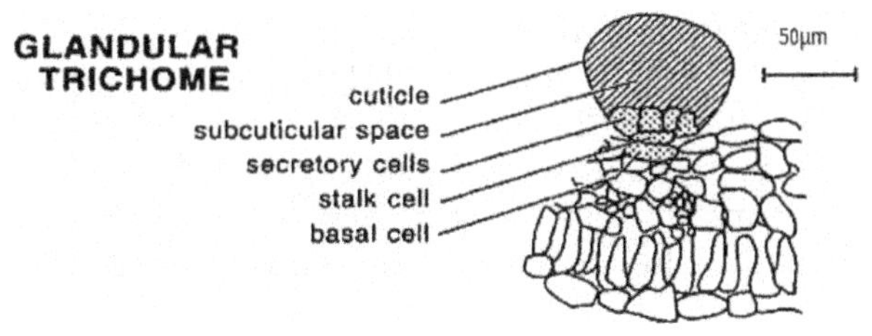

Fig. 1 Drawn from a photograph by Gershenzon et al. [6]

any terpenoid content. According to these authors, folding allows the quadratic subdivision of the cuticle due to the presence of 4 secreting cells and it is an index of non-maturation. Until the filling of the cuticular cavity by essential oils, this division tends to disappear.

In this study, we tried to evaluate the effect of water stress and AMF by *G. intraradices* on the essential oil synthesis, in terms of yield and chemical composition and also the morphology of the secretory glands of these essential oils, in an aromatic plant, basil*O. gratissimum* L, which is widely used in the traditional pharmacopoeia and cosmetics industry in Morocco .

Methods

Plant material (germination and culture)

The basil (*O. gratissimum* L.) seeds were disinfected by a passage in ethanol 95% for 1 min 30 s and immersed in mercury hypochlorite solution (1%) for 3–4 min. Seeds were then rinsed several times in sterile distilled water before being placed on an agar medium or on a filter paper moistened with sterile distilled water in Petri dishes. The petri dishes were placed in an oven at 26 °C, in the dark, for seed germination.

After germination, seedlings were transplanted into plastic pots (3 kg capacity) containing the growth substrate, with 50–60 plants per pot, and were grown in the greenhouse at temperature between 26 and 34 °C. The basil plants (M or NM) are watered daily. After the 50th day, the plants are subjected to different water regimes:

a. Water regime unstressed (NS): the plants are not deprived of water throughout their growth
b. Application of water stress (S): plants are deprived of water for 2 weeks.

Estimation of mycorrhizal root infection

The identification of mycorrhizal root infection is realized by optical microscopy through a technique not vital staining with trypan blue (TB), described by Hayman [9] revealing the set of fungal biomass.

Root samples, taken at random, are thoroughly rinsed to remove the adhering substrate. Then, the root fragments are digested in a solution of 10% potassium hydroxide (KOH) for 45 min at 90 °C in an oven, in order to empty the cell of their cytoplasmic contents which confer coloring. Then, the roots are thoroughly rinsed with distilled water and placed in the lactophenol trypan blue solution (0.5%) at 90 °C in the oven for 15 min.

Estimation of mycorrhization

We used the technique described by Trouvelot et al. [32]. This method allows to judge the state of mycorrhization and reflects the potential of the symbiotic system. Colored roots were cut into fragments of approximately 1 cm length. Thirty random fragments are assembled and crushed between slide and coverslip in lactoglycerol, with 15 fragments per slide. Endomycorrhizal infection is observed under the light microscope. Several parameters are evaluated as follows:

$$F\% = \left(\text{Number of mycorrhizal fragments}/N\right) \times 100,$$

where (F %) is the frequency of mycorrhization reflecting the importance and the percentage of fragments of infected roots, with N as the total number of root fragments observed.

(M%) is the intensity of the cortex colonization expressing the portion of the cortex colonized with respect to the entire root system, with $n5, n4, \ldots, n1$ as the number of fragments, respectively, denoted as 5, 4,..., 1,

$$\%M = (95\ n5 + 70\ n4 + 30\ n3 + 10\ n2 + n1)/N.$$

Mycorrhizal intensity

$$\%m = M \times (\text{total number})/\left(\text{number of mycorrhizal fragments}\right) = M \times 100/F.$$

(A%) is the frequency of arbuscular in roots:

$$\%A = a \times (M/100).$$

The arbuscular intensity of the mycorrhizal part is estimated as follows:

$$\%a = (100\ mA3 + 50\ mA2 + 10\ mA1)/100,$$

where $mA3$, $mA2$, *and* $mA1$% m are, respectively, assigned $A3$, $A2$, *and* $A1$ notes, and where $mA3$ = ((95$n5A3$ 70$n4A3$ + + + 30$n3A3$ 5$n2A3$ $n1A3$ +)/mycorrhizal number) × 100/m.

Similarly for A2 and A1.

Relative water contents (RWC)

The relative water content (RWC) is measured on the full development of 7th or 8th leaf using the following formula according to Bandurska [1]:

$$\text{RWC\%} = 100 \times [(\text{FW} - \text{DW})/(\text{WT} - \text{DW})],$$

where FW represents the weight of fresh leaf material, WT the the weight of fresh material from the turgid leaf which was submerged in distilled water for 4 h, and DW represents the dry weight of the sheet material placed in an oven at 70 °C for 24 h.

Environmental scanning electron microscope

The observations were performed using a Quanta 200 (FEI company) Environmental Scanning Electron Microscope . The microscope is equipped with a tungsten electron gun. The analyses are carried out under a partial pressure of water vapor.

Specifications/capabilities

It is capable of functioning under high-vacuum, low-vacuum, and environmental vacuum modes and its specifications include field emission gun, energy-dispersive silicon drift X-ray detector, cathodoluminescence system, backscattered electron detector, peltier-cooled stage, scanning transmission detector, wet scanning transmission detector, large field detector, gaseous analytical detector, gaseous backscatter detector, low-vacuum and low-voltage cones, and gaseous secondary electron detector.

Light microscopy

Optika B-350 Trinocular Biological Microscope is connected to camera AIPTEK Z600 (resolution 1080 Megapixels) and is equipped with two different types of optical system: anti-fungus 160 mm standard system and infinity corrected system (IOS). In both cases, the field diameter of the system is 20 mm, macro- and micrometric focusing (graduated 0.002 mm) wth coaxial controls.

Extraction of essential oils

Dried aerial parts (100 g) of *O. gratissimum* plants were submitted to hydrodistillation with a Clevenger-type apparatus and extracted with 2 L of water for 180 min (until no more essential oil was obtained). The essential oil was collected, dried under anhydrous sodium sulphate, and stored at 4 °C until analysis. The essential oil yield is calculated using the following formula:

$$\text{YEO}\left(\text{ml/100 g Dm}\right) = (V/\text{Dm} \times 100) \pm (\Delta V/\text{Dm} \times 100),$$

where Y.E.O represents the essential oil yield of dry matter, V the volume of essential oils collected (ml), ΔV the reading error, and Dm represents the dry plant mass (g).

GC/MS analysis

Gas chromatography (GC) analyses were performed on a Hewlett-Packard (HP 6890) gas chromatograph (FID), equipped with a 5% phenyl methyl silicone HP-5 capillary column (30 m × 0.25 mm × film thickness 0.25 μm). The column temperature was programed at 50 °C for 5 min and then increased to 200 °C with a flow rate of 4 °C/min. Gas chromatography conditions were as follows: N2 as carrier gas (1.8 ml/min); split mode was used (Flow: 72.1 ml/min, ratio 1/50); the temperature of injector and detector was 250 °C; and the final hold time was 48 min. The machine was led by a computer system type "HP ChemStation," managing the functioning of the machine and allowing to follow the evolution of chromatographic analyses. Diluted samples (1/20 in methanol) of 1 μl were injected manually.

GC/MS qualitative analyses were performed on a Hewlett-Packard equipped with a HP-5MS (Crosslinked 5% PHME Siloxane) capillary column (30 m × 0.25 mm i.d, 0.25 μm film thickness) and coupled with a mass spectrometer (HP 5973). The temperature was programmed from 50 to 250 °C at 2 °C/min. The carrier gas was He (1.5 ml/min), and a split mode was used (Flow: 112 ml/min, ratio: 1/74.7). The different compounds were confirmed by reference to their MS identities (Library of NIST/EPA/NIH MASS SPECTRAL LIBRARY Version 2.0, built Jul 1, 2002). MS operating parameters were ionization voltage, 70 eV; ion source temperature, 230 °C; and scan mass, 35–450.

Statistical analysis

One-way analysis of variance was carried out for each parameter studied. Tukey's post hoc multiple mean comparison test was used to test for significant differences between treatments (at $p \leq 0.05\%$). Univariate analysis was used to test significant differences in treatments, accessions, and their interaction for an individual parameter. All statistical analyses were performed with IBM. SPSS statistics, Version 19. The results of each experiment (biochemical assays) were repeated three times, and 20 times for mycorrhizal intensity assays.

Results

Mycorrhizal colonization

Table 1 shows mycorrhizal colonization of basil plants subjected to continuous irrigation (NS) and water stress (*S*) after treatment with trypan blue. Mycorrhizal intensity (*M*%) and arbuscular richness (*A*%) showed no differences between the stressed plant and unstressed plant. Contrariwise, control plants which were grown in a sterilized soil showed no mycorrhizal colonization.

Table 1 Mycorrhizal colonization of basil (*O. gratissimum*) roots subjected to continuous irrigation (NS) and water stress (*S*); the values followed by different letters are significantly different ($p = 0.05$)

Mycorrhizal colonization	NS	*S*
Mycorrhizal frequency (*F*%)	78.5 ± 2.75	77.3 ± 2.44
Mycorrhizal intensity (*M*%)	35 ± 2.32	32 ± 1.88
Arbuscular richness (*A*%)	20 ± 1.12	17 ± 0.95

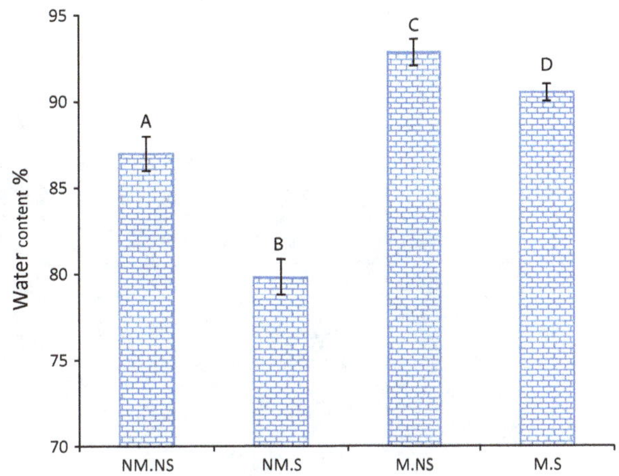

Fig. 2 Water contents in basil plant (*O. gratissimum* L.) inoculated with (*G. intraradices*) (*M*) or not (NM) and subjected to water stress (*S*) or not (NS). The values followed by *different letters* are significantly different ($p = 0.05$)

Effect of mycorrhizae and water stress on water contents

Mycorrhization and water stress influenced water contents significantly in basil plant; we recorded a significant variation between the different treatments (Fig. 2). In M.NS plants, mycorrhizal fungi increase the levels of water to up to 93%. Non-mycorrhizal plants showed the lowest levels with a dramatic decrease recorded in NM.S plants (81%). In MS plants, there has been relatively high water content (91%) reflecting the role of mycorrhizal fungi in withstanding the water deficit in stressed plants.

Microscopy observations

Observations by light microscopy of a cross section of the central portion of basil plants (*O. gratissimum*) which received normal irrigation show the presence of two kinds of peltate glands on the ventral side: glands having 4 secreting cells (Fig. 3a) and glands with 8 secreting cells (Fig. 3b, c). These glands have a swollen appearance indicating adequate irrigation of the plants.

The light microscopy also shows significant differences in the diameter, structure, and color of these structures: unstressed plants, mycorrhizal or not, have large glands having an inflated shape and are greenish (Fig. 4a, c). Contrariwise, stressed plants have glands with a reddish coloration (Fig. 4c, d) with a small structure, and a deflated aspect in NM plants (Fig. 4c). These results obtained in light microscopy are confirmed by environmental scanning electron microscopy (Fig. 5).

The mycorrhization causes a significant increase in the diameter of these cavities (Fig. 5c, d), while non-mycorrhizal plants have cavities with small diameter and this is regardless of the hydration level of plant (Fig. 5a, b). Water stress causes further deflation of these cavities

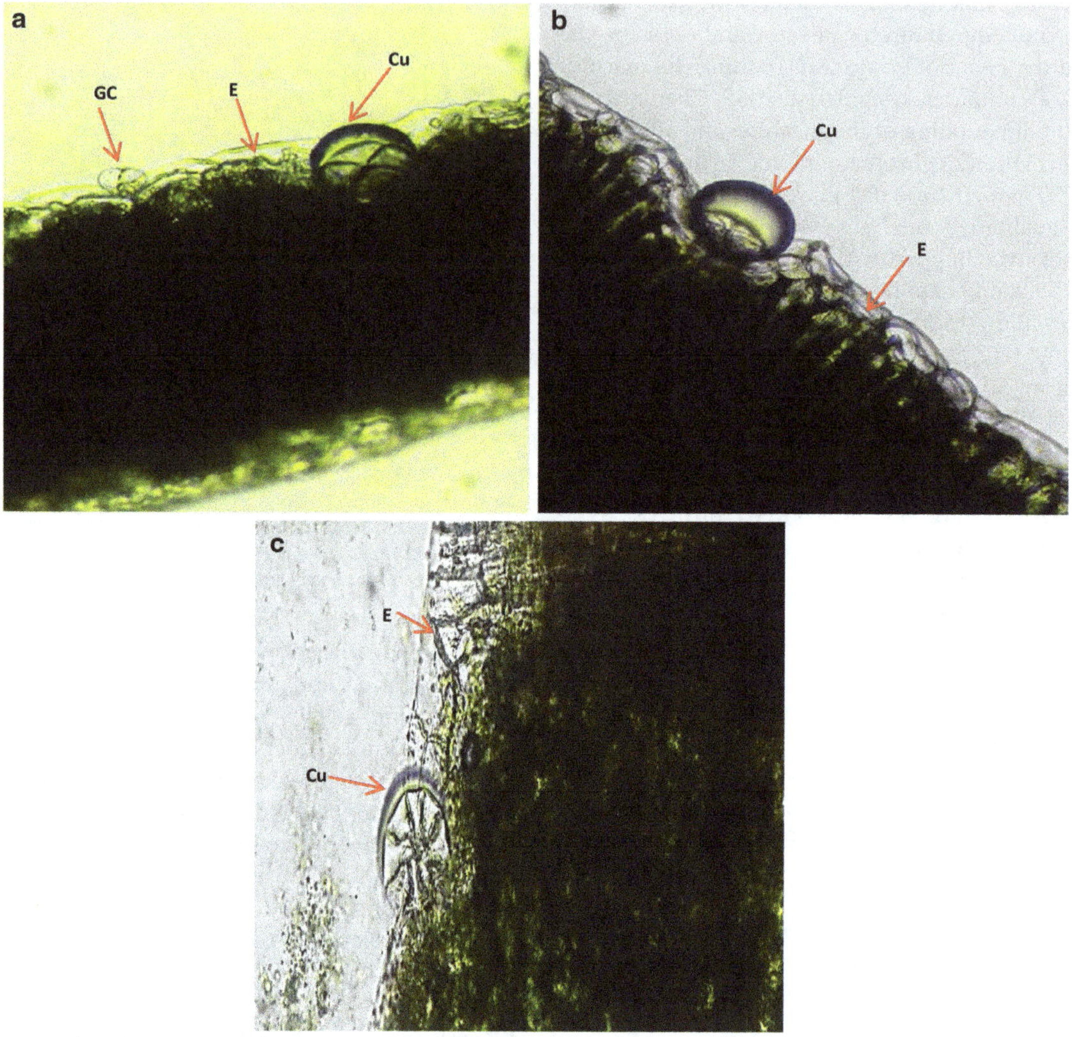

Fig. 3 Observation by light microscopy of a cross section of the central portion of a basil plant (*O. gratissimum*). NS showing two kinds of peltate glands, with 4 secreting cells (**a**) and 8 secreting cells (**b**, **c**); *E* Epidermis, *Cu* Cuticle, *GC*: capitate gland

with a more or less wrinkled appearance mainly in mycorrhizal plants. This deflation let appear at the epidermis surface a glandular excavation with a diameter as that of an inflated glandular cavity in unstressed plant (between 75 and 78 μm) (Fig. 6).

In addition, Fig. 7 shows the gland location at the basal region (a) and apical region (b) of a basil plants cultivated under optimum conditions (mycorrhized unstressed plant). It is noted that there is no difference as regards the morphology of these glands whatever the section is (basal or apical); the glands are always present in a swollen form. Contrariwise, in the plant found in stressed conditions (non-mycorrhized stressed plant), a structural difference is observed between the glands according to their location in the sheet: the glands in the basal section was found to have a swollen appearance (Fig. 8a), while those located in the apical portion was found to have a deflated aspect, with a diameter of 58 μm and leaving an excavation of a very large diameter (Fig. 6 also Fig. 8b) that can reach 80 μm or more. This difference between the two sections of the sheet can be explained by the fact that the basal section, being closest to the stem, is the first to receive water and nutrients, thus contributing to adequate supply of water and nutrients to the cells in this section and therefore a more or less normal moisture balance of the cells. On the contrary, the other section of the sheet (apical) receives less supply of these elements, which causes a reduction of volume and deflation of glands. Note that we do not see any difference in the diameter of these glands according to their location.

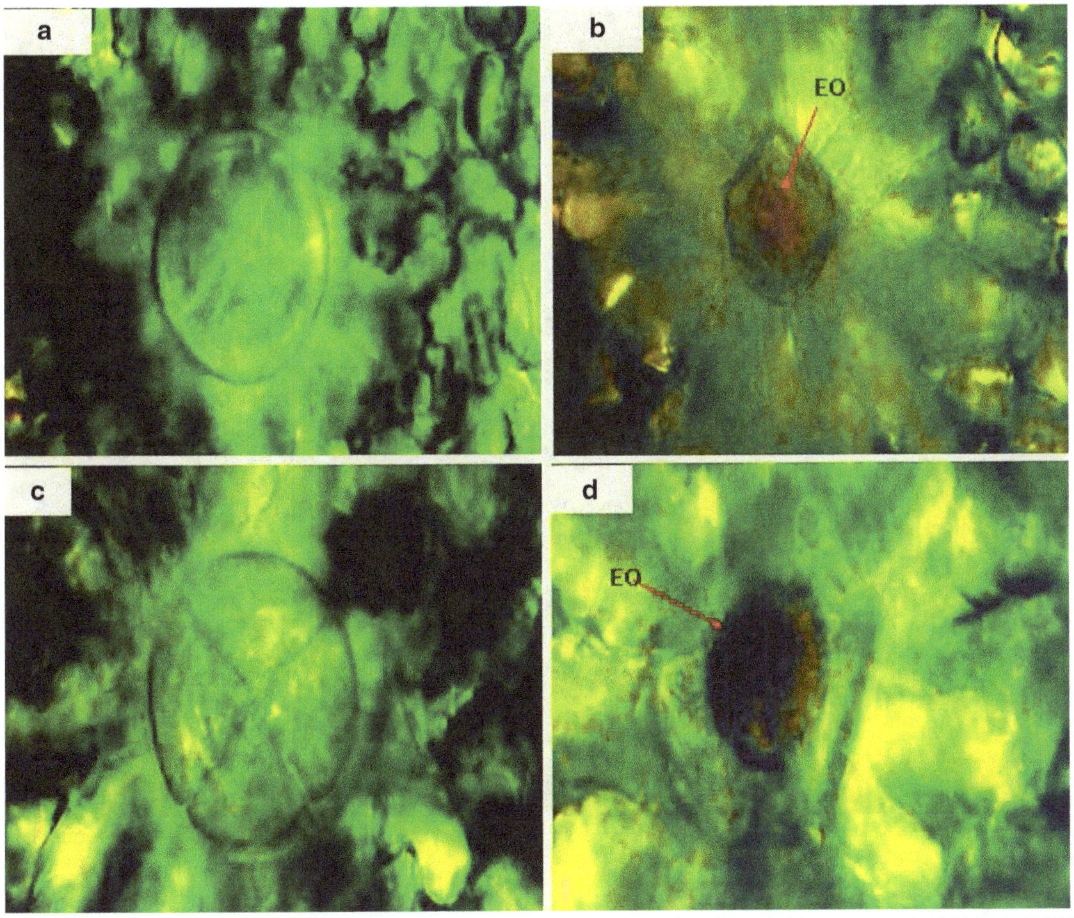

Fig. 4 Light microscopy observations of secretory glands realized at the ventral surface of basil leaves (*O. gratissimum*): NM.NS (**a**), NM.S (**b**), M.NS (**c**), and MS (**d**) (*EO*: essential oil)

Composition of basil (*O.gratissimum* L.) essential oils after inoculation with AMF and different irrigation systems

The essential oil yield is strongly changed by the mycorrhization since found a higher synthesis in both cases stress and irrigation. This increase reaches values up to 37.5% more than those found in non-mycorrhizal plants (Table 2). Moreover, water stress did not show a significant influence on the essential oil synthesis, because we noticed a slight increase in EO levels. These results can be confirmed by the increase in the number and diameter of glandular hairs in basil leaves. From the composition viewpoint, the basil essential oils have 4 categories of molecules: monoterpene hydrocarbon, oxygenated monoterpenes, oxygenated sesquiterpenes, and hydrocarbon sesquiterpenes. The hydrocarbon monoterpenes, represented by the single molecule Azulene, have very low levels irrespective of the surrounding factor.

However, oxygenated monoterpenes have the major contribution to basil essential oils, with 5 molecules whose contents changed between 87% in NM.S plants and a maximum reached 97% in M.NS plants. The methyl chavicol is the major molecule of basil essential oils, with contents ranging between a maximum of 93% in MS plants and a minimum of 84% in NM.S plants. However, its isomer (*t*-anethole) varies differently between stress and mycorrhizal conditions (Fig. 9), and water stress causes a reduction in the levels of this molecule, especially in non-mycorrhizal stressed plants (2%). However, mycorrhizal causes an increase in the levels of this molecule with a maximum recorded in M.NS plants (8%).

Hydrocarbon sesquiterpenes have the highest qualitative diversity with 9 molecules in NM.S plants and 8 molecules in the other treatments. Moreover, we note that the highest content of this family was found in the NM.S plants (8.61%) with the appearance of two molecules (germacrene-D) whose contents are 1.76% and the nearly disappearance of Aristolene (0.05%). However, this family of molecules has a very low abundance in M.NS plants with values <2% of total yield. Moreover, the oxygenated

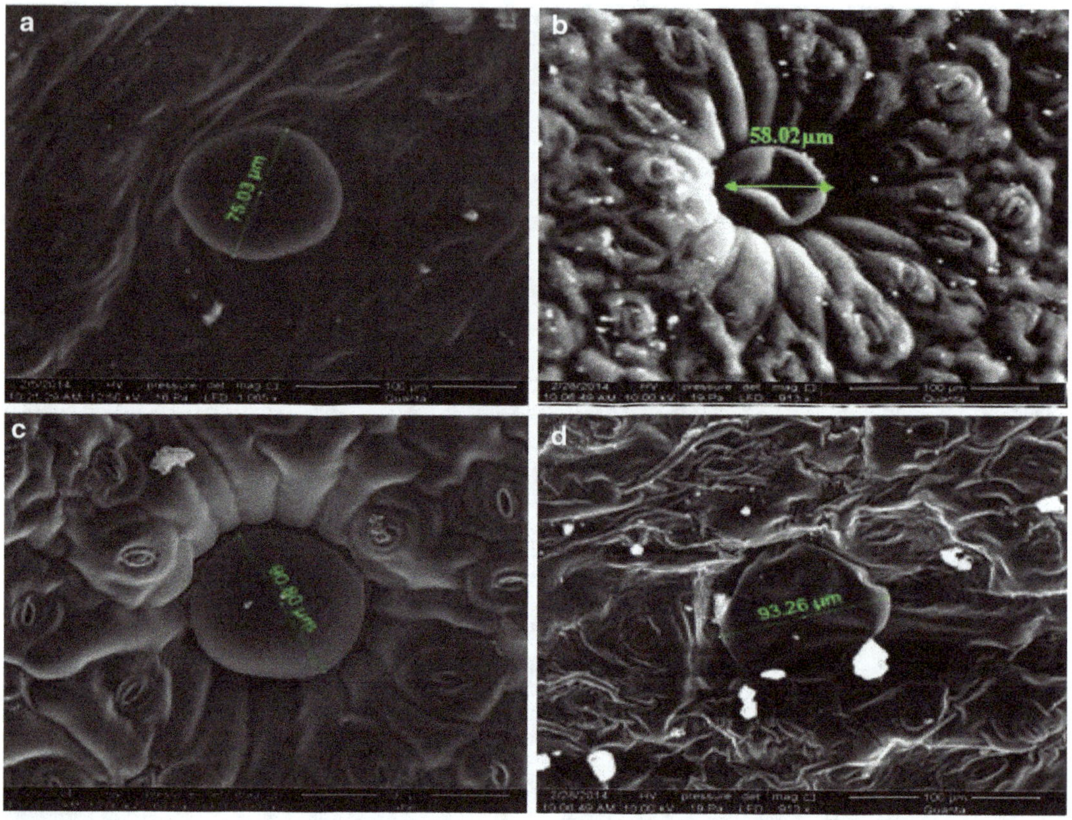

Fig. 5 Observation by environmental scanning electron microscope of the central portion of the basil leaves (*O. gratissimum*) from a plant: NM.NS (**a**), NM.S (**b**), M.NS (**c**), and MS (**d**)

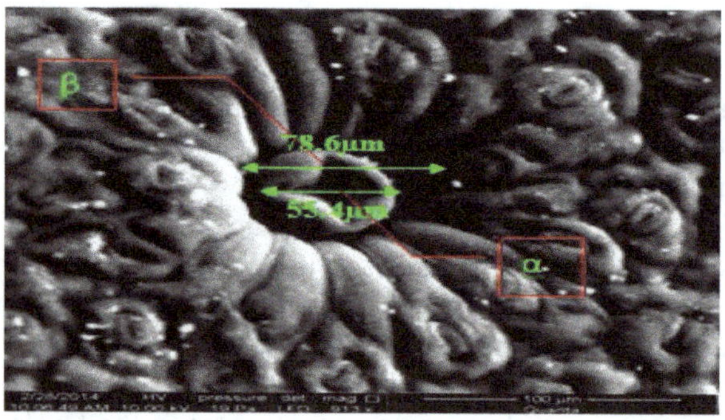

Fig. 6 Comparison of excavation glandular diameter in NM.S plant and its potential diameter in case of cuticle swelling

sesquiterpenes were represented by a single molecule (Cubenol), which appear in NM.S plants (0.2%).

Discussion

The glandular trichome morphology is largely influenced by mycorrhizal and the hydric state of plant; the observations made in this study have demonstrated the existence of two types of peltate glands according to their shape: inflated glands and deflated glands. Mycorrhization provides for plants an optimal nutrition as well as hydric or mineral, allowing the leaves to receive the water and nutrients in roughly sufficient quantities. This water inlet provides a turgor state of epidermal cells and secretory cells of the gland, causing swelling of cavities and magnification.

In NM plants, the decrease of water content in leaves causes a plasmolysis of cells following the decrease of the water content at the glands, which leads to a decrease of their volume and deflation. This decrease of volume is even more pronounced in stress condition. These results do not confirm those obtained by Copetta et al. [2, 3] who did not observe significant differences in structures between the cavities of NM plants and those M plants.

Furthermore, we were able to highlight the presence of secreting glands having 4 cells and other with 8 cells. Light microscopy observations (Fig. 3) showed that the glands with pleated and deflated appearance have a reddish color, which allows us to conclude the presence of essential oil content. These results are in agreement with those of Werker et al. [35] who showed that basil (*O. basilicum*) may have a gland with 4 or more secreting cells, and the maturation of glands is determined by

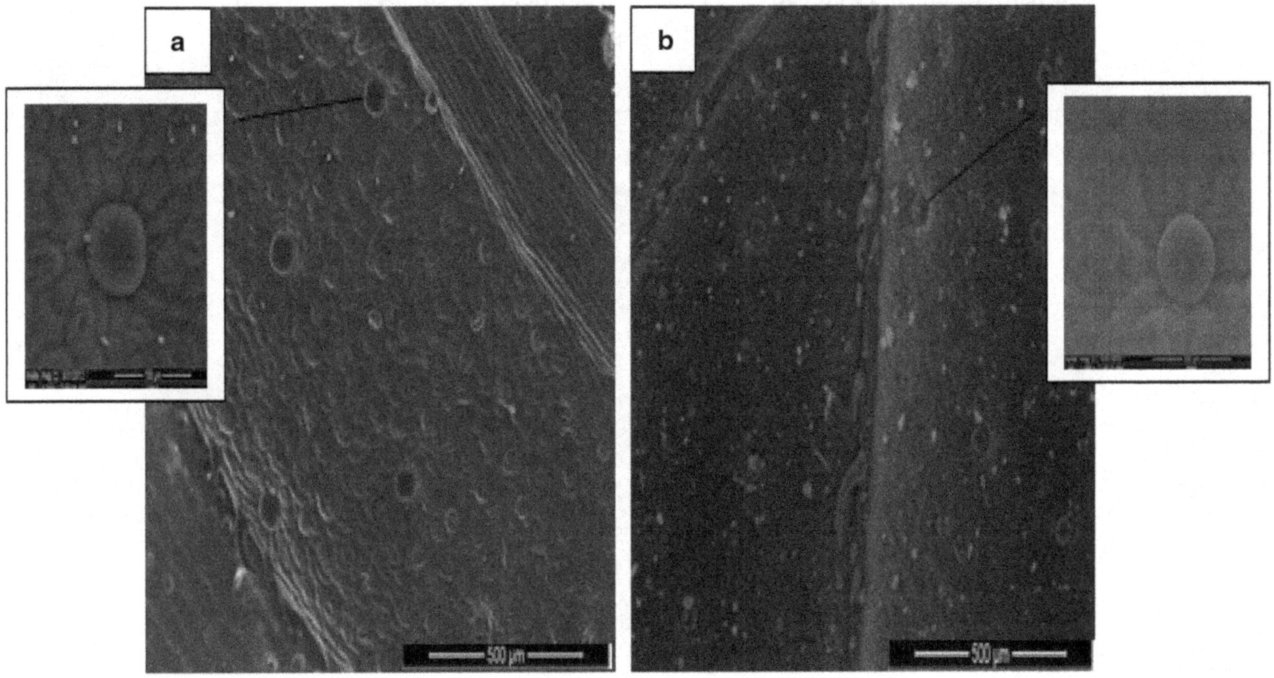

Fig. 7 Observation by environmental scanning electron microscope showed the structural variation of the secretory glands between the base portion (**a**) and apical (**b**) in M.NS plant

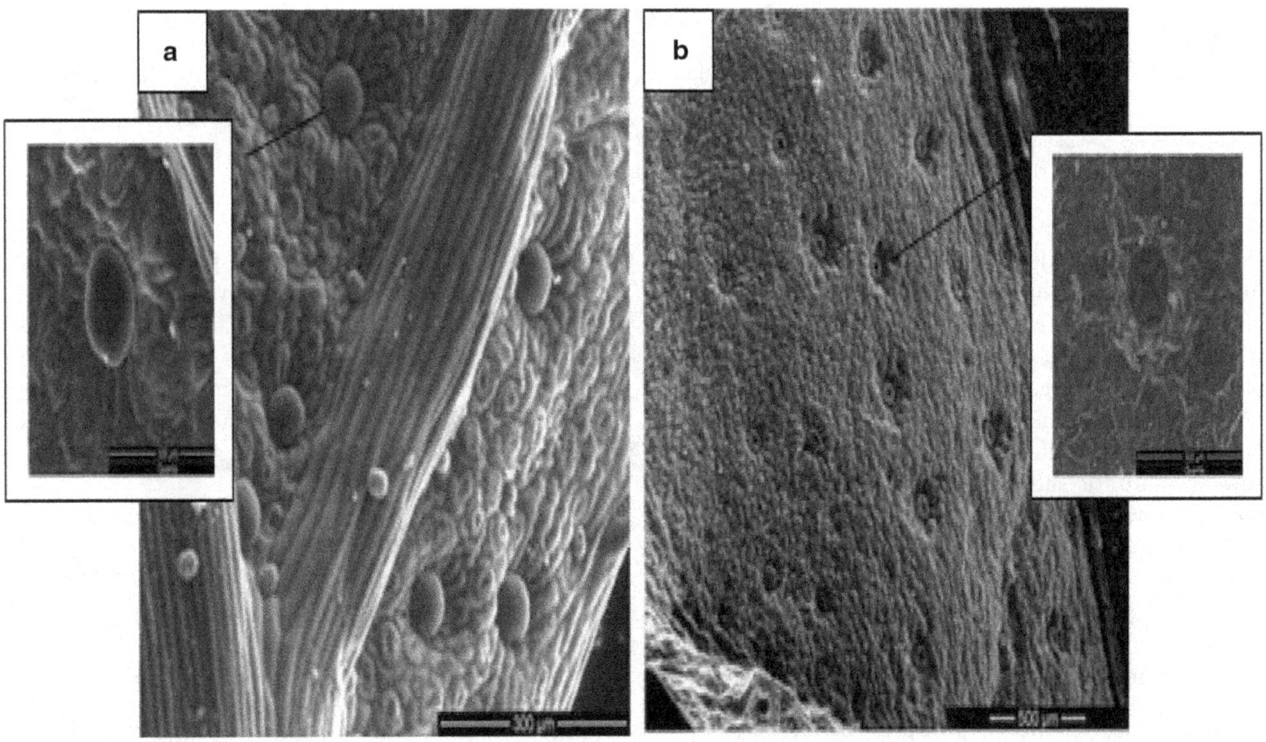

Fig. 8 Observation by environmental scanning electron microscope showed a structural variation of the secretory glands between the base portion (**a**) and apical (**b**) in NM.S plant

Table 2 Influence of mycorhization and water stress on essential oils yield and composition in *O. gratissimum* L

Treatments	**NM.NS**	**NM.S**	**M.NS**	**MS**
Oils content (%)	0.22 ± 0.03a	0.24 ± 0.02a	0.3 ± 0.1b	0.33 ± 0.08c
Components	**Peak area %**			
	Hydrocarbons monoterpenes			
Azulene	0.44 ± 0.02a	0.89 ± 0.03b	0.29 ± 0.06a	0.23 ± 0.07a
	Oxygenated monoterpenes			
Eucalyptol	2.02 ± 0.08a	0.39 ± 0.06b	0.74 ± 0.11c	0.88 ± 0.05c
α-Campholene	0.20 ± 0.05a	0.26 ± 0.03a	0.23 ± 0.02a	0.13 ± 0.02b
Methyl-chavicol	85.26 ± 0.32a	84.38 ± 0.72a	86.40 ± 0.62a	92.62 ± 0.42b
Trans anethole	4.25 ± 0.22a	2.28 ± 0.3b	8.31 ± 0.41c	2.45 ± 0.38b
Cis anéthole	0.12 ± 0.04a	0.08 ± 0.03a	1.64 ± 0.14b	0.15 ± 0.05a
Total	91.85a	87.38b	97.32c	96.23c
	Hydrocarbons sesquiterpenes			
Caryophyllene	0.21 ± 0.11a	0.36 ± 0.09a	0.11 ± 0.02b	0.15 ± 0.04b
Calarene	0.50 ± 0.15a	0.68 ± 0.12a	0.18 ± 0.14b	0.27 ± 0.11b
α-Longipinene	0.82 ± 0.18a	1.37 ± 0.09b	0.35 ± 0.18c	0.71 ± 0.16a
α-Caryophyllene	0.52 ± 0.19a	0.78 ± 0.02b	0.31 ± 0.16a	0.34 ± 0.18a
Aristolene	0.73 ± 0.14a	0.05 ± 0.2a	0.33 ± 0.08b	0.64 ± 0.1a
Germacrene-D	_	1.76 ± 0.18	_	_
β Cedrene	0.21 ± 0.11a	0.38 ± 0.09a	0.09 ± 0.02b	0.16 ± 0.08a
Aromadendrene	0.21 ± 0.13a	0.65 ± 0.15b	0.11 ± 0.09a	0.18 ± 0.09a
Cadinene	0.78 ± 0.19a	2.38 ± 0.09b	0.48 ± 0.32a	0.84 ± 0.11a
Total	3.98a	8.41b	1.96c	3.29a
	Oxygenated sesquiterpenes			
Cubenol	–	0.20 ± 0.07	–	–

One-way analysis of variance was carried out for each parameter studied. Tukey's post hoc multiple mean comparison test was used to test for significant differences between treatments (at $p \leq 0.05\%$). Univariate analysis was used to test significant differences in treatments, accessions and their interaction for an individual parameter. All statistical analyses were performed with IBM.SPSS statistics, Version 19.The results of each experiment (biochemical essays) were repeated three times and (20 times for mycorrhizal intensity essays)

The values are the average of 3 determinations, and the values followed by different letters are significantly different ($p \leq 0.05$)

the folding of the cuticle that occurs when the secretory cells stop secreting the essential oils. They also noticed that the older leaves have glands with a pleated appearance and are sunken. Contrariwise, Gang et al. [5] have not observed the presence of glands with 8 cells, and they say that the pleated glands do not present a terpenoid content. According to the same authors, the folding allows a quadratic subdivision of the cuticle as a result of the presence of 4 secreting cells, and they indicate the non-maturation of glands since at the time of filling of this cuticular cavity by essential oils, this quadratic division tends to disappear and these glands become less abundant.

In addition, according to our results, the degree of maturation of the gland is related to essential oil synthesis and cuticular cavities filled by these oils, leading to an important increase in gland diameter, regardless of the degree of fold, and also this aspect is linked to the water and nutritional status of plants.

This finding is confirmed by the position of glands in leaf. The glands in basal section have an inflated appearance and those of the apical section have a deflated aspect when the plant is under water stress condition. Moreover, the measurement of excavation diameter depends on the gland location at the leaf. The diameter of this latter can reach about 78.6 μm, diameter that gland may have when is swelling (Fig. 6b).

Effect on essential oil yield and composition

The essential oil yield is strongly influenced by the mycorrhization, since found a higher synthesis regardless the hydration level of plant (stress or irrigation). Water stress also shows a slight increase in essential oil contents. However, the combined treatment (stress and mycorrhization) can increase the EO yield about 50% more than NM.NS. These results confirm those of Copetta et al. [2, 3] which showed that this increase in mycorrhizal plants may be due to a defensive response to fungal colonization, since

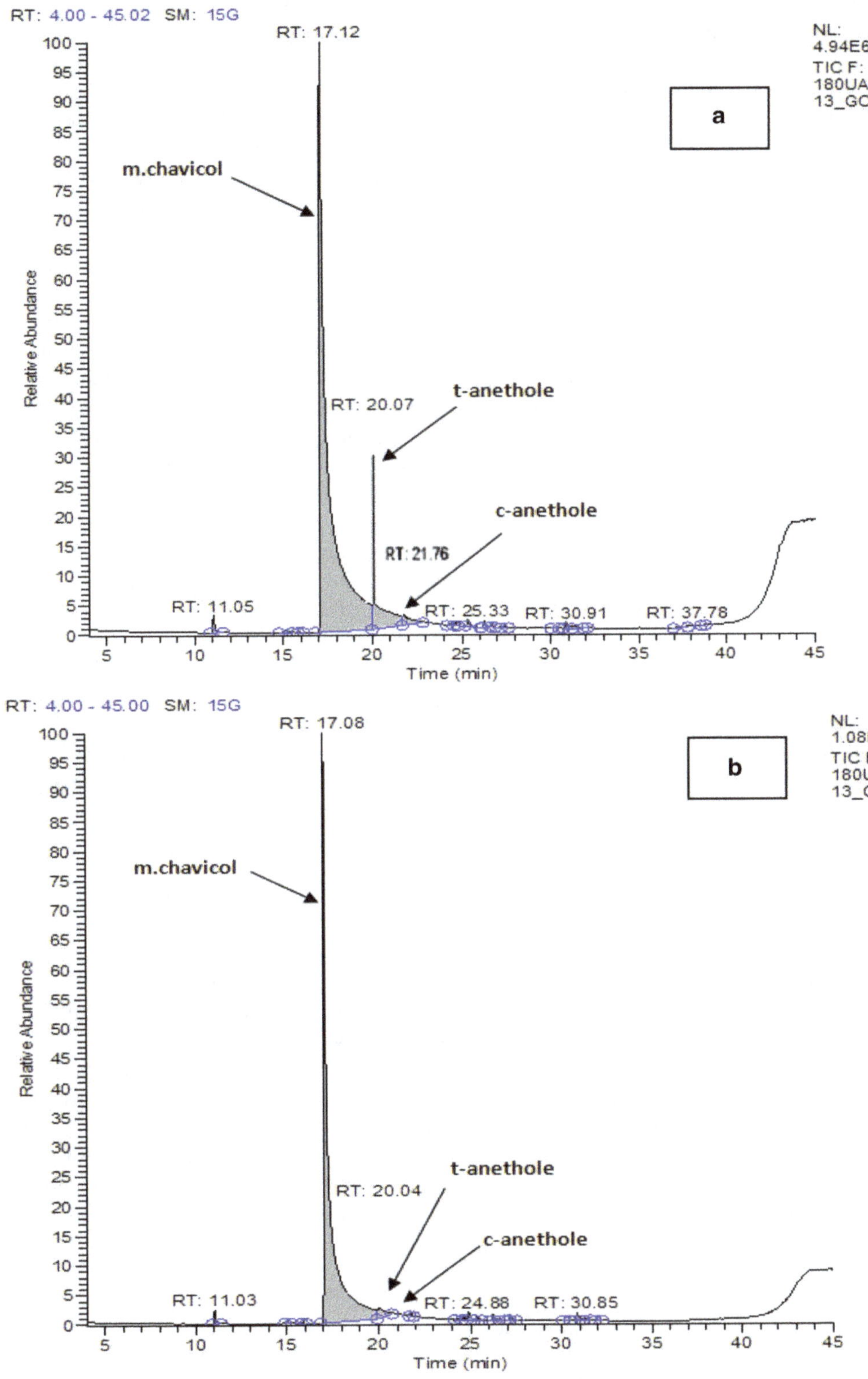

Fig. 9 Chromatograms of essential oils extracted from *O. gratissimum* L. plants: **a** M.NS, **b** MS

several molecules of these essential oils have fungicidal properties. Furthermore, the increase can also be attributed to a change in the secondary metabolism of plants and the number of gland leaves confirming the results of Zolfaghari et al. [37] and Freitas et al. [4], which recorded an increase in essential oil yields of *M. arvensis* and *Mentha piperita* L. plants inoculated with mycorrhizal fungi. The colonization by AM fungi stimulates the production of these glands probably by increasing endogenous hormone levels particularly AIA, cytokinin, and GA. These results are confirmed also by Hazzoumi et al. [10, 11].

Considering essential oil groups, we see that mycorrhization generates an increase in oxygenated monoterpene level, both in stress condition or not. By contrast, hydrocarbon sesquiterpenes decrease remarkably. These results go in the same way as those of Copetta et al. [2, 3] who have shown that basil plants (*O. basilicum* L.) inoculated by *Gigaspor arosea* and *Gigaspora margarita* induce a quantitative and not qualitative variation in the essential oil constituents, with an increase in oxygenated monoterpene level and the diminution of hydrocarbon sesquiterpenes. In the same way, Khaosaad et al. [15] showed in two species of basil (*O. vulgare* and *O. basilicum*) that the inoculation with *G. rosea* and *G. mosseae* lead to an increase in essential oil yield and changed the composition. This increase is related to the abundance of secretory glands and their maturation stage (secretory or postsecretory stage) according to Sudria et al. [31].

The influence of mycorrhization in the levels of essential oil groups is mainly due to the change in the contents of different molecules constitutive of this group. Following inoculation by *G. intraradices*, the increase is mostly influenced by the main compounds variation (m-chavicol and *t*-anethole); these results are confirmed by Copetta et al. [2] who showed a significant increase in levels of some molecules like eucalyptol and linalool in *O. basilicum* inoculated by *G. Margarita*. Moreover, in studies realized on *M. arvensis*, Khaliq and Janardhanan [16], Gupta et al. [8], and Freitas et al. [4] showed a relationship between the presence of AMF and some change in essential oil composition. Similar results were confirmed by Kapoor et al. [13] on *Coriandrum sativum*, and Nemec and Lund [22] on *Citrus Jambhiri* inoculated by *G. intraradices*.

Our results also show that water stress causes a slight increase in essential oil yield; this increase may be a result of the activation of certain enzymes involved in the synthesis of some secondary metabolites implicated in defense. These results are in agreement with those of Khalid [14] who reported an increase in essential oil yield after applying water stress on *Ocimum basilicum* L., *Ocimum americanum* L., *Cymbopogon nardus* L., and *Cymbopogon pendulm* L. plants.

Furthermore, we find that water stress causes a decrease in oxygenated monoterpene levels and a very important increase in hydrocarbon sesquiterpenes. These results are consistent with those of Simon et al. [28] who noticed an increased levels of linalool and methyl chavicol after applying water stress on basil plants (*O. basilicum*). Moreover, Khalid [14] reported high levels of methyl chavicol after applying water stress on *O. basilicum*, while in *O. americanum* L. plants, he recorded a decrease in levels of this molecule. Eucalyptol slightly decreases after water stress in *O. basilicum* and *O. americanum*, while the cadinene contents increase, which is consistent with our results.

Influence of mycorrhization in the presence or absence of water stress on the two isomers: methyl chavicol and *trans*-anethole

Essential oils (*O. gratissimum*) are characterized by m-chavicol as chemotype, with the presence of its isomer (*t*-anethole). These two isomers are different only in the double bond of the propenyl chain, and we can find this natural isomer in many EO: (*Artemisia dracunculus* L.), basil (*O. basilicum* L.), and (*Foeniculum vulgare*) [7, 18] but with different contents. This difference is mainly due to the (*O*-methyl transferase) enzyme, which uses chavicol and *trans*-anol as substrate to produce methyl chavicol and *trans*-anethole, respectively [7, 19].

This variation in the content of the two isomers and the absence of *trans*-anol suggest that in *O. gratissimum* L., the synthesis pathway of *trans*-anethole pass through direct isomerization of methyl chavicol (Fig. 10) probably through an enzyme complex and not by the transformation of *trans*-anol as already

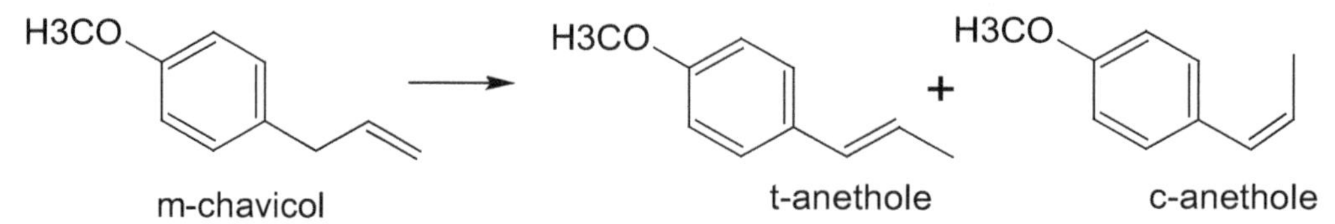

Fig. 10 Isomerization of m-chavicol to anethole according to Kishore and Kannan [17]

suggested by Gross et al. [7]. In vitro catalysis of this reaction is possible and widely used in the food industry [17, 27] Fig. 9. Low levels of cis-anethole found in our extracts can be attributed to the instability of the cis configuration and they confirm the presence of a direct isomerization.

Furthermore, our results show that these two isomers progress in opposite ways: the increase of a molecule causes a decrease of its isomer and vice versa (Table 2; Fig. 9). However, Hazzoumi et al. [11] have shown that exogenous application of GA 70 ppm on basil plants (*O. gratissimum*) causes the isomerization of m-chavicol to *t*-anethole, by the variation in the level of a specific enzyme who assured a direct isomerization of m-chavicol to *t*-anethole (changed the double bond position), and this enzyme is influenced by the plant growth regulator. Furthermore, Martín-Rodríguez et al. [21] have shown the existence of a correlation between mycorrhizal development and synthesis of GA in tomato. These authors noted an increase in endogenous GA level in the roots, and over expression of genes encoding enzymes involved in the biosynthesis of GA after mycorrhizal infection. This hypothesis can be confirmed by the results obtained in NM plants in which the isomerization phenomenon is not observed.

Conclusion

The histological study of *O. gratissimum* L. peltate glands shows clear structural differences when the plants are subject to mycorrhization or water stress. However. In case of non-mycorrhization, there is deflation of extracellular cavities and a decrease in diameter which causes an increase in the concentration of EO. Water stress leads to the same findings, but this effect is attenuated in case of mycorrhization.

These two factors also influence the synthesis of essential oils: Water stress and mycorrhizal symbiosis cause variation in yields and compostion. Mycorrhization causes increase in essential oil yield regardless of the hydric state of the plant. Furthermore, the synthesis of the majority compound, methyl chavicol, increases in mycorrhizal stressed plants (93%) and decreases in non-mycorrhizal stressed plants (84%), while its isomer, *trans*-anethole, increases in NM.NS (8%) and diminishes in NM.S plants (2%).

Abbreviations

AMF: arbuscular mycorrhizal fungi; m-chavicol: methyl chavicol; *t*-anethole: *trans*-anethole; GA: gibberellic acid; IAA: indole 3-acetic acid; EO: essential oil; M.NS: mycorrhizal unstressed; MS: mycorrhizal stressed; NM.NS: non-mycorrhizal unstressed; NM.S: non-mycorrhizal stressed.

Authors' contributions

YM and KAJ are participated in the data processing, the design of the study, performed the statistical analysis and the manuscript drafting. All authors read and approved the final manuscript.

Competing interests

The authors declare that they have no competing interests.

References

1. Bandurska H. Akumulacja wolnej proliny jako przejaw metabolicznej reakcji rooelin na dziaanie stresu wodnego. Wiad Bot. 1991;35:35–46
2. Copetta A, Lingua G, Berta G. Effects of three AM fungi on growth, distribution of glandular hairs, and essential oil production in *Ocimum basilicum* L. var. Genovese. Mycorrhiza. 2006;16:485–94.
3. Copetta A, Lingua G, Berta G, Bardi L, Masoero G (2006) Three arbuscular mycorrhizal fungi differently affect growth, distribution of glandular trichomes and essential oil composition in *Ocimum basilicum* var. Genovese. In: Proceedings of the 1st international symposium on the Labiatae: Advances in Production, Biotechnology and Utilisation, vol 723. 2006; p. 151–56.
4. Freitas MSM, Martins MA, Vieira IJC. Yield and quality of essential oils of *Mentha arvensis* in response to inoculation with arbuscular mycorrhizal fungi. Pesqui Agro pecu Bras. 2004;39:887–94.
5. Gang DR, Wang J, Dudareva N, Nam KH, Simon JE, Lewinsohn E, Pichersky E. An investigation of the storage and biosynthesis of phenylpropenes in sweet basil. Plant Physiol. 2001;125:539–55.
6. Gershenzon J, Maffei M, Croteau R. Biochemical and histochemical localization of monoterpene biosynthesis in the glandular trichomes of spearmint (*Menthaspicata*). Plant Physiol. 1989;89:1351–7.
7. Gross M, Jacob F, Nativ D, Olga L, Yael C, Einat B, Uzi R, Eli P, Efraim L. Biosynthesis of estragole and *t*-anethole in bitter fennel (*Foeniculum vulgare* Mill. var. vulgare) chemotypes. Changes in SAM:phenylpropene Omethyl transferase activities during development. Plant Sci. 2002;163:1047–53.
8. Gupta ML, Prasad A, Ram M, Kumar S. Effect of the vesicular–arbuscular mycorrhizal (VAM) fungus *Glomus fasciculatum* on the essential oil yield related characters and nutrient acquisition in the crops of different cultivars of menthol mint (*Mentha arvensis*) under field conditions. Bioresour Technol. 2002;81:77–9.
9. Hayman PE. Improved procedures for clearing and staining parasite and vesiculaire-arbiscular mycorrhizal fungi for rapid assessment of infection trans. Brit Mycol Soc. 1970;55:158–61.
10. Hazzoumi Z, Moustakime Y, Elharchli E, Joutei KA. Effect of arbuscular mycorrhizal fungi (AMF) and water stress on growth, phenolic compounds, glandular hairs, and yield of essential oil in basil (*Ocimum gratissimum* L.). Chem Biol Technol Agric. 2015;2:10.
11. Hazzoumi Z, Moustakime Y, Khalid A. Effect of gibberellic acid (GA), indole acetic acid (IAA) and benzyl-amino-purine (BAP) on the synthesis of essential oils and the isomerization of methyl-chavicol and *trans*-anethole in *Ocimum gratissimum* L. Springer Plus. 2014;3:321.
12. Hendawy SF, Khalid KA. Response of sage (*Salvia officinalis* L.) plants to zinc application under different salinity levels. J Appl Sci Res. 2005;1(2):147–55.
13. Kapoor R, Giri B, Mukerji KG. Improved growth and essential oil yield and quality in *Foeniculu mvulgare* mill on mycorrhizal inoculation supplemented with P-fertilizer. Bioresour Technol. 2004;93:307–11.
14. Khalid KhA. Influence of water stress on growth, essential oil, and chemical composition of herbs (*Ocimum* sp.). Int Agrophys. 2006;20:289–96.
15. Khaosaad T, Vierheilig H, Nell M, Zitterl-Eglseer K, Novak J. Arbuscular mycorrhiza alter the concentration of essential oils in oregano (*Origanum sp., Lamiaceae*). Mycorrhiza. 2006;16:443–46.
16. Khaliq A, Janardhanan KK. Influence of VAM Fungi on the productivity of cultivated mints. Lucknow J Medi Arom Pl Sci. 1997;19:7–10
17. Kishore D, Kannan S. Catalytic isomerization of estragole to anethole over hydrotalcites and HT-like compounds. J Mol Catal A. 2006;244:83–92.
18. Koeduka T, Baiga TJ, Noel JP, Pichersky E. Biosynthesis of *t*-Anethole in Anise: characterization of t-anol/isoeugenol synthase and an O-Methyltransferase specific for a C7–C8 propenyl side chain. Plant Physiology. 2009;149:384–94.
19. Lewinsohn E, Ziv-Raz I, Dudai N, Tadmor Y, Lastochkin E, Larkov AO, David C, Uzi R, Eli P, Eran P, Yuval S. Biosynthesis of estragole and methyl-eugenol in sweet basil (*Ocimum basilicum* L.). Developmental and chemotypic association of allylphenol *O*-methyltransferase activities. Plant Sci. 2000;160:27–35.

20. Loveys BR, Rorinson SP, Bropl JJ, Chacko EK. Mango sapburn: components of fruit sap and their role in causing skin damage. Aust J Plant Physiol. 1992;19:449–57.
21. Martín-Rodríguez JÁ, Ocampo JA, Molinero-Rosales N, Tarkowská D, Ruíz-Rivero O, García-Garrido JM. Role of gibberellins during arbuscular mycorrhizal formation in tomato: new insights revealed by endogenous quantification and genetic analysis of their metabolism in mycorrhizal roots. Physiol Plant. 2014;154(1):66–81.
22. Nemec S, Lund E. Leaf volatiles of mycorrhizal and non mycorrhizal *Citrus jambhiri* Lush. J Essent Oil Res. 1990;2:287–97.
23. Polonsky J, Bhatnagar SC, Griffiths DC, Pickett JA, Woodcock CM. Activity of quassinoids as antifeedants against aphids. J Chem Ecol. 1989;15:993–8.
24. Pozo M, Azcón-Aguilar C. Unraveling mycorrhiza-induced resistance. Curr Opin Plant Biol. 2007;10:393–8.
25. Rapparini F, Llusia J, Penuelas J. Effect of arbuscular mycorrhizal (AM) colonization on terpene emission and contents of *Artemisia annua* L. Plant Biol. 2008;10:108–22.
26. Shao HB, Chu LY, Jaleel CA, Zhao CX. Water-deficit stress-induced anatomical changes in higher plants. C R Biol. 2008;331:215–25.
27. Sharma SK, Srivastava VK, Pandya PH, Jasra RV. Solvent-free isomerization of methyl chavicol to *trans*-anethole using transition metal complexes as catalysts. Catal Commun. 2005;6:205–9.
28. Simon L, Bousquet J, Levesque RC, Lalonde M. Origin and diversification of endomycorrhizal fungi and coincidence with vascular land plants. Nature. 1993;363:67–9.
29. Singh M, Ramesh S, Kumar S, Kukreja AK. Effect of irrigation and nitrogen on herbage, oil yield and water-use efficiency in rosemary grown under semi-arid tropical conditions. J Med Aromat Plant Sci. 2000;22(1B):659–62.
30. Solinas V, Deiana S, Gessa C, Pistidda C. Reduction of the Fe(III)-desferrioxamine-B complexes by caffeic acid: a reduction mechanism of biochemical significance. Soil Biol Biochem. 1996;28(4–5):649–54.
31. Sudria C, Piñol MT, Palazón J, Cusidó RM, Vila R, Morales C, Bonfill M, Cañigueral S. Influence of plant growth regulators on the growth and essential oil content of cultured *Lavandula dentata* plantlets. Plant Cell Tissue Organ Cult. 1999;58:177–84.
32. Trouvelot A, Kough JL, Gianinazzi-Pearson V. Meesure du taux de mycorhization VA d'un système radiculaire. Recherche des méthodes d'estimation ayant une signification fonctionnelle. In: Bay G-P, Gianinazzi S, editors. Physiological aspect of mycorrizea. Paris: INRA; 1986. p. 217–21.
33. Vierheilig H, Gagnon H, Strack D, Maier W. Accumulation of cyclohexenone derivatives in barley, wheat and maize roots in response to inoculation with different arbuscular mycorrhizal fungi. Mycorrhiza. 2000;9:291–3.
34. Werker E. Trichome diversity and development. Adv Botanical Res. 2000;31:1–35.
35. Werker E, Putievsky E, Ravid U, Dudai N, Katzir I. Glandular hairs and essential oil in developing leaves of *Ocimum basilicum* L. (Lamiaceae). Ann Bot. 1993;71:43–50.
36. Zehtab-Salmasi S, Javanshir A, Omidbaigi R, Alyariand H, Ghassemi-Golezani K. Effects of water supply and sowing date on performance and essential oil production of anise (*Pimpinella anisum* L.). Acta Agron Hung. 2001;49(1):75–81.
37. Zolfaghari M, Nazeri V, Sefidkon F, Rejali F. Effects effect of arbuscular mycorrhizal fungi on plant growth and essential oil contents and composition of *Ocimum basilicum* L. Iran J Plant Physiol. 2013;3(2):643.

4

Pyritic metals sequestration on mine dumps treated with oyster mushroom (*Pleurotus ostreatus,* Jacq.et Fr.)

Johnson Masaka[1*], Marshall Mutambu[1], Rangarirayi Mhindu[1] and Kudakwashe Muringaniza[2]

Abstract

Background: Mining activities can lead to the generation of large quantities of heavy metal laden wastes which are released into the environment in an uncontrolled manner, causing widespread contamination of the ecosystem. Bioremediation is an innovative and promising technology available for removal and recovery of heavy metals from polluted environments.

Methods: Field experiments based on a Latin Square Design were employed in this study in order to determine the effectiveness of oyster mushroom (*P. ostreatus*) as a bioremediation agent for the sequestration of pyritic heavy metals from mine dumps.

Results: Compared with the background values of heavy metals in virgin forest soil, concentrations of Mn^{2+}; Cr^{2+}; Pb^{2+}; Fe^{3+}; and Cd^{2+} were substantially enriched in the study sites. The propagation of oyster mushroom in wheat straw substrate and heavy metal-polluted pyritic mine waste, reduced considerably the concentrations of mobile heavy metals in the waste by 1.2 mg kg^{-1} (98%); 2.0 mg kg^{-1} (79%); 13.1 mg kg^{-1} (93%); 7.7 mg kg^{-1} (78%); and 6.5 mg kg^{-1} (88%) for Mn^{2+}; Cr^{2+}; Pb^{2+}; Fe^{3+}; and Cd^{2+}, respectively. Results have shown that the greatest threat to the ecology in the mine dump sites is currently posed by Cd^{2+}; Fe^{3+}; and Pb^{2+}, while the least ecological threat comes from the concentrations of Mn^{2+}.

Conclusions: The mean levels of reductions in heavy metal concentrations in pyritic mine waste after mycoremedial interventions observed in the current study clearly indicate that, in bio-sorption, *P. ostreatus* is more efficient in sequestering Mn^{2+}; Fe^{3+}; and Cd^{2+}, in that order. The oyster mushroom is less efficient in the bio-sorption of Cr^{2+} and Fe^{3+} from heavy metal polluted mine waste.

Keywords: Contamination, Mine dumps, Bioremediation, Oyster mushroom

Background

Mining is one of the main income earners of the Zimbabwean economy. It contributes up to 20% of Zimbabwe's gross domestic product. Apart from its contribution to exports, the mining industry plays an important role in providing essential mineral-based raw materials to the local manufacturing industry [1]. Over the last century the generation of overburden material has grown dramatically with the increasing demand for minerals and metals in the world market and the decreasing metal ore grades being mined in Zimbabwe. As a result, metal mining activities have generated large volumes of toxic and harmful residues for long periods at mining and processing sites [2]. Since the efficiency of the extraction process is always less than 100%, a high proportion of ore that contains uneconomic material with variable amounts of metallic minerals which can be mobilized into the terrestrial environment is discarded [2–4].

Bioremediation is an innovative and promising technology available for removal and recovery of heavy

*Correspondence: johnsonmasaka@yahoo.com; johnsonmasaka@gmail.com; masakaj@msu.ac.zw
[1] Department of Land and Water Resources Management, Midlands State University, Private Bag 9055, Gweru, Zimbabwe
Full list of author information is available at the end of the article

metals from polluted environments. It involves the microbe-mediated process for clearance of contaminants [5]. The role of mushroom in biosorption of heavy metals into its biomass is well documented, and understanding its propagation on contaminated materials has numerous biotechnological implications for bioremediation of mine dump sites [6].

The microbial cell walls, which mainly consist of polysaccharides, lipids and proteins, offer many functional groups that can bind heavy metal ions, and these include carboxylate, hydroxyl, amino and phosphate groups [7]. Selection of the microbes for bioremediation of contaminated environments should be done based on an understanding of mechanisms involved in the adsorption and mobilization of heavy metals and trace elements in the soil to restore soil health [3].

With the growth of gold and nickel mining industry in the Bindura area of central Zimbabwe, there has been a considerable increase in the discharge of gold mining waste to the environment. In addition to that, the low efficiency of the mechanical and chemical gold extraction processes has caused a mine dump pile-up of discarded metalliferous mine residues, which can mobilize heavy metals into the terrestrial and aquatic environments with deleterious impacts [2, 8]. Heavy metals can pose long-term environmental and health implications because of their non-biodegradability and persistence. In view of this, the study investigates the potential of mushroom propagated on mine dumps for the biosorption of heavy metals.

Methods

Site description

Location and weather conditions

The aim of the current study was to establish the possibility of use of oyster mushroom as a green technology for the bioremediation of mine dumps contaminated with pyritic heavy metals. The study area located in the immediate vicinity of the mining town of Bindura in central Zimbabwe (17°18′S 31°20′E; 1070 m above sea level, Fig. 1) consists of gold and nickel mine waste and tailings dumps. Bindura town is spread over an area of 200 km^2, supporting a population of 46,275 [9]. The area lies in agro-ecological region two with an annual rainfall total of 800 mm. The seasons are characterized by rainy summers (November to April) and short dry winters (May to July). The average annual air temperature of the area is 23 °C with an average minimum and maximum temperature of 12 and 30 °C, respectively [10].

The Bindura mining area falls under the gold belt formations (complexes of metamorphosed basaltic and andesitic lavas and sediments) which are important sources of minerals in Zimbabwe. Pyrite is the dominant sulphide of the gold bearing rocks in the geological formations [11–13].

Bindura town is sandwiched by four major gold mines (Bindura, Trojan, Freda-Rebecca and Renny; Fig. 1). Mine waste dumps were formed as a result of disposal of spoils from the open pit mines. The mine dumps are irregularly shaped and are characterized by horizontal terraces with steep edge slopes of 5–30° and a depth of approximately 20–100 m. The primary minerals are nickel, gold and platinum [11–13].

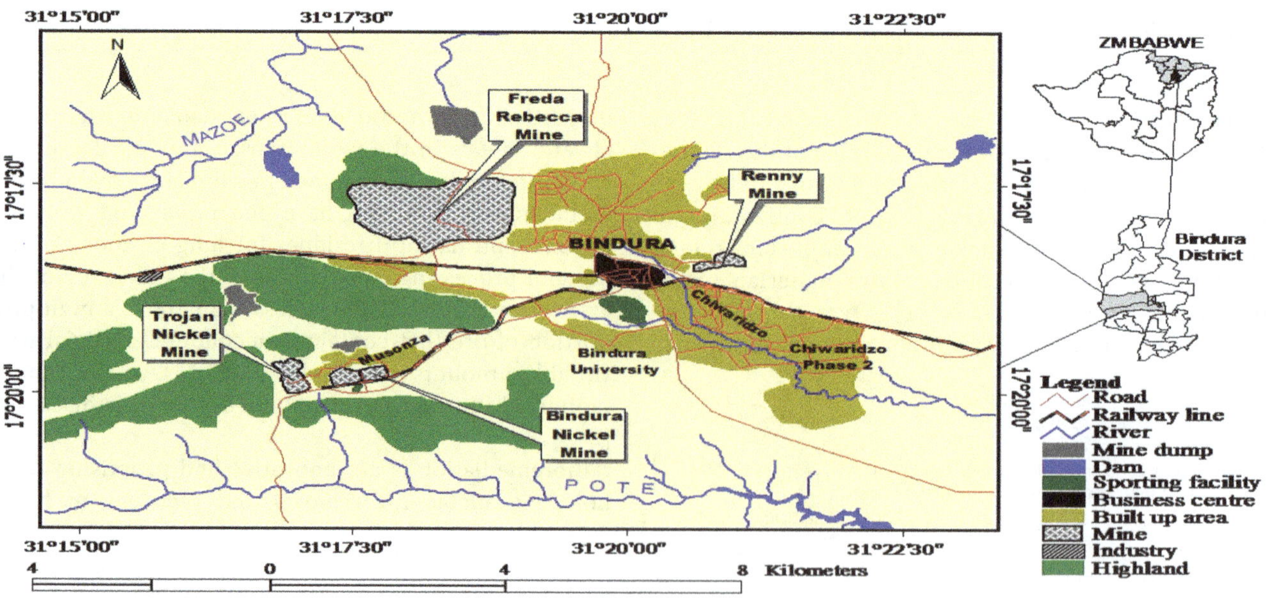

Fig. 1 Location of study sites

The four mine dumps are located in a vast area drained by the Mazowe–Pote river systems, which is a part of the greater Zambezi river catchment in Zimbabwe (Fig. 1). The mine dumps occupy micro-catchment areas that drain north-eastwards into the Mazowe and Pote rivers. The mobilization of heavy metals from the mine dumps with comparatively steel slopes (5–30°) into the river systems has a potential of contaminating the environment in the Mazowe–Pote river system. The area has a thriving mixed farming system that supplies the metropolitans of Bindura, Chinhoyi and Harare with a variety of agricultural produce.

Description of the mine dump sites bioremediation experiments

Collection of uncontaminated soil for comparison

The background contaminant concentration values were obtained in soil samples collected from a virgin forest about 5 km upstream of the mine dump sites within the Mazowe–Pote river system (Fig. 1). The soil samples were collected from 20 randomly selected sites under forest canopy using a shovel (0–20 cm depth). The samples were thoroughly mixed by hand in sterilized gloves in a 50 L plastic bucket and dried under shed. Composite samples were collected from the dried soil for laboratory analyses.

Identification of the mushrooms for bioremediation

The species *Pleurotus ostreatus* Jacq.et Fr is a common oyster edible mushroom. It is now grown commercially in Zimbabwe and around the world for food. Oyster mushrooms can also be used industrially for mycoremediation of heavy metal-polluted environments [4]. The samples of the mushroom spawn were confirmed by a mycologist in the Biological Sciences department at the University of Zimbabwe.

Preparation of the substrate and dried mushroom for the mine dump site experiments

The wheat straw substrate for the field experiments was collected from a food oyster mushroom farmer in Harare as spent mushroom compost. The dried shredded mushroom which was used as *P. ostreatus* spawn source is a product of mushroom production at the same unit. The dried mushroom fruit was placed in shredding machine before soaking in clean water to 50% moisture content for the field trial. The wheat straw substrate was hammer-milled and soaked in clean water that was mixed with formalin (2 parts water: 1 part formalin) for 24 h to eliminate wild fungus spores and bacteria. The substrate was then squeezed of its excess water to achieve 50% residual moisture content. The wheat straw substrate and shredded mushroom were then placed in separately sterilized 50 litre plastic containers and transported to the four mine dump sites for the field experiments.

Experimental design and treatments

Field experiments were employed in this study in order to determine the effectiveness of oyster mushroom as a bioremediation agent for the sequestration of pyritic heavy metals from mine dumps. The field experiments were carried out at each of the four selected mine waste dumps (Trojan, Freda-Rebecca, Renny and Bindura mines; Fig. 1).

The Latin square design was employed in the current study to control the variation in the experiments related to rows and columns in the mine dump sites. Each treatment was replicated five times. The following treatments were used in the experiments: Uncontaminated garden soil (treatment 1, control); contaminated dump site waste (treatment 2); dried oyster mushroom + contaminated dump site soil (treatment 3); sterilized wheat straw substrate + contaminated dump site soil (treatment 4); sterilized wheat straw substrate + dried oyster mushroom (mushroom spawn) + contaminated dump site soil (treatment 5).

Fifty pits arranged in the Latin square experimental design on each selected dump site were dug measuring 1 m by 1 m with a depth of 0.20 m. The entire pit was lined with 250 mμ black polyvinyl sheeting, which was at least 10 cm wider and longer than inside dimensions of the pit to allow for the plastic to be laid over the top of the walls by weights.

For treatment 1 (the control), randomly allocated pits were completely filled with uncontaminated virgin forest soil without any amendments. For treatment 2, the contaminated mine waste excavated from pits were returned into the respective pits without the oyster mushroom spores. In treatment 3, half of the excavated mine waste was thoroughly mixed with an equal amount of dried shredded mushroom using a sterilized wooden stick and returned into the pit. The dried shredded mushroom was replaced with the same amount of wheat straw substrate in treatment 4. In treatment 5, a third on dry weight basis constituted the excavated mine waste material; a third by sterilized wheat straw substrate and a third by dried shredded mushroom (with spores) were thoroughly mixed in a sterilized plastic bin and returned into the respective dump site experimental pits.

All the replicates were irrigated with drinkable water until 50% moisture level was achieved. Each replicate was covered with black plastic sheeting after adding the water in order to maintain constant moisture levels and development of the mushroom fruit bodies in treatments. The constant moisture level was maintained until the end of the study at 8 weeks.

Sampling and laboratory analyses

A non-biased systematic linear sampling method was used to collect replicate pit samples for the evaluation of the concentrations of the heavy metals at the four mine waste dump sites at the beginning and end of the mine dump experiments. The pit samples collected from a depth of 0–20 cm for each treatment at each selected mine dump site were taken to a laboratory and dried in an oven at 450 °C for 48 h. The concentrations of heavy metals (Mn^{2+}, Cr^{2+}, Pb^{2+}, Fe^{3+} and Cd^{2+}) in the samples were determined following the procedures described by Jiang et al. 2011 [14]. The heavy metal concentrations were determined using an atomic absorption spectrophotometer Varian Spectr AA200 in acetylene–air flame with GTA 100 atomiser (Labequip, Canada). The digested samples of soil were simultaneously prepared with blank sample contents of distilled water through the complete procedure, analysed and then used for correction of the analytical signals. Working standard solutions were prepared by suitable dilution (0.25, 0.5, 1, 3 and 5 ppm) of the ICP multi-element standard solution supplied by Merck Millipore (Darmstadt, Germany). The calibration was performed with a blank and standard solution samples and detection limits (LOD) were determined with a standard deviation of blank solutions. The ranges of the calibration curves were selected to match the expected concentrations of all the selected heavy metal elements of the samples studied.

The pH and moisture levels in the experimental pits were measured using a Kecheng Medidor-009 digital pH and moisture metre (Jiangsu Kecheng Optoelectronic, China).

The mushroom fruit bodies were harvested at the end of the experiment (at 8 weeks) and sun-dried until residual moisture was 50%. Dried mushroom samples were analysed using an atomic absorption spectrophotometer following the procedures described by Jiang et al. 2011 [14].

Germination percentage test for in situ experiment

Replicate pits at each dump site were planted with sugar bean seed (*Phaseolus vulgaris* L.) after the remediation experimental period had lapsed. An in-row spacing of 10 cm by an inter-row of 30 cm was used for planting the seeds and each replicate was planted with 24 bean seeds (8 seeds in the in-row × 3 seeds in the inter-row lines). The planted plots were kept moist and analysed for germination at 7 days after sowing.

Single-factor and comprehensive heavy metal pollution assessment at mine dump sites

The single-factor index method was employed to evaluate the pollution of the selected heavy metals, but research has shown that the single-factor index cannot express accurately the comprehensive impact caused by each kind of heavy metals. The Nemerow index method takes not only the extreme value but also the environmental quality index based on weighted multi-factors. The method can reflect the degree of soil pollution caused by various heavy metals. To assess the degree of heavy metal contamination, the pollution index (PI) for each heavy metal and Nemerow integrated pollution index (NIPI) [15] for the five heavy metals (Mn^{2+}, Cr^{2+}, Pb^{2+}, Fe^{3+} and Cd^{2+}) were attributed to each sample. The PI was determined as follows:

$$\mathrm{PI} = \frac{C_i}{S_i}, \tag{1}$$

where PI is the pollution index of the pollutant i at the mine dump site; C_i is the measured concentration value of the heavy element i; S_i is the concentration value of the respective metal in the uncontaminated virgin forest soil. In the current study, the Nemerow integrated (multi-factor) pollution index (NIPI) was expressed as

$$\mathrm{NIPI} = \frac{\sqrt{(P_i^2\mathrm{max} + P_i^2\mathrm{ave})}}{2}, \tag{2}$$

where NIPI is Nemerow integrated pollution index at location i; P_i max and P_i ave represent the maximum and average values of PI, respectively. Based on Nemerow multi-factor pollution index, the environmental quality is divided into five levels shown in Table 1.

The current research employed the potential (E_r^i) and comprehensive potential (RI) ecological risk indices described in detail by Hakanson [16] and Douay et al. [17] to evaluate the potential ecological risk of heavy metals. In order to establish the E_r^i and RI values, the single heavy metal pollution index was computed using the following model:

$$C_f^i = \frac{C_D^i}{C_R^i}, \tag{3}$$

where C_f^i is the contamination factor; C_D^i is the degree of contamination (that is the quantified value of a particular heavy metal in contaminated material); C_R^i is reference value, which is the background value of a particular heavy metal of virgin forest soil.

The comprehensive potential ecological risk index (RI) was calculated using the following equations:

$$E_r^i = T_f^i \times C_f^i \tag{4}$$

$$\mathrm{RI} = \sum_{i=1}^{n} E_r^i, \tag{5}$$

Table 1 Nemerow multi-factor pollution, potential and comprehensive ecological indices. Source: Hakanson [16] and Douay et al. [17]

Nemerow multi-factor pollution index, NIPI		
NIPI ranges		**Pollution grades**
≤0.7		Clean
0.7 < NIPI ≤ 1		Cautionary
1 < NIPI ≤ 2		Light pollution
2 < NIPI ≤ 3		Medium pollution
NIPI > 3		Heavy pollution
Potential (E_r^i) and comprehensive (RI) ecological harm indices rating for the heavy metal		
E_r^i range	**RI range**	**Ecological pollution grades**
<40	<150	Light ecological risk
40–79	150–299	Middle ecological risk
80–159	300–600	Strong ecological risk
160–320	>600	Very strong ecological risk
>320	<150	Extremely strong ecological risk

where E_r^i is the potential ecological risk coefficient; T_f^i is the biological toxic factor of a single element, which is determined for Mn = Fe = Cr = 2, Pb = 5 and Cd = 30 [16]. The potential (E_r^i) and comprehensive potential (RI) ecological risk ratings are divided into five levels shown in Table 1.

Statistical analyses

Treatment effects on heavy metal concentrations in mine waste and oyster mushroom fruit bodies and trends in mine dump waste pH were performed using Generalized One-Way ANOVA of completely randomized blocks in Latin square design [18]. Flux data on concentrations of heavy metals were log-transformed if needed, to normalize the distributions before the statistical analysis. Differences between treatment means were considered significant at $p \leq 0.05$ as determined by Fisher's protected least significant difference test.

Results and discussion

Concentrations of heavy metals and pH in mine dumps after growing oyster mushroom

The mine dump waste at each of the four study sites in central Zimbabwe (Fig. 1) was characterized by considerably high heavy metal concentrations (Fig. 2), low organic matter content (Table 1) and high levels of waste acidity (low pH values, Fig. 3). Compared with the background values of heavy metals in virgin forest soil, concentrations of Mn^{2+}, Cr^{2+}, Pb^{2+}, Fe^{3+} and Cd^{2+} in mine dump waste were substantially ($p < 0.05$) enriched in the study sites (Fig. 2). The propagation of oyster mushroom for the bioremediation of heavy metal pollution in pyritic mine dump waste at the selected four mining sites significantly ($p < 0.05$) reduced the concentrations of the heavy metals in the mine dump waste at the end of the study (Fig. 2). Mean concentrations of the heavy metals (Fig. 2) and pH (Fig. 3) in treatment five plots were comparatively similar to those recorded in the control replicates at the end of the experiment. The differences between pH values for the control plots (uncontaminated virgin forest soil) and the plots treated with of *P. ostreatus* (treatment 5) in the heavy metal-contaminated mine waste were 0.5–0.9 pH units.

All experimental plots on pyritic mine dump waste that did not receive viable *P. ostreatus* spawn were considerably acidic with pH values ranging between 1 and 3.2 which were significantly ($p < 0.05$) lower than the pH values of unpolluted virgin forest soil (pH 5.7–6.0) (Fig. 3). Acidity in plots that were treated with wheat straw substrate, dried oyster mushroom and water applications (treatment 5) consistently approached neutral pH values after declining by 52–84% when compared with experimental pits without oyster mushroom growth over 8 weeks.

The results indicate that the mining activities at the four study sites in the Bindura mining region of Zimbabwe have the potential to mobilize significant amounts of heavy metals into surface environments of the Mazowe–Pote river catchments (Fig. 1). In the current study, the possibility of an accumulation of heavy metals in crop plants on arable lands around the Bindura area is of great concern due to the probability of food contamination through the soil–plant root interface [15].

The variations in the concentrations of the heavy metals in the mine dump waste for the various experimental pits are attributed to the heavy metal biosorption capacity of the mushroom biomass. In a related study on concentrations of heavy metals in species of mushrooms growing in the East Black Sea region, Demirbas [19] reported a higher capability of mushroom to bio-accumulate heavy metal contaminants in their fruit bodies. The *P. ostreatus* is suspected to have initially exhibited a marked tolerance towards the elevated concentrations of the heavy metals

(See figure on next page.)

Fig. 2 Concentrations of heavy metals in mine dump waste before and after treatment with oyster mushroom. *Trt* treatment, *A/Trt* after treatments, *B/Trt* before treatments, *LSL* local safe limits, *Lsd* least significant difference, *Cv* coefficient of variation

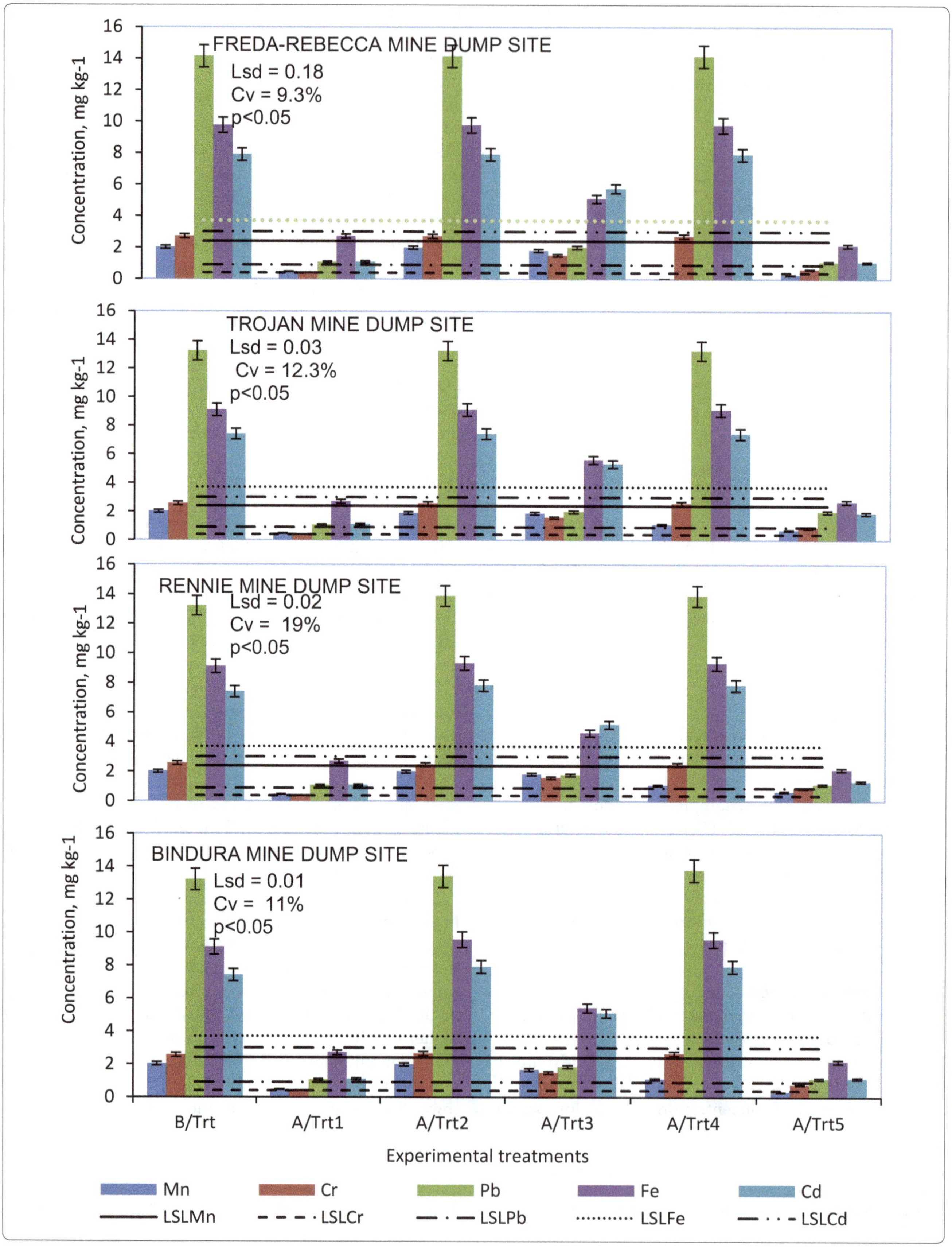
FREDA-REBECCA MINE DUMP SITE
Lsd = 0.18
Cv = 9.3%
p<0.05
TROJAN MINE DUMP SITE
Lsd = 0.03
Cv = 12.3%
p<0.05
RENNIE MINE DUMP SITE
Lsd = 0.02
Cv = 19%
p<0.05
BINDURA MINE DUMP SITE
Lsd = 0.01
Cv = 11%
p<0.05
Concentration, mg kg-1
B/Trt
A/Trt1
A/Trt2
A/Trt3
A/Trt4
A/Trt5
Experimental treatments
Mn
Cr
Pb
Fe
Cd
LSLMn
LSLCr
LSLPb
LSLFe
LSLCd

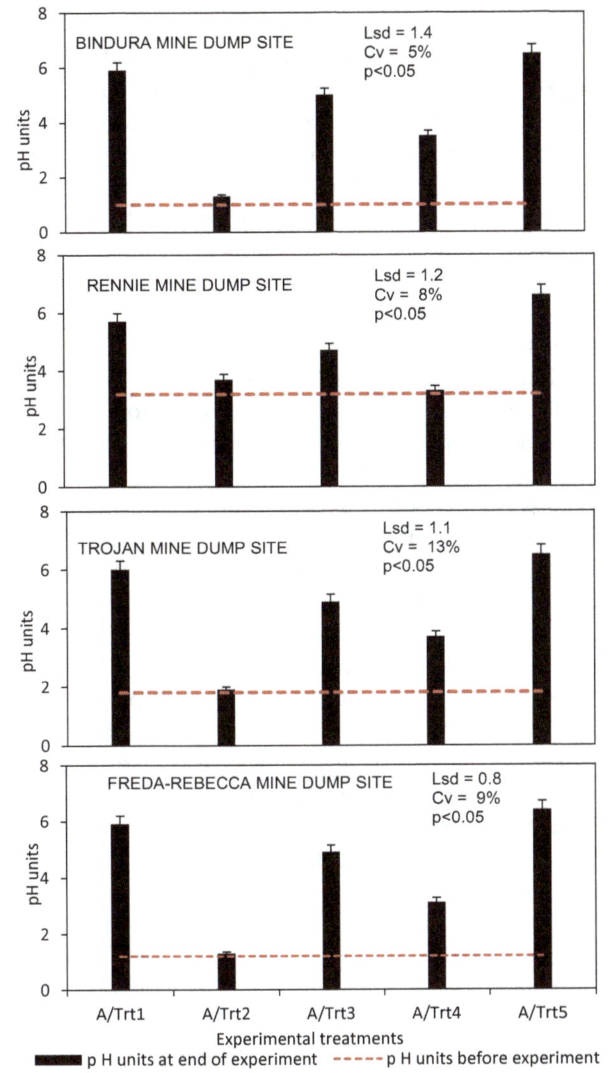

Fig. 3 pH units in mine dump waste before and after bioremediation with oyster mushroom

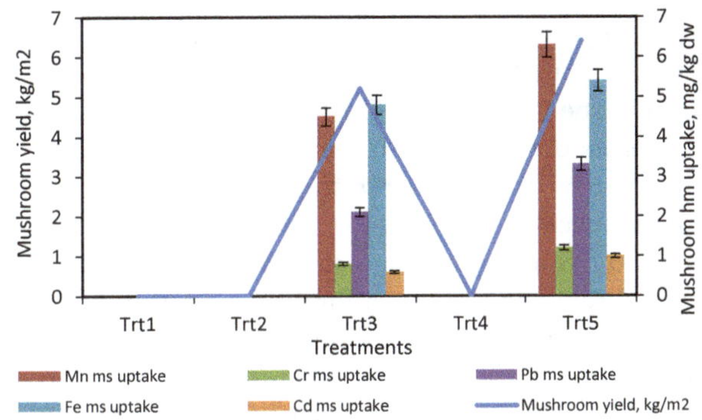

Fig. 4 Oyster mushroom yield and heavy metal uptake. *Trt* treatment, *ms* mushroom, *hm* heavy metal

(Fig. 2) and adverse conditions of high acidity (Fig. 3) in the mine dump waste recorded in this study. Related studies on mycoremediation have indicated that contaminant heavy metals can be biosorbed by mushroom bodies at binding sites found in the cellular structure [20]. In the current study, it is suggested that among the various reactive compounds associated with fungal cell walls, the extracellular polymeric substances reported by Gutierrez et al. [21] were of particular importance and are known to have significant effects on acid–base properties and metal adsorption [22–24]. The considerable reduction in the concentrations of heavy metals in the experimental plots that successfully supported vigorous growth of the basidiomycetes (Fig. 4) clearly implies that the investigated mushroom species is an efficient bio-exlusor of Mn^{2+}, Cr^{2+}, Pb^{2+}, Fe^{3+} and Cd^{2+} from their growing environments.

Considerable reductions of the heavy metal concentrations (Fig. 2) in the experimental pits that supported successful colonization and growth of the oyster mushroom coincided with comparatively near neutral pH values and had the highest mushroom fruit body yield and heavy metal uptake (Fig. 4). This clearly demonstrated that the successful propagation of the oyster mushroom in the mine dump waste has bioremediative outcomes in the decontamination of mine overburden material. It has been reported previously that pH is one of the variables influencing the bioavailability and transport of heavy metals in the waste, and heavy metal mobility decreases substantially with increasing soil pH due to the precipitation of hydroxides, carbonates or the formation of insoluble organic complexes [25]. Consequently, dump site pits that received viable oyster mushroom spawn and ground wheat straw substrate recorded significant pH increases accompanied by considerable reductions in the concentrations of the mobile heavy metals (Fig. 2).

The applications of shredded mushroom to pyritic mine waste without wheat straw substrate in treatment 3 resulted in smaller, though significant ($p < 0.05$), reductions in the mean levels of the pyritic heavy metals in the mine waste at all four sites. The absence of the wheat straw substrate in the use of *P. ostreatus* as bioremediation material for the biosorption of pyritic heavy metals in mine dump waste (treatment 3), when compared with the reductions in treatment 5, brought down the content of the selected heavy metals (except for Pb^{2+} whose concentration declined by 85%) by much smaller values of 0.2 mg kg^{-1} (11%), 1.2 mg kg^{-1} (45%), 4.9 mg kg^{-1} (48%), 2.3 mg kg^{-1} (27%) for Mn^{2+}, Cr^{2+}, Fe^{3+} and Cd^{2+}, respectively, at the end of the experiment (Fig. 2).

The mean levels of reductions in heavy metal concentrations in pyritic mine waste after mycoremedial interventions observed in the current study clearly indicate that, in biosorption, *P. ostreatus* is more efficient in sequestering Mn^{2+} (98%), Fe^{3+} (93%) and Cd^{2+} (88%), in that order. The biosorption of Cr^{2+} (79%) and Fe^{3+} (78%) from heavy metal-polluted mine waste (Fig. 2) is less efficient. The Mn^{2+}, Fe^{3+}, Cd^{2+} biosorption efficiency of *P. ostreatus* observed at the four study sites was further confirmed by the current results when concentrations of the respective heavy metals at the end of the experiment were compared with local safe limits (LSL, locally accepted safe concentration limits for a heavy metal) of the heavy metal concentrations in soil (Fig. 2). In this respect, the oyster mushroom's capacity as a biosorbent of Mn^{2+}, Fe^{3+} and Cd^{2+} from pyritic mine waste dumps was re-confirmed by the fact that their concentrations after bioremediation (treatment 5) at the four study sites were lower than the LSL by 1.92 mg kg^{-1} (20%), 1.43 mg kg^{-1} (39%) and 1.65 mg kg^{-1} (55%), respectively.

Mushroom substrate is a kind of lignocellulose material which supports the growth, development and fruiting of mushroom [2, 4, 20]. Previous studies by Demirbas [19], Gavrilescu [6], Kojta and Zhang [26] have reported that an ideal substrate will contain enough nitrogen and carbohydrate for rapid mushroom growth. In the current study, substrate conditions present in treatment 3 lacked the carbohydrate component necessary for the supply of energy and organic carbon for mushroom biomass synthesis.

The propagation of *P. ostreatus* in environments that supplied the necessary conditions for normal growth of the mushroom (mushroom + substrate + contaminated soil) was able to bring down the initial high concentrations of pyritic heavy metals. The application of mushroom + substrate to heavy metal-polluted pyritic mine waste (treatment 5), when compared with the initial levels, reduced considerably the concentrations of mobile heavy metals in the waste by 1.2 mg kg^{-1} (98%), 2.0 mg kg^{-1} (79%), 13.1 mg kg^{-1} (93%), 7.7 mg kg^{-1} (78%) and 6.5 mg kg^{-1} (88%) for Mn^{2+}, Cr^{2+}, Pb^{2+}, Fe^{3+} and Cd^{2+}, respectively, at the end of the study.

Although some heavy metals like Mn^{2+}, Cr^{2+}, Fe^{3+} and Cd^{2+} are required for growth of plants in trace amounts, they may damage plants if present beyond their maximum permissible limits [27]. Recent studies have shown that heavy metal uptake by the oyster mushroom into the cell across the cell membrane as intracellular uptake is dependent on the cell metabolism [3, 28]. This type of heavy metal uptake from mine dump waste by the oyster mushroom requires energy expenditure by the mushroom cells. Since the mushroom is heterotrophic, it uses the organic substrate as the source of energy. While sunlight provides an energy source for plants, mushrooms derive all of their energy and growth materials from their growth medium, through biochemical decomposition processes. Consequently, combinations of the viable mushroom spawn in dried mushroom and the wheat straw as source of energy for the active uptake of the heavy metals from mine dump waste in treatment 5 experimental pits were followed by considerable reductions of the heavy metal concentrations to be within safe limits (Fig. 2). The application of *P. ostreatus* as a biosorbent is not only dependent on its biosorptive capacity, but also how easy it is to regenerate and reuse oyster mushroom biomass. However, the intracellular heavy metal absorption by the mushroom cells can only mean that the absorbed heavy metals are disposed off through destructive treatment.

Treatments that received the substrate + contaminated waste showed smaller biosorption capacity of the heavy metals (treatment 3). The substrate acted as the source of nutrients for the indigenous soil fungal strains at the respective sites for the biosorption of the heavy metals. Essentially, the substrate is the organic matter (Table 2) that helps in the promotion of bio-adsorption of the heavy metals on the surface of organic matter [22]. However, the role of substrate alone without the use of the mushroom (*P. ostreatus*) was not significant enough to cause a considerable

Table 2 Organic matter content in mine dump waste before and after treatment with oyster mushroom

Trts	Organic matter content in mine dump waste before and after treatments (%)							
	Freda-Rebecca mine		Trojan mine		Renny mine		Bindura mine	
	Before trt	After trt	Before trt	After trt	Before trt	After trt	Before trt	After trt
1	9.3^a	9.4^a	9.3^a	9.3^a	9.4^a	9.4^a	9.3^a	9.3^a
2	3.2^b	3.4^b	3.1^b	3.2^b	3.4^b	3.3^b	3.1^b	3.2^b
3	3.1^b	5.3^c	3.3^b	5.8^c	3.2^b	6.1^c	3.1^b	5.9^c
4	3.3^b	4.8^d	3.4^b	4.7^d	3.1^b	4.9^d	3.0^b	4.5^d
5	3.2^b	8.1^e	3.3^b	7.8^e	3.2^b	7.9^e	3.1^b	7.8^e
Lsd	0.5	0.4	0.7	1.5	0.8	1.1	0.6	1.2
CV (%)	11.0	7.0	13.0	4.0	10.0	9.0	5.0	8.0

Trt treatments, *Lsd* least significant difference, *CV* coefficient of variation

Superscripts indicate mean separations that are significant

decrease in the heavy metal concentration levels in replicate pits that received substrate + contaminated (treatment 3).

Local safe limits, mushroom yield and uptake of heavy metals after bioremediation

Except for Mn^{2+}, results have shown that the concentrations of the selected heavy metals (Mn^{2+}, Cr^{2+}, Pb^{2+}, Fe^{3+} and Cd^{2+}) in the pyritic mine dump waste before the introduction of the bioremediations were consistently higher than the recommended LSL in soils (Fig. 2). Mean levels of the pyritic heavy metals in the mine dump waste were 2.2 mg kg^{-1} (5 times), 12.6 mg kg^{-1} (21 times), 5.4 mg kg^{-1} (3 times) and 4.4 mg kg^{-1} (2 times) in excess of the recommended LSL for Cr^{2+}, Pb^{2+}, Fe^{3+} and Cd^{2+} concentrations in mine waste before mycoremediation, respectively. In this study, mean concentrations of pyritic heavy metals in the mine dump material rated against the recommended LSL before mycoremediation followed the trend $Pb^{2+} > Cr^{2+} > Fe^{3+} > Cd^{2+}$.

If untreated, the heavy metal contaminants are able to infiltrate deep and pollute the groundwater as well as the surface water [2, 17, 29]. Heavy metals in the soil may subsequently enter the human food web through plants and can constitute some risk to the ecosystem as they tend to bio-accumulate and can be transferred from one food chain to another in the local Mazowe–Pote catchment area (Fig. 1) and beyond.

Results in Fig. 4 show that mushroom fruit body yield dynamics consistently follow trends observed in the biosorption of the selected heavy metals in treatments that received the complete conditions for the successful colonization and growth of P. *ostreatus* (treatment 5). Analyses of the results (Fig. 4) have consistently reconfirmed the trend that the macrofungus is a more efficient biosorbent of Mn^{2+} and Fe^{3+} when compared with the absorption of other heavy metals from polluted mine waste dumps. Current study results have indicated that heavy metal sequestration by *P. ostreatus* differed significantly ($p > 0.05$) across the treatments.

The bioremediation of the pyritic mine dump material through propagation of *P. ostreatus* on the waste in treatment 5 substantially reduced the content of the heavy metals to almost the same levels as those found in the uncontaminated virgin forest soil (Fig. 2). Previous studies by Aydinalp and Marinova [30] have concluded that mushroom biomass accumulations and various environmental factors such as pH (Fig. 3), organic matter (Table 2), type of substrate, geochemistry of substrate and distance from the source of pollution can affect the metal content in mushrooms.

Organic matter content at mine dump site before and after bioremediation

The propagation of oyster mushroom on mine dump waste and wheat straw substrate considerably ($p < 0.05$) increased the accumulation of organic matter in the waste (Table 2). When compared with the control (organic matter in virgin forest soil), the content of organic matter in the waste of the pits that were successfully colonized by *P. ostreatus* (treatment 5) differed by only 1.4 percentage points (9.3 and 7.9% organic matter content, respectively). Fairly low content of organic matter (3.1–3.4%) was recorded consistently in replicates that were not colonized by the myco-fungus (treatment 2).

Pollution rating for a single heavy metal at mine dump sites

A quantitative analysis of single heavy metal pollution index was conducted at each study site using the Nemerow integrated pollution index (NIPI, Eq. 2) values. A comparative analysis of the data presented in Tables 3, 4 and 5 shows that the NIPI values in the mine dump waste were highly dependent ($p < 0.05$) on the colonization and growth of the mushroom. The NIPI values for Mn^{2+}, Cr^{2+}, Pb^{2+}, Fe^{3+} and Cd^{2+} varied considerably between treatment replicates after the employment of various combinations of *P. ostreatus*, organic substrate and the contaminated mine dump material (Tables 4, 5). The NIPI values fell within the ranges of uncontaminated to heavy pollution. The lowest NIPI values of 0.11–1.65 were

Table 3 Nemerow pollution index (NIPI), potential (E_r^i) and comprehensive (RI) ecological harm indices before bioremediation

Heavy metal	Trojan mine			Freda-Rebecca mine			Renny mine			Bindura mine		
	NIPI	E_r^i	RI	NIPI	E_r^i	RI	NIPI	E_r^i	RI	NIPI	E_r^i	RI
Mn^{2+}	0.9	46.0	170.0	0.8	45.0	174.0	0.9	44.0	171.0	0.9	43.0	172.0
Cr^{2+}	3.2	60.0	289.0	3.6	62.0	286.0	3.0	63.0	290.0	3.4	60.0	278.0
Pb^{2+}	4.6	170.0	659.0	4.9	174.0	653.0	4.7	172.0	658.0	4.5	175.0	662.0
Fe^{3+}	4.1	90.0	455.0	3.9	89.0	462.0	3.8	91.0	453.0	3.7	90.0	458.0
Cd^{2+}	2.6	100.0	356.0	2.4	103.0	358.0	2.5	102.0	360.0	2.6	100.0	355.0

Table 4 Nemerow indices after bioremediation at Trojan and Freda-Rebecca mine dump sites

Trt	Trojan mine					Freda-Rebecca mine				
	Mn^{2+}	Cr^{2+}	Pb^{2+}	Fe^{3+}	Cd^{2+}	Mn^{2+}	Cr^{2+}	Pb^{2+}	Fe^{3+}	Cd^{2+}
Trt1	–	–	–	–	–	–	–	–	–	–
Trt2	0.83[a]	6.78[a]	15.72[a]	2.63[a]	2.64[a]	0.81[a]	6.76[a]	15.42[a]	2.66[a]	2.71[a]
Trt3	0.45[b]	3.75[b]	2.22[b]	1.38[b]	1.91[b]	0.48[b]	3.74[b]	2.26[b]	1.39[b]	1.79[b]
Trt4	0.68[a]	6.78[a]	15.72[a]	2.64[a]	2.64[a]	0.74[a]	6.73[a]	15.64[a]	2.43[a]	2.60[a]
Trt5	0.11[d]	1.55[d]	1.20[d]	0.58[c]	0.36[c]	0.17[d]	1.62[c]	1.30[d]	0.59[b]	0.38[c]
Lsd (5%)	0.18	1.91	0.80	0.20	0.46	0.32	1.98	0.56	0.43	1.25
Cv%	9.80	3.60	5.20	7.10	4.30	5.20	4.20	8.30	8.20	9.10

Superscripts indicate mean separations that are significant

Table 5 Nemerow indices after bioremediation at Renny and Bindura mine dump sites

Trt	Renny mine					Bindura mine				
	Mn^{2+}	Cr^{2+}	Pb^{2+}	Fe^{3+}	Cd^{2+}	Mn^{2+}	Cr^{2+}	Pb^{2+}	Fe^{3+}	Cd^{2+}
Trt1	–	–	–	–	–	–	–	–	–	–
Trt2	0.79[a]	6.72[a]	15.71[a]	2.65[a]	2.69[a]	0.76[a]	6.81[a]	15.43[a]	2.56[a]	2.70[a]
Trt3	0.44[b]	3.75[b]	2.26[b]	1.28[b]	1.93[b]	0.42[b]	3.78[b]	2.27[b]	1.36[b]	1.78[b]
Trt4	0.67[a]	6.79[a]	15.74[a]	2.63[a]	2.66[a]	0.66[a]	6.75[a]	15.59[a]	2.53[a]	2.64[a]
Trt5	0.11[c]	1.54[c]	1.21[c]	0.68[c]	0.38[c]	0.18[c]	1.65[c]	1.33[c]	0.57[c]	0.39[c]
Lsd (5%)	0.16	1.89	0.95	0.44	0.93	0.20	0.25	0.40	0.38	0.25
Cv%	4.40	7.00	6.40	6.60	9.10	4.70	5.20	8.10	7.10	8.00

Superscripts indicate mean separations that are significant

recorded in the replicates that received the bioremedial combinations of sterilized wheat straw substrate + dried oyster mushroom (treatment 5), indicating heavy metal contamination reduction to levels in ranges of clean to light pollution. When NIPI values for replicates in treatment 2 (contaminated waste without *P. ostreatus*) were compared with NIPI values in treatment 5, results indicated that the propagation of *P. ostreatus* on pyritic mine dump waste reduces pollution threats of Cr^{2+} and Pb^{2+}. The application of *P. ostreatus* as bioremediation material reduces the individual heavy metal pollution threat in mine dump waste from medium pollution (2.63–2.64) to clean and uncontaminated waste (0.36–0.58) for the Fe^{3+} and Cd^{2+} contaminants (Tables 4, 5).

The largest reduction in the NIPI was observed in the concentration of Pb^{2+} in the mine waste where initial average NIPI of 15.57 (indicating heavy pollution levels) at the four mine dump sites dropped to 1.26 (light pollution level) after bioremediation using *P. ostreatus* (Tables 4, 5). The results of the current study have shown that the use of *P. ostreatus* at mine dump site reduces the heavy metal pollution degrees in the order $Cr^{2+} > Pb^{2+} > Fe^{3+} > Cd^{2+} > Mn^{2+}$. In ascending order, this trend was reversed in the order $Mn^{2+} < Cd^{2+} < Fe^{3+} < Pb^{2+} < Cr^{2+}$ in the NIPI indices after successful propagation of the oyster mushroom in the waste (treatment 5). This implies that the greatest pollution threat from a single heavy metal was posed by Cr^{2+} after treatment of mine waste with oyster mushroom. This clearly implies that the vigorous biosorption of a particular heavy metal (Cr^{2+}) by *P. ostreatus* may not necessarily be followed by the detoxification levels of Cr^{2+} into the clean ranges of pollution rating.

It has been reported in this study that the concentration of Pb^{2+} in the mine dump waste before employment of mycoremediation scored the highest difference when compared with the LSL for the heavy metal (Fig. 2). This trend in the Nemerow pollution index for the individual heavy metal was, as expected, maintained.

Potential and comprehensive ecological risk assessment for the heavy metals

The potential (Eqs. 3, 4) and comprehensive (Eqs. 4, 5) ecological risk indices were used in the current study as diagnostic tools for pollution control purposes, and are the basis for attaching special attention for the specific contaminants Mn^{2+}, Cr^{2+}, Pb^{2+}, Fe^{3+} and Cd^{2+} within the mine dump sites.

Table 6 Potential ecological risk indices after bioremediation at Trojan and Freda-Rebecca mine dump sites

Trt	Trojan mine					Freda-Rebecca mine				
	Mn^{2+}	Cr^{2+}	Pb^{2+}	Fe^{3+}	Cd^{2+}	Mn^{2+}	Cr^{2+}	Pb^{2+}	Fe^{3+}	Cd^{2+}
Trt1	–	–	–	–	–	–	–	–	–	–
Trt2	143^a	281^a	657^a	448^a	342^a	141^a	283^a	659^a	447^a	340^a
Trt3	110^b	149^b	310^b	136^b	141^b	113^b	147^b	313^b	137^b	139^b
Trt4	142^a	279^a	655^a	449^a	336^a	142^a	277^a	659^a	451^a	336^a
Trt5	100^c	110^c	123^c	113^c	111^c	102^c	111^c	122^c	114^c	113^c
Lsd (5%)	11	23	33	18	42.1	12	22	34	16	41
Cv%	10	4	5	7	4	5	4	8	8	9

Superscripts indicate mean separations that are significant

Table 7 Potential ecological risk indices after bioremediation at Renny and Bindura mine dump sites

Trt	Renny mine					Bindura mine				
	Mn^{2+}	Cr^{2+}	Pb^{2+}	Fe^{3+}	Cd^{2+}	Mn^{2+}	Cr^{2+}	Pb^{2+}	Fe^{3+}	Cd^{2+}
Trt1	–	–	–	–	–	–	–	–	–	–
Trt2	141^a	282^a	658^a	451^a	343^a	143^a	285^a	662^a	449^a	344^a
Trt3	112^b	137^b	317^b	133^b	145^b	115^b	146^b	316^b	139^b	142^b
Trt4	141^a	281^a	656^a	453^a	339^a	144^a	273^a	663^a	453^a	338^a
Trt5	101^c	112^c	125^c	114^c	114^c	100^c	114^c	121^c	117^c	114^c
Lsd (5%)	12	24	34	19	43	13	24	36	18	43
Cv%	4	7	6	6.6	9	5	5	8	7	8

Superscripts indicate mean separations that are significant

Varying the combinations of *P. ostreatus* spawn carrying matter, wheat straw substrate with mine dump waste significantly ($p < 0.05$) influenced the degrees of potential (E_r^i) and comprehensive (RI) ecological harm indices for the selected heavy metals at the four study sites (Tables 3, 6, 7). The E_r^i and RI values ranged between 46–100 and 170–659, respectively, before treatment with oyster mushroom.

The E_r^i and RI ranges clearly indicate that the potential and comprehensive ecological risk levels were light (for Mn^{2+}) to very strong (for Pb^{2+}) before the employment of bioremediation measures in the experimental mine dump pits (Table 7). The employment of bioremediation measures in the treatment of heavy metals reduced considerably the RI values in the mine dump waste (treatments 3 and 5; Tables 3, 8). The calculated comprehensive potential ecological risk (RI) indices before the bioremediation of the waste showed that the mine dump waste was lightly (Mn^{2+}), mildly (Cr^2), strongly (Fe^{3+} and Cd^{2+}) and very strongly (Pb^{2+}) polluted (Tables 7, 8). The comprehensive potential ecological risk indices after the bioremediation using *P. ostreatus* in treatment 5 were found in the order of $Mn^{2+} < Cr^{2+} < Cd^{2+} < Fe^{3+} < Pb^{2+}$. This implies that the greatest threat to the ecology in the Mazowe–Pote River catchment areas is currently being posed by Cd^{2+}, Fe^{3+} and Pb^{2+} heavy metals from the mine dump sites. The least ecological threat comes from the concentrations of Mn^{2+} in the mine dump sites.

The RI values for the mine waste after bioremediation in treatment 5 ranged from 100 to 122, indicating a safe comprehensive ecological harm index. In plots that had not received mushroom growth, the RI values ranged from 143 for Mn^{2+} (light ecological risk) to 663 for Pb^{2+} (very strong ecological risk). The results of the study show that *P. ostreatus* was able to sequestrate the pyritic heavy metals where the combination of wheat straw substrate + dried mushroom + contaminated soil was used as bioremediation material.

Sugar bean seed germination tests on mine dumps after bioremediation

The fitness of the bio-treated mine dump waste to supply normal conditions for the germination and growth of plants was tested by planting sugar bean seed (*P. vulgaris* L.) in the experimental pits at each mine dump site. Results have shown that germination percentages in replicates pits depended considerably ($p < 0.05$; Table 9) on success of colonization and growth of oyster mushroom

Table 8 Comprehensive potential ecological harm index after the bioremediation

Treatments	Trojan mine		Freda-Rebecca mine		Renny mine		Bindura mine	
	E_r^i	RI	E_r^i	RI	E_r^i	RI	E_r^i	RI
Trt1	–	–		–		–		–
Trt2	170^a	456^a	172^a	459^a	175^a	460^a	166^a	460^a
Trt3	56^b	211^b	57^b	220^b	59^b	210^b	58^b	213^b
Trt4	182^a	453^a	182^a	458^a	181^a	457^a	180^a	454^a
Trt5	36^c	113^c	34^c	119^c	39^c	109^c	39^c	117^c
Lsd (5%)	10	15	12	9	8	11	12	8
Cv%	9	11	11	9	15	7	10	11

Superscripts indicate mean separations that are significant

Table 9 Seed germination test for bio-treated mine dump waste

Trts	Trojan mine dump			Freda-Rebecca mine dump			Renny mine dump			Bindura mine dump		
	Planted seeds	Gmd	% Gmd	Planted seeds	Gmd	% Gmd	Planted seeds	Gmd	% Gmd	Planted seeds	Gmd	% Gmd
1	24	24^a	100	24	24^a	100	24	24^a	100	24	24^a	100
2	24	0b	0	24	0^b	0	24	0^b	0	24	0^b	0
3	24	5^c	21	24	6^c	25	24	5^c	21	24	4^c	17
4	24	2^b	8	24	0^b	0	24	2^b	13	24	0^b	0
5	24	18^d	75	24	20^d	83	24	19^d	79	24	22^d	92
Lsd (5%)	–	3	–	–	2	–	–	3	–	–	2	–
CV%	12	9	10	8	7	10	9	5	5	7	4	11

Superscripts indicate mean separations that are significant

in the waste. When compared with the control (virgin forest soil), there were small, but significant ($p < 0.05$; Table 9) differences in sugar bean seed germination percentages (0–6 germinated seeds) for pits treated with dried shredded mushroom + wheat straw substrate (treatment 5). Seed germination percentages in waste that was successfully colonized by *P. ostreatus* (treatment 5) ranged from 75 to 92%. This clearly implies that the propagation of the oyster mushroom in mine dump waste is capable of detoxifying the environment by reducing the concentrations of heavy metals and increasing the pH values to be within ranges of safe ecological risks.

Disposal of bioremediation material

A major challenge for the commercialization of mycoremediation has been the disposal of contaminated mushroom biomass especially in the case of repeated propagation where large amounts of biomass may be produced. The biomass has to be stored, disposed of or utilized in an appropriate manner so as not to pose any environmental risk [25]. Controlled combustion and gasification of biomass can yield a mixture of producer gas and/or pyrogas which can be used to generate thermal and electrical energy. The bio-treated waste can be composted and compacted in order to reduce its volume before reuse. Heavy metals such as Fe and Mn are plant essential metals. The high concentrations of these metals in the harvested biomass can be diluted to acceptable concentrations by combining the biomass with clean biomass in formulations of fertilizer and fodder [2].

Conclusions

The current study presents an analysis of heavy metal pollution and the possibility for use of a basidiomycetes *P. ostreatus* in the bioremediation of mine dump waste at four selected mine dump sites around the mining town of Bindura in central Zimbabwe. The greatest threat to the ecology in the Mazowe–Pote River catchment areas is currently posed by Cd^{2+}, Fe^{3+} and Pb^{2+} heavy metals from the mine dump sites. The least ecological threat comes from the concentrations of Mn^{2+} in the mine dump sites. The considerable reduction in the concentrations of heavy metals in the experimental pits that successfully supported vigorous growth of the basidiomycetes clearly implies that the investigated mushroom species is an efficient bio-exlusor of Mn^{2+}, Cr^{2+}, Pb^{2+}, Fe^{3+} and Cd^{2+} from their growing environments. The propagation of *P. ostreatus* can be employed as a restoration measure for the soil ecosystems contaminated by heavy metals at the mine dump sites. In the biosorption of the heavy metal pollutants, *P. ostreatus* is more efficient in sequestering Mn^{2+}, Fe^{3+} and Cd^{2+}, in that order. The biosorption of Cr^{2+} and Fe^{3+} from heavy metal-polluted mine waste is less efficient. The propagation of *P. ostreatus* as bioremediation material reduces the individual heavy metal pollution threat in mine dump waste from medium pollution to clean and uncontaminated waste. The use of oyster mushroom propagation in heavy metal decontamination of mine dump waste presents an opportunity to employ renewable green technologies in the remediation of contaminated environments.

Abbreviations

17°18′S 31°20′E: degrees and minutes east of Greenwich Meridian; GTA: genesis tank atomiser; NIPI: Nemerow integrated pollution index; RI: risk index; ANOVA: analysis of variance.

Authors' contributions

JM was the main academic supervisor. The field experiment design and layouts were done under his supervision including research write-up, correcting and editing. MM was the main researcher and team member in the field experiments. He participated in all the field work, data collection and analysis; part of the research write-up was done by him. RM was involved in writing the article on the research which was a very difficult business. RM worked tirelessly in field data collection, laboratory sample analyses and guiding this research write-up. KM participated in data analyses especially statistical procedures. He was there right from the beginning up to the end of the field experiments including manuscript write-up. All authors read and approved the final manuscript.

Author details

[1] Department of Land and Water Resources Management, Midlands State University, Private Bag 9055, Gweru, Zimbabwe. [2] Department of Geography and Environmental Studies, Midlands State University, Private Bag 9055, Gweru, Zimbabwe.

Acknowledgements

The study constitutes part of the research work carried out in the Faculty of Natural Resources Management and Agriculture of the Midlands State University, Zimbabwe. The researchers acknowledge the University contribution to the research work.

Competing interests

The authors declare that they have no competing interests.

Funding

No funding from any funding agency was received for carrying out this research work.

References

1. Zimbabwe Chamber of Mines. Zimbabwe Chamber of mines report. In: Zimbabwe Chamber of Mines Editorial Management, Harare, Zimbabwe. 2014. http://www.chamberofminesofzimbabwe.com/slider/mine-entra-2014-2/. Accessed 2 Apr 2017.
2. Cortez H, Pingarrón J, Muñoz J, Ballester A, Blázquez AA, González ML, García F, Coto CO. Bioremediation of soils contaminated with metalliferous mining wastes. In: Plaza G, editor. Trends in bioremediation and phytoremediation. Thiruvananthapuram: Research Signpost Publishers; 2010. p. 230–58.
3. Kumar A, Bish BS, Josh VD, Dhewa T. Review on bioremediation of polluted environment: a management tool. IJES. 2011;1(6):1079–93.
4. Chibuike GU, Obiora SC. Heavy metal polluted soils: effect on plants and bioremediation methods. Appl Environ Soil Sci. 2014;1:1–12. doi:10.1155/2014/752708.
5. Battaglia-Brunet F, Dictor MC, Garrido F, Crouzet C, Morin D, Dekeyser K. An arsenic (III)-oxidizing bacterial population: selection, characterization, and performance in reactors. J Appl Microbiol. 2002;93:656–67.
6. Gavrilescu M. Removal of heavy metals from the environment by bioremediation. Eng Life Sci. 2004;4:219–32. doi:10.1002/elsc.200420026.
7. Smith VH, Graham DW, Cleland DD. Application of resource-ratio theory to hydrocarbon biodegradation. Environ Sci Technol. 1998;32:3386–95.
8. Strong PJ, Burgess JE. Treatment methods for wine-related and distillery wastewaters: a review. Bioremediation J. 2008;12:70–87.
9. Zimstats. Zimbabwe statistics. 2012th ed. Harare: Government Printers; 2012.
10. Mugandani R, Wuta M, Makarau A, Chipindu B. Re-classification of agro-ecological regions of Zimbabwe in conformity with climate variability and change. Afr Crop Sci J. 2012;20:361–9.
11. Bartholomew DS. Gold deposits of Zimbabwe. Zimb Geol Surv Miner Resour Ser. 1990;23:73–5.
12. Baldock JW, Styles MT, Kalbskopf S, Muchemwa E. The geology of the Harare Greenstone Belt and surrounding granitic terrain. Zimb Geol Surv Bull. 1991;94:210–3.
13. Darbyshire DPF, Pitfield PEJ, Campbell SDG. Late Archean and early Proterozoic gold-tungsten mineralization in the Zimbabwe Archean craton: Rb-Sr and Sm-Nd isotope constraints. Geology. 1996;24:19–22.
14. Jiang X, Lu WX, Zhao HQ, Yang QC, Yang ZP. Potential ecological risk assessment and prediction of soil heavy-metal pollution around coal gangue dump. Nat Hazards Earth Syst Sci. 2014;14:1599–610.
15. Yang ZP, Lu WX, Long YQ, Bao XH, Yang QC. Assessment of heavy metals contamination in urban topsoil from Changchun City, China. J Geochem Explor. 2011;108:27–38.
16. Hakanson L. An ecological risk index for aquatic pollution control: a sedimentological approach. Water Res. 1980;14:975–1001. doi:10.1016/0043-1354(80)90143-8.
17. Douay F, Pelfrene A, Planque J, Fourrier H, Richard A, Roussel H, Girondelot B. Assessment of potential health risk for inhabitants living near a former lead smelter, Part 1: metal concentrations in soils, agricultural crops, and home-grown vegetables. Environ Monit Assess. 2013;185:3665–80. doi:10.1007/s10661-012-2818-3.
18. VSN International. GenStat discovery GenStat for windows. 7th ed. Oxford: VSN International; 2003.
19. Demirbas A. Concentrations of 21 metals in 18 species of mushrooms growing in the East Black Sea region. Food Chem. 2001;75:453–7.
20. Campos JA, Tejera NA. Substrate role in the accumulation of heavy metals in sporocarps of wild fungi. Biometals. 2009;22:835–41.
21. Gutierrez A, Martinez AT, Prieto A. Structural characterization of extracellular polysaccharides produced by fungi from the genus Pleurotus. Carbohydr Res. 1996;281:143–54.
22. Das N. Heavy metals biosorption by mushrooms. NPR. 2005;4:454–9.
23. Stihi C, Radulescu C, Busuioc G, Popescui IV, Gheboianu A, Ene A. Studies on accumulation of heavy metals from substrate to edible wild mushrooms. Rom Journ Phys. 2009;56(1–2):257–64.
24. Siric I, Kos I, Bedekovic D, Kaic A, Kasap A. Heavy metals in edible mushrooms *Boletus reticulatus* Schaeff collected from Zrin mountain, Croatia. Period Biol. 2014;116(3):319–22.
25. Khaled SB, Muhammad AA. Field accumulation risks of heavy metals in soil and vegetable irrigated with sewage water in western region of Saudi Arabia. Saudi J Biol Sci. 2016;23:32–44.
26. Kojta KA, Zhang D. Survey on composition and bioconcentration potential of 12 metallic elements in King Bolete (*Boletus edulis*) mushroom that emerged at 11 spatially distant sites. J Environ Sci. 2011;46:231–46.

27. Lamrood PY, Ralegankar SD. Biosorption of Cu, Zn, Fe, Cd, Pb and Ni by non-treated biomass of some edible mushrooms. Asian J Exp Biol Sci. 2013;4:190–5.
28. Shweta K, Nupur M, Prdeep B. Mushroom as a product and their role in mycoremediation. AMB Express. 2014;4:1–29.
29. Fei L, Jing-po Y, Han L. Risk evaluation of heavy metals in soil in the sewage irrigation area: a case study of Shijiazhuang. J Chem Pharm Res. 2014;6(6):2147–52.
30. Aydinalp C, Marinova S. Distribution and forms of heavy metals in some agricultural soils. Pol J Environ Stud. 2003;12(5):629–33.

5

Densely rooted rhizosphere hotspots induced around subsurface NH_4^+-fertilizer depots: a home for soil PGPMs?

Peteh Mehdi Nkebiwe[1*], Günter Neumann[2†] and Torsten Müller[1†]

Abstract

Background: Populations of plant growth-promoting microorganisms (PGPMs) inoculated in natural soil typically decline over time due unfavourable biotic and/or abiotic factor(s). Improving subsurface root density may enhance PGPM establishment due to high concentrations of organic nutrients released as root exudates. Placing subsurface root-attracting NH_4^+-fertilizer depots may form such zones of dense localized rooting ("rhizosphere hotspots") that can enhance PGPM survival. Nevertheless, required soil conditions that favour formation of rhizosphere hotspots are unknown. This study aimed to investigate: (1) background soil N_{min} effect on NH_4^+-depot-zone root growth; (2) PGPM tolerance to high NH_4^+ concentrations (± nitrification inhibitor, DMPP); (3) ability to solubilize sparingly soluble inorganic phosphates; (4) and establishment in a subsurface NH_4^+-depot.

Methods: We conducted a greenhouse rhizobox experiment using spring wheat (*Triticum aestivum* L.) to investigate the effect of background N_{min} (0, 5, 20 and 60 mg N kg^{-1}) on root growth around a 1 g NH_4^+–N depot. The tolerance of selected PGPMs to 0, 2, 10, 50, 250, 1250 mM NH_4–N and 0, 0.1, 1 and 3 M NH_4–N ± DMPP was investigated via in vitro culture tests. Promising candidates were further tested for solubilization of tri-calcium phosphate (Ca–P), rock phosphate (RP) or sewage sludge ash (SA). One PGPM was inoculated in a ^{15}N-labelled $(NH_4)_2SO_4$ + DMPP depot in rhizobox-grown maize (*Zea mays* L.) and root colonization was measured.

Results and discussion: N_{min} 5–20 mg N kg^{-1} improved depot-zone root growth, whereas 60 mg N kg^{-1} reduced it. Tested PGPMs were tolerant to up to 1250 mM NH_4–N. *Pseudomonas* sp. DSMZ 13134 and *B. amyloliquefaciens* FZB42 (not *Trichoderma harzianum* T-22) solubilized Ca–P and RP via acidification, whereas SA was not solubilized despite marked acidification. Placed ^{15}N-labelled $(NH_4)_2SO_4$ + DMPP depot led to increased localized rooting, rhizosphere acidification, shoot ^{15}N signal, N and P concentrations and contents than homogenously applied $Ca(NO_3)_2$. Inoculation of *Pseudomonas* sp. DSMZ 13134 tended to increase shoot N and P concentrations, and shoot N content relative to the control. Higher colonization of *Pseudomonas* sp. DSMZ 13134 was measured in soil around the NH_4^+-depot than in corresponding soil zone in treatments with NO_3^-.

Conclusion: These results show the first promising effects of combining fertilizer placement and application of P-solubilizing PGPMs on plant growth.

Keywords: Fertilizer placement, Localized root growth, PGPM, Bio-effector, Biofector, P-solubilizing bacteria

*Correspondence: Mehdi.Nkebiwe@uni-hohenheim.de
†Günter Neumann and Torsten Müller contributed equally to this work
[1] Fertilisation and Soil Matter Dynamics (340 i), University of Hohenheim, 70593 Stuttgart, Germany
Full list of author information is available at the end of the article

Background

Subsurface placement of manures or mineral fertilizers produces higher crop nutrient uptake and yield than conventional surface broadcast [1–4]. Among others, intense rooting in nutrient patches and increased nutrient uptake rates favour exploitation of subsurface fertilizer depots. Therefore, providing high concentrations of nutrients that stimulate root growth at the site of root contact (e.g. NH_4^+ and $HPO_4^{2-}/H_2PO_4^-$ or $CO(NH_2)_2$ and $HPO_4^{2-}/H_2PO_4^-$, the later under optimal ammonification conditions) [2, 5] and are poorly mobile in soil [6–9], crops can more efficiently acquire fertilizer-depot nutrients, thus improving their yields [4]. Nevertheless, it has been shown that intense rooting may not occur around fertilizer depots based on manure or mineral N under greenhouse and field conditions [10]; (Müller et al. 2009, unpublished results). The reasons for this are unclear. Therefore, we hypothesize that localized root growth around a subsurface fertilizer depot depends on high nutrient concentration gradients between the depot and background soil. Suggesting that a strong localized root growth response to a subsurface fertilizer depot is more likely when nutrient concentration is high in the depot and/or low in bulk soil than the other way, Sommer [11] proposed N fertilization by subsurface application of high and toxic concentrations of NH_4^+, thus creating a strong contrast to bulk soil N concentration. Moreover, NH_4^+-toxicity within such depots inhibits microbial nitrification, resulting in more persistent NH_4^+-induced root growth intensification. This NH_4^+-based N fertilization technique has been termed controlled long-term ammonium nutrition (CULTAN) [11]. Use of stabilized NH_4^+ (e.g. nitrification inhibitor (NI), 3,4-dimethylpyrazole phosphate (DMPP) [12] will additionally inhibit nitrification around depot boarders where NH_4^+-concentrations become less toxic with time.

Nevertheless, for plant species that respond to soil nutrient patches by high turnover rates of roots within the patch (high root birth and death rates e.g. *Ambrosia artemisiifolia*), a net increase in patch root density may not be observed [13].

Another approach to directly or indirectly improve plant acquisition of applied or soil nutrients is to inoculate soil with plant growth-promoting microorganisms (PGPMs) e.g. P-solubilizing bacteria and fungi. PGPMs have been described to promote plant growth via one or a combination of mechanisms. *Pseudomonas* sp., *Bacillus* sp. or *Xanthomonas* sp. have been shown to produce the phytohormone indole acetic acid (IAA) to promote plant root growth and/or the enzyme ACC-deaminase (1-aminocyclopropane-1-carboxylate deaminase), which delays root senescence by reducing ethylene levels under stress conditions [14–17]. These PGPMs have also been shown to synthesize various antifungal metabolites (e.g. phenols and phenazines) that directly suppress root pathogenic fungi and also function as low-molecular weight iron-complexing ligands (e.g. siderophore pyoverdine) for effective competition against root pathogens for iron and other nutrients [18, 19]. Other species of *Bacillus, Pseudomonas, Enterobacter* and *Trichoderma* have been shown to release protons, organic acids and chelating metabolites (e.g. Oxalic, fumaric, citric, DL-malic, DL-lactic and succinic acids) that enhance the solubility of sparingly soluble phosphates and/or secrete enzymes like phosphatases and phytases that mineralize organic phosphates in soil making them plant-available [20–25].

In order that applied PGPMs improve plant nutrient acquisition from subsurface NH_4^+-based fertilizer depots, they must be considerably tolerant to extreme pH, high NH_4^+ and salt concentrations found within and around such depots. For depots with stabilized N, tolerance to nitrification and/or urease inhibitors is additionally required for PGPM survival and root colonization within the developing "rhizosphere hotspot".

The overall aim of this study was to employ fertilizer placement as a tool to induce dense localized rooting in soil ("rhizosphere hotspot") for improved establishment of selected soil-inoculated PGPMs thereby enhancing plant nutrient-acquisition and biomass production. Rhizosphere hotspots are proposed to be areas that favour PGPM establishment in soil due to the presence of high concentrations of organic nutrients released as root exudates. We made the following hypotheses: (1) formation of rhizosphere hotspots around NH_4^+-depots is inhibited by high background mineral N concentration (N_{min}) and promoted by low bulk soil N_{min}. (2) Certain PGPMs (also referred to as microbial bio-effectors, BEs) tolerate high concentrations of NH_4^+ ($\pm$ DMPP) that are well above the limits for NH_4^+-tolerant crop species like rice (*Oryza sativa* L.). (3) BEs are able to solubilize different sparingly soluble mineral phosphates (including recycled P fertilizers) in solid and liquid in vitro cultures. (4) Inoculation of a P-solubilizing BE around a subsurface stabilized (DMPP) concentrated ^{15}N-labelled NH_4^+ depot improves BE establishment and increases N and P uptake in maize (*Zea mays* L.).

Materials and methodology

Effect of background N_{min} on NH_4^+-depot root growth

We cultivated spring wheat (*Triticum aestivum* L. var Schirocco, KWS, 1197, 29296 Bergen, Germany) in PVC rhizoboxes (40 × 20 × 2 cm, H × W × D). The Substrate was composed of 80% C-horizon loess (P_{CAL}, 5 mg kg^{-1}; P_{total}, 332 mg kg^{-1}; pH ($CaCl_2$), 7.6; C_{org}, < 0.3%; N_{total} 0.02%) and 20% quartz sand (0.6–1.2 mm Ø) (w/w). Basic fertilization included (kg^{-1} dry soil DM): 150 mg P

$(Ca(H_2PO_4)_2)$, 200 mg K (K_2SO_4), 100 mg Mg $(MgSO_4)$, 20 µmol Fe-EDDHA (Sequestrene 138, 6% Fe), 2.6 g Zn $(ZnSO_4)$; 1 mg Cu $(CuSO_4)$; 2.2 mg Mn $(MnSO_4)$, 0.54 mg Mo (Na_2MoO_4) and 0.86 mg B (H_3BO_3). Moisture content was set to 26% (60–70% max. WHC). Treatments included increasing bulk soil N_{min} concentrations: 0, 5, 20 and 60 mg N kg^{-1} soil DM (NH_4NO_3). A stabilized NH_4^+-depot was placed 7 cm to the side and 16 cm below two wheat seedlings in each rhizobox. The NH_4^+ depot (1.0 g NH_4–N + 0.17 g NO_3–N + 0.01 g DMPP) was made by placing 3.49 g $(NH_4)_2SO_4$ in soil over which 2.5 mL solution of 1.1 M NH_4NO_3 containing DMPP(4 g L^{-1}) was pipetted. The depot was mixed in a small soil volume of 5 cm^3 (2.5 cm Ø × 2 cm depth) resulting in a high NH_4–N concentration within the depot of 78.4 mg g^{-1} soil (bulk density, 1.3 g cm^{-3}). DMPP was added at the recommended rate of 1% NH_4–N (w/w) [12]. Each rhizobox contained 2.6 kg substrate.

Before transplanting, seed surfaces were disinfected by soaking in 3% H_2O_2 for 10 min. Afterwards seeds were soaked for 10 h in 2.5 mM $CaSO_4$ and pre-germinated on filter papers that were also soaked in 2.5 mM $CaSO_4$ (48 h at 25 °C in darkness). Two healthy seedlings (distinct shoot and root length ≥ 2 cm) were transplanted to each rhizobox. Greenhouse conditions were as follows: day/night length: 8 h/16 h, daytime light intensity: 311 µmol m^{-2} s^{-1} (ALMEMO 239-3, AHLBORN), av. daily temperature 15 °C (8–21 °C) and av. daily relative humidity 42% (30–65%) (Voltcraft, DL-141 TH). Four replicates were made per treatment and were arranged in a completely randomized design. Plants were grown for 60 days (10 February–07 April 2013). Total root length visible through the root observation window of the rhizobox in the radial zones: 0–4, 4–8 and > 8 cm from the NH_4^+-depot was measured at 12, 36 and 56 days after planting. Using a fine black water-resistant marker, roots were drawn on a polyethylene sheet (40 × 20 cm) placed over the soil surface on the rhizobox window. Drawings were scanned (Epson Expression 10000 XL) and analysed for total root length using *WinRhizo Pro V. 2009c* (Regent Instruments Inc., Canada). At the end of the growth period, root length density (cm cm^{-3} of substrate) could not be appropriately measured because, as shown in Fig. 2, roots grew into the nylon sheet lining the rear of the rhizobox, making it impossible to satisfactorily harvest roots. After harvesting, shoot dry weight was measured, depot and bulk soil samples were collected and frozen for later analysis of NH_4–N and NO_3–N concentrations.

PGPM tolerance to high NH_4^+-concentrations

The PGPMs shown in Table 1 were grown on appropriate solid nutrient media with increasing concentrations of NH_4–N. For each medium, six solutions with the following NH_4^+ concentrations were prepared using deionized water: 0, 2, 10, 50, 250 and 1250 mM NH_4–N $((NH_4)_2SO_4)$. For each NH_4^+-level, the constituents per liter of medium were dissolved in the corresponding solution to a final volume of 1000 mL. The pH was then adjusted to the desired level using 1 N NaOH or HCl and finally, the medium was sterilized by autoclaving at 121 °C for 20 min.

The following solid media were prepared for each NH_4^+-level:

(1) Nutrient agar (5.0 g peptone; 3.0 g meat extract; 15.0 g agar; pH 7); (2) King's B medium (from ready-made preparation [King's B medium (Basis), Carl Roth GmbH + Co. KG]; 38 g medium and 10 mL glycerol in 990 ml deionized water; pH 7); (3) DEV medium (Deutsche Einheitsverfahren; 10.0 g peptone; 10.0 g meat extract; 5.0 g NaCl; 18.0 g agar; pH 7.2); and (4) malt extract peptone agar (30.0 g malt extract; 3.0 g soya peptone; 15.0 g agar; pH 5.6).

At 45 °C, about 20 mL of each medium was poured into a sterile petri dish. For bacterial PGPMs, inocula were prepared as diluted suspensions in 2.5 mM $CaSO_4$. For each bacterium and NH_4^+-level, 100 µL inoculum (about 50 CFUs) was pipetted onto a petri dish and evenly spread using a Drigalski spatula. For each fungus, a small fragment of pre-cultured mycelium was placed at the centre of the petri dish. There were three replicates per treatment randomly placed on the same incubator shelf (incubator: WBT, Binder, Typ 1711509900313, Tuttlingen, Germany) and cultured at 24 ± 1 °C under aerobic conditions in darkness. *Trichoderma harzianum* strains were grown for 48 h and the remaining PGPMs for 1 week.

An additional incubation test on normal NH_4^+-free agar was performed. For this, selected BEs suspended in high concentrations of NH_4–N ($(NH_4)_2SO_4$ ± DMPP) were pre-incubated at 25 ± 2 °C for 15 min or 24 h before inoculation on NH_4^+-free agar. This modification was meant to simulate addition of BEs directly into a concentrated NH_4^+-fertilizer solution ($(NH_4)_2SO_4$ ± DMPP) before placement in soil. The aim was to investigate BE viability directly after exposure to high NH_4–N ± DMPP levels. *Pseudomonas* sp. DSMZ 13134, *Bacillus amyloliquefaciens* FZB42 and *Trichoderma harzianum* T-22 were tested. After pre-incubation for 15 min, 100 µL of each bacterial inoculum suspension in 0, 0.1, 1 or 3 M $(NH_4)_2SO_4$ ± DMPP (about 50 CFUs) was inoculated on culture media and incubated at 25 ± 2 °C for 48 h. in darkness under aerobic conditions. For *T. harzianum* T-22, 100 µL 0, 0.1, 1 or 3 M $(NH_4)_2SO_4$ ± DMPP was pipetted on a fragment of pre-cultured mycelium and after 15 min, the mycelium fragment was placed on

Table 1 List of PGPMs

Nr.	Active microorganism	Product name	Supplier	Recommended solid nutrient media
1	*Bacillus atrophaeus*	–	ABiTEP GmbH, Berlin, Germany	Nutrient agar (NA)
2	*Bacillus simplex* R41	–	ABiTEP GmbH, Berlin, Germany	NA
3	Bacillus spec.	–	ABiTEP GmbH, Berlin, Germany	NA
4	*Penicillium* sp. PK 112	Biological Fertilizer DC	Bayer Crop Science, Germany	Malt extract peptone agar (MEP)
5	*Bacillus amyloliquefaciens* FZB42	RhizoVital® 42	ABiTEP GmbH, Berlin, Germany	Deutsche Einheitsverfahren Agar (DEV) and NA
6	*Pseudomonas* sp. DSMZ 13134	Proradix® WP	Sourcon Padena, Tübingen, Germany	King's B medium (KM)
7	*Trichoderma harzianum* OmG08	–	Hochschule Anhalt, Bernburg, Germany	MEP
8	*Trichoderma harzianum*	Trichoderma-WG	Bayer Crop Science, Germany	MEP
9	*Trichoderma harzianum* T50	Vitalin T50	Vitalin Pflanzengesundheit GmbH, Ober-Ramstadt, Germany	MEP
10	*Trichoderma harzianum* T-22 (only in tests ± nitrification inhibitor DMPP)	Trianum-P	Koppert Biological Systems B.V., The Netherlands	MEP

NH_4^+-free agar. This procedure was adopted for *T. harzianum* T-22 because the effect of $(NH_4)_2SO_4$ ± DMPP on its survival and growth could be best quantitatively analysed by measuring the diameter of intact pieces of growing mycelium. There were four randomly placed replicates per treatment. For 24 h BE pre-incubation, only 1 or 3 M $(NH_4)_2SO_4$ ± DMPP solutions were used because marked NH_4^+-tolerance was observed at 15 min. pre-incubation. There were only three replicates per treatment to reduce bulk and facilitate management of the experiments. For 24 h. pre-incubation, *T. harzianum* T-22 spores were used as inoculum instead of a fragment of pre-cultured mycelium to be more relevant to field application. Therefore, only the presence or absence of a growing *T. harzianum* T-22 mycelium could be recorded. Growth could not be quantified because of the absence of a distinct circular mycelium to measure the diameter. For bacteria, colony characteristics were recorded and for *Pseudomonas* sp. DSMZ 13134, colonies were observed under UV light for typical yellow-green fluorescence.

Solubilization of inorganic phosphates by PGPMs

Solid media culture

To investigate the ability of selected PGPMs to solubilize insoluble inorganic phosphates, Deubel-Muromcev solid agar containing a suspended precipitate of tri-calcium phosphate ($Ca_3(PO_4)_2$) (Ca–P) was prepared [26, 27]. 0.2 g K_2SO_4, 0.4 g $MgSO_4{\cdot}7H_2O$, 20 g Purified agar (Oxoid, No. LP 0028) were dissolved to a total volume of 1000 mL using deionized water. The pH of the solution was adjusted to 6.9 using 1 N NaOH or HCl and then it was autoclaved (121 °C, 20 min). After cooling the solution to 60 °C, 10 g D (+) Glucose monohydrate (6887.1, Roth, Karlsruhe, Germany) dissolved in 110 mL deionized water and 1 g L-asparagine monohydrate (11160, Fluka, Buchs, Germany) dissolved in 50 ml deionized water were added (Solution A). To precipitate non-soluble Ca–P in the medium, 2.2 g $CaCl_2{\cdot}2H_2O$ (2461.1000 Chem Solute Th.Geyer, Renningen, Germany) dissolved in 20 mL deionized water and 3.8 g $Na_3PO_4{\cdot}12H_2O$ (6578, E. Merck, Darmstadt, Germany) also dissolved in 20 ml deionized water, each autoclaved (121 °C, 20 min) and cooled down to 60 °C. $CaCl_2$ and Na_3PO_4 solutions were simultaneously added to solution A under continuous stirring resulting in immediate clouding of the medium through the formation of Ca–P ($Ca_3(PO_4)_2$) (Eq. 1):

$$3CaCl_2 \cdot 2H_2O + 2Na_3PO_4 \cdot 12H_2O \rightarrow Ca_3(PO_4)_2 + 6NaCl + 30H_2O. \quad (1)$$

Between 40 and 45 °C, the medium was poured into petri dishes. After solidification and cooling of the agar, suspensions of PGPMs in 2.5 mM $CaSO_4$ (1×10^4 CFU or spores mL^{-1}) were inoculated and plates were incubated at 25 °C. PGPMs included (1) *Penicillium* sp. PK 112, (2) *Pseudomonas* sp. DSMZ 13134, (3) Vitalin SP11 [combined product: *Bacillus subtilis, Pseudomonas* sp., *Streptomyces* spp., natural humic acids and seaweed extract (*Ascophyllum nodosum*)] (Vitalin Pflanzengesundheit GmbH, Ober-Ramstadt, Germany), (4), *Paenibacillus mucilaginosus* (ABiTEP GmbH, Berlin, Germany), (5) *Bacillus amyloliquefaciens* FZB 42, (6) *Trichoderman harzianum* T-22.

Additionally, two other media were prepared in which instead of precipitating Ca–P, at 60 °C under continuous stirring, finely ground (< 1 mm Ø) rock phosphate (RP, 7.6% P) or sewage sludge ash (SA, 10.3% P) was added at the rate of 0.3 g P L^{-1}. RP was chosen because it is an important low-cost sparingly soluble inorganic P fertilizer (the only mineral P fertilizer allowed in certified EU organic farming) and SA was selected as a low-cost sparingly soluble recycled P fertilizer produced from sewage treatment. Only *Pseudomonas* sp. DSMZ 13134 was inoculated on RP and SA solid media because it was one of the best Ca–P solubilizing candidates.

After two days, all petri dishes were observed for clarification of Ca–P, RP or SA along streaks of growing PGPM and scanned on a dark background.

Liquid media culture

Treatments Based on Deubel-Muromcev solid agar, liquid media with Ca–P was prepared by simply omitting addition of agar to solidify the media. Two other liquid media were prepared in which Ca–P was replaced by either rock phosphate (RP, 7.6%) or sewage sludge ash (SA, 10.3%). For each of these media Ca–P, RP or SA was the only source of P present. The liquid culture conditions, sample collection and preparation, and analysis were based on methods described by [28]. Treatments included factorial combinations of three levels of sparingly soluble P fertilizers (Ca–P, RP or SA) and four PGPM levels as microbial bio-effector (No bio-effector (No BE), *Pseudomonas* sp. DSMZ 13134 (Pro); *Bacillus amyloliquefaciens* FZB42 (Rhiz); and *Trichoderma harzianum* T-22 (T-22). There were four replicates per treatment.

Preparation of solutions Using sterile deionized water (autoclaved 121 °C, 20 min), 0.2 g K_2SO_4 and 0.4 g $MgSO_4 \cdot 7H_2O$ were dissolved to 1000 mL and pH was adjusted 6.9 (Basal solution); 45.45 g D-glucose was dissolved to 500 mL (glucose); 4 g of L-asparagine was dissolved to 200 mL (L-asparagine); 5.5 g $CaCl_2 \cdot 2\ H_2O$ was dissolved to 50 mL ($CaCl_2$); 9.5 g $Na_3PO_4 \cdot 12H_2O$ was dissolved to 50 mL (Na_3PO_4). 60 mL of Basal solution was transferred into 100-mL Erlenmeyer flasks, autoclaved and then allowed to cool down in a water bath to 80–90 °C. Glucose and L-asparagine were each heated up to 70 °C in a water bath for 30 min. $CaCl_2$ and Na_3PO_4 were each autoclaved and allowed to cool down to room temperature.

Preparation liquid media Each P form (Ca–P, RP and SA) was applied separately at the level of each experimental unit (100-mL Schott Duran incubation flask) ensuring 0.018 g P $flask^{-1}$ because it was impossible to pour out 60 mL portions with the appropriate quantity of P from one stock suspension with sedimenting RP or SA into each 100-mL flask. 1.2 mL $CaCl_2$ and 1.2 mL Na_3PO_4 were added to 60 ml Basal solution in each flask under continuous stirring to evenly disperse the Ca–P precipitate formed. The total concentration of P was 0.3 g L^{-1} and NaCl was 1.75 g L^{-1}. 0.237 g RP or 0.174 g SA, each equivalent to 0.018 g P, was weighed with a semi/micro analytical balance (Mettler AT261 DeltaRange, Mettler-Toledo GmbH, Giessen, Germany) and added under continuous stirring to 60 mL Basal solution. The RP and SA used had the following chemical composition: RP, P_{Total} 7.6%; $P_{Citric\ acid}$ 6.8%; P_{NaHCO3} 1.7%; Al 1.4%; Ca 22.5%; Fe 5.1%; Si 7.7%; SA, P_{total} 10.34%; $P_{citric\ acid}$ 4.49%; Al 9.9%; Ca 8.6%; Fe 4.2%; Cd 4.2; Cr 90.8; Cu 814; Ni 78.5; Pb 224; V 59.1; Zn 4004 ppm). Total P concentration was also 0.3 g P L^{-1}. 0.105 g of NaCl was added to each RP or SA flask to maintain the same NaCl concentration in Ca–P flasks (1.75 g L^{-1}). In each Ca–P, RP or SA flask, 6.6 mL glucose and 3 mL L-asparagine was added and swelled to mix.

Inoculation and incubation Using sterile 2.5 mM $CaSO_4$, without PGPM (Control) or with the PGPMs: *Pseudomonas* sp. DSMZ 13134, 6.6 × 10^{10} CFUs g^{-1} (Pro), *Bacillus amyloliquefaciens* FZB42, 2.5 × 10^{10} Spores mL^{-1} (Rhiz) and *Trichoderma harzianum* T-22, 1 × 10^9 Spores g^{-1} (T-22) were each sequentially diluted to produce a suspension of 1 × 10^4 CFU or spores mL^{-1}. For the non-inoculated control, 2 mL of sterile 2.5 mM $CaSO_4$ solution was added to each flask. For each PGPM treatment, 2 mL inoculum suspension was added. All flasks were randomly placed on two shelves in an incubator (GFL Typ 3032, Gesellschaft für Labortechnik mbH, Burgwedel, Germany) in darkness at 30 °C. To prevent sedimentation of Ca–P, RP or SA, flasks were continuously swelled horizontally at 125 rpm.

Measurements After 65 h of incubation, 10 mL sample of medium was collected from each flask into a sterile 50-mL falcon tube after solids could settle. Tubes were centrifuged at 5000×*g* for 15 min and supernatants were carefully poured into 30-mL polyethylene bottles and stored frozen (− 18 °C). For further analysis, samples were defrosted at room temperature, pH was measured (pH/Conductivity Meter, MPC227, Metler Toledo) and they were filtered through P-free blue-band filter papers (MN 640 d) for later analysis of total P concentration by inductively coupled plasma optical emission spectrometry (ICP-OES) (Core Facility Hohenheim, University of Hohenheim).

Acid and alkaline phosphatase activity in the samples was also measured. Phosphatase activity gave an indication of the rate of mineralization or turnover of phosphates that have been assimilated already into organic matter. Phosphatase measurement had no relevance to initial solubilization of inorganic phosphates. Based on a modified *p*-nitrophenyl phosphate method ([29], phosphatase activity was measured by dephosphorylation of 0.1 mL of 150 mM *p*-nitrophenyl phosphate in 1.5-mL Eppendorf tube, which contained 0.4 mL buffer

(200 mM Na-acetate buffer, pH 5.2 for acid phosphatase or 200 mM Na-borate buffer, pH 8.2 for alkaline phosphatase), 0.4 mL deionized water and 0.1 mL liquid culture supernatant. The mixture was incubated at 30 °C under continuous shaking. For acid phosphatase activity, incubation was stopped after 30 min, whereas for alkaline phosphatase activity, it was stopped after 45 min to ensure a detectable coloration. To stop enzyme activity, 0.5 mL of 0.5 M NaOH was added immediately after incubation. For controls or blanks, a second set of tubes containing liquid media sample and buffer was incubated as described above. After incubation, NaOH was added followed by *p*-nitrophenyl phosphate. Both sets of samples and blanks were centrifuged (14,000 rpm for 5 min). About 800 μL supernatant was pipetted into a 1 mL disposable micro cuvette and measured for absorbance of yellow coloration at 405 nm using*p*-nitrophenol (*p*-NP) standards containing volumes of *p*-NP stock (20 μg mL^{-1}), 500 μl 0.5 M NaOH and volumes of deionized water to top up to 1.5 mL. *p*-NP standards included: 0, 2, 4, 8, 12, 16 and 20 μg *p*-NP mL^{-1}.

Establishment of *Pseudomonas* sp. DSMZ13134 rhizosphere hotspots

The methodology for this experiment has been fully described in Nkebiwe et al. [30]. Using a substrate composed of 80% loess [(P_{CAL}, 5 mg kg^{-1}; P_{total}, 332 mg kg^{-1}; pH ($CaCl_2$), 7.6; $CaCO_3$, 23.3%; C_{org}, < 0.3%; N_{total} 0.02%)] and 20% sand filled in rhizoboxes as described in "Effect of background N_{min} on NH_4^+-depot root growth" section, maize (*Zea mays* L. var Colisee) was grown to investigate the establishment of *Pseudomonas* sp. DSMZ13134 in rhizosphere hotspots around stabilized NH_4^+-depots. The substrate was optimally supplied (kg^{-1} soil DM) with 150 mg P ($Ca(H_2PO_4)_2$), 150 mg K (K_2SO_4); 50 mg Mg ($MgSO_4$); 20 μmol Fe, (Sequestrene 138, 6% Fe); 2.6 mg Zn ($ZnSO_4$) and 1 mg Cu, (CuSO4). Moisture content was 18%. 2400 g of substrate was filled into each rizobox (H × W × D: 40 × 20 × 2 cm).

Treatments included two factorial combinations of two N-fertilizer levels and two BE levels. N levels were 100 mg NO_3–N kg^{-1} soil as $CaNO_3$ homogenously mixed in the substrate (NO_3^--mixed) and 100 mg NH_4–N as concentrated ^{15}N labelled stabilized $(NH_4)_2SO_4$ + DMPP (64 mg N mL^{-1}) placed as a depot 5 × 5 cm to the maize seed (NH_4^+-depot). BE levels were non-inoculated control (NoBE) and *Pseudomonas* sp. DSMZ13134 (Pro) inoculated twice by placement at 0 and 23 DAS each at the rate 1 × 10^9 CFUs kg^{-1} soil in the area 2.5 cm around the NH_4^+-depot or corresponding location for NO_3^--mixed treatments. The ^{15}N-labelled $(NH_4)_2SO_4$ fertilizer solution was enriched using ^{15}N-$(NH_4)_2SO_4$ to 10 atom%.

Concentration and contents of shoot N and P, shoot DM, rhizosphere pH, root length density and root colonization of *Pseudomonas* sp. DSMZ13134 in the NH_4^+-depot or corresponding location for NO_3^--Mixed treatments as described already [30]. ^{15}N signal of dried shoot biomass was measure by isotope ratio mass spectrometry (IRMS). During grinding of dried maize shoot samples, some contamination of non-labelled samples with ^{15}N-labelled ones occurred. Nevertheless, this did not prove to be problematic due the big differences in ^{15}N signal between labelled and non-labelled plants because of the very high labeling rate of the placed $(NH_4)_2SO_4$ + DMPP fertilizer with ^{15}N.

Results

Effect of background N_{min} on NH_4^+-depot root growth

At 12 days after planting (DAP), no root growth towards the NH_4^+-depot was observed. At 36 DAP, root length (or length per unit surface area) 4 cm around the depot was low, without differences across treatments (Fig. 1a). At 4–8 cm around the depot, root length was lower in soil with no background N fertilization (0 mg N kg^{-1}) than in those with 5, 20 and 60 mg N kg^{-1}. At > 8 cm around the depot, it increased with increasing background N from 0, 5 to 20 mg N kg^{-1} and then decreased at 60 mg N kg^{-1}. At 56 DAP in the zone > 8 cm around the depot, the same trend could be observed (Fig. 1b). Furthermore, there were no differences in root growth 0–4 cm and 4–8 cm around the depot across treatments at 56 DAP (Figs. 1b, 2). Nevertheless, the trend of increasing root growth in 0–4 or 4–8 cm zones with increasing N_{min} from 0–20 mg N kg^{-1} followed by a decrease at 60 mg N kg^{-1} could be observed. However, after washing away soil from the rhizoboxes, intense rooting induced by the NH_4^+-depot in soil areas with less toxic NH_4^+ concentrations could be observed across all background N_{min} levels (Fig. 2).

As expected, residual NO_3–N in the bulk soil increased with increasing background N fertilization from 0 to 60 mg N kg^{-1}, whereas residual NH_4–N was low and similar across treatments (Fig. 3a). Within the NH_4^+-depot, NO_3–N levels were low, whereas NH_4–N levels where high, with no differences across treatments (Fig. 3b). Shoot dry matter per rhizobox at background N fertilization of 20 mg N kg^{-1} (3.7 g) was higher than those of 0 mg N kg^{-1} (2.9 g) and 5 mg N kg^{-1} (2.6 g) but not different from that of 60 mg N kg^{-1} (3.4 g) (n = 4, tukey test, $P < 0.05$).

PGPM tolerance to high N concentrations

For all strains of *Trichoderman harzianum*, mycelia diameter showed a tendency to increase with increasing NH_4–N concentrations from 0 to 10 mM, followed by a

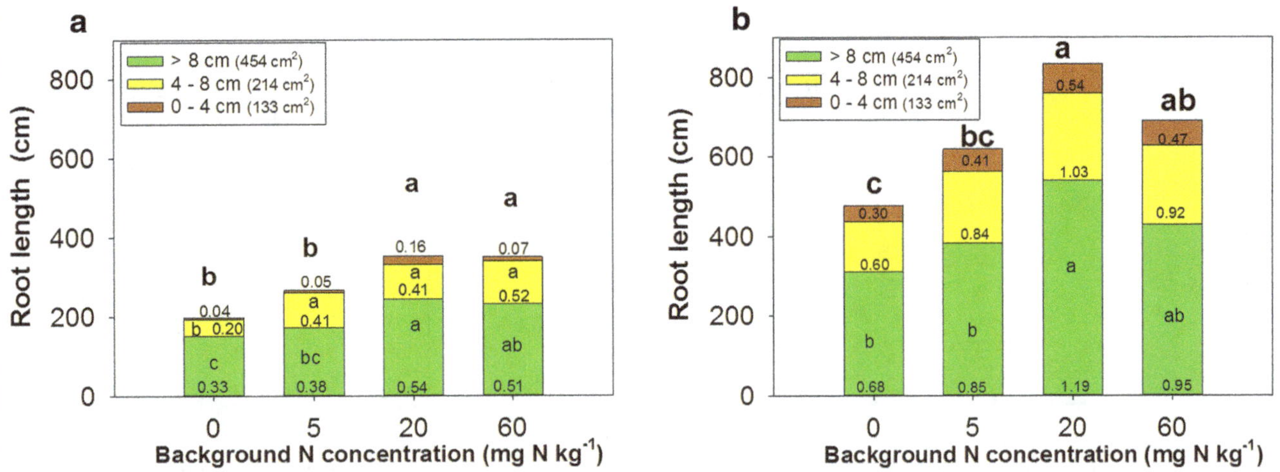

Fig. 1 Root length around the NH_4^+-depot at 36 (**a**) and 56 (**b**) days after planting. Different letters show significant difference between treatments ($P < 0.05$) (mean ± SEM, $n = 4$, One Way ANOVA, Tukey test, $\alpha = 0.05$). Bold letters: total root length; small letters: root length within radial zones around the depot only if differences were significant. Brown, yellow and green sections of bars represent root growth at radial zones 0–4 cm, 4–8 cm and > 8 cm from the depot, respectively. Numbers within bars represent root length per unit surface area on the rhizobox window (cm cm^{-2})

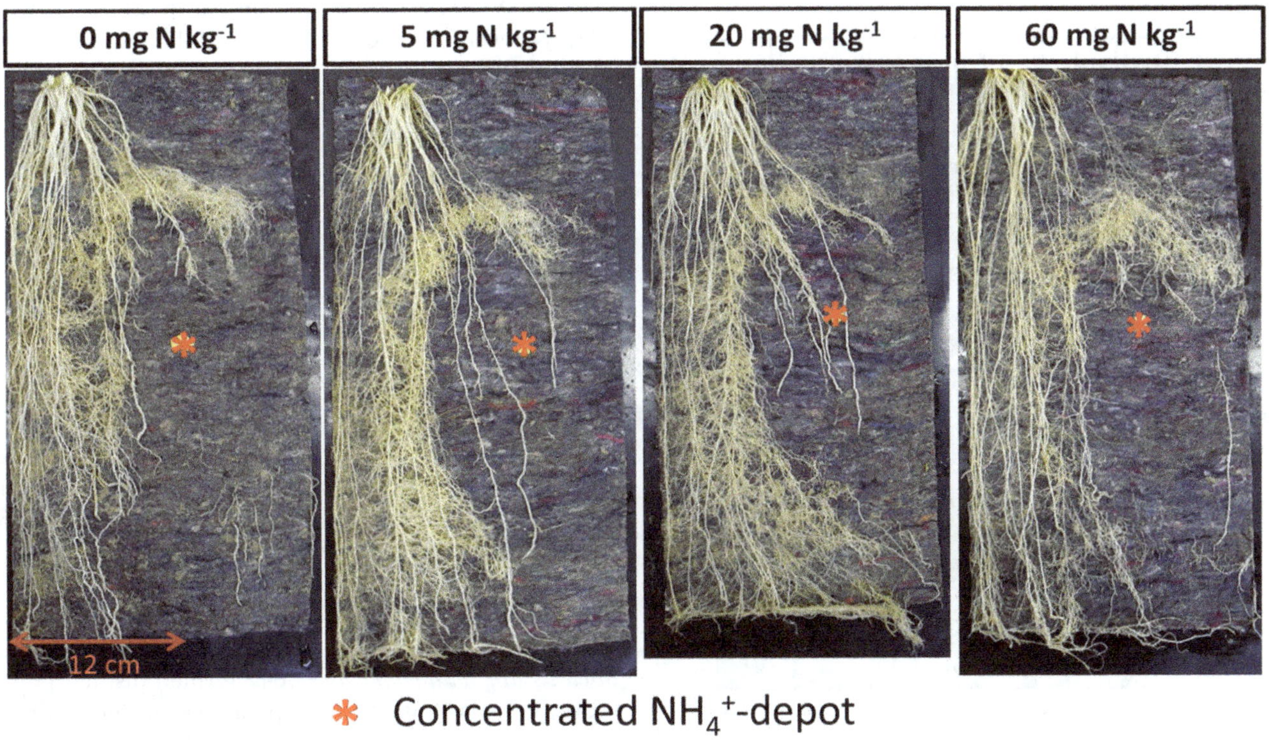

Fig. 2 Root growth inhibition within the toxic NH_4^+-depot and intensification in the outer zones with less-toxic NH_4^+ levels (56 DAP)

slight decrease at 50 mM. Mycelium diameter at 50 mM NH_4–N was higher than the one at 250 mM, which again was higher than the one at 1250 mM (Fig. 4).

After 1 week, there was some growth at 1250 mM NH_4–N for all *T. harzianum* strains meanwhile the typical green centric ring indicating conidia formation

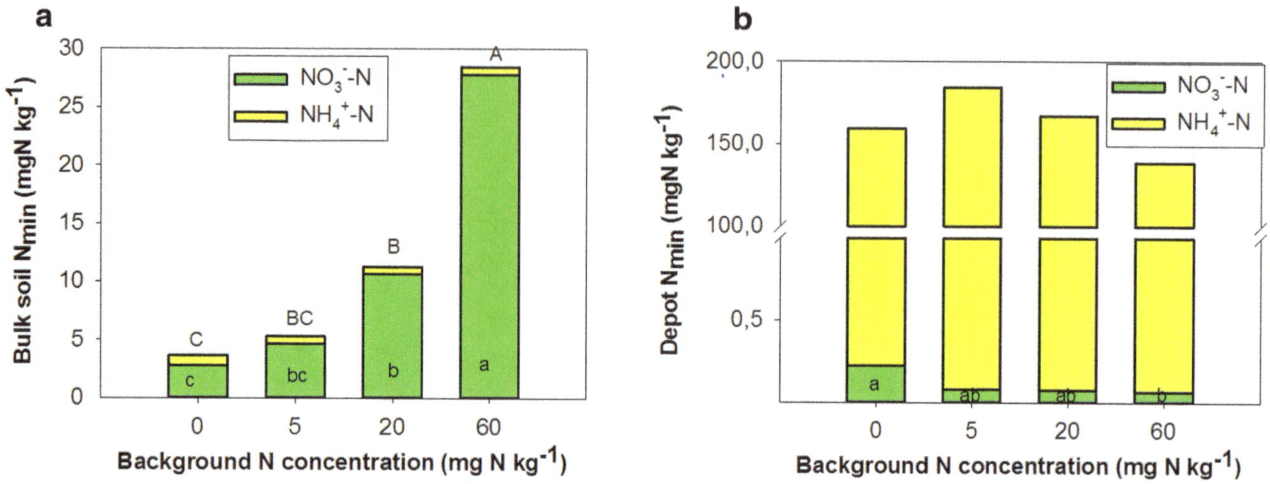

Fig. 3 Residual N_{min} in the bulk soil (**a**) and in the NH_4^+-depot (**b**) at 56 DAP. Different letters show significant difference between treatments, if differences were significant ($P < 0.05$) (mean ± SEM, $n = 4$, One-Way ANOVA, Tukey test, $\alpha = 0.05$). Data for bulk soil NH_4^+–N and total N_{min} were transformed by square roots. Block letters: difference in total N_{min} concentrations; small letters: difference in NH_4^+–N concentrations. Yellow and green section of bars represent NH_4^+–N and NO_3^-–N concentrations, respectively

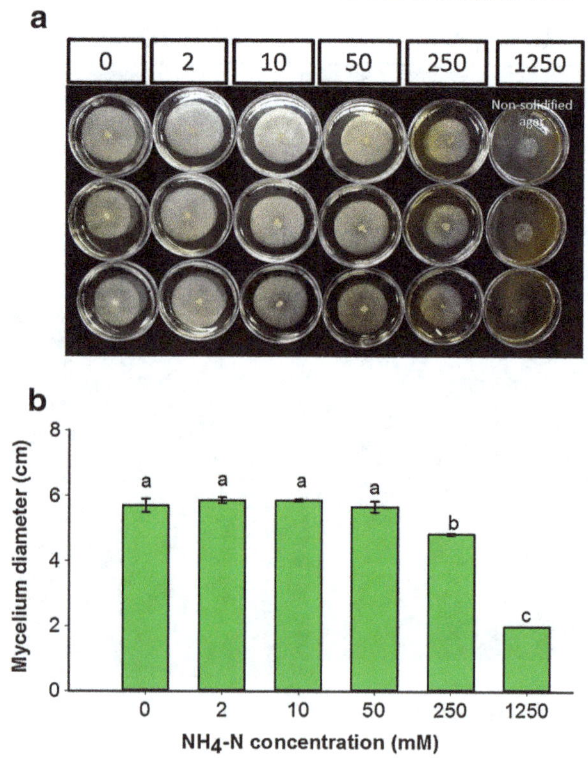

Fig. 4 Effect of NH_4^+ on growth of *Trichoderma harzianum* T-50 mycelium (**a**) and mycelium diameter (**b**) on malt extract peptone agar after 48 h. Different letters show significant difference between treatments ($P < 0.05$) (mean ± SE, $n = 3$, One-Way ANOVA, Tukey test, $\alpha = 0.05$)

at maturity [31] could be seen only in petri dishes with NH_4–N levels from 0 to 250 mM NH_4–N. Therefore, whereas 250 mM NH_4–N depressed vegetative growth only, the toxicity of 1250 mM inhibited both vegetative and generative growth.

Penicillium sp. PK 112 grew slower than *Trichoderma*. After 1 week the mycelium diameter of *Penicillium* sp. PK 112 at 10 and 50 mM NH_4–N were higher than those at 0 and 2 mM NH_4–N. It increased from 2.53 cm at 0 mM NH_4–N to 2.93 cm at 50 mM NH_4–N ($P < 0.05$), after which it showed a tendency to decrease to 2.87 cm at 250 mM NH_4–N (no significant difference). Atypical and limited growth was observed at 1250 mM NH_4–N.

For bacterial PGPMs, considerable growth could be observed at high NH_4–N levels after 1 week. For *Pseudomonas* sp. DSMZ 13134 colonies were circular, yellow, smooth, glistening, entire and convex. The typical yellow pigment which was spread outwards from each colony was intense and distinct only in agar containing 0–250 mM NH_4–N. Colonies also showed typical yellow-green fluorescence under UV light. On DEV medium, *Bacillus amyloliquefaciens* FZB42 formed distinct punctiform colonies with typical spreading bio-films, which were less spread at 250 and 1250 mM NH_4–N (Fig. 5). There were more *B. amyloliquefaciens* FZB42 colonies at 250 mM NH_4–N (97) than at 0 mM (59) ($P = 0.017$). On nutrient agar, colonies of *B. amyloliquefaciens* FZB42, *B. atrophaeus*, *B. simplex* R41 and *Bacillus* spec. were

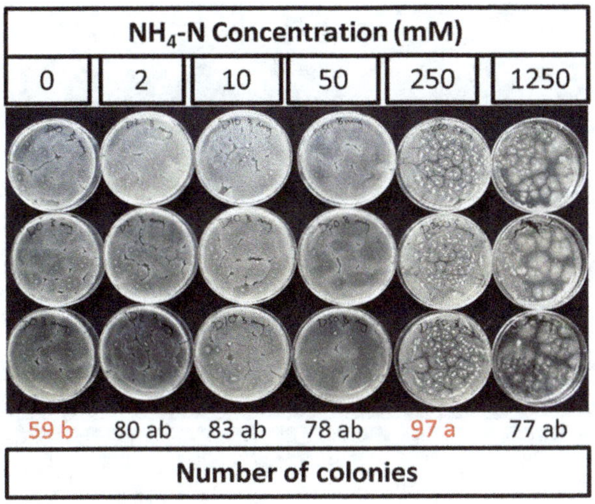

Fig. 5 *Bacillus amyloliquefaciens* FZB42 grown on DEV medium. Different letters show significant difference between treatments ($P < 0.05$) (mean ± SE, $n = 3$, One-way ANOVA, Tukey test, $a = 0.05$)

less distinct and bio-films were also observed across all NH_4–N concentrations.

For *T. harzianum* T-22 after 15-min pre-incubation of mycelium fragment in NH_4^+ ± DMPP, NH_4–N concentration ($P = 0.007$) and presence of DMPP ($P = 0.015$) affected mycelium diameter without any interaction between factors ($P = 0.575$). Irrespective of the presence of DMPP, mycelium diameter at 3 M NH_4–N (4.95 cm) was less than the one at 0 M NH_4–N (5.54 cm) ($P = 0.005$); and irrespective of NH_4–N level, mycelium diameter with DMPP (5.16 cm) was less than the one without (5.44 cm) ($P = 0.015$) (Fig. 6). After 15-min pre-incubation in 0 and 0.1 M NH_4–N, *Pseudomonas* sp. DSMZ 13134 and *B. amyloliquefaciens* FZB42 showed growth activity without any observable effect of DMPP. At 1 and 3 M NH_4–N, growth of both *Pseudomonas* sp. DSMZ 13134 and *B. amyloliquefaciens* FZB42 was suppressed (more for *Pseudomonas* sp. DSMZ 13134) and normal morphological colony characteristics were absent. Nevertheless, typical fluorescence under UV light was observed for *Pseudomonas* sp. DSMZ 13134 as well as typical bio-film formation for *B. amyloliquefaciens* FZB42 was observed.

After 24-h pre-incubation of PGPM in 1 or 3 M·NH_4–N solution with or without DMPP, growth of *T. harzianum* T-22, *Pseudomonas* sp. DSMZ 13134 or *Bacillus amyloliquefaciens* FZB42 were strongly suppressed.

Solubilization of inorganic phosphates by PGPMs

All tested PGPMs except *T. harzianum* T-22 were able to solubilize Ca–P by clarifying the agar along PGPM streaks (Fig. 7a–f). No visible changes could be observed along streaks of *Pseudomonas* sp. DSMZ 13134 on RP or SA agar (Fig. 7i). Sedimentation of RP or SA at the bottom of the petri dishes (Fig. 7g, h), suggested lack of contact between growing *Pseudomonas* on the agar surface and RP or SA at the bottom. This implied that liquid media cultures were more appropriate to test the ability of PGPMs to solubilize RP and SA.

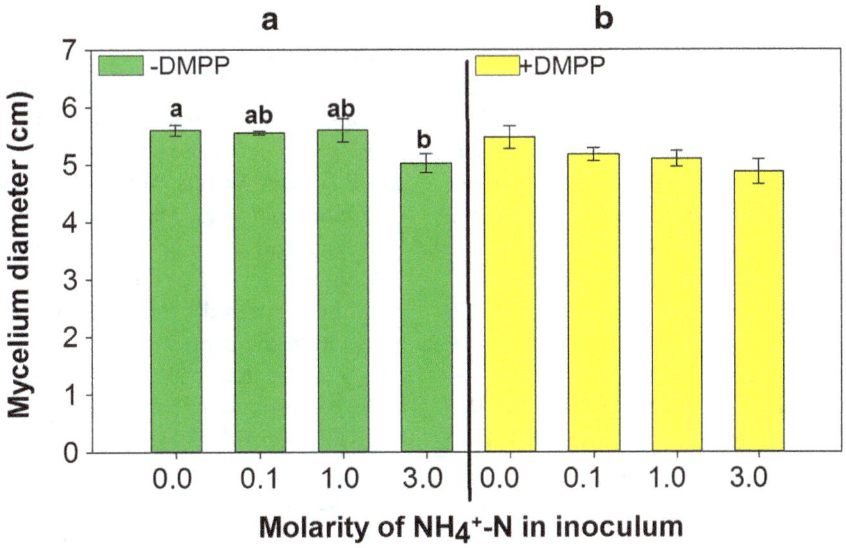

Fig. 6 Mycelium diameter of *Trichoderma harzianum* T-22 affected by 15-min pre-incubation in NH_4–N solution with DMPP (**a**) or without (**b**) or without. Block letters: difference between variants with DMPP and those without (significant effect of DMPP $P = 0.015$); Significant effect of NH_4^+ molarity ($P = 0.007$); small letters: tendency of difference between treatments ($P = 0.055$) (mean ± SE, $n = 4$, One* and Two-way ANOVA, Tukey test, $a = 0.05$)

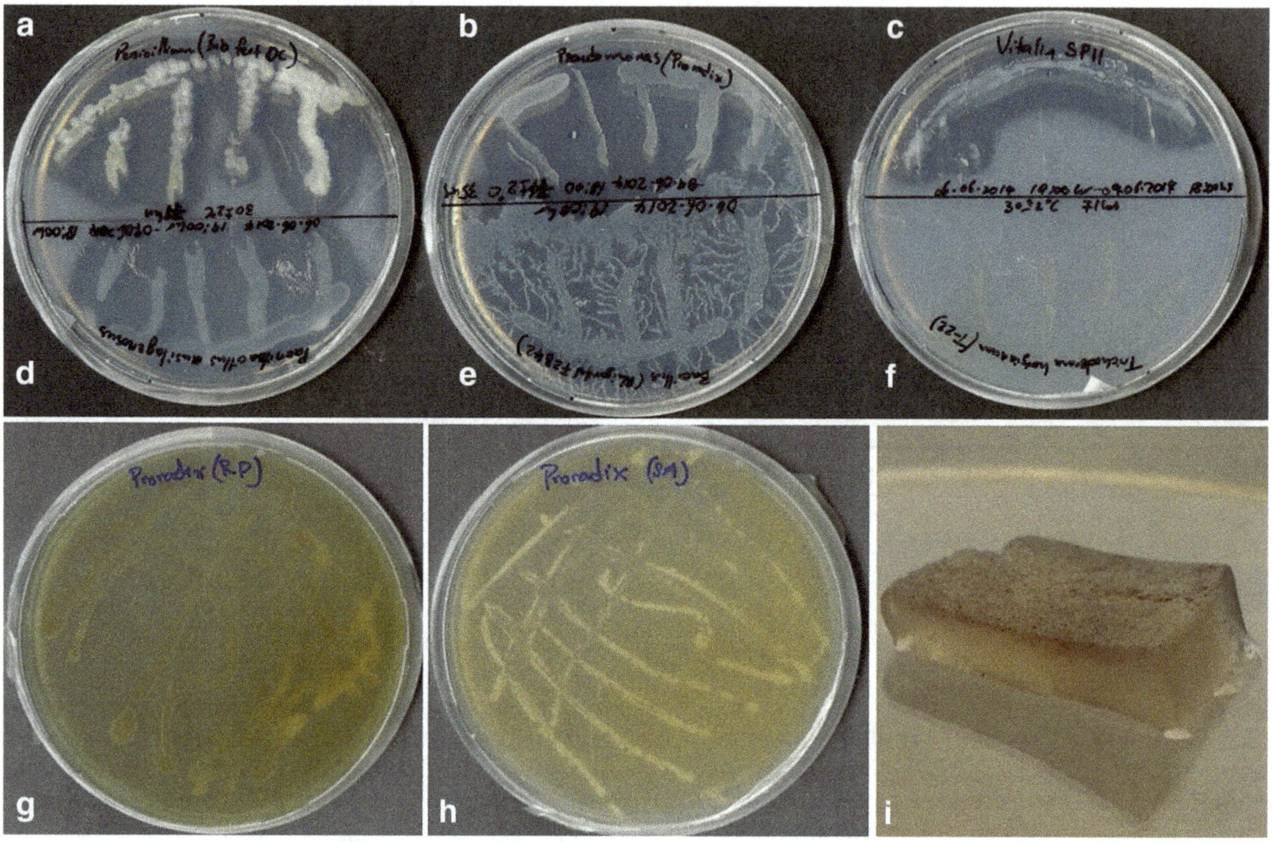

Fig. 7 PGPMs solubilize/clarify cloudy Ca–P precipitate but not rock phosphate (RP) or sewage sludge ash (SA). Cloudy Ca–P precipitate ($Ca_3(PO_4)_2$) solubilized and clarified by *Penicillium* sp. PK 112 (**a**), *Pseudomonas* sp. DSMZ 13134 (**b**), Vitalin SP11 [combined product: *Bacillus subtilis*, *Pseudomonas* sp., *Streptomyces* spp., natural humic acids and seaweed extract (*Ascophyllum nodosum*)] (**c**), *Paenibacillus mucilaginosus* (**d**), *Bacillus amyloliquefaciens* FZB 42 (**e**). But it was not solubilized by *Trichoderma harzianum* T-22 (**f**). Growth of *Pseudomonas* sp. did not lead to any visual changes with RP (**g**) or SA (**h**). A piece of agar inverted to show SA sediment (**i**)

The pH of the media after 65 h varied among and within treatments. In comparison to the starting pH, there was a reduction in the pH of the nutrient medium in the following treatments: Ca–P/Pro, RP, RP/Pro, RP/Rhiz, SA, SA/Pro, SA/Rhiz and SA/T-22 (Fig. 8a). The concentration of P in liquid media ($P_{conc.}$) was decreasing in the following order: Ca–P/Pro, RP/Pro, RP/Rhiz, RP/T-22 and RP. Very low $P_{conc.}$ were measured in the other treatments (Fig. 8b). For all treatments, pH showed a significant negative correlation with $P_{conc.}$ ($P = 0.002$; $r = -0.47$).

Increase in microbial biomass could be observed by the cloudy or murky appearance of the media for Pro and Rhiz variants and a thick mesh of mycelium for T-22 variants.

There was also much variability in acid and alkaline phosphatase activity (P_{ase}) among and within treatments (Fig. 8c, d). pH or $P_{conc.}$ did not correlate with acid or alkaline phosphatase activity.

Establishment of *Pseudomonas* sp. DSMZ13134 rhizosphere hotspots

Placement of ^{15}N-labelled stabilized $(NH_4)_2SO_4$ + DMPP had an effect on root length density (RLD), and rhizosphere pH 5 × 5 cm to seeds, shoot ^{15}N signal and shoot N and P concentrations and contents at 55DAS. Inoculation of *Pseudomonas* sp. DSMZ13134 had an effect on shoot ^{15}N signal and shoot N concentrations (Table 2) at 55DAS.

A higher number of CFUs of fluorescent *Pseudomonads* could be recovered from root rhizoplane per unit soil volume within the zone 5 × 5 cm to seed (8 cm radius) in the NH_4^+-Depot treatment (13.01 Log_{10} CFUs cm^{-3} soil) than in the corresponding zone in NO_3^--mixed treatment (12.38 Log_{10} CFUs cm^{-3} soil) (*t* test $\alpha = 0.05$, $P = 0.045$) [30].

Discussion

Root growth in the centre NH_4^+-depot was strongly inhibited because of the presence of highly phytotoxic concentrations of NH_4^+, as confirmed by N_{min} analysis of

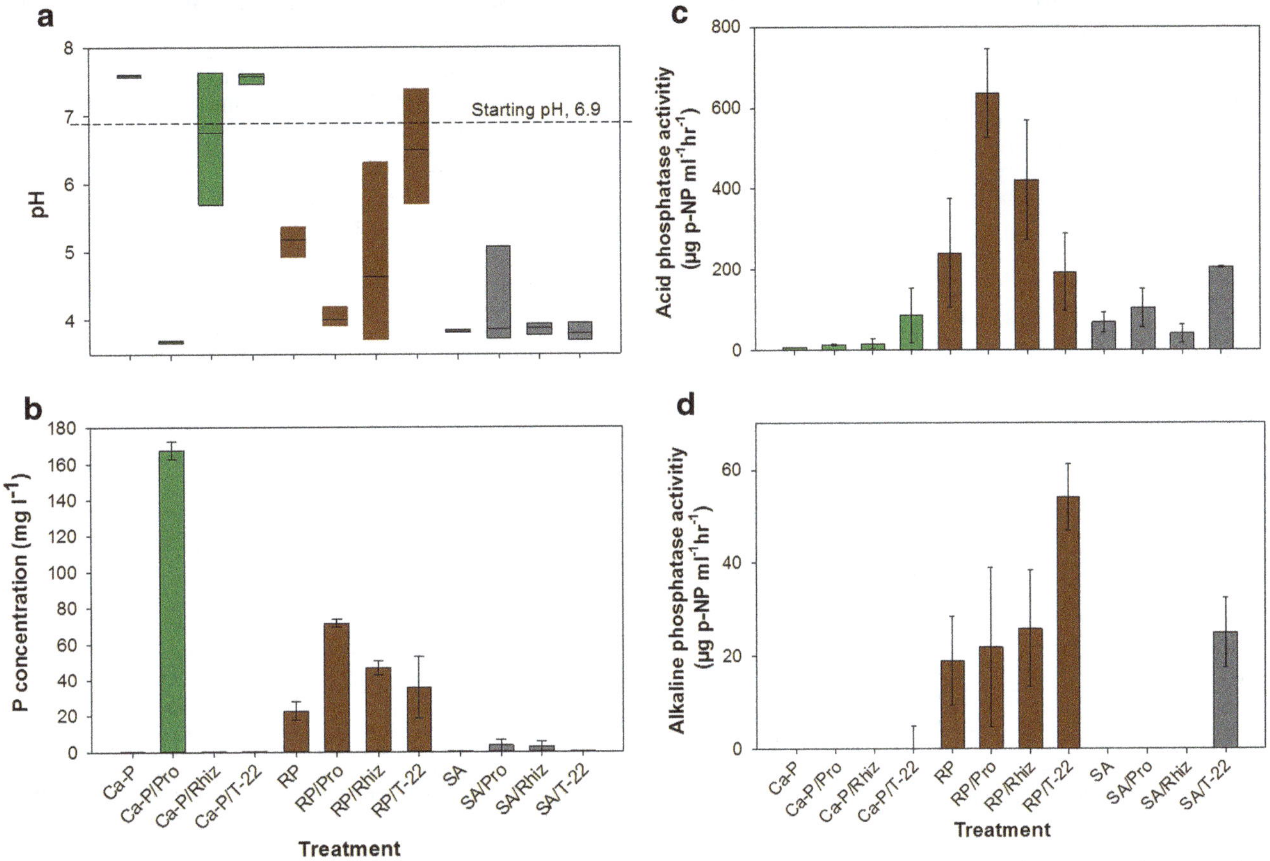

Fig. 8 Change in pH (**a**), total P concentration (**b**), acid- (**c**) and alkaline phosphatase activity (**d**) in the liquid fraction of media 65 h after onset of incubation. Ca–P, $Ca_3(PO_4)_2$; RP, rock phosphate; SA, sewage sludge ash; Pro, *Pseudomonas* sp. DSMZ 13134; Rhiz, *Bacillus amyloliquefaciens* FZB42; T-22, *Trichoderma harzianum* T-22. pH was inversely correlated to total P concentration ($P = 0.002$; $r = -0.47$)

depot soil samples. During the growth period, root length increased in all radial zones around the NH_4^+-depot. During early growth stages, root length in all zones towards the depot increased with increasing background N_{min} from 0 to 20 mg N kg^{-1} only. This was likely due to low background N_{min} levels during early growth stages, N supply to shallow root systems was inadequate for optimal root growth towards the depot. Localized root growth around the depot increased at later growth stages when the expanding N-rich boarder areas of the NH_4^+-depot became accessible to plants roots. This suggests that to optimally supply plants with N especially during early growth stages, it is effective to place starter N fertilizer close to seeds (about 5 cm below) to ensure rapid roots access [4]. However, in case of considerable nutrient mobility from depot to seeds, it is advisable to place starter N farther from seeds to avoid harmful osmotic effects and ion toxicity [32]. At 56 DAP, differences in root length occurred only in the zone > 8 cm, in which root growth was also most intense. Increasing root length in all radial zones around the depot with time and with increasing background N_{min} from 0 to 20 mg N kg^{-1} suggests that moderate N availability in the bulk soil favoured initial root growth and subsequent root growth intensification around the NH_4^+-depot. Given very high NH_4–N levels in the depot, the high background N level of 60 mg N kg^{-1} may have led to N toxicity effects resulting in reduced overall root and shoot growth.

Although background N fertilization was with an NH_4–N share of 50%, only traces of NH_4–N could be measured in the bulk soil after 56 days. This showed that rapid nitrification of NH_4–N applied in the bulk soil occurred, resulting in residual N_{min} with an NH_4–N share of only 2.4% (60 mg N kg^{-1}) to 11.9% (5 mg N kg^{-1}). This can be partially explained by the absence of a nitrification inhibitor (NI) in the NH_4NO_3 used for bulk soil N fertilization. However, in soil without background N fertilization (0 mg N kg^{-1}), low N_{min} levels (2.8 mg NO_3–N kg^{-1} and 0.87 mg NH_4–N kg^{-1}) could be measured in the bulk soil with a low NH_4–N share of 23.7%, suggesting diffusion of N from the depot into the bulk soil as NH_4^+ as well as NO_3^- after nitrification. In the treatments 5, 10,

Table 2 Effect of N-fertilizer/application method and PGPM on root growth and rhizosphere pH, shoot nutrient content and dry matter (55 days after sowing)

Source of variation	RLD 5 × 5 cm to seed (8 cm radius) (cm cm^{-3})	Rhizosphere pH 5 × 5 cm to seed (8 cm radius)	δ15 N (o/oo)$^{15}N/^{14}N$	$N_{conc.}$ (% DM)	N content (mg N $plant^{-1}$)	$P_{conc.}$ (mg P g^{-1} DM)	P content (mg P $plant^{-1}$)	Shoot DM (g $plant^{-1}$)
LS means N*BE								
$^{14}NO_3^-$-mixed*NoBE	6.4	5.62	170	1.83	114.0	2.18	13.7	6.35
$^{14}NO_3^-$-mixed*Pro	6.2	5.89	148	2.22	130.1	2.32	13.6	5.91
$^{15}NH_4^+$-depot*NoBE	10.5	4.41	25,260	2.47	137.1	2.93	16.2	5.60
$^{15}NH_4^+$-depot*Pro	9.9	4.55	21,257	2.66	133.9	3.15	15.9	5.09
Standard error	0.68	0.13	831	0.13	4.69	0.146	0.62	0.420
Two-Way ANOVA								
N	*< 0.001****	*< 0.001****	*< 0.001****	*0.002 ***	*0.014**	*< 0.001****	*0.002***	NS
$^{14}NO_3^-$	6.3	5.75	159	2.03 b	122.0 b	2.52 b	13.6 b	6.13
$^{15}NH_4^+$	10.2	4.48	23,259	2.57 a	135.5 a	3.04 a	16.0 a	5.35
BE	NS	NS	*0.036*	*0.051*	NS	NS	NS	NS
NoBE	8.4	5.01	12,715	2.15	125.7	2.56	15.0	5.97
Pro	8.0	5.22	10,702	2.44	132.0	2.73	14.7	5.50
N*BE	NS		NS	NS	NS	NS	NS	NS

P values are in italics; *NS* no significant difference, $P \geq 0.1$; $P < 0.1$ is italic; $*P < 0.5$; $**P < 0.01$; $***P < 0.001$; Means not sharing the same letters are significantly different from each other, Tukey test $\alpha = 0.05$; Factors and interaction is italic; $^{14}NO_3^-$-mixed, non-isotopically labelled $Ca(NO_3)_2$ homogenously mixed in substrate; $^{15}NH_4^+$-depot, ^{15}N-labelled stabilized $(NH_4)_2SO_4$ +DMPP placed in substrate as a depot; *BE*, bio-effector; NoBE, no bio-effector; Pro, *Pseudomonas* sp. DSMZ 13134. ANOVA on δ15N (o/oo) $^{15}N/^{14}N$ was performed on square root transformed data. RLD, root length density

20 and 60 mg N kg^{-1}, bulk soil NH_4–N concentrations were similar to the one in the treatment without background N fertilization (0 mg N kg^{-1}) after the growth period. This confirmed that bulk soil NH_4^+ at the end of the growth period partially originated from the stabilized NH_4^+-depot directly by diffusion. Given the small substrate volume of the depot, the low clay and organic matter content of the loess-dominated substrate, the high substrate moisture content and high quantity of NH_4^+ that was placed, NH_4^+ probably diffused out of the initial depot boarders after sites for NH_4^+-fixation were exhausted. This is supported by the zones around the initial depot boarders with little or no root growth likely due to the presence of phytotoxic NH_4^+ levels. This is further proven by the strong decrease in the NH_4–N concentration within the depot from the estimated rate of 78.4 mg g^{-1} soil [estimated placement rate 1 g NH_4^+–N to 5 cm^3 depot soil volume (2.5 cm Ø × 2 cm depth, bulk density, 1.3 g cm^{-3})] at the start of the experiment to the measured concentration of 0.18 mg NH_4–N g^{-1} after 56 days.

Despite considerable diffusion of NH_4^+ from the depot into the bulk soil during the experiment, low residual NH_4^+ concentrations measured in the bulk soil after 56 days may be explained by nitrification after diffusion of NH_4^+ out of the phytotoxic zones surrounding the depot resulting. With NO_3^--formation, the N-diffusion rate is expected to be more rapid that with NH_4^+ [4, 6]. High residual NH_4^+ concentrations in the depot (0.18 mg NH_4–N g^{-1}) at the end of the growth period showed that high concentrations of $(NH_4)_2SO_4$ + DMPP in a subsoil depot inhibits biological nitrification of the placed NH_4^+ by both toxicity effects of high NH_4^+ concentrations and the presence of DMPP [11, 12, 33]. This enabled placed N to persist in the depot zone as root growth stimulating and root-attracting NH_4^+. Therefore, intense localized root growth could be observed around a subsurface fertilizer depot based on high concentrations of NH_4^+ stabilized with DMPP, which led to higher N and P concentrations and contents in maize shoots (*Zea mays* L.) in comparison to variants with homogenously incorporated NO_3^--fertilizer [30].

This study also showed that soil zones around NH_4^+-depots with high root densities may be sites for effective root colonization by inoculated PGPMs. Nevertheless, in addition to cultures on selective media, molecular tracing techniques like qPCR [34] as already shown for *Pseudomonas* sp. DSMZ 13134 would provide more precision in tracing. In densely rooted soil, the establishment of PGPMs may have been promoted by high levels of root exudates that boost growth and activity of rhizosphere dwelling microorganisms [15, 35–37].

All tested fungal and bacterial PGPMs showed normal or improved growth activity on solid nutrient agar containing as high as 50 mM NH_4–N and slightly inhibited growth at 250 mM H_4–N except for *B. amyloliquefaciens* FZB42 and *Penicillium* sp. PK 112. In contrast to plant and animal cells, bacteria preferentially use NH_4^+ as a source of N and can be tolerant even to molar concentrations of NH_4–N, at which point growth limiting effects may be related more to osmotic or ionic imbalances than to direct toxic effects of NH_4^+ ions per se [38]. The threshold concentration range of 50–250 mM NH_4–N at which cultured PGPMs showed that inhibited growth is above the range for NH_4^+ toxicity in sensitive crop species like barley (*Hordeum vulgare* L.) and tolerant ones like rice (*Oryza sativa* L.) grown hydroponically [39–41]. The NH_4^+ tolerance threshold for our tested PGPMs (50–250 mM NH_4–N) is also above the range of 2–20 mM that can be common in soil solution of most agricultural soils [42]. The range is also above the NH_4–N concentrations that can be measured in soil solution a few centimeters from a subsurface NH_4^+-fertilizer band (extracted by centrifuging soil samples collected a few centimeters from a subsurface NH_4^+-fertilizer band (< 0.001–37 mM NH_4–N and 0.4–91 mM NH_4–N from a distance 4–2 cm to a fertilizer band of $(NH_4)_2SO_4$) and di-ammonium phosphate, respectively, after 5 days of contact of bands with soil, for tests on five different soil types) [43, 44].

Furthermore, all tested PGPMs showed limited growth activity at the very high NH_4–N level of 1250 mM, which could be explained by osmotic and ionic imbalances as proposed earlier. After 15-min pre-incubation of inoculum suspended in 0–3 M NH_4–N, growth of *T. harzianum* T-22 was slightly inhibited in the presence of DMPP meanwhile for *Pseudomonas* sp. DSMZ 13134 and *B. amyloliquefaciens* FZB42 after pre-incubation in 0–0.1 M NH_4–N, growth activity was not affected by DMPP. After 24-h pre-incubation in 1–3 M NH_4–N solution with or without DMPP, normal growth activity of PGPMs on solid media could no longer be achieved. This suggests that PGPMs may not be applied directly into a concentrated NH_4^+-fertilizer solution (> 50 mM NH_4–N). Alternatively, PGPMs should be inoculated in soil zones with lower NH_4–N levels in the border regions of a concentrated toxic ammonium [30] or directly into the NH_4^+-fertilizer solution if it is less concentrated.

The classical test for P solubilization on solid Ca–P media confirmed that all the tested PGPMs except *Trichoderman harzianum* T-22 were capable to solubilize inorganic P. On solid media culture, *Pseudomonas* sp. DSMZ 13134 effectively solubilized Ca–P, whereas it led to no observable changes after growth on agar with rock phosphate (RP) or sewage sludge ash (SA). However, in liquid media cultures, *Pseudomonas* sp. DSMZ 13134, *Bacillus amyloliquefaciens* FZB42 and *Trichoderma harzianum* T-22 solubilized RP more than SA. Best performing PGPMs were *Pseudomonas* sp. DSMZ 13134 and *B. amyloliquefaciens* FZB42. Increase in microbial biomass, especially for *T. harzianum* T-22, suggested that the P concentrations measured in cell-free culture-media filtrates represented underestimations of the P-solubilizing potential of tested PGPMs because measured P concentrations did not account for solubilized P that was subsequently immobilized in microbial biomass of *T. harzianum* T-22 [20]. Inoculation of T-22 did not result in considerable acidification of the Ca–P medium. Similarly to the results described by [20] (2 µg P mL^{-1} at pH 5, 48 h incubation of medium containing 0.16 g RP–P L^{-1}), inoculation of T-22 resulted in acidification of the RP medium, however, it did not lead to high P concentrations in media filtrates (35.7 µg P mL^{-1} at pH 5.7–7.4, 65 h incubation containing 0.3 g RP–P L^{-1}). This can also be explained by immobilization of solubilized P in the dense T-22 mycelium that was observed. Low pH values were measured in SA media with or without inoculated PGPM which corresponded to low P concentrations in media filtrates. In addition to immobilization of solubilized P in microbial biomass, low P concentrations in SA medium despite low medium pH may be explained by fixation of PO_4^{3-} by Al^{3+} and Fe^{3+} which may be available at low pH conditions given the high SA concentrations of Al (9.9%, 0.29 g SA–Al L^{-1}) and Fe (4.2%, 0.12 g SA–Fe L^{-1}). This explanation supports the observation that P concentrations increased with decreasing pH in the RP medium because RP had a much lower concentrations of Al (1.4%, 0.06 g RP–Al L^{-1}) and about the same concentration of Fe (5.1%, 0.20 g RP–Fe L^{-1}) as the SA medium. We recorded pH values as low as 3.7 in cell-free filtrates of culture media inoculated with *Pseudomonas* sp. DSMZ 13134 or *B. amyloliquefaciens* FZB42. This suggested that acidification by release of protons and/or low-molecular weight organic acids as already reported [20] ([24, 45–47] may have been the likely mode of action for the solubilization of the inorganic P forms. This argument is supported by the significant negative correlation between pH and P concentration in the media ($P = 0.002$; $r = -0.47$). For liquid media with RP only, the correlation was even stronger ($P < 0.0001$; $r = -0.84$). Similarly to pH and total P concentration, there was high variability in acid and alkaline phosphatase activity within and between treatments. Measured acid and alkaline phosphatase activities may only be regarded as an indicator of the turnover or mineralization of organic P [21] and not as a marker for solubilization of inorganic P. For the RP medium, there was a positive correlation between P concentration and phosphatase activity ($P = 0.015$; $r = -0.77$) indicating some turnover of organic P. Nevertheless, solubilization of inorganic P by various PGPMs in in vitro cultures does not guarantee improved P acquisition and yield of plants in soil systems containing inorganic P and inoculated with a P-solubilizing PGPM [47]. Results of in vitro Ca–P solubilization tests as a way to assess P-solubilizing potential of PGPMs and their potential to improve crop P acquisition must be critically evaluated for each soil type, and their chemical and physical properties [21]. For instance, it has been shown that maize P acquisition and biomass production after inoculation of PGPM in was strongly dependent on soil type and properties such as pH and overall fertility [48]. To increase the likelihood for identifying P-solubilizing PGPMs that are effective under field conditions, it is suggested to use the right in vitro test for PGPMs isolated from the right soil type: Ca–P or RP solubilization test for PGPMs isolated from alkaline soils; Fe–P/Al–P solubilization with PGPMs from acid soils; and phytate mineralization with those from soils rich in organic matter [21].

Several mechanisms for PGPMs to directly or indirectly improve acquisition of soil or fertilizer nutrients by crop plants such as root growth stimulation, nutrient solubilization and mineralization, and induction of resistance to biotic and abiotic stresses have been clearly described already [14–16, 18–20, 22, 25, 35, 49].

Sometimes, promised beneficial effects of several commercially available *PGPMs* are not realized even in controlled greenhouse conditions [50] or in the field [15]. This may be explained by rapid decline in PGPM populations often observed after inoculation in natural soil due to competition for nutrients, predation by other soil organisms or other unfavourable biotic or abiotic factors [37].

In our rhizobox experiment, placement of ^{15}N-labelled stabilized $(NH_4)_2SO_4$ + DMPP as a depot led to the formation of intense localized root growth around the depot (rhizosphere hotspot), which in turn led to improved N-uptake as shown by higher shoot N concentration, ^{15}N signal and N content compared to plants under homogenous supply of NO_3^-. Placement of the NH_4^+-depot also led to rhizosphere acidification, which coupled with dense localized root growth led to higher shoot P concentration and content than in plants under homogenous supply of NO_3^-. Inoculation of *Pseudomonas* sp. DSMZ13134 in soil zones in which rhizosphere hotspots developed in response to placed root-attracting NH_4^+-depots promoted the establishment of fluorescent *Pseudomonas* [30]. Inoculation of *Pseudomonas* sp. DSMZ13134 showed a trend to improve shoot N concentration and content and shoot P concentration when compare to the non-inoculated control.

Improved rhizosphere colonization of maize roots (*Zea mays* L.) around an NH_4^+-depot by rifampicin-resistant *Bacillus amyloliquefaciens* FZB42 has also been observed in a pot experiment (Mohammad et al. 2016, unpublished master thesis, Nutritional Crop Physiology, University of Hohenheim).

In addition to inoculating PGPMs in NH_4^+-based rhizosphere hotspots, inoculation of PGPM strains that are adapted or compatible to specific soil properties [21, 51], utilization of suitable nutrient additives, carrier materials and inoculation techniques may increase their survival rates PGPMs after inoculation in soil [37, 52].

Conclusions

Placement of a highly concentrated subsurface NH_4^+-depot induced the formation of rhizosphere hotspots with intense rooting around the depots with moderate bulk soil N_{min} favouring depot-zone root growth. PGPMs that readily solubilized sparingly soluble inorganic P showed considerable tolerance to high NH_4^+ concentrations. Among these NH_4^+-tolerant PGPMs, rhizosphere hotspot establishment of *Pseudomonas* sp. DSMZ13134 was tested and found to be promoted the NH_4^+-depots in rhizobox-grown maize (*Zea mays* L.) in comparison to corresponding soil zones under homogenous supply of NO_3^-. N-form effects on shoot N and P concentration and content were stronger than that of PGPM.

Abbreviations

$^{14}NO_3^-$-mixed: non-isotopically labelled $Ca(NO_3)_2$ homogenously mixed in substrate; $^{15}NH_4^+$-depot: ^{15}N-labelled stabilized $(NH_4)_2SO_4$ + DMPP placed as a depot; BE: bio-effector; Ca–P: tri-calcium phosphate ($Ca_3(PO_4)_2$); CULTAN: controlled long-term ammonium nutrition; DAP: days after planting; DMPP: nitrification inhibitor 3,4-dimethylpyrazole phosphate; ICP-OES: inductively coupled plasma optical emission spectrometry; IRMS: isotope ratio mass spectrometry; NoBE: no bio-effector; PGPM: plant growth-promoting microorganism; Pro: *Pseudomonas* sp. DSMZ 13134; Rhiz: *Bacillus amyloliquefaciens* FZB42; RLD: root length density; RP: rock phosphate; SA: sewage sludge ash; T-22: *Trichoderma harzianum* T-22.

Authors' contributions

PMN participated in conceiving and designing all studies, ran and managed all experiments, conducted all laboratory analyses (except: soil N_{min}, P concentration by ICP-OES and ^{15}N signal by IRMS), performed statistical analyses and interpretation results, drafted and participated in reviewing the manuscript, agrees to accountability. GN and TM participated in conceiving and designing all studies, participated in interpretation of results, participated in reviewing the manuscript, agree to accountability. All authors read and approved the final manuscript.

Author details

[1] Fertilisation and Soil Matter Dynamics (340 i), University of Hohenheim, 70593 Stuttgart, Germany. [2] Nutritional Crop Physiology (340 h), University of Hohenheim, 70593 Stuttgart, Germany.

Acknowledgements
We would like to express our sincere gratitude for the reliable support provided by the technical staff and student assistants of Fertilization and Soil Matter Dynamics (340 i), University of Hohenheim, Stuttgart, Germany.

Competing interests
The authors declare that they have no competing interests.

Funding
This study was part of the BIOFECTOR project funded by the European Commission within the 7th Framework Program|Grant Agmt. No. 312117.

References

1. Federolf C, Westerschulte M, Olfs H, Broll G, Trautz D. Enhanced nutrient use efficiencies from liquid manure by positioned injection in maize cropping in northwest Germany. Eur J Agron. 2016;75:130–8.
2. Jing J, Zhang F, Rengel Z, Shen J. Localized fertilization with P plus N elicits an ammonium-dependent enhancement of maize root growth and nutrient uptake. Field Crop Res. 2012;133:176–85.
3. Miller MH, Ohlrogge AJ. Principles of nutrient uptake from fertilizer bands 1. Effect of placement of nitrogen fertilizer on the uptake of band-placed phosphorus at different soil phosphorus levels1. Agron J. 1958;50(2):95–7.
4. Nkebiwe PM, Weinmann M, Bar-Tal A, Müller T. Fertilizer placement to improve crop nutrient acquisition and yield: a review and meta-analysis. Field Crop Res. 2016;196:389–401.
5. Drew MC. Comparison of the effects of a localised supply of phosphate, nitrate, ammonium and potassium on the growth of the seminal root system, and the shoot, in barley. New Phytol. 1975;75(3):479–90.
6. Barber SA. Soil nutrient bioavailability: a mechanistic approach. New York: Wiley; 1984.
7. Schenk MK, Barber SA. Phosphate uptake by corn as affected by soil characteristics and root morphology. Soil Sci Soc Am J. 1979;43(5):880–3.
8. Clarke AL, Barley KP. The uptake of nitrogen from soils in relation to solute diffusion. Aust J Soil Res. 1968;6(1):75–92.
9. Pang PC, Hedlin RA, Cho CM. Transformation and movement of band-applied urea, ammonium sulfate, and ammonium hydroxide during incubation in several Manitoba soils. Can J Soil Sci. 1973;53(3):331–41.
10. Lamb EG, Haag JJ, Cahill JF. Patch–background contrast and patch density have limited effects on root proliferation and plant performance in *Abutilon theophrasti*. Funct Ecol. 2004;18(6):836–43.
11. Sommer K. CULTAN-Düngung: physiologisch, ökologisch, ökonomisch optimiertes Düngungsverfahren für Ackerkulturen, Grünland, Gemüse, Zierpflanzen und Obstgehölze. Germany: Verlag Th. Mann; 2005.
12. Zerulla W, Barth T, Dressel J, Erhardt K, Horchler von Locquenghien K, Pasda G, Rädle M, Wissemeier A. 3,4-Dimethylpyrazole phosphate (DMPP)—a new nitrification inhibitor for agriculture and horticulture. Biol Fertil Soils. 2001;34(2):79–84.
13. Gross KL, Peters A, Pregitzer KS. Fine root growth and demographic responses to nutrient patches in four old-field plant species. Oecologia. 1993;95(1):61–4.
14. Jiang Y, Wu Y, Xu W, Cheng Y, Chen J, Xu L, Hu F, Li H. IAA-producing bacteria and bacterial-feeding nematodes promote *Arabidopsis thaliana* root growth in natural soil. Eur J Soil Biol. 2012;52:20–6.
15. Lugtenberg B, Kamilova F. Plant-growth-promoting rhizobacteria. Annu Rev Microbiol. 2009;63(1):541–56.
16. Mohite B. Isolation and characterization of indole acetic acid (IAA) producing bacteria from rhizospheric soil and its effect on plant growth. J Soil Sci Plant Nutr. 2013;13:638–49.
17. Shakir MA, Bano A, Arshad M. PGPR with ACC deaminase confer drought tolerance wheat. Soil Environ. 2012;31(1):108–12.
18. Mavrodi OV, Walter N, Elateek S, Taylor CG, Okubara PA. Suppression of *Rhizoctonia* and *Pythium* root rot of wheat by new strains of *Pseudomonas*. Biol Control. 2012;62(2):93–102.
19. Vassilev N, Vassileva M, Nikolaeva I. Simultaneous P-solubilizing and biocontrol activity of microorganisms: potentials and future trends. Appl Microbiol Biotechnol. 2006;71(2):137–44.
20. Altomare C, Norvell WA, Björkman T, Harman GE. Solubilization of phosphates and micronutrients by the plant-growth-promoting and biocontrol fungus *Trichoderma harzianum* Rifai 1295-22. Appl Environ Microbiol. 1999;65(7):2926–33.
21. Bashan Y, Kamnev AA, de-Bashan LE. Tricalcium phosphate is inappropriate as a universal selection factor for isolating and testing phosphate-solubilizing bacteria that enhance plant growth: a proposal for an alternative procedure. Biol Fertil Soils. 2013;49(4):465–79.
22. Barea JM, Toro M, Orozco MO, Campos E, Azcon R. The application of isotopic (^{32}P and ^{15}N) dilution techniques to evaluate the interactive effect of phosphate-solubilizing rhizobacteria, mycorrhizal fungi and *Rhizobium* to improve the agronomic efficiency of rock phosphate for legume crops. Nutr Cycl Agroecosyst. 2002;63:35–42.
23. Panda B, Rahman H, Panda J. Phosphate solubilizing bacteria from the acidic soils of Eastern Himalayan region and their antagonistic effect on fungal pathogens. Rhizosphere. 2016. https://doi.org/10.1016/j.rhisph.2016.08.001.
24. Pérez E, Sulbarán M, Ball MM, Yarzábal LA. Isolation and characterization of mineral phosphate-solubilizing bacteria naturally colonizing a limonitic crust in the south-eastern Venezuelan region. Soil Biol Biochem. 2007;39(11):2905–14.
25. Singh B, Satyanarayana T. Plant growth promotion by phytases and phytase-producing microbes due to amelioration in phosphorus availability. In: Satyanarayana T, Johri NB, editors. Microorganisms in sustainable agriculture and biotechnology. Dordrecht: Springer Netherlands; 2012. p. 3–15.
26. Deubel A. Einfluß wurzelbürtiger organischer Kohlenstoffverbindungen auf Wachstum und Phosphatmobilisierungsleistung verschiedener Rhizosphärenbakterien. Dissertation, Martin-Luther-Universität, Halle-Wittenberg, Germany; 1996.
27. Muromcev GS. Die lösende Wirkung einiger Wurzel-und Bodenmikroorganismen auf die wasserunlöslichen Kalziumphosphate (russ.). Agrobiologija. 1958;5:31–6.
28. de Freitas JR, Banerjee MR, Germida JJ. Phosphate-solubilizing rhizobacteria enhance the growth and yield but not phosphorus uptake of canola (*Brassica napus* L.). Biol Fertil Soils. 1997;24(4):358–64.
29. Tabatabai MA, Bremner JM. Use of *p*-nitrophenyl phosphate for assay of soil phosphatase activity. Soil Biol Biochem. 1969;1(4):301–7.
30. Nkebiwe PM, Weinmann M, Müller T. Improving fertilizer-depot exploitation and maize growth by inoculation with plant growth-promoting bacteria: from lab to field. Chem Biol Technol Agric. 2016;3(1):1–16.
31. Mendoza JLH, Pérez MIS, Prieto JMG, Velásquez JDQ, Olivares JGG, Langarica HRG. Antibiosis of *Trichoderma* spp strains native to northeastern Mexico against the pathogenic fungus *Macrophomina phaseolina*. Braz J Microbiol. 2015;46:1093–101.
32. Niehues BJ, Lamond RE, Godsey CB, Olsen CJ. Starter nitrogen fertilizer management for continuous no-till corn production contribution no. 04-099-j, k-state research and extension. Agron J. 2004;96(5):1412–8.
33. Shaviv A. Control of nitrification rate by increasing ammonium concentration. Fert Res. 1988;17(2):177–88.
34. Mosimann C, Oberhänsli T, Ziegler D, Nassal D, Kandeler E, Boller T, Mäder P, Thonar C. Tracing of two *Pseudomonas* strains in the root and rhizoplane of maize, as related to their plant growth-promoting effect in contrasting soils. Front Microbiol. 2016;7:2150.
35. Bonkowski M, Cheng W, Griffiths BS, Alphei J, Scheu S. Microbial-faunal interactions in the rhizosphere and effects on plant growth. Eur J Soil Biol. 2000;36:135–47.
36. van Overbeek LS, van Elsas JD. Root exudate-induced promoter activity in *Pseudomonas fluorescens* mutants in the wheat rhizosphere. Appl Environ Microbiol. 1995;61(3):890–8.
37. van Veen JA, van Overbeek LS, van Elsas JD. Fate and activity of microorganisms introduced into soil. Microbiol Mol Biol Rev. 1997;61(2):121–35.
38. Müller T, Walter B, Wirtz A, Burkovski A. Ammonium toxicity in bacteria. Curr Microbiol. 2006;52(5):400–6.
39. Coskun D, Britto DT, Li M, Becker A, Kronzucker HJ. Rapid ammonia gas transport accounts for futile transmembrane cycling under NH3/NH4+ toxicity in plant roots. Plant Physiol. 2013;163(4):1859–67.

40. Chen G, Guo S, Kronzucker HJ, Shi W. Nitrogen use efficiency (NUE) in rice links to NH4+ toxicity and futile NH4+ cycling in roots. Plant Soil. 2013;369(1):351–63.
41. Esteban R, Ariz I, Cruz C, Moran JF. Review: mechanisms of ammonium toxicity and the quest for tolerance. Plant Sci. 2016;248:92–101.
42. Britto DT, Kronzucker HJ. NH4+ toxicity in higher plants: a critical review. J Plant Physiol. 2002;159(6):567–84.
43. Moody PW, Aitken RL, Yo SA, Edwards DG, Bell LC. Effect of banded fertilizers on soil solution composition and short-term root-growth. 1. Ammonium-sulfate, ammonium-nitrate, potassium-nitrate and calcium nitrate. Soil Res. 1995;33(4):673–87.
44. Moody PW, Edwards DG, Bell LC. Effect of banded fertilizers on soil solution composition and short-term root-growth. 2. Mono-ammonium and di-ammonium phosphates. Soil Res. 1995;33(4):689–707.
45. Cunningham JE, Kuiack C. Production of citric and oxalic acids and solubilization of calcium phosphate by *Penicillium bilaii*. Appl Environ Microbiol. 1992;58(5):1451–8.
46. Goldstein AH. Recent progress in understanding the molecular genetics and biochemistry of calcium phosphate solubilization by Gram negative bacteria. Biol Agric Hortic. 1995;12(2):185–93.
47. García-López AM, Avilés M, Delgado A. Effect of various microorganisms on phosphorus uptake from insoluble Ca–phosphates by cucumber plants. J Plant Nutr Soil Sci. 2016;179(4):454–65.
48. Thonar C, Lekfeldt JDS, Cozzolino V, Kundel D, Kulhánek M, Mosimann C, Neumann G, Piccolo A, Rex M, Symanczik S, Walder F, Weinmann M, de Neergaard A, Mäder P. Potential of three microbial bio-effectors to promote maize growth and nutrient acquisition from alternative phosphorous fertilizers in contrasting soils. Chem Biol Technol Agric. 2017;4(1):7.
49. Richardson A, Barea J, McNeill A, Prigent-Combaret C. Acquisition of phosphorus and nitrogen in the rhizosphere and plant growth promotion by microorganisms. Plant Soil. 2009;321(1–2):305–39.
50. Schenck zu Schweinsberg-Mickan M, Müller T. Impact of effective microorganisms and other biofertilizers on soil microbial characteristics, organic-matter decomposition, and plant growth. J Plant Nutr Soil Sci. 2009;172(5):704–12.
51. Zahir ZA, Ghani U, Naveed M, Nadeem SM, Asghar HN. Comparative effectiveness of *Pseudomonas* and *Serratia* sp. containing ACC-deaminase for improving growth and yield of wheat (*Triticum aestivum* L.) under salt-stressed conditions. Arch Microbiol. 2009;191(5):415–24.
52. Marschner P. Rhizosphere biology. In: Marschner P, editor. Marschner's mineral nutrition of higher plants. 3rd ed. Academic Press: United States of America; 2012. p. 369–88.

6

Inoculation with *Azospirillum brasiliense* increases maize yield

Inocencio Junior Oliveira[1], José Roberto Antoniol Fontes[1], Bruno Fernando Faria Pereira[2] and Aleksander Westphal Muniz[1*]

Abstract

Background: Maize plants have a high nitrogen (N) demand, and thus, the fertilizers' sustainable management and the inoculation with N_2-fixing bacteria are the alternatives to reduce fertilization cost and the environmental impacts with N fertilizers. Under no tillage system, the N_2-fixing inoculant *Azospirillum brasiliense* effects were compared with mineral N fertilization for maize crop.

Results: The N fertilization treatments influenced all variables in the 2014/2015 season and the cob insertion, plant height, yield, and N concentration in leaves in the 2015/2016 season, suggesting that the N supply, either N mineral from *A. brasiliense* increased growth and yield of maize plants. The maize cultivars influenced all variables in the two seasons. There was no interaction effect of cultivars × N fertilization on the variables. Thus, the maize cultivars had similar performance regardless treatments.

Conclusions: The inoculation with *A. brasiliense* increased maize yield compared with non-N-fertilized plants and the inoculation with *Azospirillum* at sowing + mineral N at topdress provided similar grain yield as mineral N at sowing and topdress for both cultivars.

Keywords: *Zea mays* L., Inoculation, Grain yield, Mineral fertilization

Background

The maize (*Zea mays* L.), the main cereal crop in Brazil, was grown near 16 million hectares during the 2015/2016 harvest, at mean yield of 4181 kg ha^{-1}, despite its increasing global demand [1]. This low yield is explained by the high cost of inputs and low level of technology adopted by farmers. In Amazonas state, the maize yield is around only 2515 kg ha^{-1} [1].

The photosynthetic rate of maize is high, but is easily influenced by environmental stress, as the low soil N level [2, 3] is an essential element for plant cell and fundamental to increase crop yield [4]. In addition, most of tropical soils have low N availability, requiring the N fertilization as an indispensable practice [5]. In this context, inorganic fertilizers are the main N source to increase maize plant development and yield [6]. However, the inadequate management of N fertilizers has negative environmental impacts, as water body eutrophication and nitrous oxide emission [7].

The reduction of N fertilizers and their negative impacts requires urgent alternative strategies. For instance, supplementing partially the plant needs by natural processes, as biological N_2 fixation and organic phosphate solubilization [8, 9]. The commercial use of *Bradyrhizobium* is successfully implemented in Brazil and USA [10], leading together the annual N_2 fixation by soybean [11].

The beneficial effects of biofertilizers include promotion of plant growth, yield quality, nutrient mobilization, soil health, and reduced susceptibility to disease caused by environmental changes [12]. Previous studies evaluated the positive effects of diazotrophic bacteria *Azospirillum* on increasing growth, root length, yield, dry matter, and N accumulation of maize plants [6, 13–15].

*Correspondence: aleksander.muniz@embrapa.br
[1] Embrapa Amazônia Ocidental, AM 010 Road, km 29, Manaus 69010970, Brazil
Full list of author information is available at the end of the article

Besides these benefits, it has a potential to reduce the production costs [16].

In this context, we evaluated the effect of *A. brasiliense* inoculation on maize cultivated under no tillage system.

Methods

Study area

The experiment was performed in an Oxisol (Latossolo Amarelo, distrófico, muito argiloso [17]; Xanthic Hapludox [18]; Xantic Ferralsol, [19]) on an upland area on two harvests—2014/2015 and 2015/2016—under no tillage system.

The experimental area was located at km 29 of Highway AM010, in the State of Amazonas, (02°53′25″S, 59°58′06″W, 102 m a.s.l.). The climate, according to the Köppen classification, is the rainy tropical type AF, with an average temperature of 26.2 °C, and annual rainfall of 2 mm [20].

The maize was sown in December 2014 and December 2015, on millet straw desiccated with glyphosate (900 g ha^{-1} of active ingredient). The potassium and phosphorus rates used for maize crop were based in the chemical and physical properties of 0–20 cm soil layer (Table 1).

Experimental design

The experimental design was four randomized blocks in a 2 × 4 factorial combination (two cultivars of maize × four treatments with N fertilization). Five seeds were sowed per meter on 0.9-m spaced lanes. Each experimental plot size was 5.4 m (length) per 10 m (width), summing 54 m^2.

Treatments and cultivation

We used the varieties BRS Caimbé and double-cross hybrid AG 1051 in four treatments to evaluate N fertilization: (1) control (without inoculant and mineral N); (2) seeds inoculated with *A. brasiliense* bacterium; (3) N fertilization with urea (20 kg ha^{-1} at sowing and 100 kg ha^{-1} at topdress band); and (4) seeds inoculation with *A. brasiliense* + N fertilization with urea at topdress (100 kg ha^{-1} of N). Seeds were inoculated with a commercial product containing Ab-V5 and Ab-V6 strains at concentration of 10^8 cell ml^{-1}, applied at rate of 100 ml on 60,000 seeds. The inoculation was performed by adding the inoculant using an automatic pipette directly on the seeds, which were shaken in plastic bags for uniform product distribution.

The sowing was carried 1 h after inoculation. At sowing, were applied 80 kg ha^{-1} of P_2O_5 triple superphosphate and 60 kg ha^{-1} of K_2O as potassium chloride.

To control weeds, we applied nicosulfuron (50 g ha^{-1} of active ingredient) at 25 days after plant emergence, and to control armyworm [*Spodoptera frugiperda* (J.E. Smith)], we applied deltamethrin (5 g ha^{-1} of active ingredient) at 40 days after plant emergence.

Data analysis

The dependent variables were measured considering 10 central plants on each experimental plot: first cob insertion height; plant height; and total N concentration in leaves and grains. For the N concentration in leaves, plants were sampled during the flowering season and 100 grains were sampled after harvest for the N concentration determination. A 36 m^2 central area on each experimental plot was used to evaluate the variable yield of grains at 13% of moisture. All variables were subjected to analysis of variance (ANOVA), and the Tukey ($p < 0.05$) test compared the means. The data analysis was performed using SAS 9.1 statistical package.

Results

The N fertilization treatments influenced ($p < 0.01$) all variables in the 2014/2015 season and the cob insertion, plant height, yield, and N concentration in leaves in the 2015/2016 season (Table 2). The maize cultivars influenced ($p < 0.01$) all variables in the two seasons (Table 2). There was no interaction effect ($p > 0.05$) of cultivars × N fertilization on the variables. Thus, the maize cultivars had similar performance regardless treatments (Table 2).

The AG 1051 hybrid had higher cob insertion and plant height than the BRS Caimbé variety in both seasons (Figs. 1a, b, 2a, b). Considering the N fertilization treatments, only the control treatment—without N—showed lower cob insertion and plant height comparing with other treatments in both seasons (Figs. 3a, b , 4a, b), suggesting that the N supply, either N mineral from *A. brasiliense* increased growth of maize plants.

Table 1 Chemical and physical soil properties (0–20 cm) before maize sow of 2014/2015 and 2015/2016

Season	pHH_2O	SOM	N	P	K	Ca	Mg	Al	C.E.C	B.S
		g kg^{-1}		mg dm^{-3}		$cmol_c$ dm^{-3}				%
2014/2015	5.62	20.02	1.98	78	61	3.51	1.17	0	7.08	44.77
2015/2016	6.13	24.33	3.23	98	61	3.16	0.95	0	8.01	53.46

SOM soil organic matter, *N* KCl extractable nitrogen, *P and K* Mehlich-1 available phosphorus and potassium, *Ca, Mg and Al* KCl extractable calcium, magnesium and aluminum, *C.E.C* cation exchange capacity, *B.S* basis saturation

Table 2 Analysis of variance considering the influence of different N fertilization treatments, different cultivars AG 1051 e BRS Caimbé in two seasons of evaluation on first cob insertion height (FCIH), plant height (PH), grains yield (GY), N leaves concentration (NL), and N grains concentration (NG)

Sources	DF	Mean square				
		FCIH	PH	GY	NL	NG
Growing season 2014/2015						
Block	3	31.85^{ns}	149.91^{ns}	476,496.1^{ns}	2.72^{ns}	1.57^{ns}
Cultivar (C)	1	3606.13**	1199.28**	31,621,724.6**	88.05**	186.58**
Nitrogen (N)	3	227.86**	686.25**	20,983,449.3**	144.44**	13.26**
C × N	3	158.19^{ns}	482.69^{ns}	993,154.7^{ns}	2.05^{ns}	3.13^{ns}
Error	21	58.47	168.17	326617.5	7.47	0.92
CV (%)		7.3	6.4	14.0	9.7	6.1
Growing season 2015/2016						
Block	3	36.53^{ns}	6.91^{ns}	75,302.1^{ns}	4.92^{ns}	3.59^{ns}
Cultivar (C)	1	2397.58**	1928.21**	12,329,709.0**	65.35**	50.07**
Nitrogen (N)	3	214.96**	871.44**	27,503,747.9**	191.12**	5.45^{ns}
C × N	3	40.77^{ns}	107.19^{ns}	625,594.3^{ns}	4.54^{ns}	1.57^{ns}
Error	21	21.92	51.99	257,929.3	3.91	3.52
CV (%)		4.1	3.3	11.4	7.7	12.3

DF degrees of freedom

** Significant at *F* test, $p < 0.01$

^{ns} Not significant at *F* test ($p > 0.05$)

The AG 1051 hybrid had higher yield than the BRS Caimbé for both seasons (Figs. 1c, 2c). The treatments mineral N fertilization at sowing + mineral N fertilization at topdress, and the seed inoculation + mineral N at topdress had similar yield in both seasons. These treatments showed yield grain higher than the other N fertilization treatments. In addition, the yield of inoculation with *A. brasiliense* treatment was higher than control treatment in both seasons (Figs. 3c, 4c).

Considering the N concentration in leaves and grains, the AG 1051 hybrid showed higher concentration than BRS Caimbé variety in both seasons (Figs. 1d, e, 2d, e). The treatments mineral N at sowing + mineral N at topdress and seed inoculation + mineral N at topdress had the same response and were superior to the other treatments, considering the N concentration in leaves (Figs. 3d, 4d), as well as the N concentration in grains had this same behavior in 2014/2015 season (Fig. 3e), while in 2015/2016 season, the N concentration in grains was not different among the N fertilization treatments (Fig. 4e).

Discussion

We demonstrated that the *Azospirillum* seed inoculation at sowing increased N concentration in plants and growing variables compared with control treatment. Moreover, the reported yield was higher than the average of yield maize at Amazonas state [15] confirmed that *Azospirillum* seed inoculation build up plants' height. As well known, the positive effects of N on plant development are due to its function on the roots and shoots growth, composing protein, enzymes, coenzymes, nucleic acids, phytochromes, photosynthetic pigments, etc. Consequently, increasing the photosynthetically active area of plants and photoassimilates synthesis translocated to the grains [21] providing a higher yield. The demonstrated benefits of *Azospirillum* are explained by its successful association with grasses [15, 16, 22]. These microorganisms called diazotrophics reduce the atmospheric N_2 to ammonia (NH_3)-breaking triple bonds by the nitrogenase enzyme, process with high demand of energy as ATP. After this reduction, the NH_3 is converted to ammonium (NH_4^+), assimilated as glutamine by the vegetal cells [23]. Besides the N_2-fixing, these bacteria also (i) synthesize phytohormones [24]; (ii) solubilize inorganic phosphate [25, 26]; (iii) stimulate root system development [27], and (iv) increase resistance of plants by stress and their own N biological fixation [26].

Moreover, the inoculation with *Azospirillum* at sowing + mineral N at topdress provided similar grain yield as mineral N at sowing and topdress for both maize cultivars. Previous studies generally reported the benefits of *Azospirillum*. The combination of 200 ml ha^{-1} of *A. brasiliense* with 200 kg ha^{-1} of N increased the physiological response of the maize crop [6]. Gains on maize grains yield with this inoculation are reported from 7% [28] to 30% [29, 30]. Lana et al. [28] reported that inoculated plants' yield was equivalent to those topdressed with 100 kg ha^{-1} of mineral N.

The N leaves' concentrations on plants treated with inoculant at sowing + mineral N fertilization were higher

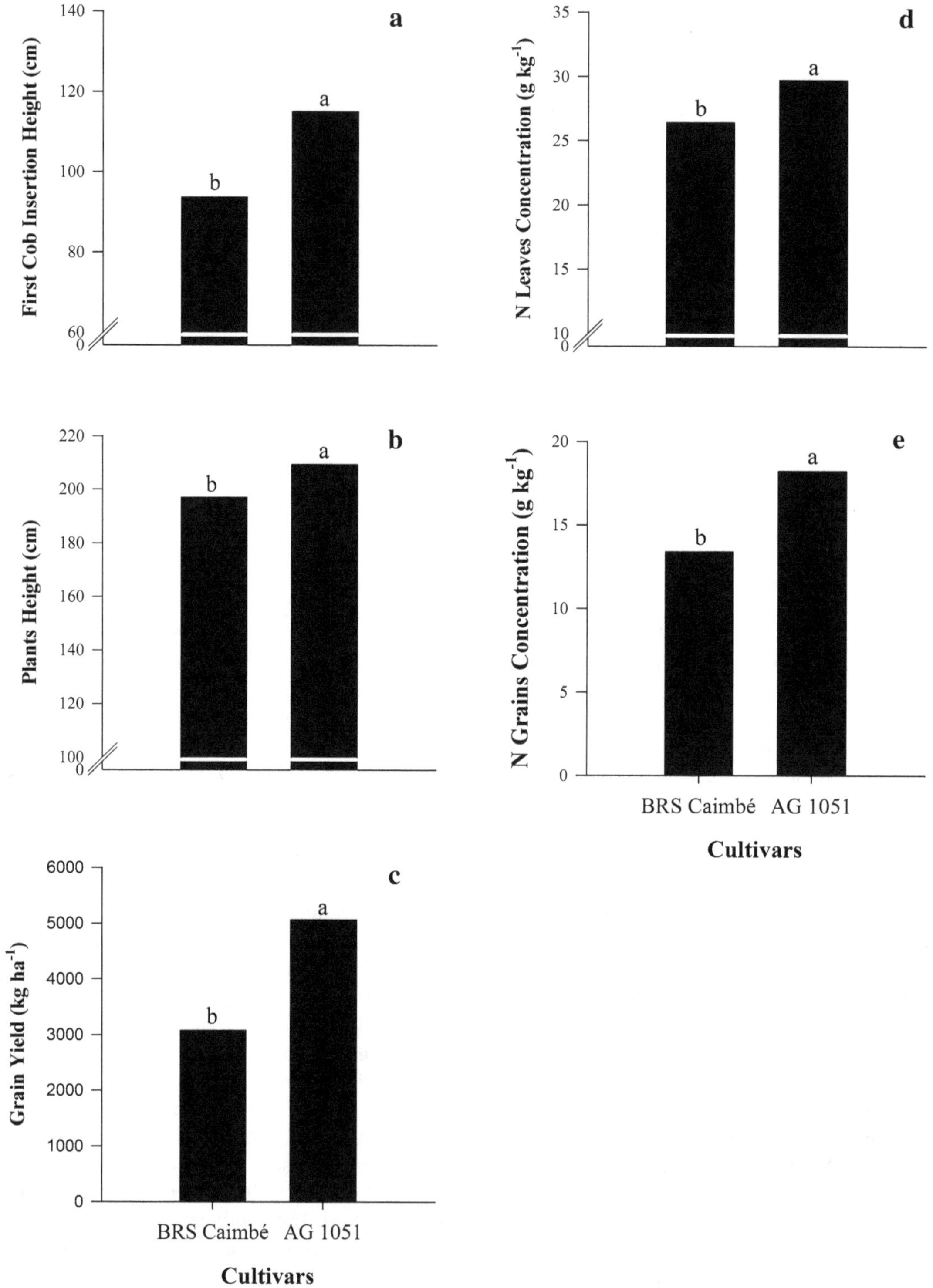

Fig. 1 Effects of BRS Caimbé and AG 1051 maize cultivars on first cob insertion height (FCIH) (**a**), plants height (PH) (**b**), grain yield (GY) (**c**), N leaves concentration (NL) (**d**), and N grains concentration (NG) (**e**) in 2014/2015 season. Different letters indicate differences ($p < 0.05$) between maize cultivars by Tukey test

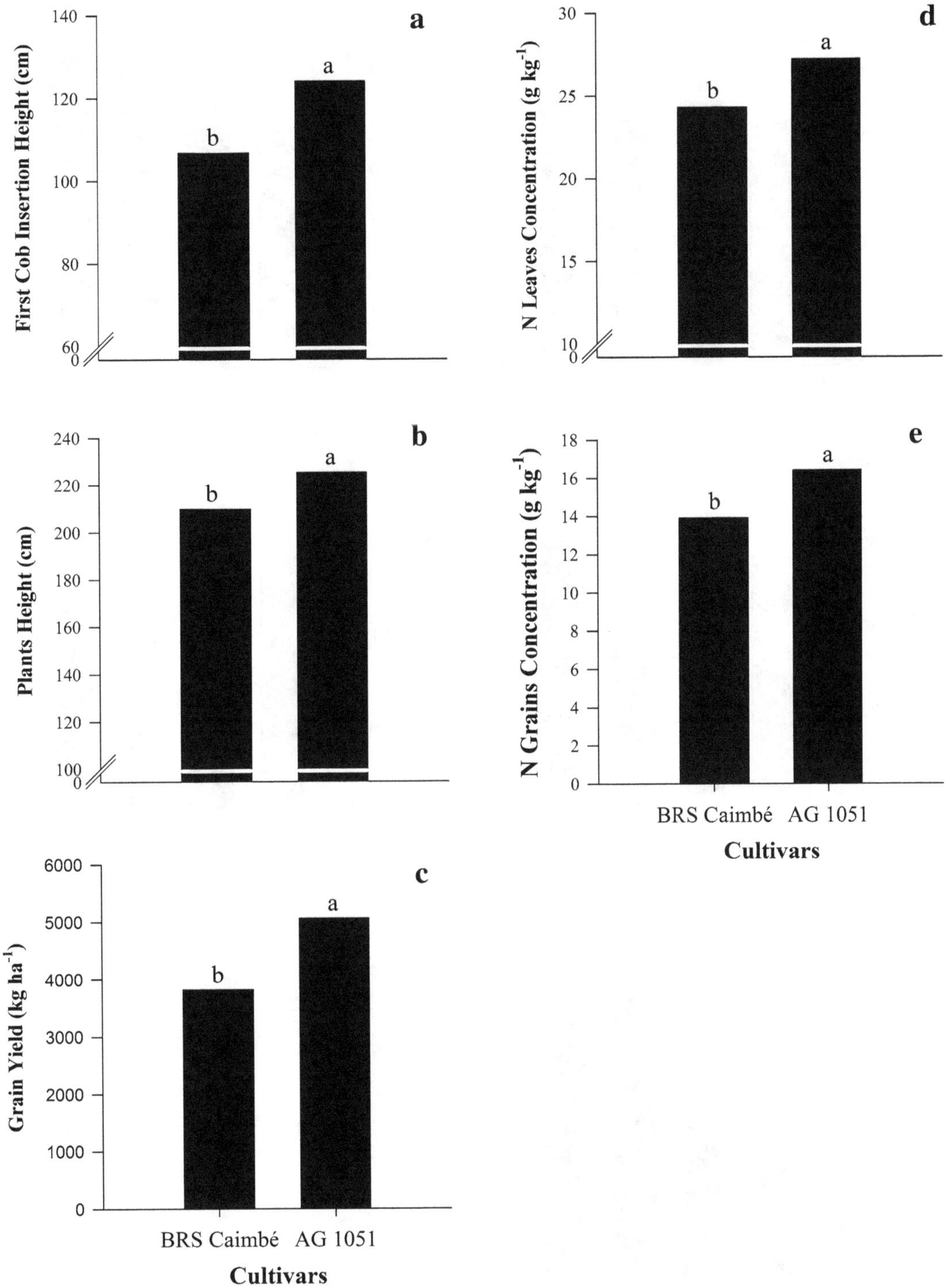

Fig. 2 Effects of BRS Caimbé and AG 1051 maize cultivars on first cob insertion height (FCIH) (**a**), plants height (PH) (**b**), grain yield (GY) (**c**), N leaves concentration (NL) (**d**), and N grains concentration (NG) (**e**) in 2015/2016 season. Different letters indicate differences ($p < 0.05$) between maize cultivars by Tukey test

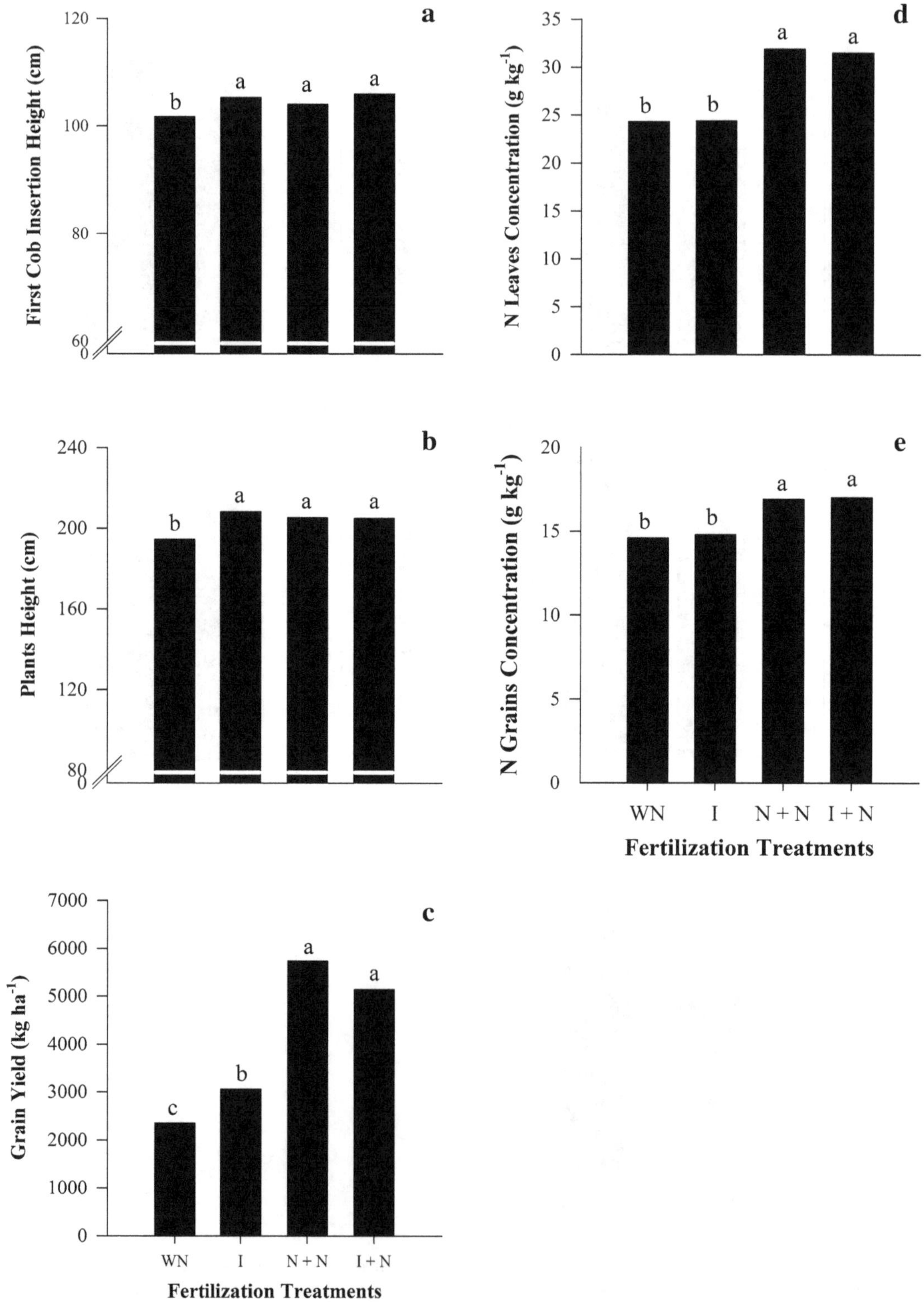

Fig. 3 Effects of N fertilization treatments without N (W N), inoculation with *Azospirillum brasiliense* (I), N fertilization as urea at sowing and at top-dress (N + N), and seeds inoculation + N fertilization at sowing (I + N) on first cob insertion height (FCIH) (**a**), plants height (PH) (**b**), grain yield (GY) (**c**), N leaves concentration (NL) (**d**), and N grains concentration (NG) (**e**) in 2014/2015 season. Different letters indicate differences ($p < 0.05$) among N fertilization treatments by Tukey test

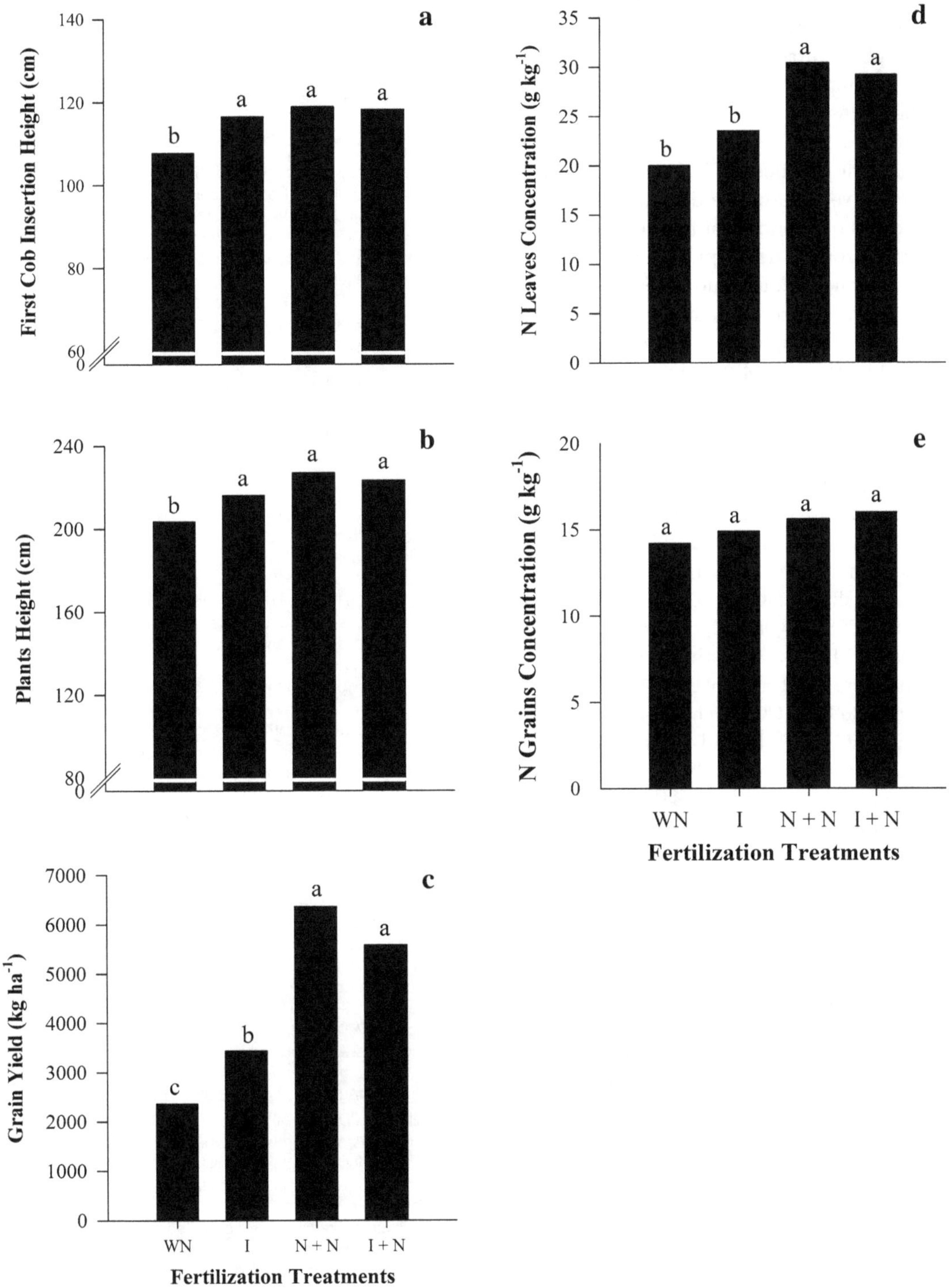

Fig. 4 Effects of N fertilization treatments without N (W N), inoculation with *Azospirillum brasiliense* (I), N fertilization as urea at sowing and at top-dress (N + N) and seeds inoculation + N fertilization at sowing (I + N) on first cob insertion height (FCIH) (**a**), plants height (PH) (**b**), grain yield (GY) (**c**), N leaves concentration (NL) (**d**), and N grains concentration (NG) (**e**) in 2015/2016 season. Different letters indicate differences ($p < 0.05$) among N fertilization treatments by Tukey test

than the sufficiency range for maize crop, from 27.5 to 32.5 g kg^{-1} [31], confirming the positive effects of the inoculant on the plant nutrition.

Considering the high economical and environmental costs of the industrial N fixation and the increasing demand for food, new technologies are essential to the best management practices of N fertilization. In this context, an alternative is associate diazotrophic bacteria N_2-fixing with crops of economical importance, increasing the vegetative growth, development, and yield [32]. Besides, it is an option to reduce and potentiate the effect of N fertilizers in the maize crop, since they are diazotrophic and thus can complement the quantity of the nutrient required by the plants via biological fixation of the atmospheric N (BNF) [15].

The high cost of chemical fertilizers and concerns for a sustainable agriculture are hightening the interest on inoculant bacteria [30]. We reported an alternative to cultivate maize without mineral fertilization at sowing, saving 20 kg ha^{-1} of N. Considering cultivated 11,600 ha $year^{-1}$ of maize in Amazonas state [1], a reduction of 232 Mg of N shall occur, or 515 Mg of fertilizer as urea, saving yearly US$ 339,680.00, considering the local costs in Manaus—AM. Moreover, the contribution of *A. brasiliense* would represent a gain of 0.103 Mg C ha^{-1}, corresponding to 0.309 Mg CO_2-eq ha^{-1} [22].

Comparing the AG 1051 hybrid cultivar advantages over BRS Caimbé variety in all studied variables including yield was previously reported [33, 34]. Hybrids can be synthesized crossing lines with complementary genetic pools, allowing best exploitation of heterosis, as suggested by [35].

Conclusions

The inoculation with *A. brasiliense* increased the maize yield compared with non-N-fertilized plants, and the inoculation with *Azospirillum* at sowing + mineral N at topdress provided similar grain yield as mineral N at sowing and topdress for both cultivars of maize. Therefore, *A. brasiliense* can be a sustainable alternative to the N supply for the maize crop of tropical regions, in addition to saving costs on maize production, especially with chemical N fertilizers.

Authors' contributions

All authors have contributed equally to the work. All authors read and approved the final manuscript.

Author details

[1] Embrapa Amazônia Ocidental, AM 010 Road, km 29, Manaus 69010970, Brazil. [2] Universidade Federal do Amazonas, Av. Gen. R. Otávio Jordão Ramos 3000 - Coroado I, FCA, DEAS, Manaus 69077-000, Brazil.

Acknowledgements

The authors wish to thank Luiz de Queiroz Agricultural Studies Foundation by the administrative support and Embrapa Western Amazon for their support in the development of field experiments.

Competing interests

The authors declare that they have no competing interests.

Funding

The project was funded by Agrisus Foundation (Project 1077-12).

References

1. Conab. Acompanhamento da Safra Brasileira: grãos, sétimo levantamento, abril 2017. Conab: Brasília. 2017.
2. Boussadia O, Steppe K, Zgallai H, El Hadj SB, Braham M, Lemeur R, Van Labeke MC. Effects of nitrogen deficiency on leaf photosynthesis, carbohydrate status and biomass production in two olive cultivars "Meski" and "Koroneiki". Sci Hortic. 2010;123:336–42.
3. Wasaya A, Tahir M, Ali H, Hussain M, Yasir TA, Sher A, Ijaz M, Sattar A. Influence of varying tillage systems and nitrogen application on crop allometry, chlorophyll contents, biomass production and net returns of maize (*Zea mays* L.). Soil Tillage Res. 2017;170:18–26.
4. Tabuchi M, Abiko T, Yamaya T. Assimilation of ammonium ions and reutilization of nitrogen in rice (*Oryza sativa* L.). J Exp Bot. 2007;58(9):2319–27.
5. Wei S, Wang X, Shi D, Li Y, Zhang J, Liu P, Zhao B, Dong S. The mechanisms of low nitrogen induced weakened photosynthesis in summer maize (*Zea mays* L.) under field conditions. Plant Physiol Biochem. 2016;105:118–28.
6. Morais TP, Brito CH, Ferreira AS, Luz JMQ. Aspectos morfofisiológicos de plantas de milho e bioquímico do solo em resposta à adubação nitrogenada e à inoculação com *Azospirillum brasilense*. Rev Ceres. 2015;62(6):589–96.
7. Qiao C, Liu L, Hu S, Compton JE, Greaver TL, Li Q. How inhibiting nitrification affects nitrogen cycle and reduces environmental impacts of anthropogenic nitrogen input. Glob Change Biol. 2015;21(3):1249–57.
8. Das K, Prasanna R, Saxena AK. Rhizobia: a potential biocontrol agent for soilborne fungal pathogens. Folia Microbiol. 2017;62(5):425–35.
9. Delaporte-Quintana P, Grillo-Puertas M, Lovaisa NC, Teixeira KR, Rapisarda VA, Pedraza RO. Contribution of *Gluconacetobacter diazotrophicus* to phosphorus nutrition in strawberry plants. Plant Soil. 2017;439:1–13.
10. Alves BJR, Boddey RM, Urquiaga S. The success of BNF in soybean in Brazil. Plant Soil. 2003;252(1):1–9.
11. Herridge DF, Peoples MB, Boddey RM. Global inputs of biological nitrogen fixation in agricultural systems. Plant Soil. 2008;311(1–2):1–18.
12. Kumar A, Verma J. Does plant–microbe interaction confer stress tolerance in plants: a review. Microbiol Res. 2018;207:41–52.
13. Marini D, Guimarães VF, Dartora J, Lana MC, Pinto Júnior AS. Growth and yield of corn hybrids in response to association with *Azospirillum brasilense* and nitrogen fertilization. Rev Ceres. 2015;62:117–23.
14. D'Angioli AM, Viani RAG, Lambers H, Sawaya ACHF, Oliveira RS. Inoculation with *Azospirillum brasilense* (Ab-V4, Ab-V5) increases *Zea mays* root carboxylate-exudation rates, dependent on soil phosphorus supply. Plant Soil. 2017;410(1–2):499–507.
15. Picazevicz AAC, Kusdra JF, Moreno ADL. Maize growth in response to *Azospirillum brasilense*, *Rhizobium tropici*, molybdenum and nitrogen. Rev Bras Eng Agrícola e Ambient. 2017;21(9):623–7.
16. Silveira APD, Sala VMR, Cardoso EJBN, Labanca EG, Cipriano MAP. Nitrogen metabolism and growth of wheat plant under diazotrophic endophytic bacteria inoculation. Appl Soil Ecol. 2016;107:313–9.
17. Solos E. Sistema Brasileiro de Classificação de Solos. 2nd ed. Rio de Janeiro: Embrapa; 2000. p. 412.
18. Soil Survey Staff. Soil taxonomy. 2nd ed. Washington: U.S. Department of Agriculture; 1999. p. 869.
19. FAO. Soil map of the world. Paris: UNESCO; 1988.
20. Scarazatti B. Boletim Agrometeorológico 2014: Estação Agroclimatológica da Embrapa Amazônia Ocidental, km 29 da Rodovia AM-010. Documentos 125. Manaus: Embrapa Amazônia Ocidental; 2016. p. 40.
21. Mengel K, Kirkby EA. Principles of plant nutrition. 5th ed. Dordrecht: Kluwer Academic Publishers; 2001.
22. Hungria M, Nogueira MA, Araujo RS. Inoculation of *Brachiaria* spp. with the plant growth-promoting bacterium *Azospirillum brasilense*: an environment-friendly component in the reclamation of degraded pastures in the tropics. Agric Ecosyst Environ. 2016;221:125–31.

23. Sangoi L, Silva LMM, Mota MR, Panison F, Schmitt A, Souza NM, Giordani W, Schenatto DE. Desempenho agronômico do milho em razão do tratamento de sementes com *Azospirillum* sp. e da aplicação de doses de nitrogênio mineral. Rev Bras Cienc do Solo. 2015;39(4):1141–50.
24. Cassán F, Perrig D, Sgroy VLV, Okon V, Vanderleyden J, Boussadia O, Steppe K, Zgallai H. Basic and technological aspects of phytohormone production by microorganisms: *Azospirillum* sp. as a model of plant growth promoting rhizobacteria. In: Maheshwari DK, editor. Bacteria in agrobiology: plant nutrient management, vol. 123(7). Berlin: Springer; 2011. p. 141–82.
25. Verma SC, Ladha JK, Tripathi AK. Evaluation of plant growth promoting and colonization ability of endophytic diazotrophs from deep water rice. J Biotechnol. 2001;91(2–3):127–41.
26. Gray EJ, Smith DL. Intracellular and extracellular PGPR: commonalities and distinctions in the plant-bacterium signaling processes. Soil Biol Biochem. 2005;37(3):395–412.
27. Lambrecht M, Okon Y, Broek AV, Vanderleyden J. Indole-3-acetic acid: a reciprocal signalling molecule in bacteria–plant interactions. Trends Microbiol. 2000;8(7):298–300.
28. Lana MC, Dartora J, Marini D, Hann JE. Inoculation with *Azospirillum*, associated with nitrogen fertilization in maize. Rev Ceres. 2012;59(3):399–405.
29. Cavallet LE, Pessoa CDP, Helmich JJ, Helmich PR, Ost CF. Corn productivity in response to nitrogen application and seed inoculation with *Azospirillum* spp. Rev Bras Eng Agric e Ambient. 2000;4(1):129–32.
30. Hungria M, Campo RJ, Souza EM, Pedrosa FO. Inoculation with selected strains of *Azospirillum brasilense* and *A. lipoferum* improves yields of maize and wheat in Brazil. Plant Soil. 2010;331(1):413–25.
31. Malavolta E, Vitti GC, Oliveira SA. Avaliação do estado nutricional das plantas: princípios e aplicações. 2nd ed. Piracicaba: Potafos; 1997.
32. Baldani JI, Caruso L, Baldani VLD, Goi SR, Dobereiner J. Recent advances in BNF with non-legume plants. Soil Biol Biochem. 1997;29(5–6):911–22.
33. Mendes FF, Guimarães LJM, Souza JC, Guimarães PEO, Pacheco CAP, Machado JRA, Meirelles WF, Silva AR, Parentoni SN. Adaptability and stability of maize varieties using mixed model methodology. Crop Breed Appl Biotechnol. 2012;12:111–7.
34. Oliveira IJ, Atroch AL, Dias MC, Guimarães LJ, Guimarães PEO. Selection of corn cultivars for yield, stability, and adaptability in the state of Amazonas, Brazil. Pesqui Agropecuária Bras. 2017;52(6):455–63.
35. Hallauer AR, Carena MJ, Miranda Filho J. Quantitative genetics in maize breeding. 2nd ed. Ames: Iowa State University; 2010. p. 663.

7

Identification and antimicrobial activity of most representative secondary metabolites from different plant species

F. Lelario[1], L. Scrano[2*], S. De Franchi[2], M. G. Bonomo[1], G. Salzano[1], S. Milan[1], L. Milella[1] and S. A. Bufo[1]

Abstract

Background: The plant kingdom constitutes an enormous reservoir of bioactive molecules, generally used by plants to prevent or to protect themselves from pathogens' attacks. To date, several primary or secondary plant metabolites have been already proven to exert antibiotic activities; nonetheless, researchers are still continuing to lavish great efforts to identify and characterize new natural molecules one by one. Aiming at the replacement of synthetic chemical products, the bioactivity of plant extracts should be assessed case by case, and active substances should be tested as individuals to obtain accurate information on the real usefulness of plant metabolites. In this work major glycoalkaloids obtained from *Solanum nigrum*, glucosinolates from *Armoracia rusticana,* and cannabinoids from *Cannabis sativa* were identified. The antimicrobial activity of crude extracts and pure components against Gram+ (*Bacillus cereus* (A1I), *Bacillus thuringiensis* (B7I2), and *Bacillus amyloliquefaciens* (A5TI)) and Gram− bacteria (*Pseudomonas orientalis* (A14-1II), and *Stenotrophomonas maltophilia* (B9TIII)), employed as model organisms, was tested.

Result: Major glycoalkaloids, glucosinolates, and cannabinoids were identified in crude plants' extracts using high-resolution LC–ESI-FTICR/MS. From antimicrobial assays useful information towards a few of biological activities of crude extracts and individual components were obtained.

Solanum nigrum extracts revealed inhibition activity on all bacteria tested as well as the main active glycoalkaloids, solamargine and solasonine, which were found to be active even when tested individually. At assayed concentrations, *A. rusticana* extract was active towards a few of the microorganisms tested, confirming that the activity of glucosinolates can be referred only partially to the mother molecules, while biological efficiency of such kind of compounds is mainly due to their enzymatic breaking off, where myrosinase converts them into isothiocyanates and/or thiocyanates. Hemp-type *C. sativa* extract showed antimicrobial activity only against Gram+ bacteria, but the main individual components tested showed always a limited bioactivity.

Conclusion: Promising results were obtained, but tests performed in vitro are only the first step of a wider investigation as required for an extensive application. Further research efforts are necessary to demonstrate the efficiency of natural substances in different target environments.

Keywords: Glycoalkaloids, Glucosinolates, Cannabinoids, *Solanum nigrum*, *Armoracia rusticana*, *Cannabis sativa*, LC–ESI-FTICR/MS, Antimicrobial activity, Gram+ bacteria, Gram− bacteria

*Correspondence: laura.scrano@unibas.it
[2] Department of European Cultures (DICEM), University of Basilicata, 85100 Potenza, Italy
Full list of author information is available at the end of the article

Background

Natural products were developed and used to relieve sickness by the dawn of human history: before the "Synthetic Era", indeed, 80% of medicine, drugs, and pesticides were obtained from roots, barks, and leaves (fluid extracts) of plants, contributing to the diseases' restraint [1, 2]. Despite this success, natural products' research has endured a global decline due to the production difficulties and small quantities obtained [3]. The necessity of using drugs and pesticides in high quantity (linked also to the population growth) has required large amounts of products, which only laboratory synthesis could realize. In this process, however, some negative aspects attributable to a large use of synthetic substances were not considered, such as the organic balance alteration, pollution of various environmental systems, resistance induction, and genetic changes in living beings in a very short period. The current trend is the return, when possible, to the preceding system of diseases' treatment, re-emphasizing the use of metabolic constituents produced by several plant, bacterial, and fungal species [4]. Due to the scientific awareness surrounding the use of natural substances instead of synthetic ones, in recent years many researchers have undertaken studies on the occurrence of secondary metabolites in plants that are widely used in almost all geographical areas and on their possible bioactivity.

Secondary metabolites are organic molecules, not involved in the normal growth and development of an organism, whose functions are largely unknown, although they seem involved in the organism defence [5, 6].

The recognition of the biological properties of thousands of these molecules has increased interest in this field for new drugs, antibiotics, insecticides, fungicides, and herbicides research and brought about a re-evaluation of bacteria, fungi, and plant role, especially in the ecological context. Terpenes (gums, resins, carotenoids, etc.), phenols (lignin, flavones, anthocyanins, tannins, etc.) as well as alkaloids are just some of the substances currently used in different application fields, giving rise to the world's growing attention due to their widespread use and to the concurrent preservation of both human health and environment.

Each family, genus, and species of several plants produce a characteristic mixture of substances that can be present in active form or in prodrug state and are used for taxonomic classification [7]. Normally, they are activated when wounding or infection in the vegetal body occurs [8]. These compounds can be active as single components or strengthen their activity due to synergistic action with other chemical compounds co-synthesized in the plant cells [9].

This work deals with the identification of more representative glycoalkaloids and glucosinolates in crude extracts obtained from *Solanum nigrum* and *Armoracia rusticana*, respectively, and of several cannabinoids extracted from *Cannabis sativa*. Moreover, the antimicrobial activity of crude extracts and some pure components against a limited number of Gram+ and Gram− bacteria, employed as model organisms, was investigated as preliminary assessment of their bioactivity.

Methods

Chemicals

Solamargine and solasonine standards were purchased from Glycomix (UK); sinigrin monohydrate was obtained from Sigma-Aldrich (Steinheim, Germany); Δ9-tetrahydrocannabinol (THC), cannabidiol (CBD), cannabinol (CBN), and cannabidiolic acid (CBDa) were purchased from HPC Standards GmbH (Cunnersdorf, Germany). Methanol, acetonitrile, and formic acid were obtained from Carlo Erba (Milan, Italy). Ultrapure water was produced using a Milli-Q RG system from Millipore (Bedford, MA, USA).

Standard preparation

Glycoalkaloids: stock solutions (1 mg L^{-1}) of pure standards in ultrapure water acidified with 1% acetic acid were prepared as reported elsewhere [10, 11] and kept in the darkness at +4 °C. Glucosinolates: stock solutions (1 mg L^{-1}) of pure standards in methanol/water (70/30, v/v) were kept in the darkness at +4 °C [12–15]. Cannabinoids: stock solutions (1 mg L^{-1}) of pure compounds in ethanol were prepared and kept in the darkness at −20 °C. Solutions prepared as above were used for analytical purposes.

For antimicrobial assays, solutions of standard compounds were prepared when necessary before each test using sterilized ultrapure water, which was also employed to dilute them up to the required concentration.

All glass apparatus were heat sterilized by autoclaving for 60 min at 121 °C before use. Aseptic handling materials and laboratory facilities were used throughout the study to maintain sterility.

Plant samples

Black nightshade (*S. nigrum*) unripe berries (glycoalkaloids), horseradish (*A. rusticana*) roots (glucosinolates), and hemp (*C. sativa*) flowers (cannabinoids) were the vegetal materials used to obtain the extracts.

Green unripe berries of black nightshade (60 berries harvested from 15 different plants in a homogeneous experimental cultivation) and horseradish root (15 plants harvested from a homogeneous experimental cultivation) were supplied, respectively, from a greenhouse located in Metaponto village (Italy) and in a field established at the Institute of Plant Genetics-National Research Council close to Policoro village (Italy). The voucher specimens of both plants were deposited at the Herbarium Lucanum (HLUC in Index Herbariorum) with the ID Code 2320 and the ID Code 9197 for *S. nigrum* and *A. rusticana*, respectively.

Two different flower samples from experimental fields located in southern Italy and derived from the registered *C. sativa* accession "Eletta campana" were supplied with the courtesy of Eletta campana S.r.l. company.

The "Eletta campana" cannabis accession has been bred and grown during the past century both in insular and peninsular areas of Italy, mainly for industrial production purposes [16, 17].

Sample #1: flowers—obtained as representative population—of the "Eletta campana" chemical phenotype, 30 flowers from 30 plants.

Sample #2: field selection of flowers used in varietal improvement schemes to select plants with a higher concentration of THC and CBD compared to the average composition of the "Eletta campana" accession, 30 flowers from 30 selected plants.

Extraction, purification, and preliminary tests

Glycoalkaloids and glucosinolates extractions were made in five replicates following previously published methods [10–15].

Cannabinoids were extracted in five replicates for each sample of hemp flowers crushed using liquid nitrogen, sieved, and lyophilized. Ultrasound-assisted extraction (USAE) was carried out using absolute ethanol as solvent according to published methods [18, 19]. Samples were centrifuged by using the Hettich Zentrifuge, MIKRO220R (Germany) for 12 min at 2400*g* and filtered (PTFE filters 0.20 μm) to clarify the liquid phase.

The replicates were brought together to constitute a representative sample of the total material collected. Subsequently, analytical determinations and antimicrobial assays were conducted.

Aiming at the determination of the minimal inhibitory concentration (MIC), preliminary tests were performed using initial concentration ranges of main representative compounds under investigation: 0.005–1 mM, for black nightshade unripe berries; 1–5 mM, for horseradish extract; 0.001–0.3 mM, for cannabis extracts. MIC assays were performed according to the European Committee for Antimicrobial Susceptibility Testing (EUCAST) and Mann and Markham methods [20, 21].

Analytical procedure

All analytical experiments were performed using a Surveyor LC system coupled to an ESI-FTICR mass spectrometer (Thermo Fisher Scientific, Bremen, Germany), equipped with a 20 W CO_2-laser IRMPD (Synrad, Mukilteo, WA, USA), emission wavelength 10.6 μm. Glucosinolates and glycoalkaloids LC separations were performed at ambient temperature by using the same chromatographic conditions reported elsewhere [11, 14, 22, 23]. For separation and identification of cannabinoids, a new optimized LC–ESI-FTICR/MS method was used.

Mass spectrometric conditions were optimized by direct infusion of standard solutions. The instrument was tuned to facilitate the ionization process and to achieve the highest sensitivity. The ESI-FTICR mass spectra obtained were used to characterize the ionization behaviour of the compounds.

Data acquisition and analyses were accomplished using the Xcalibur software package (version 2.0 SR1 Thermo Electron), and total ion current (TIC) acquisition; data were collected in full MS scan mode and processed post-acquisition to identify the compounds of interest. In addition to accurate mass determination and retention times, extensive structural information was obtained by MS/MS fragmentation performance of the compounds investigated (data not shown). The chromatographic raw data were imported, elaborated, and plotted by SigmaPlot 10.0 (Systat Software, Inc., London, UK).

Analytical determination was performed to know, in detail, the composition of crude extracts and subsequently to permit bioactivity testing of pure standards, using about the same concentrations present in the extracts.

Antimicrobial activity assays

The different extracts were tested against five bacterial strains of the culture collection stored in the Department of Sciences, University of Basilicata, Potenza, Italy. Three Gram+ bacteria [*Bacillus cereus* (A1I), *Bacillus thuringiensis* (B7I2), and *Bacillus amyloliquefaciens* (A5TI)], and two Gram− bacteria [*Pseudomonas orientalis* (A14-1II) and *Stenotrophomonas maltophilia* (B9TIII)] were employed as screening microorganisms for this study. All strains were maintained as freeze-dried stocks in reconstituted (11% w/v) skim milk, containing 0.1% (w/v) ascorbic acid, and routinely cultivated in optimal growth conditions.

Antimicrobial activities of all tested extracts were determined by agar well diffusion assay [24].

For each strain, a subculture in a specific broth (PCb) was obtained from the active stock culture by 1% (v/v) inoculum and incubated overnight at 30 °C. A volume of 200 μL for each subculture was used to inoculate the agar media (to achieve a final concentration of 10^9 CFU L^{-1}) and distributed into Petri plates. Each extract (60 μL) was poured into wells (5 mm Ø) bored in the agar plates, and then the plates were incubated at 30 °C. The organic solvent was used as negative control, while the antibiotic was used as positive control. The experiments were performed in triplicate and the antimicrobial activity of each extract was expressed as mean diameter (mm) of the zone of inhibition (ZoI) produced by the respective extract after 24 h of incubation. A value of ZoI < 10 mm was stated to indicate a low antimicrobial activity; 11 < ZoI < 15 mm, a middle antimicrobial activity; and ZoI > 16 mm, a high antimicrobial activity.

Extracts producing an inhibition zone were screened to determine the minimum inhibitory concentrations and evaluate the antimicrobial effectiveness of each extract against different bacterial strains by means of the agar well diffusion method [24]. The medium inoculated with the strain subculture was distributed into Petri plates, and different concentrations of extracts, ranging from 1 to 100 mg L^{-1}, were poured into wells bored in the agar plates and the plates were incubated for 24 h. After incubation, the MIC was determined as the lowest concentration of the extract inhibiting the growth of bacterial strains. The MIC values were obtained in triplicate tests.

Results were compared by analysis of variance (ANOVA) and Bonferroni post hoc test using GraphPad Prism 6 software, version for Windows.

Results

Analytical outcomes

Identification of main glycoalkaloids

Figure 1 shows the LC–FTICR/MS separation in positive ion mode of an aqueous extract of black nightshade berries. Analysis of the extracts revealed the presence of two main glycoalkaloids identified by accurate *m/z* values of protonated species, comparison with authentic standard, and on the basis of IRMPD fragmentation in the ICR cell of precursor ions $[M+H]^+$. In the insets, the mass spectra of two main peaks corresponding to solasonine found at *m/z* 884.50079 ($C_{45}H_{74}NO_{16}$, exact *m/z* 884.50021) and solamargine found at *m/z* 868.50476 ($C_{45}H_{74}NO_{15}$, exact *m/z* 868.50530) are reported. Both compounds were identified with a mass error lower than 1 ppm, which indicates a very good mass accuracy. In the IRMPD MS spectra (data not shown), several common loss from sugar moiety and product ions were observed. Ions generated from fragmentation of B-ring or E-ring of aglycons were diagnostically useful for establishing their membership in the general family of glycoalkaloids [25]. The other intense peak in the TIC (Fig. 1) can be due to a derivative

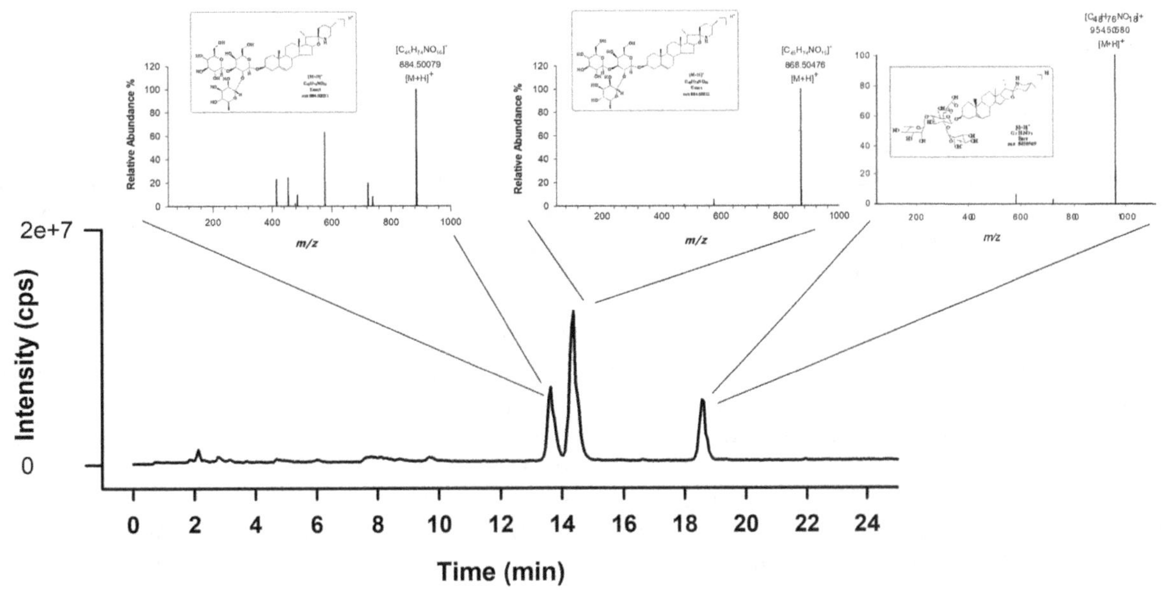

Fig. 1 LC/ESI-FTICR TIC chromatogram acquired in positive mode of a black nightshade berries' extract. Mass spectra of three main peaks, corresponding to solasonine (found at *m/z* 884.50079), solamargine (found at *m/z* 868.50476), and malonyl-solamargine (found at *m/z* 954.50525) along with their molecular structures are reported in the insets

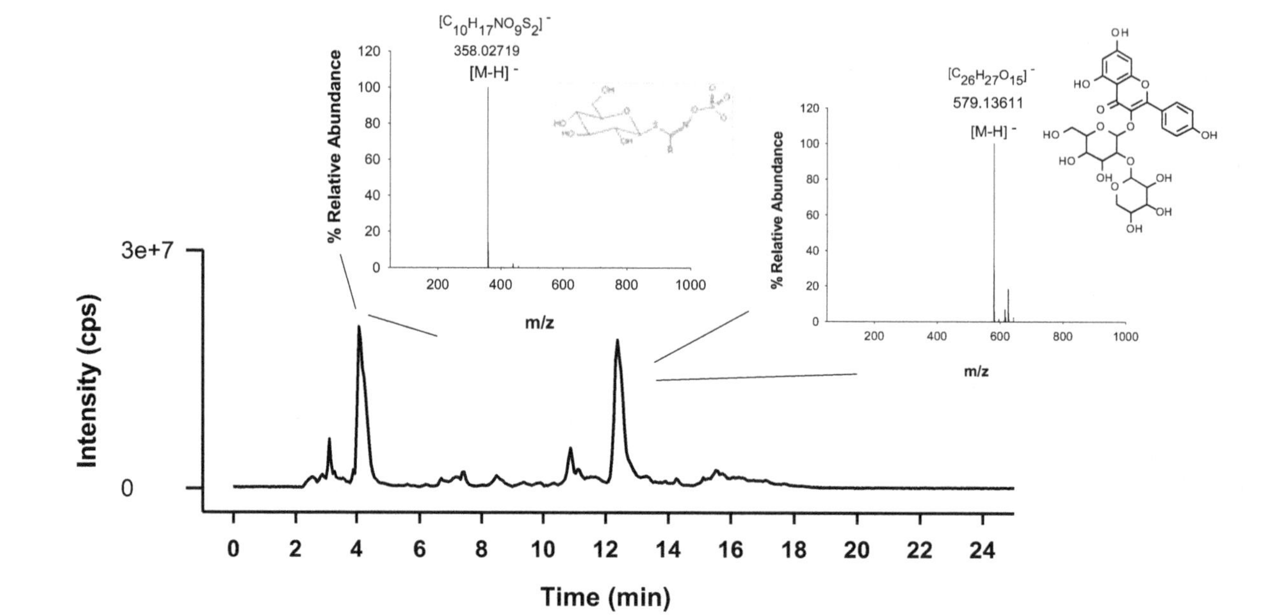

Fig. 2 LC/ESI-FTICR TIC chromatogram acquired in negative ion mode of root crude extract of horseradish. Mass spectra of the main peaks corresponding to sinigrin (found at *m/z* 358.02747) and rustoside (found at *m/z* 579.13611), and their molecular structures are reported in the insets

compound of solamargine, the malonyl-solamargine at *m/z* 954.50525 ($C_{48}H_{75}O_{18}N$, exact *m/z* 954.50569).

Quantitative analysis revealed that black nightshade berries extract contains a high amount of solamargine and solasonine (1.35 and 1.52 g kg^{-1} dry weight, respectively) and a small concentration of other minor known glycoalkaloids, confirming results obtained by Ventrella et al. [13].

Identification of glucosinolates

The identification of GLSs was based on the study of characteristic fragments of these compounds in IRMPD MS/MS spectra, and on the measure of accurate masses observed using LC/ESI-FTICR/MS, according to Agneta et al. [12, 22]. In Fig. 2, the total ion chromatogram (TIC) acquired in negative ion mode of a horseradish root extract is shown. The qualitative and quantitative analyses of this extract confirmed the occurrence of a high amount of sinigrin (2.04 g kg^{-1} dry weight), which accounts for more than 90% of the total GLS, and of the other 16 GLSs in trace quantity [12, 22]. In the inset of Fig. 2, the mass spectrum of the peak corresponding to sinigrin, found at *m/z* 358.02747 ($C_{10}H_{17}NO_9S_2$, exact *m/z* 358.02720, error 0.8 ppm) is shown. By accurate high-resolution mass analysis, the peak eluting at 12.5 min was excluded to be a glucosinolate, but was found to be rustoside, also known as kaempferol 3-lathyroside, which is a member of the class of compounds known as flavonoid-3-*O*-glycosides normally derived from horseradish.

GLSs exhibited $[M-H]^-$ as the precursor ion that corresponds to easy deprotonation of the sulphate group. Moreover, the dissociation of $[M-H]^-$ precursor ion yielded abundant product ions, which gave much information on the structure of the side chain and were of great value for a correct assignment of known and unknown GLSs. Typical fragments of GLS with nominal *m/z* 97, 195, 241, 259, and 275, which correspond to the fragment ions $HSO4^-$, $C_6H_{11}O_5S^-$, $C_6H_9O_8S^-$, $C_6H_{11}O_9S^-$, and $C_6H_{11}O_8S_2^-$, respectively, were found in the spectrum examined (data not shown). Other characteristic fragments, such as $[M\text{-}80\text{-}H]^-$, $[M\text{-}162\text{-}H]^-$, $[M\text{-}178\text{-}H]^-$, $[M\text{-}196\text{-}H]^-$, and $[M\text{-}242\text{-}H]^-$, were very informative for correct molecular identification of GLSs [15].

Identification of cannabinoids

Using optimized reversed-phase liquid chromatography (RP-HPLC) coupled to electrospray ionization in positive mode (ESI^+) and Fourier transform ion cyclotron resonance (FTICR)/MS, together with tandem mass spectrometry (MS^n) studies performed using IRMPD and collisional induced dissociation (CID), it was possible to separate and quantify four known cannabinoids (THC, CBD, CBDa, and CBN), useful for the chemotype definition and the classification of *C.*

Abbreviation	Name	Molecular structure
CBD	Cannabidiol	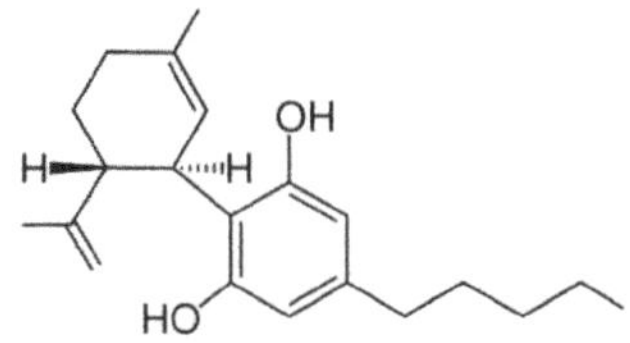
CBDa	Cannabidiolic acid	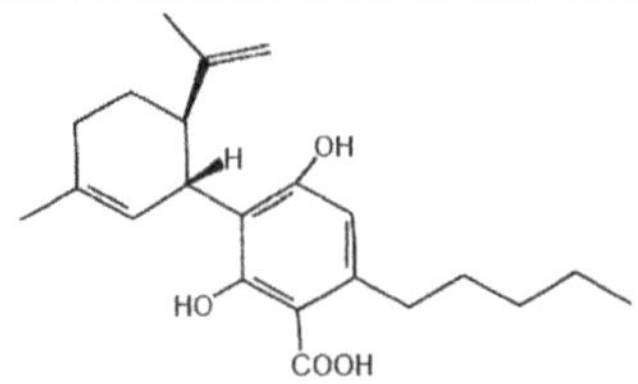
CBDV	Cannabidivarin	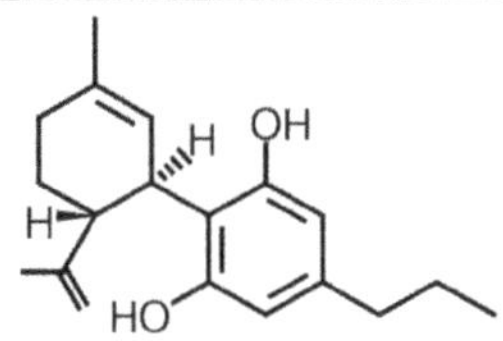
CBN	Cannabinol	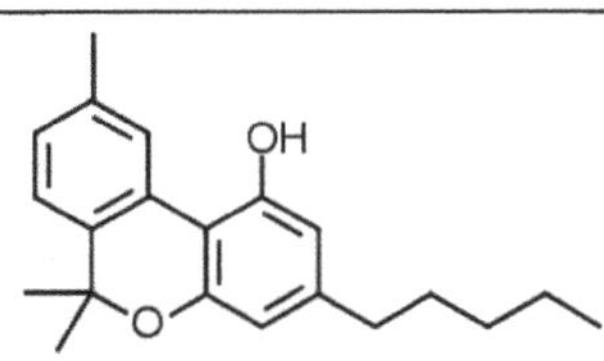
THC	Δ9-tetrahydrocannabinol	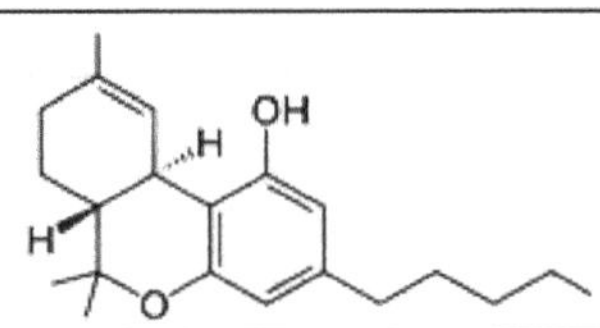

Scheme 1 Common names and molecular structures of cannabinoids detected

sativa (Scheme 1). The total ion current (TIC) chromatogram (Fig. 3) revealed the occurrence of three main cannabinoid peaks assigned to CBD, THC, and CBDa; in the insets, the mass spectra of these peaks are shown: CBD, found at *m/z* 315.23159 ($C_{21}H_{30}O_2$, exact *m/z* 315.23184, error − 0.8 ppm); THC, found at *m/z* 315.23148 ($C_{21}H_{30}O_2$, exact *m/z* 315.23186, error − 1.2 ppm); CBDa, found at *m/z* 359.22126 ($C_{22}H_{30}O_4$, exact *m/z* 359.22169 error − 1.2 ppm). CID and IRMPD fragmentation of precursor ions $[M+H]^+$ generates several common species that are diagnostically useful for establishing their identity (data not shown). The wide peak at 7.4 min corresponds to cannabidivarin (CBDV, $C_{19}H_{26}O_2$, *m/z* $[M+H]^+$ 287.20056), a non-psychoactive cannabinoid homologue of CBD with the side chain shortened by two methylene bridges; it is not useful to determine the chemotype in *Cannabis* plant destined for human consumption or industrial transformation (Scheme 1).

The quantification of secondary metabolites (THC, CBD, CBN, CBDa) was performed in parallel through low-resolution mass spectra, selected reaction monitoring (SRM), and high-resolution total ion current (TIC).

As described in Table 1, the chemovar analysed as sample #1 does not exceed the THC limit (0.2%) recommended by the European Union regulations [26, 27], confirming previous findings [28].

The ([THC] + [CBN])/[CBD] ratio (phenotypic index) of samples was used to assess the chemical phenotype (chemotype) of the specific accession [29].

The high content of cannabidiol (CBD) suggests that the "Eletta campana" accession can be defined as an industrial hemp having a ratio [CBD]/[THC] > 10 (CBD-prevalent chemotype) [30]. In our case, the CBN

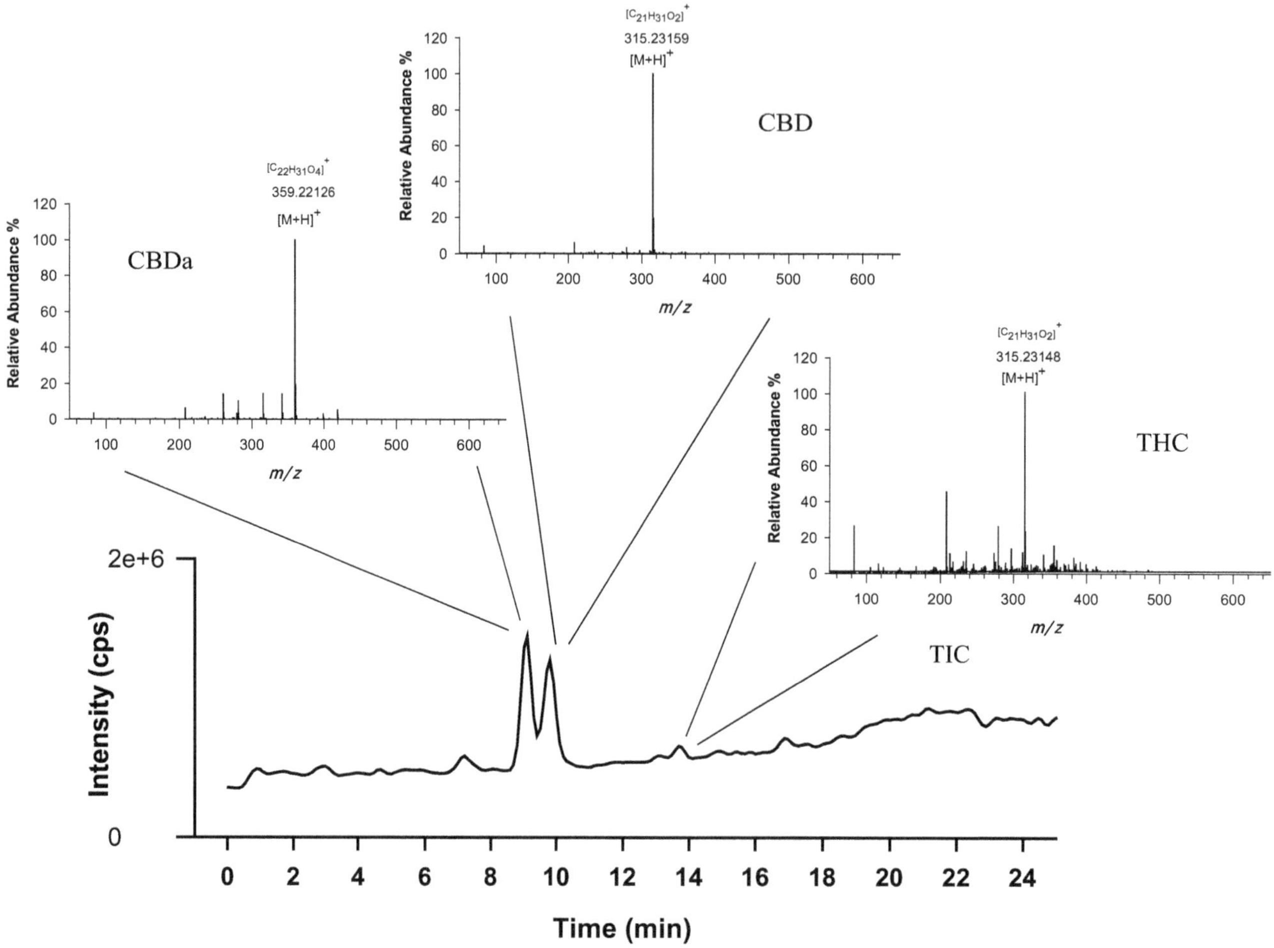

Fig. 3 Total ion current (TIC) obtained using LC–ESI-FTICR MS in positive ionization of an ethanol extract obtained from representative flowers of *C. sativa* accession "Eletta campana". The insets show the mass spectra corresponding to peaks of CBD (found at *m/z* 315.23159), THC (found at *m/z* 315.23148), and CBDa (found at *m/z* 359.22126). CBN (occurring at 12.3 min) is not of significant concentration (see Table 1). The wide peak at 7.4 min corresponds to cannabidivarin (CBDV, $C_{19}H_{26}O_2$, found at *m/z* 287.20056) a homologue of CBD with the side chain shortened by two methylene bridges

Table 1 Composition of the two samples of *C. sativa* analysed: Δ9-tetrahydrocannabinol (THC), cannabidiol (CBD), and (CBN + THC)/CBD ratio identifying the chemotype

Samples	THC	CBD	CBN	(CBN + THC)/CBD
1	0.021% (2.1 ± 0.5 g kg^{-1} dw)	0.781% (78.1 ± 1.3 g kg^{-1} dw)	0.008% (0.81 ± 0.12 g kg^{-1} dw)	0.037
2	0.050% (5.0 ± 0.8 g kg^{-1} dw)	0.967% (96.7 ± 1.7 g kg^{-1} dw)	0.010% (0.98 ± 0.28 g kg^{-1} dw)	0.061

concentration is not significant for the chemical definition of cannabis quality.

Data reported in Table 1 for the sample #2 indicate that the field selection of plant flowers was able to discriminate a group of plants with a higher content of analysed cannabinoids.

Antimicrobial activity assays

The antimicrobial activities and MICs were evaluated against selected bacterial strains giving different results depending on the type of plant under observation.

Statistics of the antimicrobial activity data (diameters of ZoI) confirmed that the diameter ranges chosen

Table 2 Antimicrobial activity and MIC of *S. nigrum*, *A. rusticana*, and *C. sativa* extracts tested

Bacteria	*S. nigrum*		*A. rusticana*		*C. sativa* Sample #1[a]		*C. sativa* Sample #2[b]	
	ZoI (mm)[c]	MIC (mg L^{-1})	ZoI (mm)[c]	MIC (mg L^{-1})	ZoI (mm)[c]	MIC (mg L^{-1})	ZoI (mm)[c]	MIC (mg L^{-1})
Gram+								
Bacillus cereus (A1I)	15.2 ± 1.0^{A}	5 ± 1^{H}	NI	–	37.0 ± 1.0^{C}	5 ± 1^{H}	36.5 ± 1.0^{C}	5 ± 1^{H}
Bacillus thuringiensis (B7I2)	13.5 ± 0.8^{B}	10 ± 1^{K}	NI	–	37.5 ± 0.9^{C}	10 ± 1^{K}	37.5 ± 0.7^{C}	5 ± 1^{H}
Bacillus amyloliquefaciens (A5TI)	15.0 ± 0.9^{A}	10 ± 1^{K}	NI	–	34.5 ± 0.9^{D}	5 ± 1^{H}	34.0 ± 0.9^{D}	5 ± 1^{H}
Gram–								
Pseudomonas orientalis (A14-1II)	13.5 ± 1.0^{B}	10 ± 1^{K}	NI	–	NI		NI	
Stenotrophomonas maltophilia (B9TIII)	15.0 ± 1.0^{A}	10 ± 1^{K}	NI	–	NI		NI	

Assays were performed in triplicate and results are the mean of three values ± standard deviation

NI no inhibition zone was observed

Different letters in superscript to numerical data indicate significant differences (p value < 0.05)

[a] Sample #1 is composed of flowers representative of the "Eletta campana" cannabis accession

[b] Sample #2 was obtained from a field mass selection of "Eletta campana" flowers having a higher THC and CBD content

[c] ZoI: diameter of inhibition zone obtained after 24 h of incubation in agar well diffusion assays. ZoI < 10 mm: low antimicrobial activity; 11 < ZoI < 15 mm: middle antimicrobial activity; ZoI > 16 mm: high antimicrobial activity

Table 3 Antimicrobial activity and MIC of pure compounds as components *of S. nigrum*

Bacteria	Solamargine		Solasonine		Solamargine/solasonine (1:1 v/v)	
	ZoI (mm)[a]	MIC (mg L^{-1})	ZoI (mm)[a]	MIC (mg L^{-1})	ZoI (mm)[a]	MIC (mg L^{-1})
Gram+						
Bacillus cereus (A1I)	15.0 ± 1.0^{A}	5 ± 1^{H}	14.0 ± 0.9^{A}	5 ± 1^{H}	15.0 ± 1.0^{A}	5 ± 1^{H}
Bacillus thuringiensis (B7I2)	12.0 ± 0.9^{B}	20 ± 1^{L}	12.0 ± 0.8^{B}	40 ± 1^{M}	13.0 ± 1.0^{B}	40 ± 1^{M}
Bacillus amyloliquefaciens (A5TI)	12.0 ± 0.9^{B}	20 ± 1^{L}	12.0 ± 1.0^{B}	40 ± 1^{M}	14.5 ± 1.0^{A}	20 ± 1^{L}
Gram–						
Pseudomonas orientalis (A14-1II)	13.0 ± 0.8^{B}	40 ± 1^{M}	13.0 ± 0.7^{B}	40 ± 1^{M}	13.5 ± 1.0^{B}	40 ± 1^{M}
Stenotrophomonas maltophilia (B9TIII)	15.0 ± 0.9^{A}	10 ± 1^{K}	14.0 ± 0.8^{A}	10 ± 1^{K}	15.0 ± 1.1^{A}	10 ± 1^{H}

Assays were performed in triplicate and results are the mean of three values ± standard deviation

Different letters in superscript to numerical data indicate significant differences (p value < 0.05)

[a] ZoI: diameter of inhibition zone obtained after 24 h of incubation in agar well diffusion assays. ZoI < 10 mm: low antimicrobial activity; 11 < ZoI < 15 mm: middle antimicrobial activity; ZoI > 16 mm: high antimicrobial activity

(< 10 mm; 11–15 mm, > 16 mm) were able to well discriminate significant differences among the antimicrobial activities (Tables 2, 3, 4).

Solanum nigrum and *C. sativa* extracts demonstrated a certain antimicrobial activity, while *A. rusticana* did not reveal any activity against bacteria in this research.

The Gram– bacteria, *P. orientalis* and *S. maltophilia*, were sensitive only to the *S. nigrum* extract, showing a middle inhibition diameter of 13.5 and 15 mm, respectively; moreover, this extract proved a middle antimicrobial activity against all Gram+ bacteria (inhibition zone ranging from 13.5 to 15.2 mm) (Table 2).

Both flower samples of *C. sativa* showed a similar effect on Gram+ bacteria with a high antimicrobial activity; these extracts were more effective against *B. thuringiensis* and *B. cereus* with 37.5 and 37.0 mm diameter of inhibition zone, respectively, while *B. amyloliquefaciens* was slightly less sensitive (Table 2).

The active extracts of *S. nigrum* and *C. sativa* were subjected to determine MIC by the agar well diffusion method against the respective susceptible bacterial species (Table 2). The results obtained indicated that Gram+ and Gram– bacterial species tested were sensitive to different extracts in a similar way with an MIC of 5–10 mg L^{-1}. The more effective extracts were the

Table 4 Antimicrobial activity and MIC of standard pure compounds as components of *C. sativa*

Bacteria	THC		CBD		CBD/THC	
	Zol (mm)[a]	MIC (mg L^{-1})	Zol (mm)[a]	MIC (mg L^{-1})	Zol (mm)[a]	MIC (mg L^{-1})
Gram+						
Bacillus cereus (A1I)	8.5 ± 1.2^A	> 60	12.0 ± 1.0^B	> 60	14.5 ± 1.4^C	> 60
Bacillus thuringiensis (B7I2)	9.5 ± 1.3^A	> 60	13.0 ± 1.1^B	> 60	14.5 ± 1.3^C	> 60
Bacillus amyloliquefaciens (A5TI)	8.0 ± 0.8^A	> 60	11.0 ± 0.9^B	> 60	15.0 ± 1.3^C	> 60
Gram−						
Pseudomonas orientalis (A14-1II)	NI	–	NI	–	NI	–
Stenotrophomonas maltophilia (B9TIII)	NI	–	NI	–	NI	–

Assays were performed in triplicate and results are the mean of three values ± standard deviation

NI no inhibition zone was observed, *THC* Δ9-tetrahydrocannabinol, *CBD* cannabidiol

Different letters in superscript to numerical data indicate significant differences (p value < 0.05)

[a] Zol: diameter of inhibition zone obtained after 24 h of incubation in agar well diffusion assays. Zol < 10 mm: low antimicrobial activity; 11 < Zol < 15 mm: middle antimicrobial activity; Zol > 16 mm: high antimicrobial activity

two samples of *C. sativa* with the higher antimicrobial ability and a low inhibitory concentration (Table 2).

The antimicrobial activity of standard pure components of the plants was investigated to understand whether the activity observed in our experiments was due to the synergistic action of more than one constituent in the extracts [31].

In the case of *S. nigrum*, solamargine, solasonine, and the solamargine/solasonine mixture (1:1 v/v) were tested. All bacteria were sensitive to both components with a middle antimicrobial activity ranging from 12 to 15 mm (Table 3).

Among Gram+ bacteria, *B. cereus* was the most sensitive (with an MIC of 5 mg L^{-1}) compared to the other two species, *B. thuringiensis* and *B. amyloliquefaciens*, which were inhibited at higher concentrations (ranging from 20 to 40 mg L^{-1}). Gram− bacteria, instead, showed the same behaviour in the presence of standard pure compounds (Table 3).

The *C. sativa* components were able to inhibit only the Gram + bacteria tested; THC showed a low antimicrobial activity, while CBD and the CBD/THC mixture (1:1 v/v) proved a middle activity, underlining a stronger effect when the mixture was used (Table 4); nevertheless, the bacterial species appeared not very sensitive to the standard pure components, requiring an inhibitory concentration of > 60 mg L^{-1}.

Discussion

The antimicrobial activity and MIC were evaluated against selected bacterial strains of significant environmental and health concern, used as model of target organisms. *B. cereus* is an endospore-forming Gram+ bacterium that can cause food poisoning. Capable of adapting to a wide range of environmental conditions, it is distributed widely in nature and is commonly found in the soil as a saprophytic organism. As a soil bacterium, *B. cereus* can spread easily to many types of foods such as vegetables, eggs, meat, and dairy products, and is known to cause 2–5% of food-borne intoxications due to its secretion of emetic toxins and enterotoxins. Food poisoning occurs when food is left without refrigeration for several hours before it is served. The remaining spores of contaminated food from heat treatment grow well after cooling and are the source of food poisoning. In addition, *B. cereus* is an opportunistic human pathogen and is occasionally associated with infections, causing periodontal diseases and other more serious infections [32]. The availability of natural substances active towards this microorganism, but well tolerated by the human body, could be useful to increase the food storage time. *B. thuringiensis* is a Gram+, rod shaped, and aerobic spore-forming soil bacterium producing crystalline proteins (endotoxins) that have insecticidal properties; on the other hand, this bacterial species synthesizes several enzymes and toxins that give them a wide adaptation to natural habitats [33]. The intrinsic resistance and adaptability of this bacterium makes it an ideal model for the tests performed in this research. *B. amyloliquefaciens* is a non-pathogenic Gram+ soil bacterium. Similar to other *Bacillus* species, it is capable of producing endospores allowing it to survive for extended periods of time. The species also shows some antifungal properties, which are influenced by environmental nitrogen availability. It synthesizes a natural antibiotic protein active against other photogenic *Bacillus* spp. and is used in agriculture, aquaculture, and hydroponics to fight root pathogens [34]. For this reason, it was important to verify if it was inhibited by secondary metabolites produced by plants. *P. orientalis* is a Gram−, rod-shaped bacterium placed in the *P. fluorescens* group. It shows antagonistic activity against several phytopathogenic bacteria [35] and as *B. amyloliquefaciens* can be inhibited by secondary metabolites produced by plants. *S. maltophilia* is an aerobic, non-fermentative, Gram− bacillus possessing flagella in a multitrichous formation, and is found naturally in the rhizosphere. However, it is also the third most common nosocomial pathogen with multi-drug resistance that targets immune-compromised patients in hospitals, making it important in medical bacteriology [36].

Antimicrobial trials have demonstrated that black nightshade extract is active on assayed microorganisms. Solamargine and solasonine, main components of black nightshade extract, were very active also when tested individually or as binary mixture. Glycoalkaloids, which are produced by widely cultivated Solanaceae plants, are confirmed to be bioactive substances useful for different applications acting as cellular membrane disrupting factors or inhibitors of acetylcholinesterase activity [6, 11, 13].

Horseradish (*A. rusticana*) extract was not active towards any of the tested microorganisms at assayed concentrations as expected for glucosinolates in the absence of the enzymatic reaction needed for the production of active derivatives [37].

Hemp (*C. sativa*) extract showed antimicrobial activity only against Gram+ bacteria, as Gram− bacteria seem to be more resistant to the secondary metabolites contained in the extract [38, 39]. The different compositions of the samples analysed did not influence their bioactivity.

The main components tested, either as individual compounds or as CBD–THC mixture, showed a bioactivity about three times lower compared to the raw extract; the rationale behind such a behaviour could be that antibacterial properties were due to the synergistic effect of many components (such as terpenoids, carboxylic moieties, and simple or complex phenols) present in the extract even if the prenyl moiety of cannabinoids has been highlighted as effective in antimicrobial activity [40–43].

The results of the antimicrobial activity of natural extracts against both Gram+ and Gram− microorganisms of this study are in agreement with the research work of Tajkarimi et al. [44]. Gram+ bacteria tend to be more sensitive to the antimicrobial properties of natural extracts [45, 46], while Gram− are less susceptible to the antibacterial action of natural substances since they possess an outer membrane surrounding the cell wall, which restricts diffusion of hydrophobic compounds through their lipopolysaccharide covering. In addition, Gram− microorganisms generally present higher MICs than the Gram+ ones. Numerous researchers investigating the same topic agree that these compounds are, in most cases, slightly more active against Gram+ than Gram− bacteria. However, Wilkinson et al. [47] remarked that some exceptions may occur, since *Aeromonas hydrophila* (Gram−) appears to be one of the most sensitive species to the action of essential oils obtained from different natural essences (thyme, cinnamon, bay, clove, almond, etc.). Moreover, the authors have postulated that individual components of natural extracts exhibit different degrees of activity due to their chemical composition, which can vary according to the geographical origin and harvesting period.

It should also be emphasized that biopharmaceuticals and biopesticides may have different action mechanisms than those of the conventional synthetic products, even if the corresponding compounds are similar. A low or inadequate dosage could cause failure of protection, which could lead to the abandoning of natural products in favour of conventional methods. Therefore, using bioactive substances efficiently requires specific knowledge of the agent and the target disease for optimizing their application time, doses, and rates.

Conclusion

In conclusion, the bioactivity of plant extracts to replace synthetic chemical products should be assessed case by case, and active substances should be tested as individuals to obtain more extended information on the real applicability of plant metabolites against pathogens. Promising results were obtained, and glycoalkaloids antimicrobial activity was confirmed herein, in line with previous reports against insects [6, 11, 13]. But tests performed in vitro are only the first step of a deeper research aimed at extending the use of natural substances to combat plant or animal diseases. Further research efforts are necessary to demonstrate plant secondary metabolite efficiency in the target environments, to better understand their biological activities and to develop actions strategies of such complex mixtures usage.

Abbreviations

CBD: cannabidiol; CBDa: cannabidiolic acid; CBN: cannabinol; CID: collisional induced dissociation; ESI: electrospray ionization; EUCAST: European Committee for Antimicrobial Susceptibility Testing; GLS: glucosinolate; IRMPD: infrared multiple photon dissociation; LC–FTICR/MS: liquid chromatography–Fourier transform ion cyclotron resonance/mass; MIC: minimal inhibitory concentration; PCb: potato and carrot broth; RP-HPLC: reversed-phase liquid chromatography; SRM: selected reaction monitoring; THC: Δ9-Tetrahydrocannabinol; TIC: total ion current; USAE: ultrasound-assisted extraction; ZoI: zone of inhibition.

Authors' contributions

All authors of this research paper have directly participated in the planning, execution, or analysis of this study and read and approved the final manuscript. FL performed analytical MS determinations and wrote the relative paragraphs in "Methods". MGB carried out microbial assays and wrote the relative paragraphs in "Methods". SM prepared crude extracts and surrogates and wrote the relative paragraphs in "Methods". SDeF selected all plant species tested for the experiments and provided the registration of plants' materials. GS discussed the bioactivity results. LM, SAB, and LS planned this research and discussed a part of the results. LS and SAB revised the whole manuscript. All authors read and approved the final manuscript.

Author details
[1] Department of Sciences, University of Basilicata, 85100 Potenza, Italy.
[2] Department of European Cultures (DICEM), University of Basilicata, 85100 Potenza, Italy.

Acknowledgements
We are grateful to the Company "Eletta campana S.r.l." for providing the flower samples of their registered *C. sativa* accession.

We would like to thank the Herbarium Lucanum for the prompt supply of the voucher codes relating to the deposited plants of *S. nigrum* and *A. rusticana*.

Competing interests
The authors declare that they have no competing interests.

Funding
BIOMON Project, granted by SAIPEM S.p.A.

References

1. Newman DJ, Cragg GM, Snader KM. Natural products as sources of new drugs over the period 1981–2002. J Nat Prod. 2003;66(7):1022–37.
2. Cragg GM, Kingston DG, Newman DJ. Anticancer agents from natural products. Boca Raton: CRC Press; 2011.
3. McChesney JD, Venkataraman SK, Henri JT. Plant natural products: back to the future or into extinction? Phytochemistry. 2007;68(14):2015–22.
4. Tripathi P, Dubey NK. Exploitation of natural products as an alternative strategy to control postharvest fungal rotting of fruit and vegetables. Postharvest Biol Technol. 2004;32(3):235–45. https://doi.org/10.1016/j.postharvbio.2003.11.005.
5. Agostini-Costa TD, Vieira RF, Bizzo HR, Silveira D, Gimenes MA. Secondary Metabolites. In: Sasikumar D, editor. Chromatography and its applications. London: InTech; 2012. https://doi.org/10.5772/35705.
6. Chowński S, Adamski Z, Marciniak P, Rosiński G, Büyükgüzel E, Büyükgüzel K, Falabella P, Scrano L, Ventrella E, Lelario F, Bufo SA. A review of bioinsecticidal activity of Solanaceae alkaloids. Toxins. 2016;8(3):60. https://doi.org/10.3390/toxins8030060.
7. Croteau R, Kutcahn TM, Lewis NG. Natural products. In: Buchanan B, Gruissem W, Jones R, editors. Biochemistry and molecular biology of plants. Rockville: American Society of Plant Physiologists; 2000. p. 1250–318.
8. Osbourn AE. Saponins and plant defense—a soap story. Trends Plant Sci. 1996;1:4–9.
9. Braga LC, Leite AAM, Xavier KGS, Takahashi JA, Bemquerer MP, Chartone-Souza E, Nascimento MAA. Synergic interaction between pomegranate extract and antibiotics against *Staphylococcus aureus*. Can J Microbiol. 2005;51(7):541–7. https://doi.org/10.1139/w05-022.
10. Cataldi TRI, Lelario F, Bufo SA. Analysis of tomato glycoalkaloids by liquid chromatography coupled with electrospray ionization tandem mass spectrometry. Rapid Commun Mass Spectrom. 2005;19(21):3103–10.
11. Adamski Z, Halamunda J, Marciniak P, Nawrocka M, Ziemnicki K, Lelario F, Scrano L, Bufo SA. Effect of various xenobiotics on hatching success of *Spodoptera exigua* eggs as compared to a natural plant extract. J Toxicol Environ Health Part A. 2009;72:1132–4.
12. Agneta R, Rivelli AR, Ventrella E, Lelario F, Sarli G, Bufo SA. Investigation of glucosinolate profile and qualitative aspects in sprouts and roots of horseradish (*Armoracia rusticana*) using LC–ESI–hybrid linear ion trap with Fourier transform ion cyclotron resonance mass spectrometry and infrared multiphoton dissociation. J Agric Food Chem. 2012;60(30):7474–82.
13. Ventrella E, Marciniak P, Adamski Z, Rosiński G, Chowański S, Falabella P, Bufo SA. Cardioactive properties of Solanaceae plant extracts and pure glycoalkaloids on *Zophobas atratus*. Insect Sci. 2015;22(2):251–62.
14. Bianco G, Lelario F, Battista FG, Bufo SA, Cataldi TRI. Identification of glucosinolates in Capers by LC–ESI using a hybrid linear ion trap with Fourier-transform ion cyclotron resonance mass spectrometry (LC–ESI-LTQ-FTICR-MS) and infrared multiphoton dissociation. J Mass Spectrom. 2012;47(9):1160–9. https://doi.org/10.1002/jms.2996.
15. Cataldi TRI, Lelario F, Orlando D, Bufo SA. Collision-induced dissociation of the A+ 2 isotope ion facilitates glucosinolates structure elucidation by electrospray ionization-tandem mass spectrometry with a linear quadrupole ion trap. Anal Chem. 2010;82(13):5686–96.
16. Barbieri R, Tedeschi P. Eletta Campana e T4, nuove cultivar di canapa per l'ambiente campano. Sementi Elette. 1968;14(6):412–7.
17. De Meijer EPM. Fibre hemp cultivars: a survey of origin, ancestry, availability and brief agronomic characteristics. J Int Hemp Assoc. 1995;2(2):66–73.
18. Kim HK, Verpoorte R. Sample preparation for plant metabolomics. Phytochem Anal. 2010;21(1):4–13.
19. Mandal V, Mohan Y, Hemalatha S. Microwave assisted extraction: an innovative and promising extraction tool for medicinal plant research. Pharmacogn Rev. 2007;1(1):7–18.
20. European Committee for Antimicrobial Susceptibility Testing (EUCAST) of the European Society of Clinical Microbiology and Infectious Diseases (ESCMID). Determination of minimum inhibitory concentrations (MICs) of antibacterial agents by broth dilution. Clin Microbiol Infect. 2003;9(8):1–7. https://doi.org/10.1046/j.1469-0691.2003.00790.x.
21. Mann CM, Markham JL. A new method for determining the minimum inhibitory concentration of essential oils. J Appl Microbiol. 1998;84:538–44.
22. Agneta R, Lelario F, De Maria S, Möllers C, Bufo SA, Rivelli AR. Glucosinolate profile and distribution among plant tissues and phenological stages of field-grown horseradish. Phytochemistry. 2014;106:178–87.
23. Lelario F, Bianco G, Bufo SA, Cataldi TRI. Establishing the occurrence of major and minor glucosinolates in *Brassicaceae* by LC–ESI hybrid linear ion-trap and Fourier-transform ion cyclotron resonance mass spectrometry. Phytochemistry. 2012;73(1):74–83.
24. Russo D, Bonomo MG, Salzano G, Martelli G, Milella L. Nutraceutical properties of *Citrus clementina* juices. Pharmacologyonline. 2012;1:84–93.
25. Lelario F, Labella C, Napolitano G, Scrano L, Bufo SA. Fragmentation study of major spirosolane-type glycoalkaloids by collision-induced dissociation linear ion trap and infrared multiphoton dissociation Fourier transform ion cyclotron resonance mass spectrometry. Rapid Commun Mass Spectrom. 2016;30(22):2395–406.
26. Commission Regulation (EEC) No 1164/89 of 28 April 1989, laying down detailed rules concerning aid for fibre flax and hemp. OJ L 121, 29.4.1989:4–10.
27. Commission Regulation (EC) No 2860/2000 of 27 December 2000 amending Regulation (EC) No 2316/1999 laying down detailed rules for the application of Council Regulation (EC) No 1251/1999 establishing a support system for producers of certain arable crops, to include flax and hemp grown for fibre, specifying the rules on set-aside areas and amending the base areas for Greece and Portugal.
28. Avico U, Pacifici R, Zuccaro P. Variations of tetrahydrocannabinol content in cannabis plants to distinguish the fibre-type from drug-type plants. Bull Narc. 1985;37(4):61–5.
29. De Meijer EPM, Van Soest LJM. The CPRO *Cannabis germplasm* collection. Euphytica. 1992;62(3):201–11.
30. De Meijer EP, Bagatta M, Carboni A, Crucitti P, Moliterni VC, Ranalli P, Mandolino G. The inheritance of chemical phenotype in *Cannabis Sativa* L. Genetics. 2003;163(1):335–46.
31. Russo EB, Taming THC. Potential cannabis synergy and phytocannabinoid-terpenoid entourage effects. British J Pharmacol. 2011;163(7):1344–64.
32. Bottone EJ. *Bacillus cereus*, a volatile human pathogen. Clin Microbiol Rev. 2010;23(2):382–98. https://doi.org/10.1128/CMR.00073-09.

33. Keshavareddy G, Kumar ARV. *Bacillus thuringiensis*. In: Omkar O, editor. Ecofriendly pest management for food security. London: Academic Press; 2016. p. 443–73. https://doi.org/10.1016/B978-0-12-803265-7.00014-2.
34. Priest F, Goodfellow M, Shute L, Berkeley R. *Bacillus amyloliquefaciens* sp. nom., nom. rev. Int J Syst Bacteriol. 1987;37:69–71.
35. Dabboussi F, Hamze M, Elomari M, Verhille S, Baida N, Izard D, Leclerc H, et al. Taxonomic study of bacteria isolated from Lebanese spring waters: proposal for *Pseudomonas cedrella* sp. nov. and *P. orientalis* sp. nov. Res Microbiol. 1999;150(5):303–16. https://doi.org/10.1016/S0923-2508(99)80056-4.
36. Ryan RP, Monchy S, Cardinale M, Taghavi S, Crossman L, Avison MB, Berg G, van der Lelie D, Dow JM. The versatility and adaptation of bacteria from the genus *Stenotrophomonas*. Nat Rev Microbiol. 2009;7:514–25.
37. Sotelo T, Lema M, Soengas P, Cartea ME, Velasco P. In vitro activity of glucosinolates and their degradation products against brassica-pathogenic bacteria and fungi. Appl Environ Microbiol. 2015;81(1):432–40.
38. Van Klingeren B, Ten Ham M. Antibacterial activity of Δ9-tetrahydrocannabinol and cannabidiol. Antonie Van Leeuwenhoek. 1976;42(1–2):9–12.
39. Gershenzon J, Dudareva N. The function of terpene natural products in the natural world. Nat Chem Biol. 2007;3(7):408–14.
40. Ali EM, Almagboul AZ, Khogali SM, Gergeir UM. Antimicrobial Activity of *Cannabis sativa* L. Chin Med. 2012;1:61–4. https://doi.org/10.4236/cm.2012.31010.
41. Appendino G, Gibbons S, Giana A, Pagani A, Grassi G, Stavri M, Rahman MM. Antibacterial cannabinoids from *Cannabis sativa*: a structure-activity study. J Nat Prod. 2008;71(8):1427–30. https://doi.org/10.1021/np8002673.
42. Pollastro F, Taglialatela-Scafati O, Allara M, Munoz E, Di Marzo V, De Petrocellis L, Appendino G. Bioactive prenylogous cannabinoid from fiber hemp (*Cannabis Sativa*). J Nat Prod. 2011;74(9):2019–22. https://doi.org/10.1021/np200500p.
43. Docimo T, Consonni R, Coraggio I, Mattana M. Early phenylpropanoid biosynthetic steps in *Cannabis sativa*: link between genes and metabolites. Int J Mol Sci. 2013;14(7):13626–44. https://doi.org/10.3390/ijms140713626.
44. Tajkarimi MM, Ibrahim SA, Cliver DO. Antimicrobial herb and spice compounds in food. Food Control. 2010;21(9):1199–218. https://doi.org/10.1016/j.foodcont.2010.02.003.
45. Compean KL, Ynalvez RA. Antimicrobial activity of plant secondary metabolites: a Review. Res J Med Plants. 2014;8:204–13. https://doi.org/10.3923/rjmp.2014.204.213.
46. Sasso S, Scrano L, Bonomo MG, Salzano G, Bufo SA. Secondary metabolites: applications on cultural heritage. Commun Appl Biol Sci. 2013;78(2):101–8.
47. Wilkinson JM, Hipwell M, Ryan T, Cavanagh HMA. Bioactivity of *Backhousia citriodora*: antibacterial and antifungal activity. J Agric Food Chem. 2003;51(1):76–81. https://doi.org/10.1021/jf0258003.

8

Structural characterization of carbon and nitrogen molecules in the Humeome of two different grassland soils

Marios Drosos[1*], Davide Savy[1], Michael Spiteller[2] and Alessandro Piccolo[1*]

Abstract

Background: The surface layers of two German grassland soils, a silt loam soil from Dortmund (Soil A) and a sandy loam soil from Hannover (Soil B), were subjected to the Humeomics fractionation to identify the molecular composition of the soil Humeome.

Methods: The separated Humeomic fractions were analysed by high performance size exclusion chromatography hyphenated to a high-resolution electrospray-Orbitrap Mass Spectrometry (MS). Empirical formulae were obtained by mass data to describe the carbon and nitrogen molecular structures in the soil organic matter (SOM) of both soils.

Results: Results showed 175 masses for Soil A and 139 for Soil B distributed among the Humeomic fractions. Masses were classified according to molecular weight, unsaturation degree, oxygenation, and presence of nitrogen in the formula. The molecular information obtained with MS was consistent with the physical–chemical properties and environmental condition of the two grassland soils: (i) nitrogenated compounds were generally more numerous and more relatively abundant in the Humeome of Soil B, which was more anoxic and lower in C/N ratio than Soil A, (ii) highly oxygenated compounds were more numerous and abundant in the more oxic Soil A, (iii) most unsaturated formulae were comparatively more abundant in Soil A than Soil B, in line with differences in environmental conditions. The empirical formulae were then matched with their molecular structures, based on the application of ChemSpider and PubChem databases, and were found to be distributed into 16 specific chemical groups. The Van Krevelen plots built on the resulting carbon and nitrogen molecular structures provided a visual comparison of the Humeome of the two soils. The organosoluble fractions in both soils were dominated by aliphatic amides and saccharide ethers, while the hydrosoluble fractions comprised mainly aromatic amides, heterocyclic nitrogen compounds, and saccharide ethers.

Conclusions: Even though the two soils contained different compounds, 66 molecules were found to be common. Based only on differences in soil texture and oxic conditions, these findings indicate the existence of similar molecular dynamics in the stabilization of organic matter in these two grassland soils. The main common group in the Humeome of both soils was the saccharide ethers, which were additionally bound to aromatic compounds in the hydrosoluble fractions. Our detailed molecular study of the Humeome of these grassland soils confirms the potential of the Humeomic procedure to assess not only the carbon and nitrogen molecular composition in SOM, but also the mechanism of their long-term persistence.

Keywords: Carbon, Nitrogen, Grassland soils, Molecular structure, SOM, Humic matter, Humeomics, Sequential chemical fractionation, ESI-Orbitrap-MS

*Correspondence: drosos.marios@gmail.com; alessandro.piccolo@unina.it
[1] Centro Interdipartimentale di Ricerca sulla Risonanza Magnetica Nucleare per l'Ambiente, l'Agroalimentare e di Nuovi Materiali (CERMANU), Università di Napoli Federico II, Via Università 100, 80055 Portici, NA, Italy
Full list of author information is available at the end of the article

Background

Soil organic matter (SOM) plays a fundamental role in soil fertility [1]; it contributes to its physical–chemical stability [2] and interacts with heavy metals and organic hexogenous materials, including those of hazardous nature [3]. Its composition and molecular structure has baffled scientists for decades [4], although recent evidence of its supramolecular rather than macropolymeric character [5] gained a general consensus [6–10]. By this understanding, SOM contains a variety of small heterogeneous relatively small compounds which self-assemble in superstructures by weak interactions, and may thus be potentially analytically determined.

This complex supramolecular matrix was shown to have intermolecular interaction of different strength, and may only be simplified using a step-wise extraction procedure named Humeomics that involves increasingly severe extraction conditions [11]. Humeomics enabled the successful characterization of a large number of compounds [12] in purified humic substance's (HS) extracts such as humic acids (HA) [11], its size fractions [13] and humin [14]. In particular, the Humeomic procedure even improved the detection of molecular components of soil organic carbon (SOC) in the unextractable organic matter by both mass spectrometry (MS) [15] and nuclear magnetic resonance (NMR) experiments [16], and enabled the characterization of compounds rich in quaternary carbon. These findings are relevant in light of the recent interest in the characterization of highly unsaturated SOC [17–19] driven by the availability of ultra-high-resolution mass spectrometers.

Conversely, highly unsaturated soil organic nitrogen (SON) is still largely uncharacterized, even though its existence and relevance has been postulated [20]. Moreover, the role of SON in the supramolecular arrangement of the soil Humeome is still unknown, and this uncertainty is reflected in the technical difficulty of its measurement and the consequent poor knowledge on its vital plant nutrition capacity [21–23]. Further molecular understanding of the dynamics of SOC and SON would be also important for their broad ecological relevance since their recognized transport to natural waters and atmosphere [24, 25].

The presence of more types of N-containing organic constituents in soil has led to classify SON according to apparent molecular weight, chemical nature and origin. SON categories include readily extractable hydrophilic aminoacids [26] generally associated with LMW fraction of SON, while polyphenol-like matter are attributed to the HMW fraction [27]. Furthermore, heterocyclic SON has been investigated by pyrolysis [20] and later by X-ray methods [28], but both methods do not provide accurate structural information. This type of SON may be originally aromatic in nature, or possibly induced by combustion [29]. Furthermore, proteins may be another source for SON [30], which is assumed to resist degradation due to encapsulation in SOM hydrophobic domains [31], thus extending the SON lifespan [32]. Nevertheless, the molecular characterization of single components of SON has been only recently attempted and mainly focused only on DON and sediment characterization [33].

A positive correlation between fine soil texture and SON abundance was inferred [34], arguing that a physical protection is provided by finer soil particles that enable physical preservation of even the most labile N-containing compounds. Conversely, in coarser soils, molecular recalcitrance is prevalently due to a chemical mechanism, thus resulting in a greater preservation of non-nitrogenated compounds, such as lignins [34]. Furthermore, it has been shown that the presence of degradative enzymes from microbial biomass affects SOC and SON composition [35]. In a recent application of Humeomics to an Italian silt loamy agricultural soil, it was found that the main molecular components of the soil Humeome were heterocyclic nitrogen compounds (HN) and amides (AD) [36]. However, more experimental work on the stability of SON is required to confirm the importance of its heterocyclic components in soils.

While much work by Fourier transform ion cyclotron (FT-ICR)-MS is available on the characterization of environmental DOM [37] that on SOC and SON is generally lacking due to the extreme complexity of terrestrial materials [38]. Conversely, the Orbitrap-MS technique has been successfully applied for the accurate characterization of masses in humic matter and Humeomic fractions separated from soils [36, 39, 40]. In fact, despite the relatively lesser mass resolution, Orbitrap-MS can successfully enable the assignment of individually detected masses within Humeomic fractions to specific molecular formulae by improving data processing [41]. The molecular characterization of the soil Humeome rather than that of bulk SOM enables a detailed description of SOM molecular composition and lay stronger foundations for the molecular fingerprinting of soil humus [36, 42], thus representing a much more accurate alternative to current methods such as UV–Vis spectroscopy [43], MS [44], outdated degradative methods [45], and even poorly sensitive NMR spectroscopy [46]. The molecular investigation of the soil Humeome not only opens up to the correlation between SOM composition and its origin, but also enables to identify specific molecular biomarkers for different soil ecosystems, inasmuch as it is already attempted for DOM [47, 48].

The objective of this study was hence, to reach a deeper molecular knowledge of the humus composition in two

Table 1 Structural, geographical and selected physico-chemical properties of soils

	Soil A	Soil B
Origin	Dortmund Barop, An der Palmweide, Germany	Hannover Fuhrberger Feld marsh, Germany
Latitude	51.5149	52.3705
Longitude	7.4660	9.7332
Altitude (m)	96	57
Soil classification (USDA)	Haplic Cambisol	Gleyic Podzol
Clay (< 2 μm) (%)	16.00	2.40
Silt (< 50–2 μm) (%)	73.00	29.00
Sand (2000–50 μm) (%)	11.00	68.60
Textural class	Silt loam	Sandy loam
pH (water, ratio 1:2.5)	6.74	5.00
C_{org}. (%)	2.88	1.88
N_{tot}. (%)	0.13	0.10

Table 2 Weight yields (mg) of Humeome fractions obtained from two grassland soils and relative (%) percentage of molecules common in both soils for each Humeome fraction

Fraction	Soil A	Soil B	Relative (%) percentage of molecules common in both soils	
			Soil A	Soil B
ORG1	535	382	56.5	62.0
ORG2	1100	878	68.7	60.0
ORG3	15	46	89.7	87.2
AQU2	729	684	55.4	78.2
AQU3	9	496	24.1	85.7
AQU4	8	1	57.8	83.0
RESOM	138	321	59.5	20.4
Total Humeome	2534	2808		

different grassland soils for both its organic carbon and nitrogen forms.

Methods

Soil samples

Two surface-layered (0–10 cm) German grassland soils were collected from the Barop area in Dortmund (Soil A) and from the Fuhrberger Feld area, in Hannover (Soil B). The soils were air-dried, and sieved with a 2-mm sieve. Their characteristics are reported in Table 1.

Soil Humeome fractionation

The applied soil Humeome fractionation was applied on two replicates for each soil, as described earlier [36, 39] and briefly below. Weight yields of the separated Humeome fractions from both soils are reported in Table 2.

Unbound fraction (ORG1)

The dried soil was weighed in an amount corresponding to 1.0 g of its content in soil organic carbon (SOC) calculated from its % of OC. Soil samples were subjected to the first step by extracting the unbound organosoluble fraction (ORG1) with a dichloromethane and methanol 2:1 solution, under stirring for 8 h at room temperature. The supernatant was centrifuged (2000 rpm for 5 min), filtered through GF-C filters (Whatman), and rotoevaporated.

Weakly ester-bound fractions (ORG2 and AQU2)

The soil residue was then transferred in Teflon tubes (Nalgene) and added with 100 mL of a Boron trifluoride (Acros Organics) solution at 12% w/v in methanol and heated for 8 h at 90 °C. After cooling, the supernatant was separated and transferred in a liquid/liquid extraction funnel where addition of a 1:1 water and chloroform mixture allowed separation of the weakly ester-bound compounds in the organosoluble (ORG2) and hydrosoluble (AQU2) fractions, which were, respectively, rotoevaporated, dialysed (Spectrapore 1000 Da cut-off membranes) and freeze-dried.

Strongly ester-bound fractions (ORG3 and AQU3)

The soil residue was suspended in a KOH 1 M (Sigma-Aldrich) solution in methanol for 2 h under refluxing. The supernatant was separated and placed in a liquid/liquid extraction funnel where addition of a 1:1 water and chloroform mixture allowed separation of the strongly ester-bound compounds in the organosoluble (ORG3) and hydrosoluble (AQU3) fractions. The former was rotoevaporated while the latter was dialysed (Spectrapore 1000 Da) and freeze-dried.

Ether-bound fraction (AQU4)

Hydroiodic Acid (Acros Organics), 47% w/v in deionized water (Milli-Q) was added to the residue, heated for 72 h at 70 °C. The supernatant was neutralized to pH 7 with a $NaHCO_3$ (Carlo Erba) solution and I_2 the evolved reduced to HI by $Na_2S_2O_3$ (Carlo Erba). This ether bound (AQU4) fraction was dialysed (Spectrapore 1000 Da) and freeze-dried.

Residual organic matter (RESOM)

The remaining unextracted humic matter (RESOM) was isolated from the soil residue by 0.5 M NaOH and 0.1 M $Na_4P_2O_7$ (Carlo Erba) solution. The supernatant was dialysed (Spectrapore 1000 Da) and freeze-dried.

Liquid chromatography ESI-Orbitrap mass spectrometry

Each Humeome fraction (ORG1, ORG2, ORG3, AQU2, AQU3, AQU4 and RESOM) was dissolved in a 1 mL water and acetonitrile 1:1 solution for a 0.4 g L^{-1} final concentration and added with either 10 µL of LC–MS Grade 25% NH_3 (Romil) solution for the ORG and RESOM fractions, or 10 µL of LC–MS Grade 37% HCl (Romil) solution for the AQU fractions. Samples were injected (50 µL) in a Rheodyne loop in a HPSEC system connected to the LC/MS system. HPSEC comprised a Phenomenex Bio-Sep SEC-S 2000 column (300 × 7.8 mm) and precolumn (30 × 7.8 mm), both thermostatted at 30 °C. A Dionex P 580 pump ensured a 0.3 mL min^{-1} elution of a 55/45 A/B solution (A: 5 mM CH_3COONH_4 in Milli-Q water and 5% CH_3CN, pH 7; B: 100% CH_3CN). Mass spectra were obtained with a LTQ Orbitrap (Thermo Electron, Waltham, MA) and positive ESI, 100–1000 *m/z* mass range, and 1.0 s scan time. N_2 was the sheath gas (45 AU) and He was the collision gas (7.99 AU). Spray voltage was set at 4.00 kV, spray current at 2.05 µA, capillary temperature at 260 °C, and capillary voltage at 14.93 V.

Data analysis

Masses were translated into molecular formulae using the Excalibur bundled software (Thermo) and setting the following constraints: $P < 1$; $1 > C > 60$; $1 > H > 120$; $0 > O > 60$; $0 > N > 20$ and 5 ppm error. In most cases only one molecular formula fulfilled all requirements. In case of multiple assignments, priority was given to the formulae which could be included in an already existing series of increasing homologues with matching double bond equivalent (DBE) rates. By these criteria, no assignment was ambiguous. The most probable chemical structure for each empirical formula was found by the ChemSpider (http://www.chemspider.com), and the PubChem (https://pubchem.ncbi.nlm.nih.gov/) databases, allowing the assignment to class-specific groups. Each mass peak was integrated for both replicates and means were obtained for all areas under curve (AUC). Semiquantitative assessment (%) was carried out for each peak dividing its AUC by the sum of all AUCs within the specific Humeome fraction. Relative standard deviations were for all data within 10%. Results are reported in Additional file 1: Tables S1–S14.

Results and discussion

The two soils of this study were diverse in texture, pH, and organic carbon content (Table 1). In particular, the silt loam Soil A had a neutral pH while the sandy loam hydromorphic Soil B was slightly acidic. The former showed a larger percentage of organic carbon and higher C/N ratio than the latter.

Compounds in the Humeome and their structural characterization

HPSEC-ESI(+)-Orbitrap-MS analysis enabled the characterization of 175 distinct empirical formulae in Soil A and 139 in Soil B. They were arranged in following three categories: (i) presence (or absence) of N atoms in the empirical formula, henceforth abbreviated in N and no-N; (ii) high or low degree of unsaturation (high or low double bonding equivalent = DBE), abbreviated as HDBE and LDBE; (iii) high or low degree of oxygenation per C atom, abbreviated in HOx and LOx, respectively. Presence of N in a formula was objectively determined, while for the other two criteria the discriminating factors were chosen arbitrarily as it follows: high unsaturation when $DBE \geq 6$ and low unsaturation when $DBE \leq 5$; high oxygenation when C/O ratio in formulae was ≥ 0.5 and low oxygenation when ≤ 0.5. In case of DBE greater than 10, the abbreviation UHDBE for ultra high unsaturation was used, and formulae showing absence of O atoms were designated by the NOx (no oxygen) abbreviation.

Finally, the empirical formulae were classified into 16 specific chemical groups after identifying their molecular structure based on chemical software programs and databases: aliphatic amides (Aliphatic AD), aromatic amides (Aromatic AD), glycosidic amides (Glycosidic AD), amines (AM), heterocyclic nitrogen compounds (HN), dicarboxylic acids (DA), hydroxylic acids (HA), aliphatic esters (ES), aliphatic ethers (ET), oligosaccharides (saccharide ET), heterocyclic oxygen compounds (HO), phenolic acids (PA), phenolic esters (PE), phenols (PH), steroids (SE), and sterols (ST).

Nitrogenated formulae with low unsaturation and oxygenation degree (N, LDBE, LOx)

All Humeomic fractions of both soils revealed a number of structures characterized by 12 or more C atoms, few unsaturations and typically 4 or occasionally more O atoms. These molecules included mostly a single N atom, even though up to four were found in some formulae. Since these compounds were nearly completely saturated, and given that these humic molecules had likely plant or bacterial origin, it may be inferred that membrane lipids or biopolyesters, such as cutin and suberin, were the most probable precursor for this class of compounds [49]. The main constituents of these biomolecules are either long-chain mono- or di-oic acids, with the chain bearing one or more unsaturations and/or one or more O substituents in the ω or mid-chain position. This suggests that the molecules designated as N, LDBE, LOx revealed (Additional file 1: Tables S1–S14) may derive from the break-down of lipids, cutin and suberin after cell death, and have included nitrogen in the formula at a later stage. ChemSpider and PubChem databases showed that the main compounds fitting this category were aliphatic AD, and to a lesser extent, some AM and HN.

Nitrogenated formulae with low unsaturation and high oxygenation degree (N, LDBE, HOx)

Molecular structures characterized by the C_{10-29} chain were found in Soil A, and to a much lesser extent in number and amount, in Soil B. These nitrogenated compounds showed few double bonds, but contained several O atoms and were designated as N, LDBE, HOx. The origin of these molecules may be from either a more oxygenated aliphatic cutin and suberin material or a product of lignin breakdown, (Additional file 1: Tables S1–S14). It is inferred that a slow oxidation process may have occurred over time in these compounds and enriched the structure with hydroxyl groups. In fact, the evident anoxic conditions of Soil B have yielded smaller amounts of these structures than the oxic Soil A (Table 3). The application of chemical databases revealed that the main compounds fitting this category were mainly glycosidic AD, and to a much lesser extent, AM.

Nitrogenated formulae with high- or ultra-high-unsaturation and low- or no-oxygenation degree (N, HDBE/UHDBE, LOx/NOx)

Although the low degree of oxygenation of such unsaturated and poorly oxygenated compounds should have suggested a low hydrosolubility of compounds in this category, they remained in the water phase during the extraction due to positively charged nitrogen. Moreover, the much greater number of nitrogen atoms in the UHDBE formulae strongly implied also the presence of heterocyclic N. In case of HDBE formulae, their marked hydrosolubility in the presence of fewer N atoms may be explained with the presence of one or more amino groups instead of heterocycles. While the different substitution pattern may indicate that HDBE and UHDBE formulae may be chemically different, also the chemical databases of both these categories showed a different complexation degree of HN and aromatic AD (Additional file 1: Tables S1, S4–S7 and S11–S14).

Nitrogenated formulae with high unsaturation and oxygenation degree (N, HDBE, HOx)

Hydrosolubility of this group of molecular structures is a consequence of the presence of several oxygen and nitrogen atoms in these formulae. It is likely that these compounds derived from a domain of saccharides bearing different methylation degree, to which they are bound with glycosidic bonds. The chemical databases suggested an abundance of HN, aromatic AD and glycosidic AD compounds (Additional file 1: Tables S4–S7 and S11–S13).

Table 3 Weight yields (%) per type of compound found in the Humeomic fractions

Soil	Class of compound	Fraction						
		ORG1	ORG2	ORG3	AQU2	AQU3	AQU4	RESOM
Soil A	N, LDBE, LOx	49.10	57.54	84.74	0.73	29.72	6.18	2.34
	N, LDBE, HOx	12.70	4.94	3.46	0.02	5.51	–	–
	N, HDBE, LOx	–	0.50	–	55.25	9.53	7.51	8.48
	N, HDBE, HOx	–	–	–	34.68	4.47	0.44	0.26
	N, UHDBE, LOx/NOx	–	–	–	1.47	5.92	13.89	19.43
	Non-N, LDBE, LOx	–	0.69	–	–	4.67	–	0.13
	Non-N, LDBE, HOx	–	2.91	0.63	6.73	20.17	2.01	–
	Non-N, HDBE, LOx	9.16	6.94	0.87	1.12	7.25	10.94	–
	Non-N, HDBE, HOx	29.10	26.48	10.30	–	12.76	59.03	69.36
Soil B	N, LDBE, LOx	75.09	23.16	58.77	–	2.78	62.59	0.56
	N, LDBE, HOx	–	0.83	–	–	1.40	1.78	0.37
	N, HDBE, LOx	–	–	–	69.93	–	–	–
	N, HDBE, HOx	–	–	–	8.47	5.16	7.08	–
	N, UHDBE, LOx/NOx	–	–	–	11.37	7.85	3.76	78.27
	Non-N, LDBE, LOx	2.71	30.21	0.33	1.43	–	10.1	–
	Non-N, LDBE, HOx	–	–	–	0.12	9.87	1.07	–
	Non-N, HDBE, LOx	–	–	–	6.34	8.52	1.07	–
	Non-N, HDBE, HOx	22.20	45.80	40.90	2.34	64.42	12.55	20.80

N nitrogenated, *LDBE/HDBE/UHDBE* low/high/ultra high double bond equivalent, *LOx/HOx/NOx* low/high/no oxygenation

Non-nitrogenated formulae with low unsaturation and oxygenation degree (non-N, LDBE, LOx)

Based on the empirical formulae of this category, the most plausible attribution involves a nearly saturated hydrophobic chain such as those present in compounds commonly found in the MS spectra of humic acids and humin [11, 50]. The great abundance of these oxygenated formulae in Soil B would appear odd, given the anoxic character of this soil. However, it is more likely that this category of compounds was also originally present in Soil A, but underwent oxidation over time, as implied by the prominent abundance of highly oxygenated formulae (non-N, LDBE, HOx) found therein (Table 3). Additionally, the low unsaturation of these formulae is consistent with the less oxidizing environment of the marsh where Soil B was sampled. The application of ChemSpider and PubChem databases suggested the structures of aromatic compounds such as PH, PE and ST, but also of aliphatic compounds such as ES, DA and HA (Additional file 1: Tables S2, S5, S7–S11 and S13).

Non-nitrogenated formulae with low unsaturation and high oxygenation degree (non-N, LDBE, HOx)

As already mentioned, this category of formulae was more numerous and abundant in Soil A than in Soil B, owing to the greater availability of oxygen in the former, which translated into smaller unsaturation and oxygenation (Table 3). The chemical databases for this category indicated structures for aromatic compounds such as PH, PE, and HO. However, a number of structures were also present as ES and ET, and in large amount, as saccharide ET (Additional file 1: Tables S2–S6 and S11–S13).

Non-nitrogenated formulae with high unsaturation and low oxygenation degree (non-N, HDBE, LOx)

These formulae with large degree of unsaturation were found in both soils, although with different distributions. The structures of these molecules entailed a degree of unsaturations that plausibly implied an aromatic nature. Despite high unsaturation and low oxygenation degree, ChemSpider and PubChem databases again assigned the structures to aromatic compounds such as HO, SE, ST, PH, PA and PE (Additional file 1: Tables S1-S3, S5–S6 and S11–S14).

Non-nitrogenated formulae with high unsaturation and high oxygenation degree (non-N, HDBE, HOx)

Compounds characterized by a high degree of unsaturation and oxygenation without N atoms in their formula were revealed by the high-resolution Orbitrap-MS as the most ubiquitous category of compounds. As in previous categories, the chemical databases suggested aromatic compounds, such as PH, PE and HO, also in this category. However, the main constituents were saccharide ET, which they were linked to aromatic moieties, as HO or phenols, in the hydrosoluble fractions (Additional file 1: Tables S1–S3 and S5–S14), thus forming sugar ethers combined to flavonoids or flavanols.

Van Krevelen plots of soil Humeomes

The graphical representation of the MS data for both soils is shown in Van Krevelen diagrams which were built by plotting O/C versus H/C ratios of all detected analytes in any fraction (Fig. 1). Although different components were found in the two soils, they still contained 66 common molecules. It is, thus, noteworthy that despite their geographical distance (about 200 km) and diverse properties (Table 1), similar biochemical and physical–chemical mechanisms are active in stabilizing specific molecules in the two soils.

In the case of the unbound fraction (ORG1), both soils showed that aliphatic AD and saccharide ET were the main chemical groups. However, Soil A comprised 12.7% of glycosylated AD compounds (N, LDBE, HOx formulae in Additional file 1: Table S1), which were totally absent in Soil B (Fig. 2).

The weakly ester-bound organosoluble fraction (ORG2) was similar to ORG1 and contained mainly aliphatic AD and saccharide ET. Nevertheless, the relative percentage of these two groups was inverted in the two soils: Soil A had a larger amount of aliphatic AD, while Soil B was richer in saccharide ET (Fig. 3). Yet, the oxygenated molecules of Soil A were more oxidized, whereas the reduced anoxic molecules of Soil B had a more lipidic character.

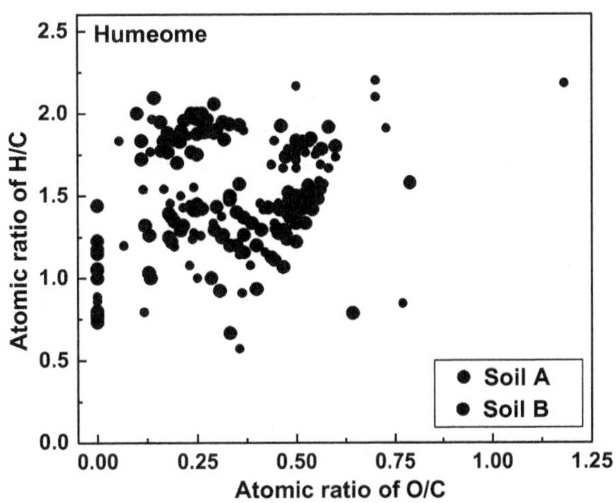

Fig. 1 Van Krevelen plot of total Humeome comprising all molecular structures identified in Soil A (red) and Soil B (grey)

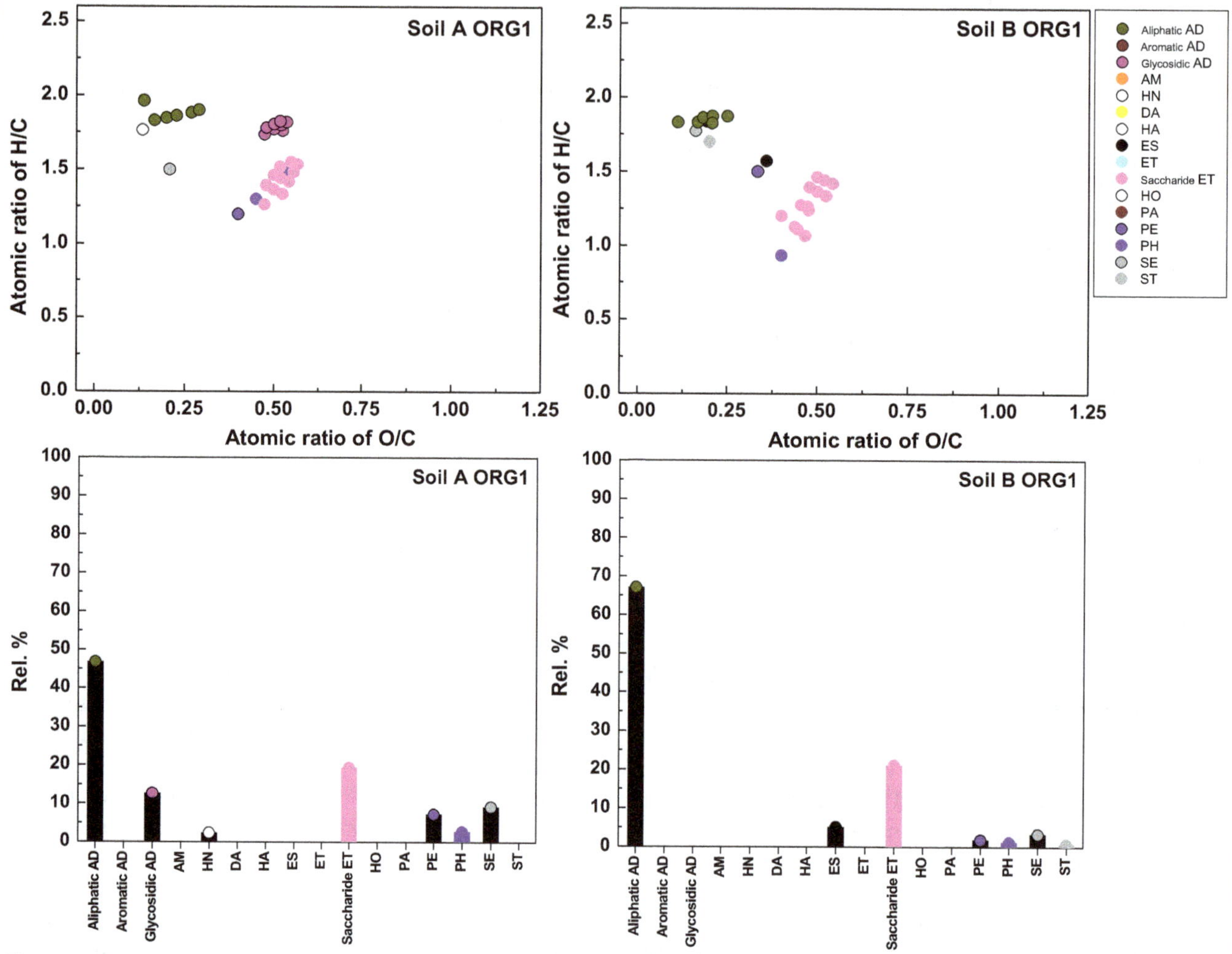

Fig. 2 Van Krevelen plots of molecular structures identified in fraction ORG1 for both soils and their relative percentage distribution in different classes

Aliphatic AD and saccharide ET were again the main compounds found for the strongly ester-bound organosoluble fraction (ORG3). However, the abundance of AM was much larger in Soil A than in Soil B (Fig. 4). It is worth to mention that the silt loam soil (Soil A) had more organosoluble contents than the sandy loam soil (Soil B). This indicates a greater stability of SOM in Soil A, with a consequent larger resistance to oxidation. Another important outcome was that the percentage of common molecules found in both soils was rising from ORG1 to ORG3 (Table 2). This result indicated that the similarity of the two grassland soils is more related to the recalcitrant part of SOM than to the labile fraction.

Soil B showed a larger components' abundance in the hydrosoluble fractions, possibly due to a greater binding capacity of the larger surface for this silt loam soil than for the sandy loam Soil A. However, the hydrosoluble weakly ester-bound fraction (AQU2) was found similar in quantitative abundance in both soils, with aromatic AD and HN compounds being the main representative groups (Fig. 5). Soil A contained partially more oxidized ethers and more reduced aromatic AD than Soil B, that showed instead a larger amount of reduced HN compounds.

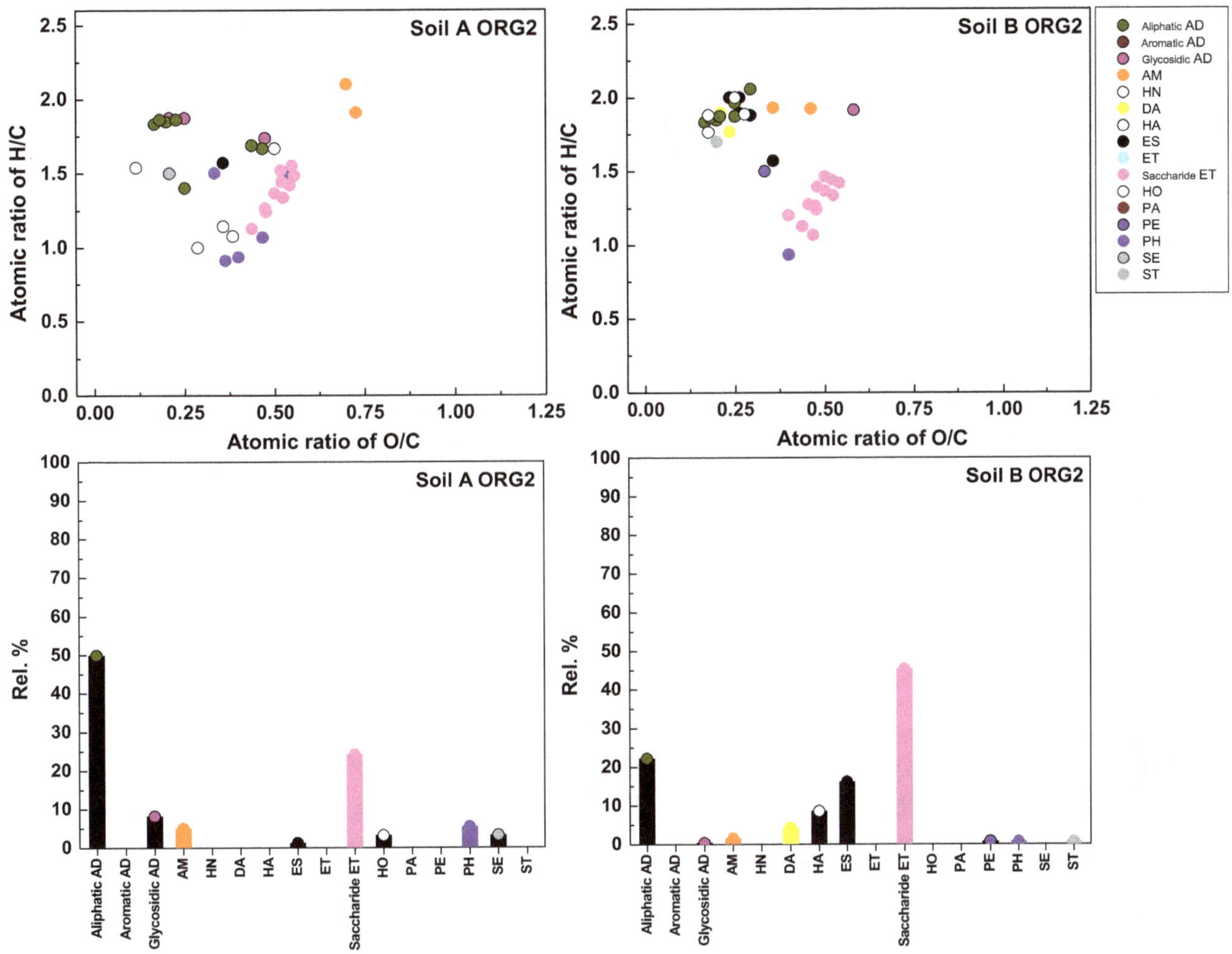

Fig. 3 Van Krevelen plots of molecular structures identified in fraction ORG2 for both soils and their relative percentage distribution in different classes

In the strongly ester-bound fraction (AQU3), the main difference was the greater percentage of Saccharide ET and more reduced HN compounds in Soil B than in Soil A (Fig. 6). Nevertheless, it is important to underline that the mass yield of AQU3 in Soil B was almost 55 times larger than in Soil A (Table 2).

The ether-bound fraction (AQU4) showed a significantly small yield for both soils (Table 2). Nevertheless, Soil B still revealed a larger content of reduced HN compounds than in Soil A and a complete absence of Aromatic AD (Fig. 7).

Finally, the residual organic matter (RESOM) showed similar Van Krevelen plots for both soils (Fig. 8), although the percentage of aromatic AD and saccharide ET as the main groups of compounds in the fraction, was inverted in the two soils. In fact, aromatic AD was more abundant in Soil B, whereas saccharide ET was predominant in Soil A.

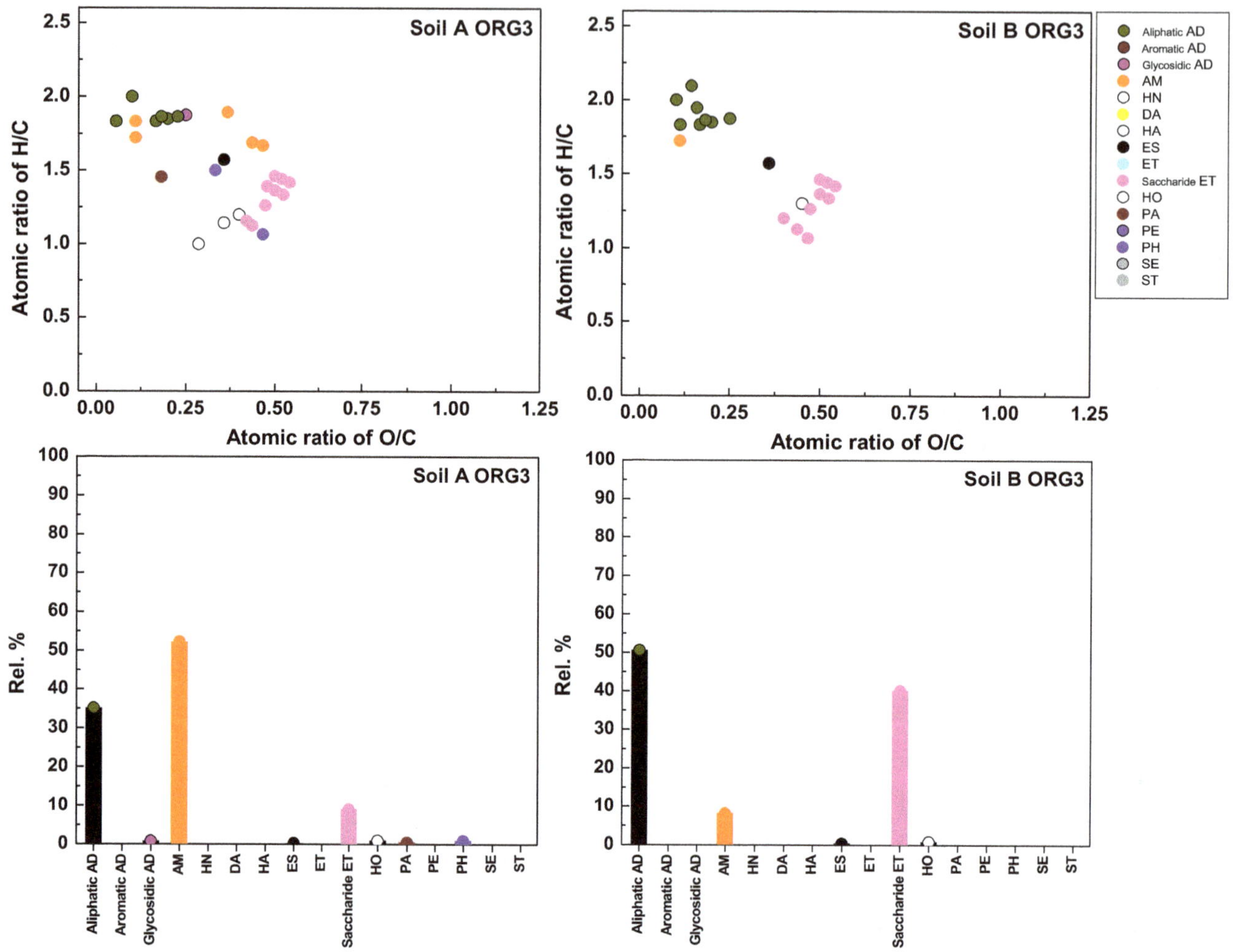

Fig. 4 Van Krevelen plots of molecular structures identified in fraction ORG3 for both soils and their relative percentage distribution in different classes

Conclusions

Two different grassland soils were fractionated here by the Humeomics procedure and the resulting fractions subjected to HPSEC-ESI-Orbitrap-MS characterization. A considerable number of masses was detected and classified according to nitrogenation, oxygenation and unsaturation.

The different properties and environmental conditions of the two grassland soils were reflected in their molecular composition. In particular, the detected masses and their MS relative quantitative assessment indicated a greater nitrogenation of compounds for the hydromorphic sandy loam Soil B and larger oxygenation of components in the oxic silt loam Soil A. The Van Krevelen diagrams built on MS data were proved to be a useful fingerprinting tool to graphically reveal the differences in molecular composition of SOC and SON between the two soils.

The compositional differences among Humeomic fractions were rationalized by taking into account the fact

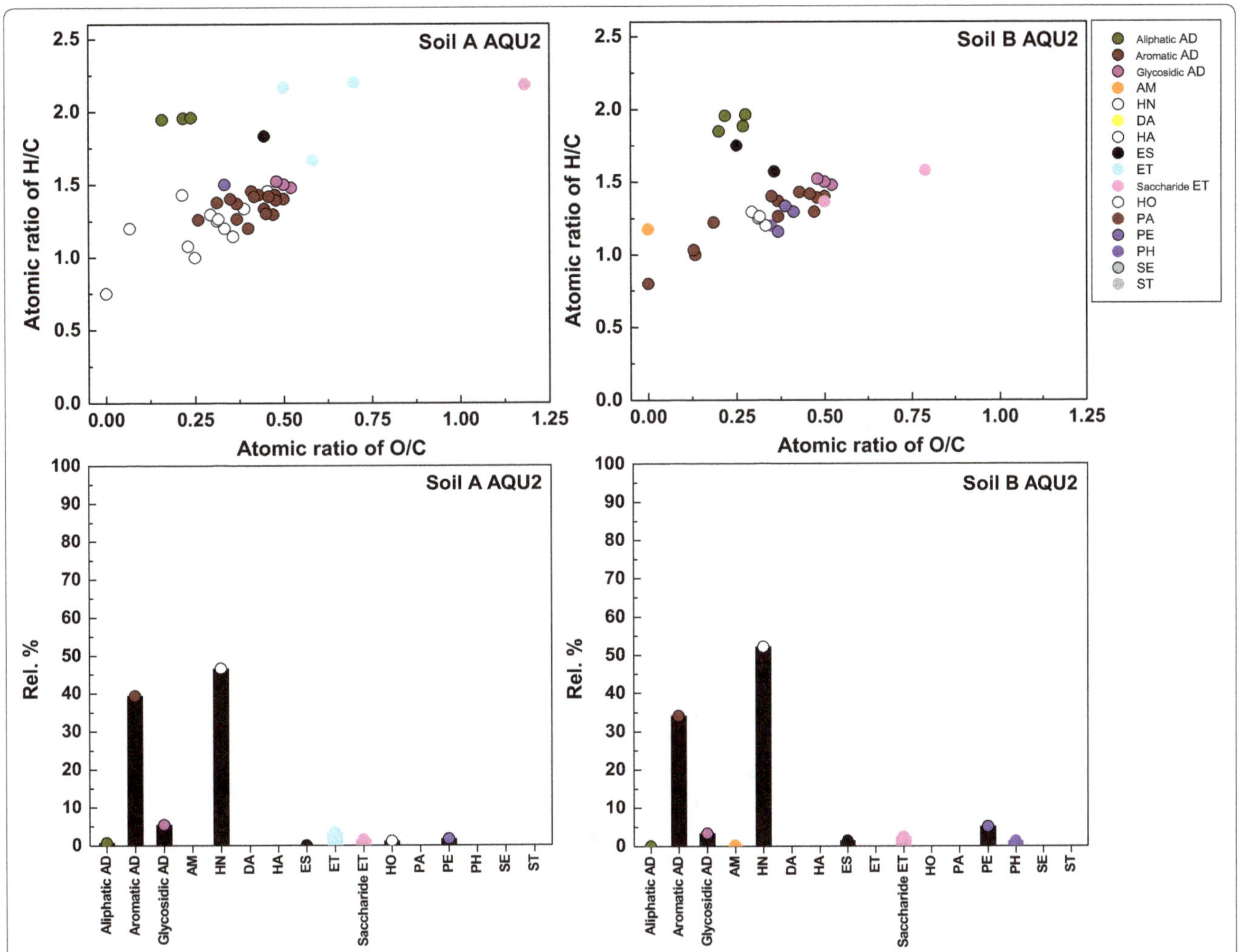

Fig. 5 Van Krevelen plots of molecular structures identified in fraction AQU2 for both soils and their relative percentage distribution in different classes

that carbon–carbon bonds were unaffected, whereas C–O and C–N bonds were subjected to cleavage by hydrolytic reactions of increasing strength (non-covalent > ester > ether/amide). This step-wise approach highlighted the patterns by which some formulae lost the same chemical groups after a specific cleavage, thereby revealing the specific type of chemical bond existing in the humic molecules. This approach, although time consuming, provided unprecedented information on the chemical structure of previously unknown compounds, including the heterocyclic nitrogenated molecules.

It is to be underlined that even though molecular formulae were distributed in statistical categories according to their degree of oxygenation and DBE, most of these categories always resulted into a mixture of different types of compounds. This suggests that interpretations of Van Krevelen results should not be based only on the empirical formulae stemming out from software programs accompanying high-resolution spectrometers, but they should be rather referred to the corresponding molecular structures, as further supported by chemical databases in silico [51]. In fact, careful structural

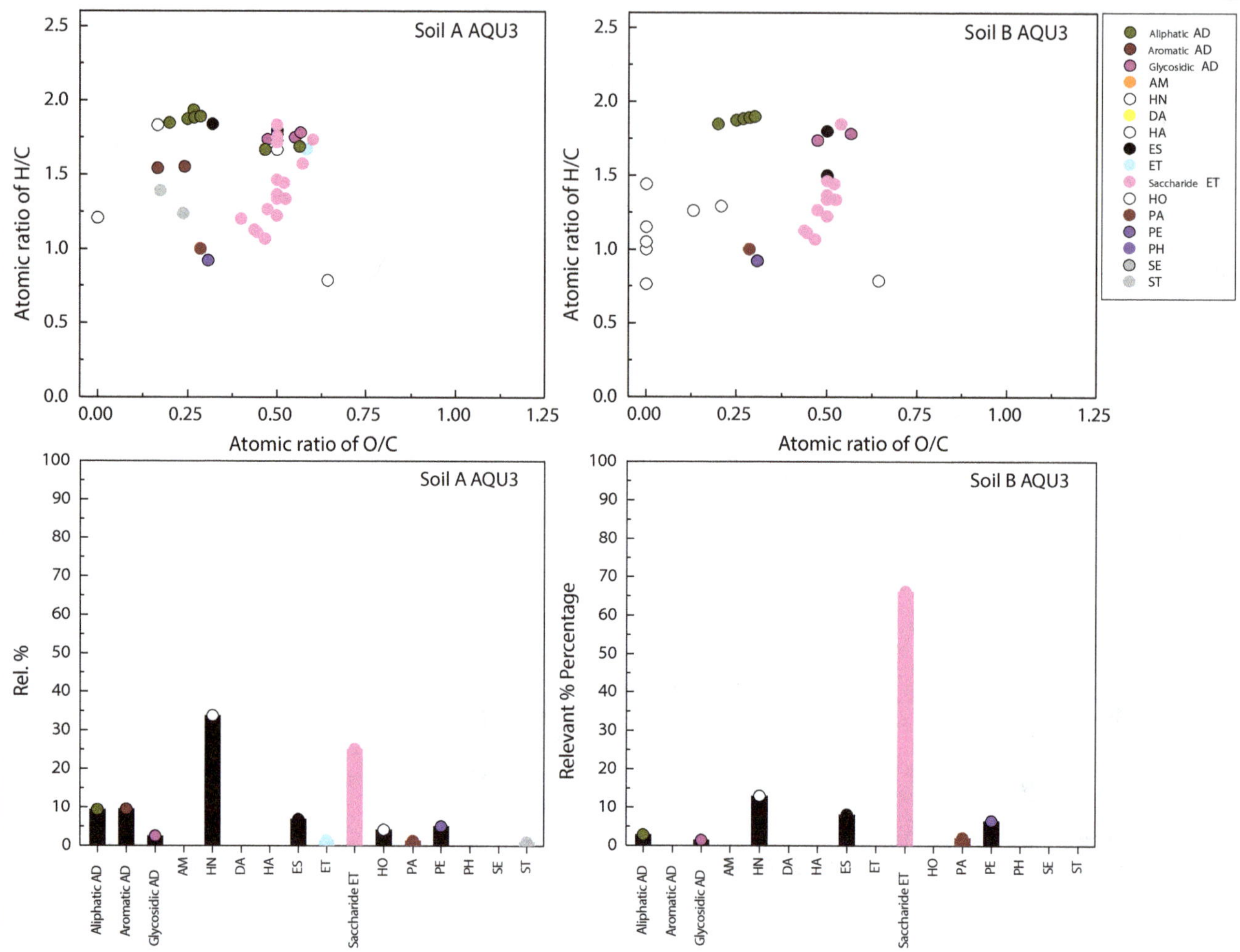

Fig. 6 Van Krevelen plots of molecular structures identified in fraction AQU3 for both soils and their relative percentage distribution in different classes

interpretation by ChemSpider and PubChem databases of empirical formulae obtained from Orbitrap mass spectra revealed that organosoluble fractions of both soils were dominated by aliphatic AD and saccharide ET, while aromatic AD, HN and saccharide ET were predominant in the hydrosoluble fractions. Hence, saccharide ET was found to be the class of compounds common in all types of fractions in both soils, although these components were bound to aromatic moieties in the hydrosoluble fractions. Therefore, these findings indicate that saccharide ET compounds, such as flavanon glycosides, are recalcitrant in grassland soils and represent a stabilized component of SOM.

The mass ratio of the organosoluble to hydrosoluble extracts from the parental soils was 1.9 for the silty Soil A, dropping to 0.9 for the sandy Soil B (Table 2). This difference was due to the high amount of labile ORG1 content present in Soil A, while Soil B was found more abundant in molecules held together by strongly bound esters (AQU3–ORG3) and organominerals (RESOM). However, the relative percentage of compounds being common between the two soils was found to increase

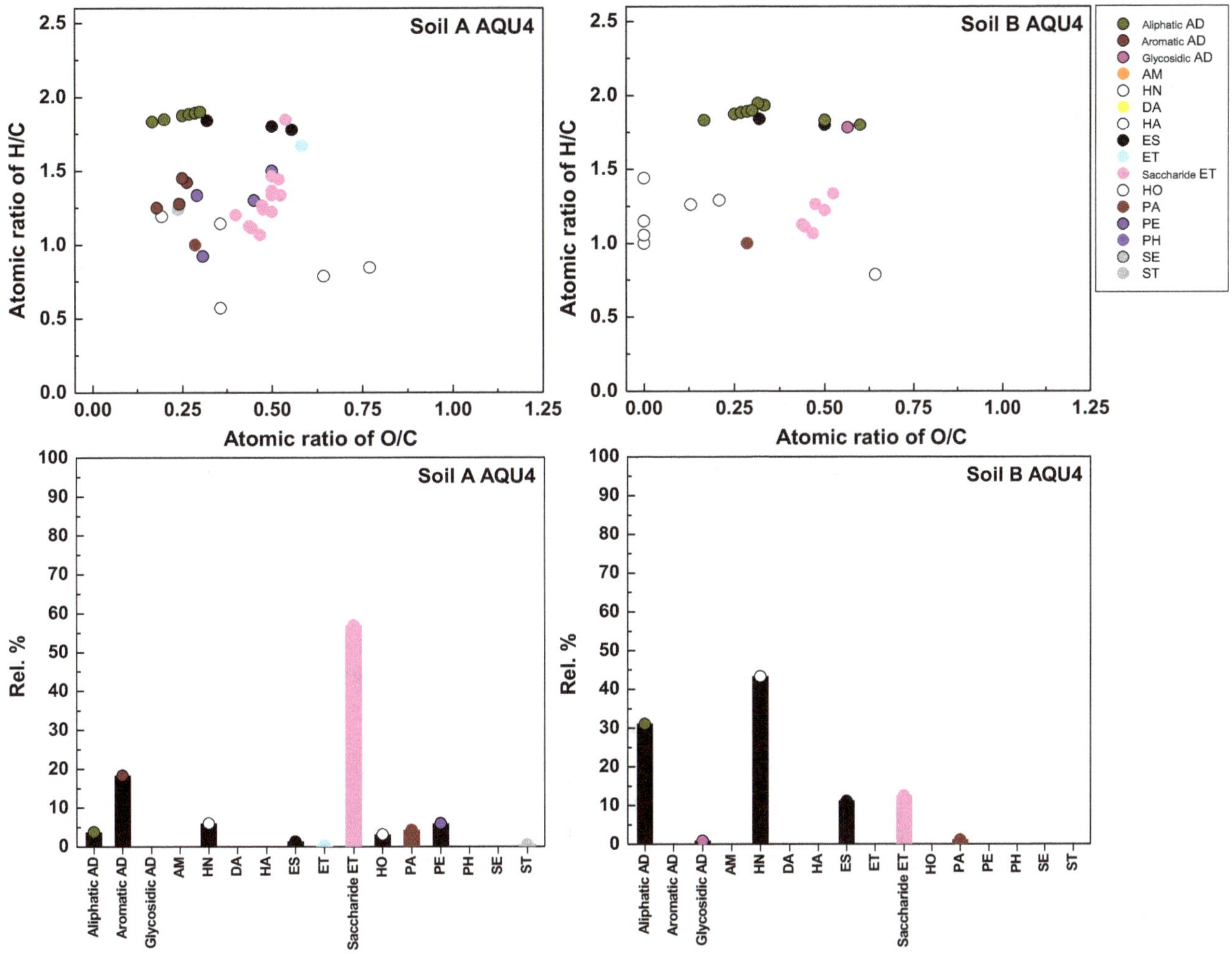

Fig. 7 Van Krevelen plots of molecular structures identified in fraction AQU4 for both soils and their relative percentage distribution in different classes

from ORG1 to ORG3, whereas no such an absolute trend was revealed for the hydrosoluble fractions AQU2-AQU4 and RESOM (Table 2). This result suggests that the stabilization of organic matter in soils is mainly due to its organosoluble fraction, while the hydrosoluble components may be more easily subjected to abiotic and biotic transformations. This conclusion is in line with the supramolecular nature of SOM that implies a stabilization of humus by a progressive accumulation of lipophilic molecules in soil, due to their physical–chemical separation from aqueous media [52]. However, differences in the organosoluble to hydrosoluble mass ratios between the two soils may be related to the varying association of humic molecules with either silt or sand components, thereby forming SOM aggregates of different complexity and stability [53]. Therefore, a further investigation on the humeome of soil aggregates of different dimension and stability may enlarge our knowledge on the mechanisms of SOM sequestration.

The findings of this work indicate that a detailed study of the Humeome of two grassland soils not only

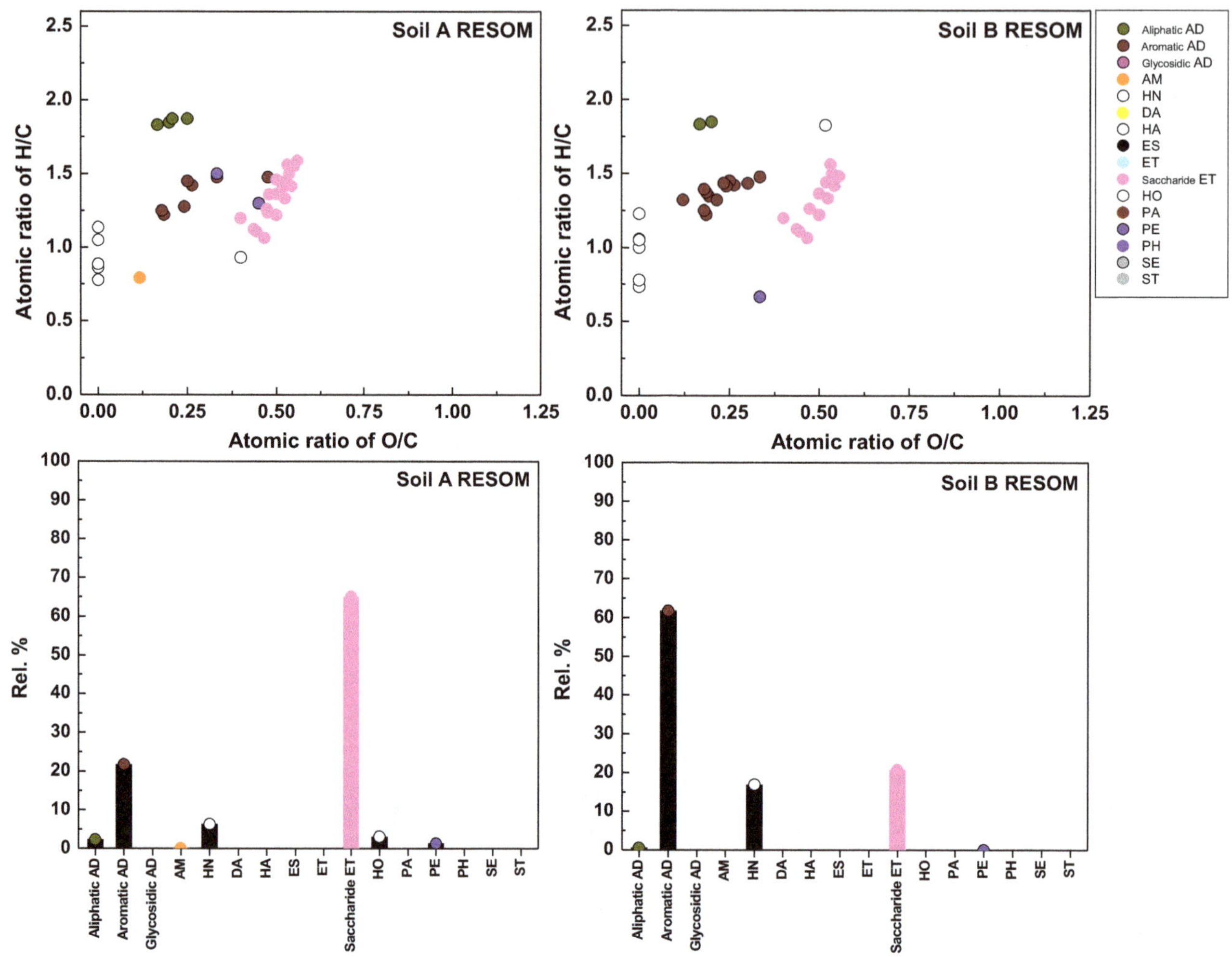

Fig. 8 Van Krevelen plots of molecular structures identified in fraction RESOM for both soils and their relative percentage distribution in different classes

provided unprecedented information on the specific carbon and nitrogen molecular structures in SOM, but also revealed that similar mechanisms are active in the two soils for the long-term stabilization of organic carbon.

Authors' contributions

All authors have contributed equally to the work. All authors read and approved the final manuscript.

Author details

[1] Centro Interdipartimentale di Ricerca sulla Risonanza Magnetica Nucleare per l'Ambiente, l'Agroalimentare e di Nuovi Materiali (CERMANU), Università di Napoli Federico II, Via Università 100, 80055 Portici, NA, Italy. [2] Institut für Umweltforschung (INFU) der Fakultät Chemie und Chemische Biologie, Technische Universität Dortmund, Otto Hahn Straße 6, 44221 Dortmund, Germany.

Acknowledgements

The collaboration of Drs. Antonio Nebbioso and Sebastian Zühlke in obtaining the Orbitrap mass spectra at INFU is gratefully appreciated.

Competing interests

The authors declare that they have no competing interests.

Funding

Not applicable.

References

1. Chendev YG, Sauer TJ, Ramirez GH, Burras CL. History of east European chernozem soil degradation; protection and restoration by tree windbreaks in the Russian steppe. Sustainability. 2015;7:705–24.
2. Ouni Y, Ghnaya T, Montemurro F, Abdelly C, Lakhdar A. The role of humic substances in mitigating the harmful effects of soil salinity and improve plant productivity. Int J Plant Prod. 2014;8(3):353–74.
3. Fukushima M, Tatsumi K. Degradation pathways of pentachlorophenol by photo-Fenton systems in the presence of iron (III), humic acid, and hydrogen peroxide. Environ Sci Technol. 2001;35(9):1771–8.
4. Hayes MH, Clapp CE. Humic substances: considerations of compositions, aspects of structure, and environmental influences. Soil Sci. 2001;166(11):723–37.

5. Piccolo A. The supramolecular structure of humic substances. Soil Sci. 2001;166(11):810–32.
6. Baalousha M, Motelica-Heino M, Galaup S, Le Coustumer P. Supramolecular structure of humic acids by TEM with improved sample preparation and staining. Microsc Res Tech. 2005;66(6):299–306.
7. Peuravuori J. NMR spectroscopy study of freshwater humic material in light of supramolecular assembly. Environ Sci Technol. 2005;39(15):5541–9.
8. Fedotov GN, Dobrovol'skii GV, Shoba SA. Microphase stratification in supramolecularhumic matrix as a factor forming the nanostructure of soil gels. Dokl Biol Sci. 2011;437(1):128–30.
9. Schmidt MWI, Torn MS, Abiven S, Dittmar T, Guggenberger G, Janssens IA, Kleber M, Kögel-Knabner I, Lehmann J, Manning DAC, Nannipieri P, Rasse DP, Weiner S, Trumbore SE. Persistence of soil organic matter as an ecosystem property. Nature. 2011;478:49–56.
10. Lehmann J, Kleber M. The contentious nature of soil organic matter. Nature. 2015;528:60–8.
11. Nebbioso A, Piccolo A. Basis of a humeomics science: chemical fractionation and molecular characterization of humicbiosuprastructures. Biomacromol. 2011;12(4):1187–99.
12. Drosos M, Nebbioso A, Piccolo A. Humeomics: A key to unravel the humusic pentagram. In: Zanella A, editor. Humusica: towards a unified classification of humus systems, humus manual. Appl Soil Ecol; 2017. https://doi.org/10.1016/j.apsoil.2017.07.027.
13. Nebbioso A, Piccolo A. Advances in humeomics: enhanced structural identification of humic molecules after size fractionation of a soil humic acid. Anal Chim Acta. 2012;720:77–90.
14. Nebbioso A, Vinci G, Drosos M, Spaccini R, Piccolo A. Unveiling the molecular composition of the unextractable soil organic fraction (humin) by humeomics. Biol Fertil Soil. 2015;51(4):443–51.
15. Nebbioso A, Piccolo A, Lamshöft M, Spiteller M. Molecular characterization of an end-residue of humeomics applied to a soil humic acid. RSC Adv. 2014;4(45):23658–65.
16. Nebbioso A, Mazzei P, Savy D. Reduced complexity of multidimensional and diffusion NMR spectra of soil humic fractions as simplified by Humeomics. Chem Biol Technol Agric. 2014;1(1):1.
17. Kramer RW, Kujawinski EB, Hatcher PG. Identification of black carbon derived structures in a volcanic ash soil humic acid by Fourier transform ion cyclotron resonance mass spectrometry. Environ Sci Technol. 2004;38(12):3387–95.
18. Hockaday WC, Grannas AM, Kim S, Hatcher PG. Direct molecular evidence for the degradation and mobility of black carbon in soils from ultrahigh-resolution mass spectral analysis of dissolved organic matter from a fire-impacted forest soil. Org Geochem. 2006;37(4):501–10.
19. DiDonato N, Chen H, Waggoner D, Hatcher PG. Potential origin and formation for molecular components of humic acids in soils. Geochim Cosmochim Acta. 2016;178:210–22.
20. Schulten HR, Sorge-Lewin C, Schnitzer M. Structure of "unknown" soil nitrogen investigated by analytical pyrolysis. Biol Fertil Soil. 1997;24(3):249–54.
21. Bhogal A, Murphy DV, Fortune S, Shepherd MA, Hatch DJ, Jarvis SC, Gaunt JL. Distribution of nitrogen pools in the soil profile of undisturbed and reseeded grasslands. Biol Fertil Soils. 2000;30:356–62.
22. Hannam KD, Prescott CE. Soluble organic nitrogen in forests and adjacent clear cuts in British Columbia, Canada. Can J For Res. 2003;33:1709–18.
23. Chen CR, Xu ZH. On the nature and ecological functions of soil soluble organic nitrogen in forest ecosystems. J Soils Sediment. 2006;6:63–6.
24. Chen HY, Chen LD, Chiang ZY, Hung CC, Lin FJ, Chou WC, Gong GC, Wen LS. Size fractionation and molecular composition of water-soluble inorganic and organic nitrogen in aerosols of a coastal environment. J Geophysical Res: Atmos. 2010. https://doi.org/10.1029/2010JD014157.
25. Nebbioso A, Piccolo A. Molecular characterization of dissolved organic matter (DOM): a critical review. Anal Bioanal Chem. 2013;405(1):109–24.
26. Gao J, Helmus R, Cerli C, Jansen B, Wang X, Kalbitz K. Robust analysis of underivatized free amino acids in soil by hydrophilic interaction liquid chromatography coupled with electrospray tandem mass spectrometry. J Chromatogr A. 2016;1449:78–88.
27. Chen CR, Xu ZH. Analysis and behavior of soluble organic nitrogen in forest soils. J Soil Sediment. 2008;8(6):363–78.
28. Leinweber P, Jandl G, Eckhardt KU, Kruse J, Walley FL, Khan MJ, Blyth RIR, Regier T. Nitrogen speciation in fine and coarse clay fractions of a Cryoboroll-new evidence from pyrolysis-mass spectrometry and nitrogen K-edge XANES. Can J Soil Sci. 2010;90(2):309–18.
29. Knicker H, González-Vila FJ, Polvillo O, González JA, Almendros G. Fire-induced transformation of C-and N-forms in different organic soil fractions from a Dystric Cambisol under a Mediterranean pine forest (Pinus pinaster). Soil Biol Biochem. 2005;37(4):701–18.
30. Zang X, Hatcher PG. A Py–GC–MS and NMR spectroscopy study of organic nitrogen in Mangrove Lake sediments. Org Geochem. 2002;33(3):201–11.
31. Zang X, van Heemst JD, Dria KJ, Hatcher PG. Encapsulation of protein in humic acid from a histosol as an explanation for the occurrence of organic nitrogen in soil and sediment. Org Geochem. 2000;31(7):679–95.
32. Gleixner G, Poirier N, Bol R, Balesdent J. Molecular dynamics of organic matter in a cultivated soil. Org Geochem. 2002;33(3):357–66.
33. McKee GA, Hatcher PG. Alkyl amides in two organic-rich anoxic sediments: a possible new abiotic route for N sequestration. Geochim Cosmochim Acta. 2010;74(22):6436–50.
34. Grandy AS, Strickland MS, Lauber CL, Bradford MA, Fierer N. The influence of microbial communities, management, and soil texture on soil organic matter chemistry. Geoderma. 2009;150(3):278–86.
35. Khan KS, Mack R, Castillo X, Kaiser M, Joergensen RG. Microbial biomass, fungal and bacterial residues, and their relationships to the soil organic matter C/N/P/S ratios. Geoderma. 2016;271:115–23.
36. Drosos M, Nebbioso A, Mazzei P, Vinci G, Spaccini R, Piccolo A. A molecular zoom into soil Humeome by a direct sequential chemical fractionation of soil. Sci Tot Environ. 2017;586:807–16.
37. Hertkorn N, Benner R, Frommberger M, Schmitt-Kopplin P, Witt M, Kaiser K, Kettrup A, Hedges JI. Characterization of a major refractory component of marine dissolved organic matter. Geochim Cosmochim Acta. 2006;70:2990–3010.
38. Piccolo A, Spiteller M, Nebbioso A. Effects of sample properties and mass spectroscopic parameters on electrospray ionization mass spectra of size-fractions from a soil humic acid. Anal Bioanal Chem. 2010;397:3071–8.
39. Drosos M, Piccolo A. The molecular dynamics of soil humus as a function of tillage. Land Degrad Dev. 2018. https://doi.org/10.1002/ldr.2989.
40. Nebbioso A, Piccolo A. Modification of chemical and conformational properties of natural organic matter by click chemistry as revealed by ESI-Orbitrap mass spectrometry. Anal Bioanal Chem. 2015;407:8515–23.
41. Hawkes J, Dittmar T, Patriarca C, Tranvik LJ, Bergquist J. Evaluation of the Orbitrap mass spectrometer for the molecular fingerprinting analysis of natural dissolved organic matter (DOM). Anal Chem. 2016;88:7698–704.
42. Piccolo A, Spaccini R, Drosos M, Vinci G, Cozzolino V. The molecular composition of humus carbon: recalcitrance and reactivity in soils. In: Carlos G, Nannipieri P, Hernandez T, editors. The future of soil carbon, its conservation and formation. Cambridge: Academic Press; 2018. https://doi.org/10.1016/B978-0-12-811687-6.00004-3.
43. Tahiri A, Richel A, Destain J, Druart P, Thonart P, Ongena M. Comprehensive comparison of the chemical and structural characterization of landfill leachate and leonarditehumic fractions. Anal Bioanal Chem. 2016;408(7):1917–28.
44. Katsumi N, Yonebayashi K, Okazaki M, Nishiyama S, Nishi T, Hosaka A, Watanabe C. Characterization of soil organic matter with different degrees of humification using evolved gas analysis-mass spectrometry. Talanta. 2016;155:28–37.
45. Rumpel C, Kögel-Knabner I, Bruhn F. Vertical distribution, age, and chemical composition of organic carbon in two forest soils of different pedogenesis. Org Geochem. 2002;33(10):1131–42.
46. Šmejkalová D, Spaccini R, Piccolo A. Multivariate analysis of CPMAS^{13}C-NMR spectra of soils and humic matter as a tool to evaluate organic carbon quality in natural systems. Eur J Soil Sci. 2008;59:496–504.
47. Koch BP, Witt M, Engbrodt R, Dittmar T, Kattner G. Molecular formulae of marine and terrigenous dissolved organic matter detected by electrospray ionization Fourier transform ion cyclotron resonance mass spectrometry. Geochim Cosmochim Acta. 2005;69:3299–308.
48. Roth VN, Dittmar T, Gaupp R, Gleixner G. Ecosystem-specific composition of dissolved organic matter. Vadose Zone J. 2014. https://doi.org/10.2136/vzj2013.09.0162.

49. Kolattukudy PE. Biopolyester membranes of plants: cutin and suberin. Science. 1980;208(4447):990–1000.
50. Grasset L, Amblès A. Aliphatic lipids released from a soil humin after enzymatic degradation of cellulose. Org Geochem. 1998;29(4):893–7.
51. Fedoros EI, Orlov AA, Zherebker A, Gubareva EA, Maydin MA, Konstantinov AI, Krasnov KA, Karapetian RN, Izotova EI, Pigarev SE, Panchenko AV, Tyndyk ML, Osolodkin DI, Nikolaev EN, Perminova IV, Anisimov VN. Novel water-soluble lignin derivative BP-Cx-1: identification of components and screening of potential targets in silico and in vitro. Oncotarget. 2018;9(26):18578–93.
52. Piccolo A. In memoriam Prof. F.J. Stevenson and the Question of humic substances in soil. Chem Biol Technol Agric. 2016;3:23. https://doi.org/10.1186/s40538-016-0076-2.
53. Six J, Elliot ET, Paustian K. Soil macroaggregate turnover and microaggregate formation: a mechanism for C sequestration under no-tillage agriculture. Soil Biol Biochem. 2000;32:2099–103.

Effect of increasing inoculum ratio on energy recovery from chicken manure for better use in Egyptian agricultural farms

Ouahid Elasri[1*†], Mahmoud Salem[2†], Mohamed Ramdani[3], Oulya Zaraali[4] and Latrach Lahbib[5]

Abstract

Background: Poultry production is one of the fastest growing agricultural activities in Egypt. This activity is delivering large quantities of chicken manure. This organic waste can have a negative impact on the environment and public health in the vicinity of the Egyptian agricultural farms. In this study, we treated this organic waste by anaerobic digestion for achieving the maximum production of biogas with a high percentage of methane. For this, we investigated the different ratio of bacterial inoculum and manure chicken content.

Methods: Seven anaerobic mixtures with different inoculums (1:1, 1:2, 1:3, 1:4, 1:5, 1:6 and 1:7) were tested. All digesters are placed in a constant temperature 35 °C. Biogas and methane production during 50 days of incubation.

Results: The best ratio in terms of biogas produce is the low ratio (1:7) with a production of 73.3 Nml/gVS. But from a qualitative point of view, the largest ratio (1:1) was delivering a biogas with the highest calorific value in terms of methane content with 69.9% of biogas.

Conclusions: The energy valorization of chicken manure depends on the waste/inoculum ratio. The maximum biogas produced is using the lowest ratio of waste and inoculum (1:7) but the highest methane percentage in the biogas is reached by the highest of ratio (1:1). Thus, from a qualitative point of view, the high percentage of methane is obtained for the highest ratio 1:1, however, from a quantitative point of view, the largest production is obtained from the lowest ratio 1:7.

Keywords: Anaerobic digestion, Biogas, Manure chickens, methane, ratio, Calorific value

Background

Poultry production has been one of the fastest growing agricultural activities in Egypt. The poultry sector employment represents approximately 6% of Egypt's 23.7 million labor force in 2003 and more than 15% of the agricultural work force [1]. In 2014, this country has produced 140 millions chicken heads [2]. The growth of this Egyptian activity leads to increasing quantities of chicken manure in farms and breeding units. Therefore, this large production of chicken meat produces large quantities of organic wastes.

This waste must be strictly managed according to the governmental regulations because it delivers high content of nitrogen and pathogenic bacteria, mainly staphylococci and enterobacteriaceae [3, 4]. These quantities of bacteria and nitrogen can be lost through lixiviation or runoff and will be retrieved in groundwater and surface water, causing diseases or epidemics [5, 6]. It is involved too in the production of greenhouse gas, with the big emission factor (0.023 kg $head^{-1}$ $year^{-1}$) [7]. Thus, the chicken manure has a negative impact on the environment and public health.

In other hand, Egypt was exposed to repeated electricity failures because of a combination of increasing demand, natural gas supply shortages, old infrastructure, insufficient generation and transport infrastructure. In addition to that, this country has become dependent on

*Correspondence: elasriouahid@yahoo.fr
†Ouahid Elasri and Mahmoud Salem contributed equally to this work
[1] Biochemistry and Biotechnology Laboratory, Mohamed First University, Oujda, Morocco
Full list of author information is available at the end of the article

hydrocarbon imports. Consequently, this energy deficit has a negative impact on economical trade balance and the country budget [8].

In front of these two problems, the progressive production of chicken's manure and the energy deficit is an ideal situation to develop a green technology that combines valorization of waste and production of energy. Among the current technologies, anaerobic digestion is gaining more importance across the world, thanks to its efficiency as a green treatment [9]. This process is based on the degradation of various organic wastes and substrates in hermetically closed digester [10, 11]. These organic wastes are partially converted by microorganisms to biogas [12, 13].

Biogas is a renewable alternative fuel for internal combustion engines that has several advantages over conventional fuels, including lower costs, reduced levels of harmful emissions and being carbon neutral [14]. Thus, the biogas is delivering energy by cogeneration motor able to supply the farms and the breeding units in green energy [15].

In this work, we treated this Egyptian organic waste by anaerobic digestion. Precisely, we have studied the different ratios of waste and inoculum for producing the ideal quantity of biogas favoring a large high calorific value. This study is the first attempt to study the ratio of waste and inoculum in the Egyptian area.

Methods

Chicken manure and inoculum

Chicken manure contains manure droppings and bedding material. Taken from a commercial poultry farm housing with 6000 broilers per cycle, with an annual manure production 14 tons at Sharqia governorate in Egypt. This organic waste is collected in a 20-l plastic bucket with tight lids, and immediately stored in a freezer at − 20 °C in the laboratory.

The inoculum used in this work is obtained after 50 days of anaerobic digestion of cattle manure at mesophilic condition (35 °C) in our laboratory. All the organic matters in this inoculum are exhausted and the bacteria are very activated [16]. The compositions of chicken manure and inoculum are summarized in Table 1.

Laboratory experiments

In this work, eight anaerobic digestion tests were conducted: 1 control (inoculum only) and 7 experimental (waste with different quantity of inoculum). These tests are based on completely randomized design. The use of a control test allowed us to set up the deduction of the amount of biogas production by the residual organic matter in the inoculum alone. The experimental tests have included the combination of seven different ratios (1:1, 1:2, 1:3, 1:4, 1:5, 1:6 and 1:7) of poultry manure and inoculum, respectively, without dilution with water (Table 2). The tests were prepared with a balance to allow the mass balance by grams, and we assumed an inoculum density of 1 [17]. Consequently, each experimental test had different C/N ratio ranged from 16 to 17.5 and a total solid content ranged from 14 to 32%. Each test was replicated twice.

Table 1 Characterization of substrates and inoculum

Parameter	Unit	Chicken manure	Inoculum
Water content	g	26.89 ± 0.4	95.88 ± 0.3
Total solid	g/100 g	73.11 ± 0.2	04.12 ± 0.1
Volatile solid	g/100 g	81.43 ± 0.2	77.78 ± 0.1
Total carbon	g/100 g	47.23 ± 0.2	45.11 ± 0.1
Total Kjeldahl nitrogen	g/100 g	02.99 ± 0.3	01.70 ± 0.1
C/N ratio	–	16:1	27:1
pH	–	08.89 ± 0.2	08.75 ± 0.1

Mean values originated from three independent replications with standard deviations

Digesters

The eight types of digesters used in this work are conical flasks of 500 ml. Each digester was sealed with a rubber stopper and connected via its biogas outlet to a 100-ml burette glass collecting apparatus (Fig. 1). The fabricated biogas collecting device has a valve which was used to run-off the process.

All digesters are placed in automatically controlled temperature water bath, with a constant temperature of 35 °C. The agitation of these digesters is manually performed for 30 s only once a day for two reasons, (i) to achieve a contact between the microorganisms and waste, and (ii) to prevent scum formation. The dry anaerobic digestion is performed during 50 days per batch mode.

Measurement and analysis of biogas production

Every day, we followed the biogas production by moving acidified saline water in a graduated and an inverted burette connected to the digester [18]. The volumes of biogas produced were adjusted to the standard conditions of pressure and temperature [19].

The content of methane (CH_4) in biogas product was determined as follows. A known volume of the biogas produced (V_1) in a digester was syringed out and injected into serum bottle which contained 20 g l^{-1} of KOH solution (Fig. 2). This serum bottle was shaken manually for 3–4 min so that all the CO_2 and H_2S was absorbed by KOH solution [20]. The volume of the remaining biogas (V_2) which was 99.9% CH_4 in the serum bottle was determined by means of a syringe. The percentage of the report (V_2/V_1) provided the content of CH_4 in this biogas production [21].

Table 2 Experimental protocol for preparing the eight digesters

Ration waste/ inoculum	Poultry manure (g)	Inoculum (g)	Water (g)	C/N ratio	Initial TS (%)
1:1	40±0.1	40±0.1	0	16.13	32.66±0.2
1:2	40±0.1	80±0.1	0	16.44	27.44±0.2
1:3	40±0.1	120±0.1	0	16.74	20.89±0.2
1:4	40±0.1	160±0.1	0	17.02	17.85±0.2
1:5	40±0.1	200±0.1	0	17.28	17.00±0.2
1:6	40±0.1	420±0.1	0	17.53	15.41±0.2
1:7	40±0.1	280±0.1	0	17.76	14.78±0.2

Mean values originated from three independent replications with standard deviations

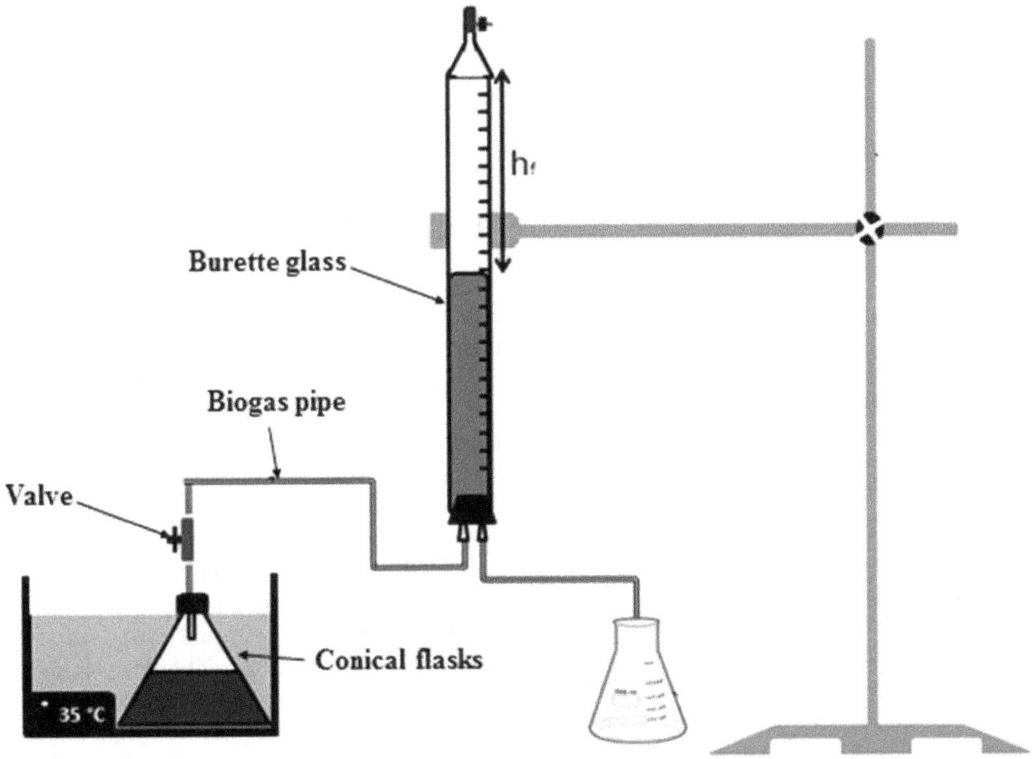

Fig. 1 Measurement of biogas production, (hf: Moving the measurement solution)

Chemical analyses

Total solids (TS) and volatile solids (VS) are considered as the most important qualitative and quantitative criteria for an organic waste treated by anaerobic digestion. The TS was performed according to the standard protocol which consists of drying the fresh matter (FM) at 105 °C to a constant weight. The VS is a gravimetric method based on the mass loss of the dry sample (sample from the determination of TS) in a muffle furnace at 550 °C for 6 h [22].

We measured the pH in digesters with a pH meter (WTW 720 handheld pH meter, Germany), which was calibrated by both standards (pH=4 and pH=7). The pH was determined two times: The start of tests (pH_i) to verify the beginning of the anaerobic digestion and at the end of tests (pH_f), i.e., after the 50 days period of anaerobic digestion.

Statistical analysis

All data collected of this work were subjected to statistical analysis using Statistical Package for the Social Sciences (SPSS). We started these analyses with the Levene test to determine the homogeneity of the variance

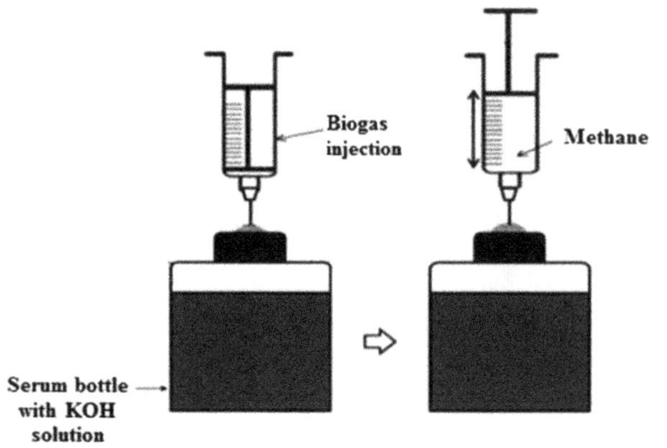

Fig. 2 Determination of the quality of biogas [20]

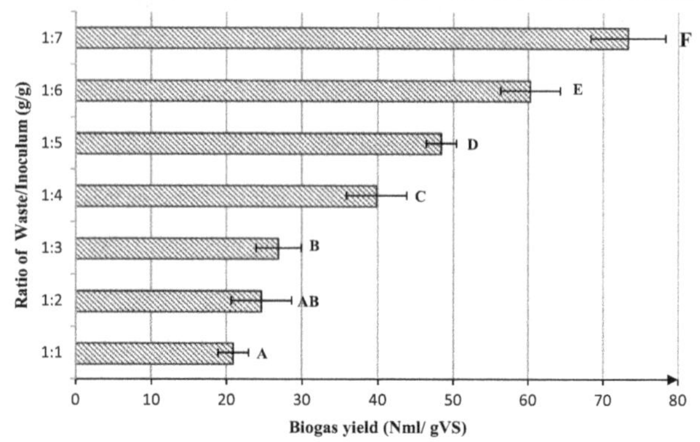

Fig. 3 Comparison of the biogas potential produced by different ratios. Mean values originated from three independent replications. Horizontal bars represent standard deviations. Means with different letters indicate significant differences (Duncan's Test $p < 0.05$)

(Homoscedasticity). Then, we performed one-way analysis of variance (ANOVA) to determine the significant differences for the different parameters (biogas, methane yield and pH). Moreover, the Duncan's multiple range tests were used to establish which ratio is significantly important.

Results and discussion

Statistical analysis

Table 3 illustrates the analysis of variance of methane production, biogas yield and pH. The variances are homogeneous (verified by the Levene test). Thus, we opt for the ANOVA test. This last test shows that the *p* values are all less than 0.001 ($p < 0.001$), which allows us to conclude that the differences observed at the level of methane production, the biogas yield and the fluctuation of the pH are statistically significant.

Comparison of the biogas potential produced

After 50 days of incubation, we observed that all ratios of waste and inoculum did produce biogas. The lowest among these ratios (1:7) is the most important ratio in terms of the quantitative amount of biogas produced 73.3 Nml/gVS compared to other ratios (Fig. 3). As well as the comparison of means by Duncan test showed that there has a significant increase in quantity of biogas in the ratio (1:7).

We also note that the largest C/N ratio (17.76) belongs to this ratio (1:7). Therefore, this biogas is correlated with an excellent C/N ratio. Thus, the latter provides an excellent supply of carbon and nitrogen for the bacterial arsenal [23]. The increase of the waste and inoculum ratio (1:1) leads to two types of inhibition: The first type is by production of ammonia because the introduction of a large amount of nitrogenous substrate in the digester. Abouelenien et al. (2010) confirmed that this ratio of chicken manure and inoculum (1:1) produces a large fraction of ammonium causing a low production of biogas [24]. This ammonia (NH_4^+) product at a deamination during digestion anaerobic, it is able to diffuse passively into the inoculum bacteria mainly the methanogens thanks to its hydrophobic character, creating a proton imbalance [9], leading to a loss of potassium [25] and blocking enzymatic reactions [26]. The second type due to the accumulation of volatile fatty acids (VFAs) resulting from the biodegradation anaerobic of this waste. This is because of the transport and transformation's mechanisms from liquid phase to solid and biological degradation of VFAs is becoming difficult [24, 25]. Hwu et al. [29] recommended that the transportation limitations were causing sludge floatation resulting from adsorption of the long chain of fatty acids into cell walls. Finally, the

Table 3 Analysis of variance of methane production, biogas yield and pH

	Degrees of freedom	Sum of squares	Mean squares	*F*
Methane production	6	1405.9	234.3	101.68***
Biogas yield	6	7011	1168.4	160.9***
pH	6	1.736	0.289	62.12***

*** $p < 0.001$

percentage of waste having access to the cells of inoculum has restricted the production of biogas [30].

On the other hand, this low ration (1:7) contains a small fraction in TS (14.78%) which leads to a low approval in organic matter and nitrogen (Table 2). Thus, this ratio does not exhibit inhibition by ammonium or volatile fatty acids. The lower ratio of waste and inoculum indicates also a lower quantity of ammonia and VFAs within the microorganism present in digester. Thus, the existence of a higher inoculum quantity than organic waste dedicated to the valorisation by anaerobic digestion may dilute the ammonia and VFAs [24–31]. This low ratio also indicated a higher adaptation of microorganism to the substrate content resulting in a higher biogas yield [32].

Some researchers states that this low biogas production may be due to imbalances of hydrolytic, fermentative, and acetogenic bacteria, and methanogenic archaea [33]. Generally, these imbalances are caused by improper C/N ratio, accumulation of VFAs, and high content of total ammonia–nitrogen and free ammonia concentration [34, 35]. Finally, the 1:7 ratio is the optimal ratio for the production of biogas from chicken droppings. On the other hand, the important inoculum load reduces the risk for the installation of inhibition. These results are similar to work such as Neves at 2004, which confirmed that the low ratio is better for anaerobic digestion of other organic waste and substrate [36].

Comparison of the biogas quality produced

The comparison of means by Duncan test showed that the ratio (1:1) has a significant content of methane in biogas produced. Thus, we noted that the high ratio (1:1) has the highest methane content (69.9%) compared with the other ratio studied (Fig. 4).

Therefore, the ratio 1:1 is the most important qualitative ratio because it produces a biogas with a large high calorific value of order (6.9 kWh/m^3). We have already noticed that this large ratio is producing the lowest production of biogas but with an excellent quality.

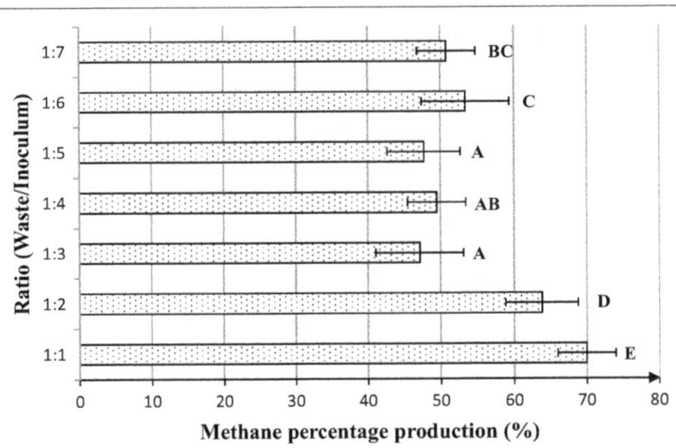

Fig. 4 Comparison of the methane percentage in biogas produced by different ratios of waste/inoculum. Mean values originated from three independent replications. Horizontal bars represent standard deviations. Means with different letters indicate significant differences (Duncan's test $p < 0.05$)

This big production of methane in low biogas production is indicated by the occurrence of few survivals of the anaerobic consortium that are able to digest the accumulated long chain of fatty acids and to reproduce the methane [30]. Nazaitulshila [37] described the same result, by studying the fat and oils where the increase in the ratio to 1:1 makes it possible and has recorded a large percentage of methane. The other researcher such as Chynoweth [38] reported that a maximal methane yields were obtained with ratio of waste and inoculum between 1:2 and 1:1 in anaerobic batch digestion of herbaceous, woody feedstock and municipal wastes [39]. Based on a volatile solid, Li et al. [40] have demonstrated that higher cumulative methane production was observed for the 1:1 ratio of waste and inoculum, when kitchen waste was used as a substrate. Finally, the high ratio of waste and inoculum (1:1) influences methane yields for this organic waste.

The evolution of pH

We observed that all the ratios in this work showed a decrease of the pH level at the end of the anaerobic digestion except for the high ratio (1:1) which showed a pH increase (Fig. 5). The decrease of pH in the ratios (1:2–1:7) is due to the production of organic metabolites in four steps of anaerobic digestion of this waste [41]. On the other hand, the increase of pH in the ratio 1:1 is due to ammonia–nitrogen (NH_3–N) concentration could contribute to the alkalinity and could help to maintain the pH value [42]. But the excess of ammonium can inhibit the destruction of organic compounds, the production of volatile fatty acids and methanogenesis activities [43]. Therefore, this result explains the low production of biogas by the high ratio (1:1).

The ratio 1:7 shows a slight decrease in pH from 7 to 6.93 but this low fall remains in the optimal interval of the biogas production [44]. Finally, the treatment of chicken manure is done according to the intended use, if we want a large production of biogas with a low-calorific value is used the ratio 1:7 but if we need a biogas of small quantity and with a large calorific value we used the large 1:1 ratio.

Conclusion

The treatment of Egyptian manure chicken by anaerobic digestion is feasible and positively recommended. The use of an optimum waste/inoculum ratio is essential for the best energetic valorization of this type of organic waste because the ratio of waste/inoculum is impacting

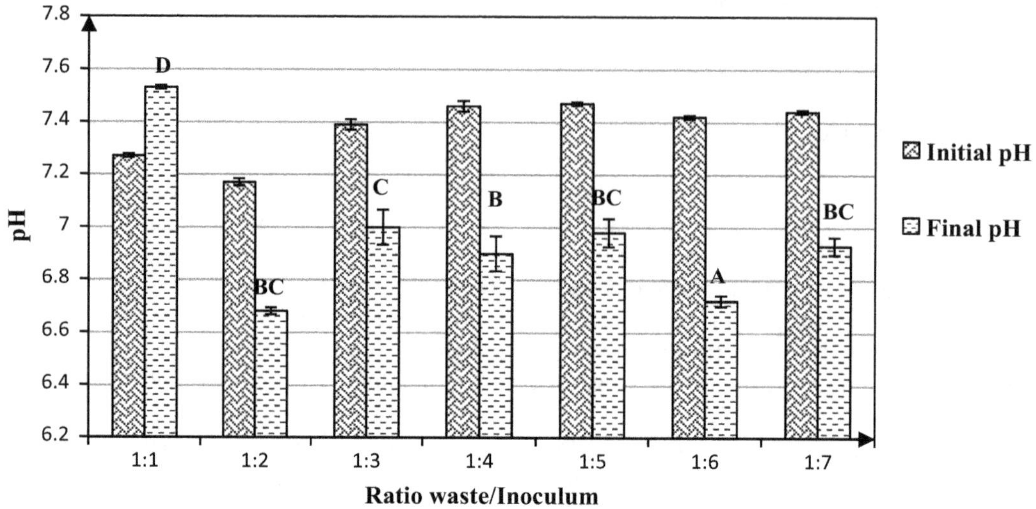

Fig. 5 Comparison of the evolution of pH by different ratios of waste/inoculum. Mean values originated from three independent replications. Horizontal bars represent standard deviations. Means with different letters indicate significant differences (Duncan's test $p < 0.05$)

the biogas and methane production. The maximum biogas volume produced is using the lowest ratio of waste and inoculum (1:7) but the highest methane percentage in the biogas is reached by the highest ratio (1:1). Therefore, from a qualitative point of view, the high percentage of methane is obtained for the highest ratio 1:1; however, from a quantitative point of view, the largest production is obtained from the lowest ratio 1:7. The lesson learnt is that the ratio waste/inoculum was clearly the most important parameters for the treatment of chicken manure and the related production of green energy in Egyptian agricultural farms.

Abbreviations

C/N: Carbon-to-nitrogen ratio; KOH: Potassium hydroxide; CH_4: Methane; TS: Total solids; VS: Volatile solids; FM: Fresh matter; pH: Potential of hydrogen; VFAs: Volatile fatty acids; p: Probability value.

Authors' contributions

OE, MS designed the experiments; MS performed the experiments; OE, MS, MR, OZ and LL wrote this manuscript. All authors read and approved the final manuscript.

Author details

[1] Biochemistry and Biotechnology Laboratory, Mohamed First University, Oujda, Morocco. [2] Agricultural Engineering Research Institute, Agricultural Research Center, Nadi El Said St., Giza, Egypt. [3] Ecology, Water and Environment Laboratory, Mohamed First University, Oujda, Morocco. [4] Applied Biology and Pathology Laboratory, Abdelmalek Essaâdi University, Tetouan, Morocco. [5] Faculty of Sciences Semlalia, Cadi Ayyad University, Marrakech, Morocco.

Acknowledgements

We deeply thank Mrs Hicham Elasri and A. Fauzi Taybi for their technical support to produce this work.

Competing interests

The authors declare that they have no competing interests.

Funding

Not applicable.

References

1. Hosny FA. Poultry sector-country review (Egypt). FAO Animal Production and Health Division: Italy; 2006.
2. FAOSTAT, Food and Agriculture Organization of the United Nations. Statistics division, live animals, Egypt, Stocks, Chickens. http://www.fao.org/faostat/en/#data/QA. Accessed 30 July 2018.
3. Arifin B, Bono A, Janaun J. The transformation of chicken manure into mineralized organic fertilizer. J Sustain Sci Manag. 2006;1(1):58–63.
4. Elasri O, Afilal ME. Potential for biogas production from the anaerobic digestion of chicken droppings in Morocco. Int J Recycl Org Waste Agric. 2016;5:195–204.
5. Ganoulis J. Risk analysis of wastewater reuse in agriculture. Int J Recycl Org Waste Agric. 2012;1(1):1–9.
6. Kostadinova G. Sanitary hygienic assessment of drinking water from underground source at a pig farm. Agric Sci Technol. 2013;5(4):448–54.
7. Jun P, Gibbs M, Gaffney K. CH_4 and N_2O emissions from livestock manure. In: Penman J, Kruger D, Galbally I, editors. Good practice guidance and uncertainty management in national greenhouse gas inventories. Hayama: IPCC Task Force on National Greenhouse Gas Inventories; 2002. p. 321–81.
8. Radwan AF, Hanafy AA, Elhelw M, El-Sayed AE-HA. Retrofitting of existing buildings to achieve better energy-efficiency in commercial building case study: hospital in Egypt. Alex Eng J. 2016;55(4):3061–71.
9. Chen Y, Cheng JJ, Creamer KS. Inhibition of anaerobic digestion process: a review. Bioresour Technol. 2008;99(10):4044–64.
10. El Asri O, Ramdani M, Latrach L, Haloui B, Ramdani M, Afilal ME. Comparison of energy recovery after anaerobic digestion of three Marchica lagoon algae (*Caulerpa prolifera*, *Colpomenia sinuosa*, *Gracilaria bursa-pastoris*). Sustain Mater Technol. 2017;11:47–52.
11. El Asri O, Ramdani M, Latrach L, Haloui B, Mohamed R, elamin Afilal M. Energetic valorization of Nador lagoon algae and proposal to use it as a means of elimination of the eutrophication in this lagoon. Ecol Eng. 2017;103:236–43.
12. Budzianowski WM. A review of potential innovations for production, conditioning and utilization of biogas with multiple-criteria assessment. Renew Sustain Energy Rev. 2016;54:1148–71.
13. Perlatti B, Forim MR, Zuin VG. Green chemistry, sustainable agriculture and processing systems: a Brazilian overview. Chem Biol Technol Agric. 2014;1(1):5.

14. Kim Y, Kawahara N, Tsuboi K, Tomita E. Combustion characteristics and NOX emissions of biogas fuels with various CO_2 contents in a micro co-generation spark-ignition engine. Appl Energy. 2016;182:539–47.
15. Calise F, Cremonesi C, di Vastogirardi GN, d'Accadia MD. Technical and economic analysis of a cogeneration plant fueled by biogas produced from livestock biomass. Energy Procedia. 2015;82:666–73.
16. Van Aarle IM, Perimenis A, Lima-Ramos J, de Hults E, George IF, Gerin PA. Mixed inoculum origin and lignocellulosic substrate type both influence the production of volatile fatty acids during acidogenic fermentation. Biochem Eng J. 2015;103:242–9.
17. Perimenis A, et al. Metabolic profile of mixed culture acidogenic fermentation of lignocellulosic residues and the effect of upstream substrate fractionation by steam explosion. Biomass Convers. Biorefinery. 2015;6(1):25–37.
18. Elasri O, Mahaouch M, elamin Afilal M. The evaluation and the development of three devices for measurement of biogas production. Phys Chem News. 2015;75:75–85.
19. Kafle GK, Kim SH, Sung KI. Ensiling of fish industry waste for biogas production: a lab scale evaluation of biochemical methane potential (BMP) and kinetics. Bioresour Technol. 2013;127:326–36.
20. El Asri O, Hafidi I, elamin Afilal M. Comparison of biogas purification by different substrates and construction of a biogas purification system. Waste Biomass Valorization. 2015;6(4):459–64.
21. Ergüder T, Güven E, Demirer G. Anaerobic treatment of olive mill wastes in batch reactors. Process Biochem. 2000;36(3):243–8.
22. APHA. Standard methods for the examination of water and wastewater. 21st ed. Washington, D.C: American Public Health Association and Water Environment Federation; 2005.
23. Bouteleux C, et al. *Escherichia coli* behavior in the presence of organic matter released by algae exposed to water treatment chemicals. Appl Environ Microbiol. 2005;71(2):734–40.
24. Abouelenien F, Fujiwara W, Namba Y, Kosseva M, Nishio N, Nakashimada Y. Improved methane fermentation of chicken manure via ammonia removal by biogas recycle. Bioresour Technol. 2010;101(16):6368–73.
25. Yenigün O, Demirel B. Ammonia inhibition in anaerobic digestion: a review. Process Biochem. 2013;48(5–6):901–11.
26. Wittmann C, Zeng A-P, Deckwer W-D. Growth inhibition by ammonia and use of a pH-controlled feeding strategy for the effective cultivation of *Mycobacterium chlorophenolicum*. Appl Microbiol Biotechnol. 1995;44(3–4):519–25.
27. Rinzema A, Boone M, van Knippenberg K, Lettinga G. Bactericidal effect of long chain fatty acids in anaerobic digestion. Water Environ Res. 1994;66(1):40–9.
28. Zaccheria F, Mariani M, Ravasio N. The use of rice bran oil within a biorefinery concept. Chem Biol Technol Agric. 2015;2(1):23.
29. Hwu C-S, Tseng S-K, Yuan C-Y, Kulik Z, Lettinga G. Biosorption of long-chain fatty acids in UASB treatment process. Water Res. 1998;32(5):1571–9.
30. Pereira MA, Sousa DZ, Mota M, Alves MM. Mineralization of LCFA associated with anaerobic sludge: kinetics, enhancement of methanogenic activity, and effect of VFA. Biotechnol Bioeng. 2004;88(4):502–11.
31. Mata-Alvarez J, Macé S, Llabrés P. Anaerobic digestion of organic solid wastes. An overview of research achievements and perspectives. Bioresour Technol. 2000;74(1):3–16.
32. Cirne DG, Paloumet X, Björnsson L, Alves MM, Mattiasson B. Anaerobic digestion of lipid-rich waste—effects of lipid concentration. Renew Energy. 2007;32(6):965–75.
33. Brown D, Li Y. Solid state anaerobic co-digestion of yard waste and food waste for biogas production. Bioresour Technol. 2013;127:275–80.
34. Hansen KH, Angelidaki I, Ahring BK. Anaerobic digestion of swine manure: inhibition by ammonia. Water Res. 1998;32(1):5–12.
35. Li Y, Park SY, Zhu J. Solid-state anaerobic digestion for methane production from organic waste. Renew Sustain Energy Rev. 2011;15(1):821–6.
36. Neves L, Oliveira R, Alves MM. Influence of inoculum activity on the biomethanization of a kitchen waste under different waste/inoculum ratios. Process Biochem. 2004;39(12):2019–24.
37. Nazaitulshila R, Idris A, Harun R, Wan Azlina WAKG. The influence of inoculum to substrate ratio on the biochemical methane potential of fat, oil, and grease in batch anaerobic assays. Energy Sources Part Recovery Util Environ Eff. 2015;37(6):590–7.
38. Chynoweth DP, Turick CE, Owens JM, Jerger DE, Peck MW. Biochemical methane potential of biomass and waste feedstocks. Biomass Bioenergy. 1993;5(1):95–111.
39. Yoon Y-M, Kim S-H, Shin K-S, Kim C-H. Effects of substrate to inoculum ratio on the biochemical methane potential of piggery slaughterhouse wastes. Asian Australas J Anim Sci. 2014;27(4):600–7.
40. Li C, Champagne P, Anderson BC. Evaluating and modeling biogas production from municipal fat, oil, and grease and synthetic kitchen waste in anaerobic co-digestions. Bioresour Technol. 2011;102(20):9471–80.
41. Kalyuzhnyi SV. Batch anaerobic digestion of glucose and its mathematical modeling. II. Description, verification and application of model. Bioresour Technol. 1997;59(2–3):249–58.
42. Li Y, Zhang R, Chen C, Liu G, He Y, Liu X. Biogas production from co-digestion of corn stover and chicken manure under anaerobic wet, hemi-solid, and solid state conditions. Bioresour Technol. 2013;149:406–12.
43. Krylova NI, Khabiboulline RE, Naumova RP, Nagel MA. The influence of ammonium and methods for removal during the anaerobic treatment of poultry manure. J Chem Technol Biotechnol. 1997;70(1):99–105.
44. Zhou J, et al. Biogas production and microbial community shift through neutral pH control during the anaerobic digestion of pig manure. Bioresour Technol. 2016;217:44–9.

10

Potential alteration of iron–humate complexes by plant root exudates and microbial siderophores

Assunta Nuzzo[1*], Antonio De Martino[2], Vincenzo Di Meo[2] and Alessandro Piccolo[1*]

Abstract

Background: Two humic acids extracted from a volcanic soil (HA_1) and a leonardite (HA_2) were used to form insoluble complexes with iron metals. To simulate the rhizospheric processes that displace iron from complexes of humic molecules and solubilize the soil humeome, the insoluble iron–humates were treated with a solution of siderophore deferoxamine mesylate (DFOM) and a mixture of citric, oxalic, tartaric and ketoglutaric acids, at different concentrations and contact times.

Results: Results showed that the removal of iron from humic complexes varied depending on the concentration of the extractants and extraction times. At large concentration, the mixture of organic acids was generally a better extractant than the siderophore, probably because of their lower solution pH. However, at smaller concentrations, the extracting capacity of the DFOM solution and the mixture of organic acids was generally similar. Differences in iron extractability between the two iron–humates were attributed to both the humic molecular composition and the steric hindrance of the extractants. Iron was more easily released from complexes formed with HA_1 rich in aliphatic C than from those of HA_2 rich in aromatic C, possibly because the more flexible conformational structure of HA_1 was more accessible to the bulky DFOM than the rigid conformation of iron complexes made by the largely aromatic HA_2.

Conclusions: This work provided evidence that iron–humate complexes may be potentially used to enhance iron nutrition of plants, whose exudates rich in organic acids, together with the siderophores produced by rhizospheric microbes, can displace iron from complexes and enhance its solubility. Concomitantly, the same process may solubilize humic molecules from the same iron–humates, thereby enhancing the concentration of bioactive humic matter in the soil solution.

Keywords: Siderophore, Iron, Humic complexes, Humic acids, Organic acids, Rizosphere, Iron chlorosis, Organic ligands, Iron extraction, Plant root exudates

Background

Iron (Fe) is an essential microelement for the growth and life of plants. Iron chlorosis is a widespread disease that affects plant growth and reduces the yield of many crops [1]. Iron deficiency results in a decrease of chlorophyll in plant leaves, leading to a progressive yellowing of plants, starting from the younger leaves and proceeding to the older parts of plants.

The causes of plant iron deficiency can be different, ranging from nutritional disorder to infections caused by microorganisms. Soil pH plays a key role in iron supply to plants [2], since iron solubility is enhanced in acidic soils and depressed in alkaline soils. In high pH calcareous soils, iron availability can be dramatically reduced [3], and iron chlorosis is widely diffused [4], because soluble Fe may be as low as 10^{-10} M, while a normal requirement for plant growth is $\geq 10^{-7}$ M [5].

*Correspondence: assunta.nuzzo@unina.it; alessandro.piccolo@unina.it
[1] Centro Interdipartimentale di Ricerca sulla Risonanza, Magnetica Nucleare per l'Ambiente, l'Agroalimentare ed i Nuovi Materiali (CERMANU), Università di Napoli Federico II, Via Università, 100, 80055 Portici, NA, Italy
Full list of author information is available at the end of the article

Fertilization with synthetic chelates is the most common practice to solve iron chlorosis, and ethylenediamine-*N, N′*-bis (*o*-hydroxyphenyl) acetic acid (EDDHA) is one of the most efficient iron chelating agents [6]. However, a natural plant survival process is the exudation in the rhizosphere of a variety of natural organic ligands, which can form soluble complexes with Fe^{3+}, thus favoring the mobilization of Fe from soil oxides/hydroxides [7, 8] or Fe–humates [9–11]. Microbial siderophores [12], plant root exudates including phytosiderophores [13], organic acids [14, 15] and phenolic compounds [16, 17] belong to the natural organic ligands that facilitate iron uptake by plants. It is reported that more than 95% of the total plant-available Fe in the soil solution may be represented by this organic Fe pool [18].

An important Fe source in soil is represented by the insoluble Fe complexes with humic substances (HS). Iron complexation by humic substances is attributed to the oxygen-containing functional groups (carboxylic, phenolic and carbonyl) and, to a lesser extent, nitrogen- or sulfur-containing functions [19]. This iron can be made soluble by displacing humic molecules through a ligand exchange mechanism. In fact, the stability constants of Fe complexes with organic acids [20] and with siderophores [21] are generally much larger than those calculated for iron–humate complexes [22].

HS can be defined as ubiquitous natural compounds arising from the chemical and biological degradations of plant and animal residues [23]. A general consensus regards HS as supramolecular associations of heterogeneous and relatively small molecules, which are held together in only apparently large molecular sizes by weak forces, such as hydrogen and hydrophobic bonds [24], and whose conformations can be disrupted by the action of weak organic acids [25]. Based on their solubility in aqueous solutions as a function of pH, HS can be divided into three fractions: humic acids (HA) is the fraction that is soluble under alkaline pH conditions, fulvic acids (FA) is the fraction soluble in water at all pH values, and humin is the insoluble fraction.

Soil iron oxides were recently shown to preferentially sequester oxidized and highly unsaturated organic molecules, rich in aromatic rings, polyphenols and carboxyl groups [26]. The formation of Fe–O–C bonds in stable [27] and large-sized organometallic amorphous complexes [28] had likely the effect to enhance the recalcitrance of the soil humeome that is intended as the totality of humic molecules present in soil. In fact, organic compounds bound to iron were extracted by Humeomics technique, a sequential chemical fractionation developed to characterize the molecular composition of a humified matter, only after stable molecular components had been progressively uncovered by first hydrolyzing ester and ether fractions [29].

The binding to iron enables the persistent fixation of humic molecules in soil, including the nitrogen-containing compounds [30], and, thus the formation of organo-Fe complexes can be reckoned as a major mechanism of molecular stabilization of soil organic carbon. Finally, it is plausible to believe that any plant- or microbe-mediated process that displaces iron from complexes with SOM exposes humic molecules to solubilization and enhances their bioactivity in soil.

A more detailed understanding of the ability of natural organic ligands to displace Fe from complexes with HS may be thus important not only to promote a more efficient iron uptake by crops, but also to solubilize in the rhizosphere humic compounds that may act as plant biostimulants. The aim of this work was hence to assess the solubilization of iron from Fe complexes with two different humic acids by various concentrations of a siderophore and a mixture of organic acids at different times of extraction.

Methods

Humic matter

HS were extracted from: (1) a volcanic soil (Typic Dystrandept) sampled in the caldera of Lake Vico near Rome (Italy), and (2) a North Dakota Leonardite (Mammoth Chem. Co.). Original materials were shaken overnight and in a 0.5 M NaOH and 0.1 M $Na_4P_2O_7$ solution under N_2 atmosphere [31]. The resulting humic acids were precipitated by lowering the pH to 1 with 6 M HCl, purified by three cycles of dissolution in 0.1 M NaOH and precipitation in 6 M HCl, treated with a 0.5% (v/v) HCl–HF solution for 36 h, dialyzed against deionized water until chloride-free (Spectrapore 3 dialysis tubes, 3500 Mw cutoff) and freeze-dried. They were then redissolved in 0.5 M NaOH solution and passed through a strong cation exchange resin (Dowex 50) to further eliminate divalent and trivalent metals and freeze-dried again. An aliquot of samples was subsequently suspended in deionized water and titrated (VIT 90 Videotitrator, Radiometer, Copenhagen, Denmark) to pH 7 with a CO_2-free solution of 0.5 M NaOH. After having reached a constant pH 7, the resulting sodium–humates were left under titration for 2 h, filtered through a Millipore 0.45 μm and freeze-dried. Total acidity, carboxylic and phenolic groups of humic samples were determined according to the methods described by Schnitzer [32]. The elemental composition (C, N, H) of humates was carried out by an EA 1108 Elemental Analyzer (Fisons Instruments).

Solid-state nuclear magnetic resonance (NMR) spectroscopy

A 300 MHz Bruker Avance spectrometer, equipped with a 4 mm wide-bore MAS probe, was used to run solid-state spectra of humic samples. Each fine-powdered sample (5 mg) was packed into a 4 mm zirconium rotor, stoppered by a Kel-F cap and spun at a rate of 13,000 ± 1 Hz. In particular, ^{13}C NMR spectra were acquired through the cross-polarization magic-angle spinning (CPMAS) technique, by using 2 s of recycle delay, 1 ms of contact time, 30 ms of acquisition time and 4000 scans.

Synthesis of iron–humate complexes

Humic samples were added with a 0.8 N $FeCl_2$ solution to completely neutralize, with the Fe^{2+} cations the total acidity of humic solutions (Table 1), thereby precipitating the formed iron–humate complexes. The insoluble complexes were separated from the supernatant by centrifugation, dialyzed and freeze-dried. The total Fe content in complexes was measured by atomic absorption spectrometer (AAS) (Perkin Elmer, Analyst 700) following mineralization of the iron–humates by a HNO_3–$HClO_4$ treatment at 300 °C.

The solid iron–humate complexes were then treated with the following extracting solutions: (1) a mixture of citric, oxalic, tartaric and ketoglutaric acids (Merck KGaA, Germany) and (2) an aqueous solution of the commercially available deferoxamine mesylate (DFOM) (Merck KGaA, Germany), mimicking the naturally occurring siderophore deferoxamine B. An extraction experiment was based on the concentration of 10^{-3} M for both extracting solutions, to simulate the concentration of organic acids in maize root exudates reported in literature [33]. Other extraction experiments were conducted with solution concentrations of an order of magnitude either above or below 10^{-3} M.

For each extractant, 20 mL of solution was added to 50 mg of iron–humate sample placed in screwed-cap plastic vials and shaken on a rotating shaker for 1, 2, 4 and 16 h. The suspensions were then centrifuged and the supernatants separated. A second extraction was conducted by adding to the residue of the first extraction 20 mL more of the extractant and the vials shaken again for the same periods of time. All Fe extractions from complexes were carried out in triplicate. The concentration of iron in the supernatants following the consecutive extractions was determined by AAS.

Results and discussion

Elemental composition, total, carboxyl and phenolic acidities of HA used in this study are reported in Table 1. The humic acid extracted from volcanic soil (HA_1) showed a smaller C content and a larger amount of acidic functional groups than the humic material extracted from leonardite (HA_2). HA contain multiple functional groups that can interact with metal ions through complex formation, as carboxyl and phenolic groups. The larger content of the total acidic functional groups found in HA_1 in comparison to HA_2 may explain the greater amount of Fe complexed by the former material (Table 1) and the establishment of a network of intra- and inter-molecular iron complexes.

Further chemical differences between HA_1 and HA_2 were shown by ^{13}C-CPMAS-NMR spectroscopy (Fig. 1). The NMR spectra were divided into the following intervals: 0–105 ppm (aliphatic C); 105–165 ppm (aromatic C); 165–190 (carboxyl C), and the relative percent distribution of signal areas in the different resonance intervals is reported in Table 2. HA_1 showed a larger content of aliphatic than aromatic carbons, contrary to HA_2, in which the aromatic components were instead greatly prevalent. Moreover, the content of carboxyl carbons was larger in HA_2 than in HA_1.

The percentages of iron extracted from the humic complexes after two consecutive extractions by either DFOM solution or the mixture of organic acids at different concentrations are reported in Tables 3 and 4. The amount of iron extracted varied with the molecular composition of humic materials, the concentration of the extracting solution and the time of extraction. In particular, the amount of iron recovered from HA_1 by extraction with the mixture of organic acids at the concentration of 0.01 M was significantly larger than that recovered with the same concentration of DFOM solution, regardless of the number of extractions and shaking time. These results indicate that the mixture of organic acids was significantly more efficient than the DFOM siderophore in displacing iron from complexes with HA (Table 3).

Table 1 Chemical properties of humic acids

Sample	*C* (%)	*H* (%)	*N* (%)	Ash (%)	Acidity (meq g^{-1})			Complexed Fe (%)
					Total	Carboxyl	Phenolic	
HA_1	50.8	4.2	2.3	1.0	10.7	4.7	6.0	7.57
HA_2	64.2	3.5	1.5	1.5	9.3	4.4	4.9	6.83

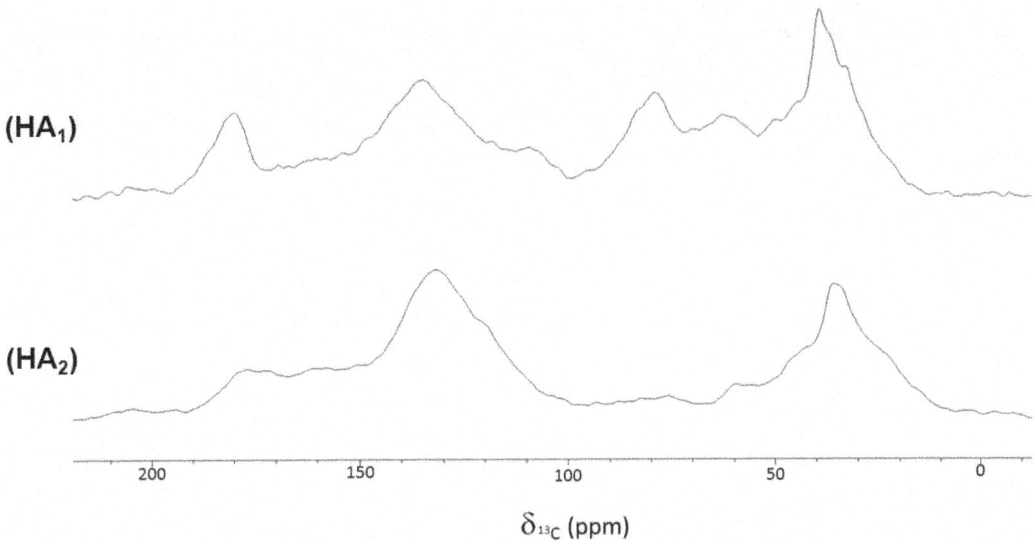

Fig. 1 ^{13}C-CPMAS-NMR spectra of humic acids extracted from a volcanic soil (HA_1) and a leonardite (HA_2)

Table 2 Relative carbon distribution (%) in different regions of chemical shift (ppm) in ^{13}C-CPMAS-NMR spectra of humic samples

Sample	Aliphatic C (0–105 ppm)	Aromatic C (105–165 ppm)	Carboxyl C (165–190 ppm)
HA_1	78.1	16.0	10.2
HA_2	51.7	39.4	16.6

Conversely, the Fe-removing capacity was similar for both extractants at the 0.001 M concentration in both the first and second extractions, except for a slightly larger Fe amount extracted by the organic acids after 16 h of shaking. A similar behavior was observed for the 0.0001 M concentration at all shaking times in the first extraction, whereas in the second extraction the mixture of organic acids was again capable of removing a significantly larger Fe amount than the DFOM solution.

In the case of HA_2, the capacity of both extractants to displace iron from the iron–humate complexes was significantly smaller than for HA_1 (Table 4). The largest iron-removing capacity was also shown in HA_2 by the mixture of organic acids at the 0.01 M concentration in both the first and second extractions and at increasing extraction times. The difference in iron-removing

Table 3 Fe recovered (%) from iron–humate complexes formed with HA_1, after a first (I) and a second (II)[a] extraction after different shaking times (*h*), with a solution of a DFOM siderophore and a mixture of four organic acids: citric, oxalic, tartaric and ketoglutaric acids (Mix) at different concentrations

Extractant	*h*	Concentration					
		0.01 M		0.001 M		0.0001 M	
		I	II	I	II	I	II
DFOM	1	24.3 ab	13.9 a	18.5 bcd	11.4 a	6.1 cde	2.1 a
	2	32.8 b	20.4 ab	21.6 defg	12.3 ab	6.4 cde	1.9 a
	4	53.6 c	22.6 ab	24.1 defgh	15.3 abc	6.9 de	1.9 a
	16	57.6 c	26.4 ab	24.9 efgh	19.6 cdef	6.7 de	1.9 a
Mix	1	68.4 cde	53.9 c	14.4 abc	12.6 ab	6.3 cde	5.0 bc
	2	77.2 def	65.3 cd	25.5 fgh	15.1 abc	7.0 de	5.5 bcd
	4	86.1 ef	95.2 f	28.5 hi	19.1 cde	7.3 e	6.5 cde
	16	88.2 f	78.9 def	33.1 i	26.6 gh	6.8 de	4.4 b

All values are average of three replications. Values followed by the same letter within columns are not significantly different by Tukey's test at the level of $P \leq 0.05$

[a] Fe recovered after the second extraction was calculated on the basis of iron left in the solid iron–humates after the first extraction

Table 4 Fe recovered (%) from iron–humate complexes formed with HA_2, after a first (I) and a second (II)[a] extraction at different hours of shaking (*h*), with a solution of the siderophore DFOM and a mixture of four organic acids: citric, oxalic, tartaric and ketoglutaric acids (Mix) at different concentrations

Extractant	*h*	Concentration					
		0.01 M		0.001 M		0.0001 M	
		I	II	I	II	I	II
DFOM	1	1.3 a	1.8 a	1.0 a	0.8 a	0.7 a	0.9 ab
	2	3.1 a	2.4 a	2.2 ab	1.1 a	1.1 abc	0.7 a
	4	4.0 a	2.7 a	2.6 ab	1.1 a	1.3 bcde	1.2 bcd
	16	9.2 a	7.4 a	3.8 bc	3.3 bc	1.8 f	1.7 ef
Mix	1	26.6 b	25.4 b	5.4 cd	6.5 de	1.1 abc	1.5 cdef
	2	42.0 c	38.1 c	6.9 de	7.7 e	1.5 cdef	2.3 g
	4	53.5 d	37.9 c	7.0 de	10.3 f	1.6 def	2.5 gh
	16	66.7 e	23.3 b	15.8 g	15.5 g	2.8 ih	3.1 i

All values are average of three replications. Values followed by the same letter within columns are not significantly different by Tukey's test at the level of $P \leq 0.05$

[a] Fe recovered after the second extraction was calculated on the amount of iron left in the complexes after the first extraction

Table 5 Values of pH of a siderophore solution (DFOM) and a mixture of four organic acids: citric, oxalic, tartaric and ketoglutaric acids (Mix) at different concentrations

Extractant	Concentration		
	0.01 M	0.001 M	0.0001 M
DFOM	4.44	4.87	5.39
Mix	1.98	2.68	3.44

capacity between the mixture of organic acids and the DFOM solution remained significant at the 0.001 M concentration, whereas it disappeared at the 0.0001 M concentration except for the longest shaking time (16 h) of the first extraction and all shaking times of the second extraction.

These findings suggest that the Fe displacement from humic complexes is governed by different factors. The pH of the extracting solutions is an important parameter that varies with the concentration of solutions (Table 5). In fact, the pH for the mixture of organic acids was significantly lower than for the DFOM solution at all the applied concentrations. This difference in pH values may partially account for the larger Fe recovery observed for the mixture of organic acids with respect to the siderophore solution. Moreover, the presence of different carboxyl functional groups in the mixture of organic acids, which are most effectively metal complexing agents, may further explain the greater Fe extracting capacity of the organic acids mixture as compared to the siderophore.

Other important factors that control the process of alteration of iron–humate complexes are the molecular composition of the humic acids and the possible steric hindrance of the two extractants that may prevent a direct contact with the iron-complexing sites. In fact, the large aliphatic characteristics of HA_1 extracted from soil may confer a conformational flexibility that may favor an easier iron displacement from the humic-complexing groups than for HA_2, whose prevalent aromatic composition and consequent abundance of π–π stacking interactions is instead conducive to a more rigid conformational structure that prevents the accessibility of complexed iron by the extracting solutions. In fact, a physical inhibition to reach the iron-complexing sites may explain the significant reduction of Fe extracted by DFOM when passing from HA_1 to HA_2. Moreover, it may be conceivable that the bulky siderophore molecule, with respect to the simpler structures of organic acids, was sterically hindered from reaching the iron complexed by the HA_2 functional groups.

Conversely, the bulky character of the DFOM molecule may have been less determinant in displacing iron from the more sterically flexible Fe–humate complex in HA_1, thereby accounting the Fe extraction mainly to the large stability of the iron–siderophore complex [21]. In fact, the stability constants of complexes formed between Fe and DFOM are greater than for the iron–humates complexes and for the complexes with the organic acids [20]. Therefore, the large stability of the DFOM–iron complex was the thermodynamic drive that determined the quantitative iron extraction from the iron–humates complexes.

Conclusions

This study showed that iron complexed by humic acids can be displaced and solubilized by the action of microbially produced siderophores or root-exuded organic acids. The major drive for the iron removal from the iron complexes is attributed to the stability constants with the new ligands, which are larger than those between iron and the humic acids. However, the extent of Fe displacement was also dependent on both the molecular composition and consequent conformational flexibility of humic materials and the steric hindrance of ligands. Complexes made by humic acids rich in aromatic carbon were less potentially susceptible to release iron to organic ligands present in the soil solution than complexes formed with alkyl-rich humic acids. In fact, the highly aromatic humic acids are liable to be more conformationally rigid and inhibit the accessibility of the bulky siderophore molecules to the iron-complexing sites.

These findings suggest not only that iron–humate products may be usefully employed to increase plant availability of iron metals in alkaline soils, but also that the humic molecules may be displaced from the iron complexes that stabilize organic matter in soils [30] and play a bioactive role in the rhizosphere. These results hence indicate that both plant-exuded organic acids and microbial siderophores play a fundamental role in regulating the dynamics and turnovers of humic molecules in soil, thereby ultimately controlling both the nutrition of crops and the stability of soil organic matter.

Abbreviations

HA_1: humic acid extracted from a volcanic soil; HA_2: humic acid extracted from a leonardite; DFOM: deferoxamine mesylate; Fe: iron; EDDHA: ethylenediamine-*N*, *N'*-bis (*o*-hydroxyphenyl) acetic acid; HS: humic substances; HA: humic acids; FA: fulvic acids; SOM: soil organic matter; NMR: nuclear magnetic resonance; CPMAS: cross-polarization magic-angle spinning; AAS: atomic absorption spectrometer; Mix: a mixture of four organic acids: citric, oxalic, tartaric and ketoglutaric acids.

Authors' contributions

All authors have contributed equally to the work. All authors read and approved the final manuscript.

Author details

[1] Centro Interdipartimentale di Ricerca sulla Risonanza, Magnetica Nucleare per l'Ambiente, l'Agroalimentare ed i Nuovi Materiali (CERMANU), Università di Napoli Federico II, Via Università, 100, 80055 Portici, NA, Italy. [2] Dipartimento di Agraria, Università di Napoli Federico II, Via Università, 100, 80055 Portici, NA, Italy.

Acknowledgements

The authors thank the European Community's Seventh Framework Programme (FP7/2007–2013) for providing funding.

Competing interests

The authors declare that they have no competing interests.

Funding

This work was partially supported by the project "BIOFECTOR" funded by the European Community's Seventh Framework Programme (FP7/2007–2013).

References

1. Caballero R, Ordovás J, Pajuelo P, Carmona E, Delgrado A. Iron chlorosis in gerber as related to properties of various types of compost used as growing media. Commun Soil Sci Plant Anal. 2007;38:2357–69.
2. Marsh HV Jr, Evans HJ, Matrone G. Investigations of the role of iron in chlorophyll metabolism I. Effect of iron deficiency on chlorophyll and heme content and on the activities of certain enzymes in leaves. Plant Physiol. 1963;38:632.
3. Chen Y, Barak P. Iron nutrition of plants in calcareous soils. Adv Agron. 1982;35:217–40.
4. de Santiago A, Delgado A. Predicting iron chlorosis of lupin in calcareous spanish soils from iron extracts. Soil Sci Soc Am J. 2006;70:1945–50.
5. Lindsay WL, Schwab AP. The chemistry of iron in soils and its availability to plants. J Plant Nutr. 1982;5:821–40.
6. García-Marco S, Torreblanca A, Lucena JJ. Chromatographic determination of Fe chelated by ethylenediamine-*N*-(*o*-hydroxyphenylacetic)-N'-(p-hydroxyphenylacetic) acid in commercial EDDHA/Fe^{3+} fertilizers. J Agric Food Chem. 2006;54:1380–6.
7. Cesco S, Römheld V, Varanini Z, Pinton R. Solubilization of iron by water-extractable humic substances. J Plant Nutr Soil Sci. 2000;163:285–90.
8. Tomasi N, De Nobili M, Gottardi S, Zanin L, Mimmo T, Varanini Z, Römheld V, Pinton R, Cesco S. Physiological and molecular aspects of Fe acquisition by tomato plants from natural Fe complexes. Biol Fertil Soil. 2013;49:187–200.
9. Gerke J. Solubilization of Fe(III) from humic-Fe complexes, humic/Fe-oxide mixtures and from poorly ordered Fe-oxide by organic acids—consequences for P adsorption. J Plant Nutr Soil Sci. 1993;156:253–7.
10. Colombo C, Palumbo G, Sellitto VM, Rizzardo C, Tomasi N, Pinton R, Cesco S. Characteristics of insoluble, high molecular weight Fe–humic substances used as plant Fe sources. Soil Sci Soc Am J. 2012;76:1246–56.
11. Colombo C, Palumbo G, He J-Z, Pinton R, Cesco S. Review on iron availability in soil: interaction of Fe minerals, plants, and microbes. J Soils Sediments. 2014;14:538–48.
12. Neilands JB. Iron absorption and transport in microorganisms. Annu Rev Nutr. 1981;1:27–46.
13. Takagi S, Nomoto K, Takemoto T. Physiological aspect of mugineic acid, a possible phytosiderophore of graminaceous plants. J Plant Nutr. 1984;7:469–77.
14. Römheld V. Existence of two different strategies for the acquisition of iron in higher plants. In: Winkelmann G, van der Helm D, Neilands JB, editors. Iron transport in animal, plants and microorganisms. Weinheim: VCH Chemie; 1987. p. 353–74.
15. Jones DL, Darrah P, Kochian L. Critical evaluation of organic acid mediated iron dissolution in the rhizosphere and its potential role in root iron uptake. Plant Soil. 1996;180:57–66.
16. Cesco S, Neumann G, Tomasi N, Pinton R, Weisskopf L. Release of plant-borne flavonoids into the rhizosphere and their role in plant nutrition. Plant Soil. 2010;329:1–25.
17. Cesco S, Mimmo T, Tonon G, Tomasi N, Pinton R, Terzano R, Neumann G, Weisskopf L. Plant-borne flavonoids released into the rhizosphere: impact on soil bio-activities related to plant nutrition. A review. Biol Fertil Soil. 2012;48:123–49.
18. Van Hees PAW, Lundstrom US. Equilibrium models of aluminium and iron complexation with different organic acids in soil solution. Geoderma. 2000;94:201–21.
19. Evangelou VP, Marsì M. Composition and metal ion complexation behaviour of humic fractions derived from corn tissue. Plant Soil. 2001;229:13–24.
20. Lindsay WL. Chemical equilibria in soils. New York: Wiley; 1979. p. 449.
21. Cline GR, Powell PE, Szaniszlo PJ, Reid CPP. Comparison of the abilities of hydroxamic acid, synthetic and other natural organic acids to chelate iron and other ions in nutrient solution. Soil Sci Soc Am J. 1982;46:1158–64.

22. Stevenson FJ. Humus chemistry: genesis, composition, reactions. 2nd ed. New York: Wiley; 1994.
23. Humus Piccolo A, Conservation Soil. In: Piccolo A, editor. Humic substances in terrestrial ecosystems. Amsterdam: Elsevier; 1996. p. 225–64.
24. Piccolo A. The supramolecular structure of humic substances: a novel understanding of humus chemistry and implications in soil science. Adv Agron. 2002;75:57–134.
25. Piccolo A, Conte P, Cozzolino A. Chromatographic and spectrophotometric properties of dissolved humic substances compared with macromolecular polymers. Soil Sci. 2001;166:174–85.
26. Lv J, Zhang S, Wang S, Luo L, Cao D, Christie P. Molecular-scale investigation with ESI-FT-ICR-MS on fractionation of dissolved organic matter induced by adsorption on iron oxyhydroxides. Environ Sci Technol. 2016;50:2328–36.
27. Yang F, Zhao L, Gao B, Xu X, Cao X. The interfacial behavior between biochar and soil minerals and its effect on biochar stability. Environ Sci Technol. 2016;50:2264–71.
28. Nuzzo A, Sánchez A, Fontaine B, Piccolo A. Conformational changes of dissolved humic and fulvic superstructures with progressive iron complexation. J Geochem Explor. 2013;129:1–5.
29. Drosos M, Nebbioso A, Mazzei P, Vinci G, Spaccini R, Piccolo A. A molecular zoom into soil humeome by a direct chemical sequential fractionation of soil. Sci Tot Environ. 2017;586:807–16.
30. Drosos M, Piccolo A. The molecular dynamics of soil humus as a function of tillage. Land Degrad Dev. 2018;29:1792–805.
31. Piccolo A. Characteristics of soil humic substances extracted with some organic and inorganic solvents and purified by the HCl–HF treatment. Soil Sci. 1988;146:418–26.
32. Schnitzer M. Organic matter characterization. In: Page AL, editor. Methods of soil analysis. 2nd ed. Madison: American Society of Agronomy; 1982.
33. Kraffczyk I, Trolldener G, Beringer H. Soluble root exudates of maize: influence of potassium supply and rhizosphere microorganisms. Soil Biol Biochem. 1984;16:315–22.

11

Biodegradable mulching spray for weed control in the cultivation of containerized ornamental shrubs

Matteo Giaccone[1], Chiara Cirillo[1], Pasquale Scognamiglio[1], Maurizio Teobaldelli[1], Alessandro Mataffo[1], Adriano Stinca[2], Antonio Pannico[1], Barbara Immirzi[3], Gabriella Santagata[3], Mario Malinconico[3] and Boris Basile[1*]

Abstract

Background: Weed control represents a major issue in plant cultivation in containers. Manual weed control is very expensive and the use of chemical herbicide or plastic mulch films has a large environmental impact. The aim of this study was to test the efficacy of an experimental biodegradable chitosan-based mulching spray in controlling weed growth in containers. This research also studied the effect of this mulch on the growth of *Viburnum lucidum* Mill. plants to test for possible phytotoxic effects.

Results: The study compared a total of six treatments derived from three types of weed control (no weed control; herbicide, oxadiazon; mulching spray) applied in containers either filled only with the sterile substrate or filled with the sterile substrate and then artificially inoculated with seeds of the weed species [*Sonchus asper* (L.) Hill subsp. *asper* and *Epilobium montanum* L.]. The mulch controlled the weeds effectively for more than 2 months after its application even under severe weed infestation. The mulching spray controlled the emergence of *S. asper* more efficiently than *E. montanum* plants, probably because the latter has a stronger capacity to penetrate the mulch film during emergence.

Conclusions: Three months after its application, the mulch started to degrade and this allowed some weeds to emerge in the containers, but, in general, the mulch performed better than the herbicide. The chitosan-based mulch did not have any negative effective on the growth of *V. lucidum* plants.

Keywords: Chitosan, *Viburnum lucidum*, *Sonchus asper*, *Epilobium montanum*, Plant biomass

Background

Weed control represents a major issue in plant cultivation [1]. Indeed, weeds can compete with cultivated plants for resources (light, water, nutrients, etc.) [2] and can represent a secondary host for crop pests [3]. In container cultivation, most of these negative effects may be amplified compared to open-field cultivations because of the reduced availability of growing media. Furthermore, the high fertility of the growth substrate often enhances weed growth rate compared to open-field cultivations [4]. Padgett and Frazier [5] estimated that manual weed control in nursery container cultivation requires 1542 h of labor per hectare. Currently, weed control is mainly chemical, but the most widely used herbicides, including those with broad-spectrum activity, cannot guarantee complete coverage against weeds [6]. This lack of full coverage forces operators to perform expensive manual weeding before marketing [1]. In addition, herbicides are phytotoxic [7] and dangerous for both the environment [8] and people [9]. Nowadays, plastic sheets are largely used for mulching in both open-field and container cultivations [10]. However, there is large concern about the use of plastic in agriculture, because of environmental issues [11]. It was estimated that around 700,000 tons of plastic are used annually for agriculture [12] and these

*Correspondence: boris.basile@unina.it
[1] Department of Agricultural Sciences, University of Naples Federico II, Portici, Naples, Italy
Full list of author information is available at the end of the article

materials are often incorrectly disposed of, being dumped in marginal areas or even burned in bonfires.

Lately, researchers have focused on the development of plastic materials with low environmental impact [10, 13]. This has allowed the introduction of biodegradable plastic sheets in open-field crops [14] that, depending on their specific weight can be fully biodegraded in the soil at the end of the plant growing cycle. Similarly, in container cultivation, discs made of different materials (coconut, geotextile fiber, and recycled paper) were introduced to cover container surfaces [10, 15]. Other techniques involve the use of plant materials, such as pine bark [13]. Immirzi et al. [16] developed sprays, made of sodium alginate, able to create a film on the soil resistant to weed stem penetration. The modality of applying this type of spray appeared to be suitable for container cultivation and also for use in nurseries. Sodium alginate-based sprays are soluble in water, but when calcium or other divalent cations is added to a water solution of sodium alginate, this compound gives rise to an insoluble gel. The strong cross-linked network occurring between the divalent cations and sodium alginate may cause the alginate to stiffen and withdrawn, causing surface cracking of the mulching which makes it more vulnerable to weed emergence [16]. Another way of developing these mulching sprays is the substitution of alginate with chitosan. Chitosan is a cationic carbohydrate biopolymer derived from chitin, the second most abundant polysaccharide present in nature after the cellulose. The main sources of chitin are the shell wastes of shrimp, lobsters, and crabs [17]. Due to its film-forming capacity, chitosan may be a suitable candidate when developing films for mulching applications in agricultural activities [18] especially in the case in which traditional films (polyethylene) or biodegradable ones cannot be used [19]. Chitosan readily dissolves in dilute solutions of most organic acids, like acetic acid, while it is insoluble in water. This feature is suitable for chitosan-based film applications on the soil; the water resistance of the films, in fact, could ensure their permanence on the soil for the required cultivation time, especially in such a case where it is not possible to use preformed traditional plastic coverages. Anifantis et al. [20] studied soil temperature under a traditional (LDPE black film), a biodegradable (polylactic film black) and two different black sprayable mulching materials (chitosan cellulosic fibers and galactomannans/agarose black) used in the cultivation of sunflowers in greenhouses. This study suggested that the sprayable mulching materials did not improve heat storage in the soil, whereas this effect is significant under traditional mulching films. Little is known about the performance of mulching sprays based on chitosan in weed control. The aim of this research was to study the effectiveness of an experimental mulching spray based on chitosan in controlling weeds during pot cultivation of an ornamental shrub.

Methods

Location and plant material

The trial was carried out at the experimental fields of the Department of Agricultural Sciences (University of Naples Federico II), located in Portici (Naples, Southern Italy; 40°49′11″N–14°20′28″E). On 01 July 2010, 1-year-old plants of *Viburnum lucidum* Mill. were transplanted in 5.5 L plastic containers (top diameter, 24 cm) filled with a steam-sterilized substrate mix containing 40% of coconut fiber, 40% of peat, and 20% of perlite. Before filling the pots, the substrate was homogeneously mixed with a controlled-release fertilizer $15-8-12+2MgO$ containing B, Cu, Fe, Mn, Mo, and Zn as trace elements (Basacote Plus, COMPO EXPERT Italia Srl) at the dose of 2 kg/m^3 (110 g/pot). No additional fertilization was provided to the plants throughout the rest of the experiment. The containers were placed on plastic mulching sheets. Drip-irrigation was provided daily with one emitter per container (4 L/h each)—located about 5 cm above the substrate—until the first drops of water came out from the pot drainage holes (substrate saturation). No pesticide application was required throughout the experiment.

Experimental design

The study compared a total of six treatments (Table 1) derived from three types of weed control (no weed control, NC; herbicide, H; mulching spray, M) applied in containers either filled with the sterile substrate and then artificially inoculated with seeds of the weed species (IS) or filled only with the sterile substrate that was left non-inoculated (NIS). Therefore, the six treatments were: non-inoculated substrate–no weed control (NIS–NC), non-inoculated substrate–herbicide (NIS–H), non-inoculated substrate–mulching spray (NIS–M), inoculated substrate–no weed control (IS–NC), inoculated substrate–herbicide (IS–H), and inoculated substrate-mulching spray (IS–M). The weed species selected were *Sonchus asper* (L.) Hill subsp. *asper* (*Asteraceae*) and *Epilobium montanum* L. (*Onagraceae*), because they are the most widespread in nurseries in the Campania region (A. Stinca, unpublished observation). The experiment had a complete randomized block design with six treatments and four replications. The experimental unit consisted of five plants (i.e., a total of 120 plants = 5 plants × 6 treatments × 4 replications).

Characteristics of the seed inoculum

Seed inoculum was prepared by mixing 50 achenes of *S. asper* subsp. *asper* and 50 seeds of *E. montanum* with 250 mL of sterile substrate. This mix was added to each inoculated container on top of the sterile substrate. The propagules were collected from plants in nature, air-dried and stored at 4 °C. Seed inoculum was applied in the same date of plant transplant (1 July 2010).

Characteristics of the herbicide

The herbicide used was a granular formulation of oxadiazon at 2% (Ronstar® Bayer Crop Science, Monheim, Germany) that is widely used in the nursery [21, 22]. Herbicide was distributed in each container according to the recommended doses (50 kg/ha) on the same day of plant transplant.

Characteristics of the mulching spray

The mulching spray was prepared as follows: 1.5 g of chitosan at 75% deacetylation degree was dissolved in 100 ml of acetic acid solution (3% vol), then 1.5 g of polyglycerol, 1.5 g of cellulosic fibers (used after sonication process) and 0.2 g of carbon black were added. To study the mechanical properties of the chitosan-based film, a film was prepared by spraying the mulching solution on polystyrene Petri dishes. The obtained film before testing on soil (thickness of 150 microns, a width of 4 mm, and a length of about 28 mm) was characterized by a Young modulus of 1301 ± 41 MPa, a stress at break of 101 ± 6 MPa, a strain at break of $10 \pm 1\%$, and a biodegradability 8% higher than cellulose (unpublished results; biodegradability test was performed by Novamont following ASTM D5988-96 normative). Mechanical tests were performed on six specimens. The product was applied in the containers with a compressed air spray gun (outgoing hole diameter of 3.1 mm) in a quantity of 2 L/m^2 to create on the substrate a film with an estimated thickness of around 150 µm [20]. Before the spray treatments, the soil was moistened up to saturation to avoid any percolation of the sprays. The mulching spray was applied on 1 July 2010.

Similarly, the films were prepared to perform a puncture test. To follow the duration on the soil of the mulching materials, a different preformed disk of chitosan-based films was put on the soil in the same condition of the experimental plot. Six discs were picked up every 2 weeks and performed puncture test on them. This test was carried out with an Instron machine, (model 4301, Instron, Canton, MA, USA) [16, 23]. Briefly, the samples were cut with a punch cutter with a diameter of 42 mm and were trapped in cups fixed on the underside of the Instron instrument. They underwent the action of a force exerted by a spherical dart linked to a steel rod fixed on the upper side of the apparatus. The dart, moving down at a fixed rate of 2 mm/min, penetrated the sample until its rupture. The applied load was recorded as a function of the displacement; the parameters obtained were normalized with respect to the half of the area of the sphere and with respect to an estimated value of the displacement of 10 mm.

Table 1 Treatments used in the experiment and their abbreviations

Treatment	Abbreviation
Non-inoculated substrate–no weed control	NIS–NC
Non-inoculated substrate–herbicide weed control	NIS–H
Non-inoculated substrate–mulching weed control	NIS–M
Inoculated substrate–no weed control	IS–NC
Inoculated substrate–herbicide weed control	IS–H
Inoculated substrate–mulching weed control	IS–M

Measurements

Shoot growth of *V. lucidum* plants was measured on five days: 01, 14, and 29 July; 29 August; and 23 October 2010. The length of all the shoots produced by each plant was measured in all containers.

Weed infestation in the containers was evaluated on the same days when vegetative measurements of *V. lucidum* plants were carried out. In all containers, each live weed plant was counted and identified according to Pignatti [24] and Tutin et al. [25, 26]. On the last three measuring dates (29 July, 29 August, and 23 October), four containers per treatment (one per block) were sampled and all the weed plants collected, separated by species and their dry weight measured after they were dried to constant weight in a ventilated oven (set at 60 °C).

Air temperature, relative humidity, and rainfall were measured at a weather station located about two kilometers from the experimental site.

Statistical analysis

All statistical analyses were performed with SPSS software (SPSS Chicago, IL). The significance of the effects ($P < 0.05$) of the substrate seed inoculation (SSI), the type of weed control (TWC), the time (T) and the SSI × TWC, SSI × T, TWC × T, and SSI × TWC × T interactions on the measured parameters was evaluated by three-way ANOVA using the least significant difference (LSD) as a post hoc test for mean separation ($P < 0.05$).

Results and discussion

During the trial (from July to October), mean daily air temperature ranged between 11.2 and 28.9 °C (Fig. 1), whereas the maximum air temperature was 34.1 °C (16 July). In addition, mean air relative humidity ranged between 41 and 87% and total rainfall was 333.4 mm (with a total of 39 rainy days). Under these weather conditions, the chitosan-based mulching spray was effective in controlling weed emergence even under severe seed infestation (Fig. 2 and Table 2). Indeed, on the first measurement date after treatment application (14 July), 25% and 60% of substrate surface was occupied by weeds in IS–H and IS–NC containers, respectively (Fig. 2a), whereas in the other treatments, the percentage of infestation was either very low (NIS–NC and IS–M) or zero (NIS–H and NIS–M). In IS–H and IS–NC, containers reached maximum percentage of substrate surface covered by weeds on 29 July (around 40 and 70%, respectively), whereas in the containers of the other treatments, this parameter reached the maximum on 29 August (values between 6 and 18%). In all measuring dates between 14 July and the end of the experiment, the percentage of substrate surface covered by weeds was significantly higher in IS–NC than in the other treatments, whereas IS–H containers had intermediate values (significantly higher than IS–M, NIS–NC, NIS–H, and NIS–M containers). Furthermore, from 29 July to 23 October, the cumulated total dry weight of weed plants was significantly higher in IS–NC and IS–H containers than in the other treatments (Fig. 2b and Table 2). The dry weight of weed plants in NIS–H and NIS–M containers was either zero or very low throughout the experiment. These effects included both a delay in weed emergence and a decrease in total number and biomass of weeds per container (Fig. 3 and Table 2). The seasonal pattern of the dry weight of *S. asper* plants was very similar to that already described for the total dry weight of all the weeds (Fig. 3). On 14 and 29 July, the number of live *S. asper* plants counted in IS–NC was significantly higher than in the other treatments, whereas on the same

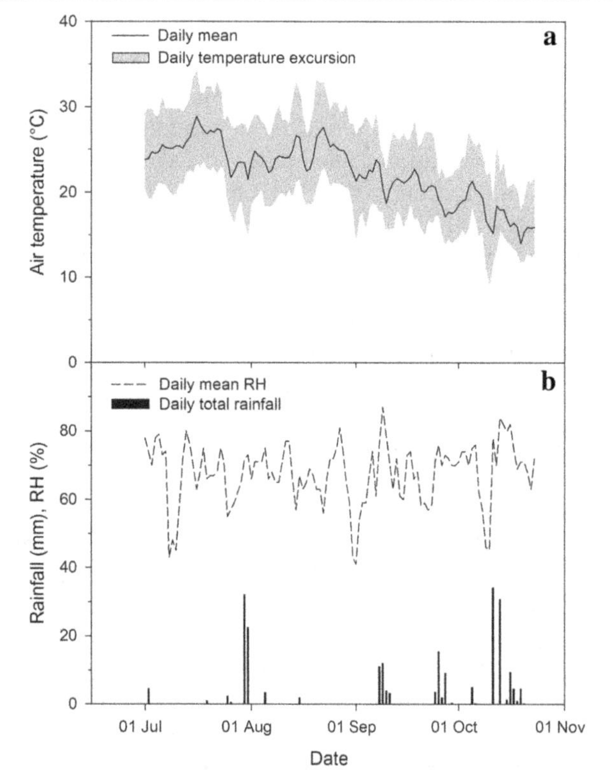

Fig. 1 Seasonal pattern of daily **a** mean, minimum and maximum air temperature and **b** relative air humidity and rainfall

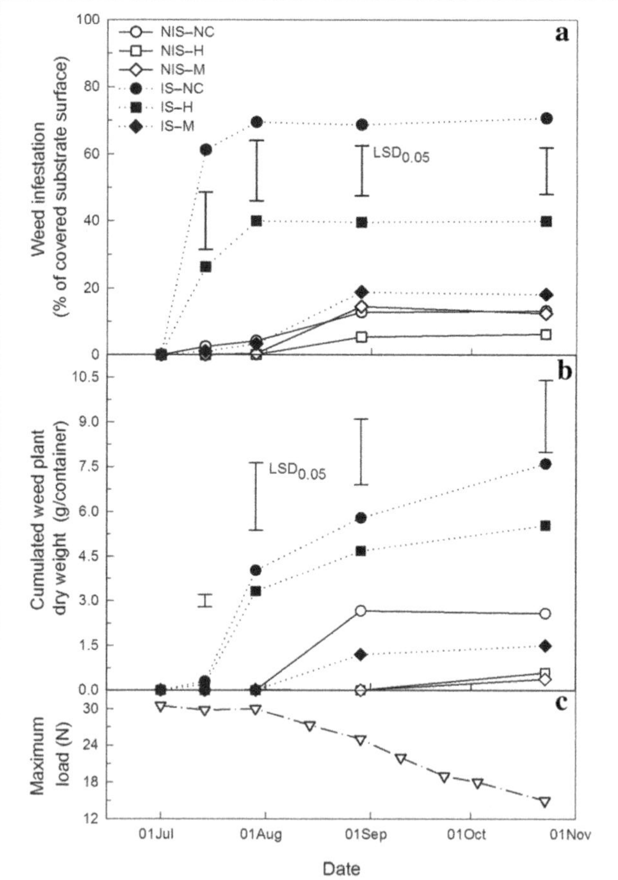

Fig. 2 Seasonal pattern of **a** the percentage of substrate surface infested by weeds and of **b** the cumulated total dry weight of weeds in containers with *Viburnum lucidum* plants grown on non-inoculated substrate (NIS) without weed control (NIS–NC), or with herbicide application (NIS–H), or with spray mulching application (NIS–M); or on substrate inoculated with *Epilobium montanum* and *Sonchus asper* subsp. *asper* seeds (IS) without weed control (IS–NC), or with herbicide application (IS–H), or with spray mulching application (IS–M); and **c** seasonal pattern of the resistance at the puncture test of mulching spray (each measure ± 3N)

days IS–H containers had an intermediate number of this weed plant (Fig. 3 and Table 2). On 9 September in IS–NC and IS–H containers, there was a sudden decrease in the number of alive *S. asper* plants, and this was followed by a second flush of plant germination in IS–NC containers. No differences were found in this parameter between the other treatments throughout the experiment. On 9 September, dry weight of *E. montanum* plants in IS–NC plants was significantly higher than the other treatments (Fig. 3), whereas on the last measuring date this parameter was highest in IS–NC and IS–H, intermediate in IS–M containers and around zero in all the other treatments. From 14 July to the end of the experiment, the number of alive *E. montanum* plants was higher in IS–NC containers than the other treatments (Fig. 3). The number of these weed plants was very similar in IS–H and IS–M containers. Even though previous studies demonstrated that the herbicide used in this study reduced by 96–97% the infestation of plants of the genus *Sonchus* in containers [27], the mulching spray appeared to be more efficient than oxadiazon in controlling these weed species (Fig. 3). The case of *E. montanum* (Fig. 3), which was controlled with similar efficiency by the herbicide and the mulch, was different. This was perhaps because the herbicide controlled *E. montanum* better than *S. asper* subsp. *asper* (Fig. 3), as previously reported [21]. The difference between the efficacy of the mulching spray in controlling the two inoculated weed species (Fig. 3 and Table 2) is quite interesting. Mulching spray fully controlled *S. asper* subsp. *asper* for more than 2 months (from 1 July to 9 September), whereas some *E. montanum* plants had already emerged in mulched containers on the second measuring date (14 July, Fig. 3). This difference in the time of emergence between weed species may be due to a greater capacity of *E. montanum* seedlings to break the polymeric structure created by the mulching sprays. This occurred even though at the beginning of the experiment the mulch was still characterized by high resistance to puncture (Fig. 2c). In this puncture test, the resistance of a dard to penetrate through the sprayable mulch was measured under controlled conditions. This test was designed to simulate the resistance of this material to weed growth. Mechanical performances play a fundamental role in mulching films, as they influence durability and functionality. The mulching films undergo, during the cultivation period, ageing processes due to weathering agents (solar radiations, high air temperature and relative humidity, etc.) and the employment of chemical substances. The prolonged exposure of films to these factors determines simultaneously progressive degradation and photooxidation processes, with a variation of physical and mechanical properties of the films [28]. In a previous study, chitosan-based spray mulch performed better in comparison to mulch based on sodium alginate [16]. Indeed, sodium alginate-based spray mulch was reported to present cracks already 1 month after application and that in this period of time some weeds grew through them [16]. The profile of the maximum load curve (Fig. 2c) is consistent with the loss of plasticizer previously reported [19]. As a result, the coating became more rigid and showed a lower resistance to rupture and displacement.

Some *S. asper* subsp. *asper* plants appeared in mulched containers on the last measuring date (23 October). This was probably (Fig. 2c) caused by the loss of resistance to puncturing on the film that became very rigid

Table 2 Significance of the effects ($P<0.05$) of the substrate seed inoculation (SSI), the type of weed control (TWC), the time (T) and the SSI × TWC, SSI × T, TWC × T, and SSI × TWC × T interactions on the measured parameters evaluated by three-way ANOVAs

Source of variation	Significance						
	% of substrate surface covered by weeds	Weed plant dry weight per container			Number of alive plants per container		Shoot length of *Viburnum lucidum*
		All weeds	*Sonchus asper*	*Epilobium montanum*	*Sonchus asper*	*Epilobium montanum*	
Substrate seed inoculation (SSI)	*<0.001*	*0.045*	*<0.001*	*<0.001*	*<0.001*	*<0.001*	*0.012*
Type of weed control (TWC)	*<0.001*	*0.001*	*0.012*	*0.020*	*0.007*	*<0.001*	0.157
Time (T)	*<0.001*	*<0.001*	*<0.001*	*<0.001*	*<0.001*	*<0.001*	*<0.001*
SSI × TWC	*<0.001*	*0.036*	*0.013*	*<0.001*	*0.002*	*<0.001*	0.053
SSI × T	*<0.001*	0.182	*<0.001*	*<0.001*	*<0.001*	*0.001*	0.670
TWC × T	*<0.001*	0.582	0.620	0.563	0.130	*0.006*	0.681
SSI × TWC × T	*<0.001*	0.729	0.630	0.489	0.168	*0.006*	*0.025*

Significant effects are reported in italic

and/or by the damage caused to the mulch structure by *E. montanum* plants. When mulch started to degrade, it also became partially vulnerable to root penetration of some weed species whose seeds were naturally disseminated on the mulch. This was particularly evident with of *Chamaesyce maculata* (L.) Small, *Erigeron sumatrensis* Retz., *Oxalis corniculata* L., and *Taraxacum officinale* Weber, that reached a maximum frequency higher than one plant per container. *Chamaesyce maculate* and *E. sumatrensis* are considered alien plants of the Italian vascular flora. The time of the first appearance of most of these weed species was after 29 July. On 23 October, the number of live *C. maculate* plants was significantly higher in NIS–NC and NIS–M containers than in the

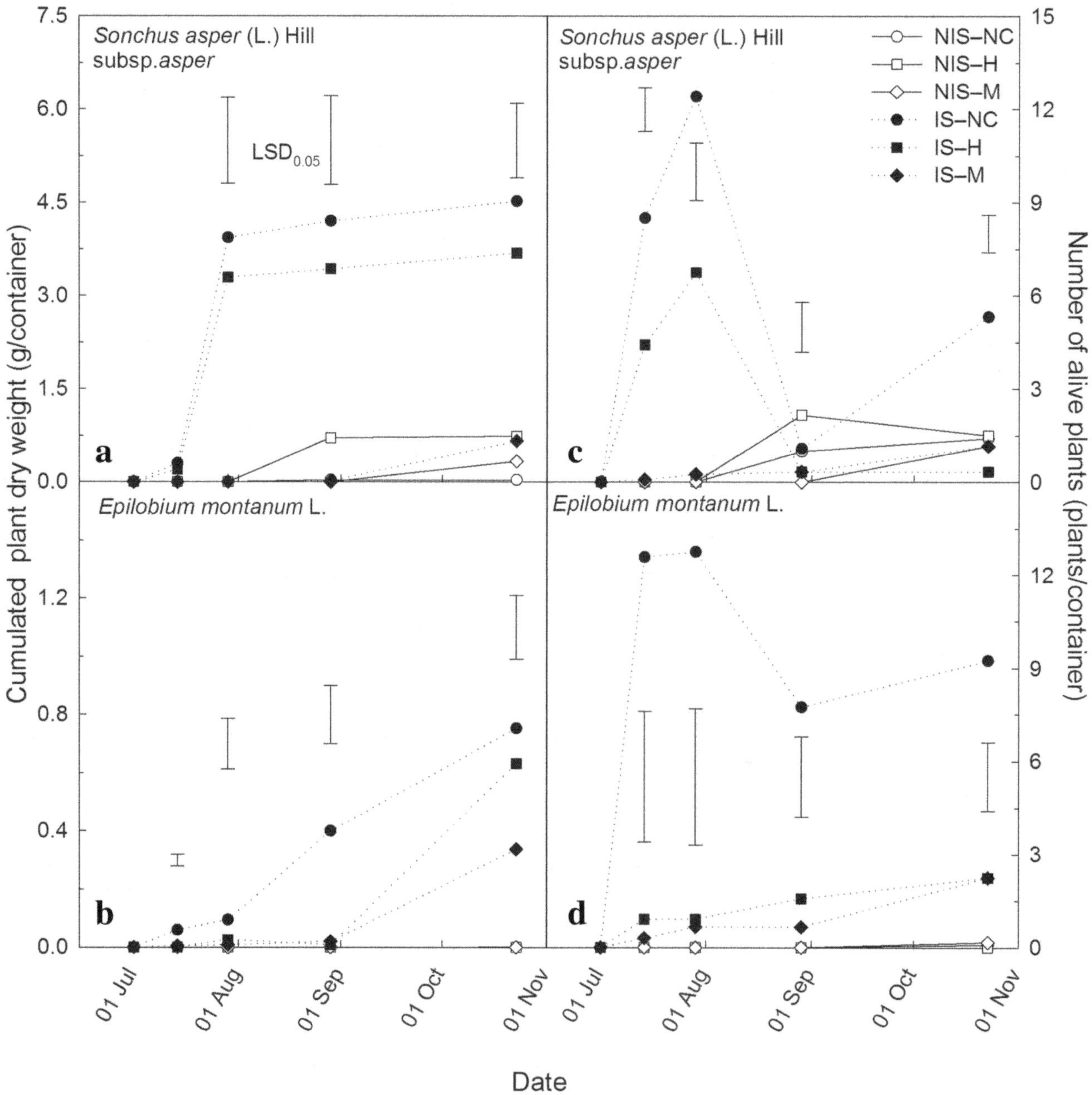

Fig. 3 Seasonal pattern of **a**, **b** dry weight and **c**, **d** number of alive plants of **a**, **c** *Sonchus asper* subsp. *asper* and **b**, **d** *Epilobium montanum* plants in containers with *Viburnum lucidum* plants grown on non-inoculated substrate (NIS) without weed control (NIS–NC), or with herbicide application (NIS–H), or with spray mulching application (NIS–M); or on substrate inoculated with *Epilobium montanum* and *Sonchus asper* subsp. *asper* seeds (IS) without weed control (IS–NC), or with herbicide application (IS–H), or with spray mulching application (IS–M)

other treatments. Similar, results were also found for *O. corniculata* (L.). On the last date of measurements, IS–M containers had the higher number of *E. sumatrensis* than the other treatments. Similarly, on the same date, NIS–M containers had the highest number of *Taraxacum officinale* plants.

In addition, the following weed species were found occasionally: *Portulaca oleracea* L. subsp. *oleracea, Veronica arvensis* L., *Poa annua* L., *Carduus pycnocephalus* L. subsp. *pycnocephalus, Cardamine hirsuta* L., *Silene gallica* L., *Mercurialis annua* L., and *Stellaria neglecta* Weihe.

In the last two measuring dates, plant growth of the cultivated shrub (*V. lucidum*) significantly decreased in inoculated containers when no weed control or herbicide was applied (IS–NC and IS–H, respectively) (Fig. 4). This negative effect on the growth of the cultivated plants can be the result of competition with weed plants for resources (light, water, and nutrients), but also of the release by weeds of root growth-inhibiting chemicals [4]. The latter is generally called allelopathy [29] and has been reported to be stronger in containers [4]. Another possible contribution to this negative effect can be due to the possible phytotoxicity of the herbicide. Different studies have reported that herbicides different from the one used here can have negative effects on the cultivated plant when used at high doses [30]. However, in this study, oxadiazon had no negative effect on *V. lucidum* plant growth when the herbicide was applied in non-inoculated containers (NIS–H treatment, Fig. 4). This suggests that oxadiazon did not have any important phytotoxic effect on the cultivated plant at least at the used recommended dose. Previous studies reported that some herbicides such as atrazine and trifluralin induce stress in the cultivated plant emphasizing the negative effects of allelopathy [31]. This can account for some of the negative effects on *V. lucidum* growth in IS–H containers. The mulching spray did not have any negative effect on the growth of the cultivated plants (Fig. 4).

In this research, the chitosan-based sprayable mulching was evaluated in open field under the typical summer conditions of a Mediterranean climate. The effectiveness of the sprayable mulching in controlling weed growth may also be affected by weather and soil microbiological conditions that may affect its degradation rate as previously reported for chitosan and for other biodegradable mulching materials [32, 33]. This may play an important role in the use of these mulching materials in open-field conditions. More studies are needed to test the performance of this chitosan-based material under different environmental and soil conditions.

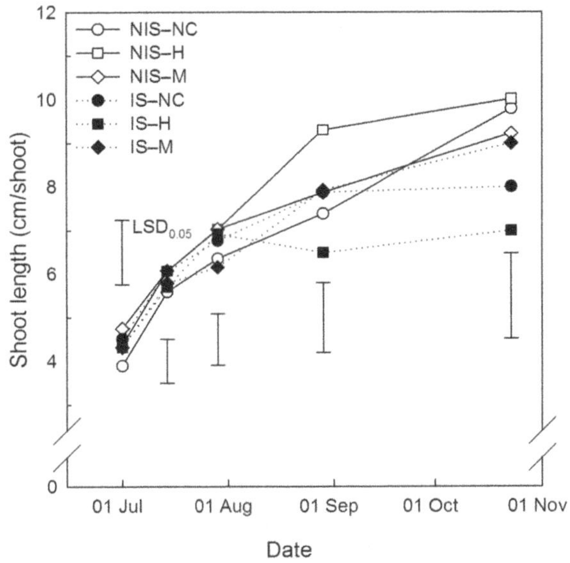

Fig. 4 Seasonal pattern of shoot growth of *Viburnum lucidum* plants grown on non-inoculated substrate (NIS) without weed control (NIS–NC), or with herbicide application (NIS–H), or with spray mulching application (NIS–M); or on substrate inoculated with *Epilobium montanum* and *Sonchus asper* subsp. *asper* seeds (IS) without weed control (IS–NC), or with herbicide application (IS–H), or with spray mulching application (IS–M)

Conclusions

For more than 2 months after its application, chitosan-based mulching spray efficiently controlled weed growth in containers even under severe weed infestation. The mulch inhibited the emergence of *S. asper* subsp. *asper* plants slightly better than *E. montanum* plants. Three months after its application, the mulch film started to degrade and this allowed the growth of some weed plants in the containers, but, in general, the mulch performed better than oxadiazon herbicide. This was also true because in the mulched containers, the cultivated *V. lucidum* plants grew better than in containers where weeds were not controlled or where the herbicide was applied.

Authors' contributions

All authors have contributed substantially to the work. All authors read and approved the final manuscript.

Author details

[1] Department of Agricultural Sciences, University of Naples Federico II, Portici, Naples, Italy. [2] Department of Environmental, Biological and Pharmaceutical Sciences and Technologies, University of Campania Luigi Vanvitelli, Caserta, Italy. [3] Institute of Polymers, Composites, and Biomaterials, CNR, Pozzuoli, Naples, Italy.

Acknowledgements

The authors would like to thank Mr. Giustino Bocchetti for the important technical assistance throughout the experiment, Dr. Claudia Rocco for the help during data collection and analysis, and Dr. Dianna Pickens (University of Naples Federico II) for her help in the English editing.

Competing interests

The authors declare that they have no competing interests.

Funding

No sources of funding were provided for this study.

References

1. Mathers HM. Novel methods of weed control in containers. HortTechnol. 2003;13:28–34.
2. Booth BD, Murphy SDD, Swanton CJ. Weed ecology in natural and agricultural systems. London: CABI Publishing; 2003.
3. Norris RF, Kogan M. Interactions between weeds, arthropod pests, and their natural enemies in managed ecosystems. Weed Sci. 2000;48:94–158.
4. Berchielli-Robertson DL, Gilliam CH, Fare DC. Competitive effects of weeds on the growth of container-grown plants. HortScience. 1990;25:77–9.
5. Padgett JH, Frazier TL. The relationship between costs and pricing of woody ornamentals. Georgia Agric Exp Stn Bull. 1962;100:34.
6. Duke SO, Powles SB. Glyphosate: a once-in-a-century herbicide. Pest Manag Sci. 2008;64:319–25.
7. Staats D, Klett J. Evaluation of weed control and phytotoxicity of preemergence herbicides applied to container-grown herbaceous and woody plants. J Environ Hortic. 1993;11:78–80.
8. Ueji U, Inao K. Rice paddy field herbicides and their effects on the environment and ecosystems. Weed Biol Manag. 2001;1:71–9.
9. Richard S, Moslemi S, Sipahutar H, Benachour N, Seralini GE. Differential effects of glyphosate and roundup on human placental cells and aromatase. Environ Health Perspect. 2005;113:716–20.
10. Chong C. Experiences with weed discs and other nonchemical alternatives for container weed control. HortTechnol. 2003;13:23–7.
11. Kyrikou I, Briassoulis D. Biodegradation of agricultural plastic films: a critical review. J Polym Environ. 2007;15:125–50.
12. Espi E, Salmeròn A, Fontecha A, Garcìa Y, Real AI. Plastic films for agricultural applications. J Plast Film Sheet. 2006;22:85–102.
13. Cochran DR, Gilliam CH, Eakes DJ, Wehtje GR, Knight PR, Olive J. Mulch depth affects weed germination. J Environ Hortic. 2009;27:85–90.
14. Bilck AP, Grossmann MVE, Yamashita F. Biodegradable mulch films for strawberry production. Polym Test. 2010;29:471–6.
15. Smith DR, Gilliam CH, Edwards JH, Olive JW, Eakes DJ, Williams JD. Recycled waste paper as a nonchemical alternative for weed control in container production. J Environ Hortic. 1998;16:69–75.
16. Immirzi B, Santagata G, Vox G, Schettini E. Preparation, characterization and field-testing of a biodegradable sodium alginate-based spray mulch. Biosyst Eng. 2009;102:461–72.
17. Lezica RP, Quesada-Alluè L. Chitin Methods in plants biochemistry. In: Dey PM, editor. Carbohydrates, vol. 2. London: Academic Press Limited; 1990. p. 443–81.
18. Malinconico M, Immirzi B, Massenti S, La Mantia FP, Mormile P, Petti L. Blends of polyvinyl alcohol and functionalized polycaprolactone. A study on the melt extrusion and post-cure of films suitable for protected cultivation. J Mater Sci. 2002;37:4973–8.
19. Santagata G, Schettini E, Vox G, Immirzi B, Mugnozza GS, Malinconico M. Biodegradable spray mulching and nursery pots: new frontiers for research. In: Malinconico M, editor. Soil degradable bioplastics for a sustainable modern agriculture. Berlin: Springer; 2017. p. 105–37.
20. Anifantis A, Canzio G, Cristiano G, De Lucia B, Russo G, Vecchietti L, Immirzi B, Malinconico M, Santagata G. Influence of the use of drip irrigation systems and different mulching materials on ornamental sunflowers in greenhouse cultivation. Acta Hort. 2012;952:385–92.
21. Lamont GP, Spohr LJ. Evaluation of oxadiazon and simazine for weed control and phytotoxicity in container-grown ornamental plants. Sci Hortic. 1988;34:93–9.
22. Hood LR, Klett JE. Preemergent weed control in container-grown herbaceous and woody plants. J Environ Hortic. 1992;10:8–11.
23. Malinconico M, Immirzi B, Santagata G, Schettini E, Vox G, Scarascia Mugnozza GM. An overview of innovative biodegradable materials for agricultural application. In: Moeller HW, editor. Progress in polymer degradation and stability research. Hauppauge: Nova Science Publishers; 2008. p. 69–114.
24. Pignatti S. Flora d'Italia 1-3. Bologna: Edagricole; 1982.
25. Tutin TG, Heywood VH, Burges NA, Moore DM, Valentine DH, Walters SM, Webb DA. Flora Europaea 1. Cambridge: Cambridge University Press; 1964–1980.
26. Tutin TG, Burges NA, Chater AO, Edmondson JR, Heywood VH, Moore DM, Valentine DH, Walters SM, Webb DA. Flora Europaea 1. 2nd ed. Cambridge: Cambridge University Press; 1993.
27. Bailey RE, Simmons JA. Oxadiazon for weed control in woody ornamentals. Weed Sci. 1979;27:396–400.
28. Feuilloley P, Cesar G, Benguigui L, Grohens Y, Pillin I, Bewa H, Lefaux S, Jamal M. Degradation of polyethylene designed for agricultural purposes. J Polym Environ. 2005;13:349–55.
29. Rice EL. Biological control of weeds and plant diseases: advances in applied allelopathy. Norman: University of Oklahoma Press; 1995.
30. Sharman KV. Herbicides for container-grown rain forest species. HortScience. 1993;28:303–5.
31. Einhellig FA. Interaction involving allelopathy in cropping systems. Agron J. 1996;88:886–93.
32. Mostafa HM, Sourell H, Bockisch FJ. Mechanical properties of some bioplastics under different soil types used as biodegradable drip tubes. Agric Eng Int CIGR J. 2010;12:12–21.
33. Li C, Moore-Kucera J, Miles C, Leonas K, Lee J, Corbin A, Inglis D. Degradation of potentially biodegradable plastic mulch films at three diverse US locations. Agroecol Sust Food. 2014;38:861–89.

12

Effects of paraffin external coating on the degradable flower pot made of sludge cake

Udomsak Kongmuang[1], Chaovayut Phornpimolthape[1] and Ikuharu Morioka[2*]

Abstract

Objectives: The researchers have tried to economically make a flower pot from sludge cake produced in paper mill industries for the purpose of decreasing its volume. The crude flower pot did not resist natural sprinkling of water or natural rain. The aims of this study were to clarify two hypotheses: (1) whether life time of a degradable flower pot on the ground would be prolonged or not by coating it with appropriate material, and (2) whether paraffin, material used for coating flower pot, would prevent the toxicity of sludge cake in it.

Methods: Sludge cake was mixed with soil. The circular plaster mold was used as a fixed mold. Three types of coating materials, paraffin wax, tapioca starch and modified tapioca starch were selected. Water durability test was performed with a water sprinkler. As the toxicological testing, leaching test and seed germination test were used using the paraffin-coated flower pot.

Results: In the water durability test, the durability was different among types of the coated flower pot. Among them, the paraffin prolonged the life time of flower pot on the ground until 5 weeks at least and may be 8 weeks. The results of leaching test using the paraffin-coated flower pot showed that three heavy metals, lead, nickel and copper, were lower than the standard in Thailand, but the paraffin had no preventive effect on the leaching of the heavy metals from the sludge cake in the flower pot. The seed germination test showed no significant difference in the number of germination of Chinese kale and the length of its leaf from those grown in the plant soil, and suggested that the paraffin-coated flower pot seemed to have no suppressive effect on the seed germination of Chinese kale.

Conclusion: The paraffin prolongs the life time of flower pot on the ground. Although it has no preventive effect on the toxicity of the sludge cake in the flower pot, the paraffin-coated flower pot seems to have no negative effect on the environment. This paraffin-coated flower pot has the possibility to reuse the sludge cake in the horticultural practice of growing the ornamental plant.

Introduction

Paper mill industries in Thailand generally produce sludge cake after wastewater treatment. This sludge cake is usually disposed into an industrial open landfill [1, 2]. Since it contains some heavy metals, especially lead [3, 4], it becomes one of the sources of pollutants to cause environmental degradation [2, 5] and health problems. To decrease its volume disposed into a landfill, many researchers have tried to recycle rigid materials from it and investigated its safety [6, 7].

The sludge is not a waste, but has a lot of fertilizing values. The application of sludge for agricultural purposes is beneficial because organic matter improves nutrients for the plant germination and growth. Thus, the sludge is used as soil amendment [8]. The sludge compost is economically reliable to grow the plant [9]. However, the sludge contains potentially toxic elements. High levels of toxic elements become the risk of health of plants and humans. When farmers plan to use the sludge, they must

*Correspondence: moriokai@wakayama-med.ac.jp
[2] Graduate School of Health and Nursing Science, Wakayama Medical University, Mikazura 580, Wakayama 641-0011, Japan
Full list of author information is available at the end of the article

confirm whether the sludge is suitable for planting or not [10].

In the previous report [11], the researchers have tried to make a flower pot for reducing the volume of sludge cake economically. They also have clarified that the new flower pot can be made from the sludge cake with clay, and that it has the possibility to have no negative effect on the environment of disposing it. However, the flower pot does not resist natural sprinkling of water or natural rain. Thus, the researchers have discussed on the necessity of coating its surface to use it for a longer time. The material used for coating it would also have possibility to prevent the environmental degeneration and health problems of the sludge cake.

The aims of this study were to clarify two hypotheses: (1) whether life time of a degradable flower pot on the ground would be prolonged or not by coating it with appropriate material, (2) whether material used for coating it would have preventive effect on the toxicity of sludge cake in it or not.

Methods

Cooperative industry

The cooperative industry is a medium-sized (227 workers) paper mill industry in Thailand, producing approximately 4500 m^3 day^{-1} wastewater in 2017. After being treated chemically using aluminum sulfate and biologically using urea, ferric chloride and phosphoric acid, and then mixed with polyacrylamide [11], the wastewater becomes sludge cake (Eurochem, Activated System). The sludge cake (about 5 m^3 day^{-1}) is disposed into the industrial landfill.

Making a flower pot

The components of making a flower pot were a binder and a fixed model [11]. The clay in the soil was used as the binder. The soil was brought from Bang Len, Nakhon Pathom Province, Thailand (Bang Len soil). This contained sand 20.68%, silt 28.30% and clay 51.02% and was classified as heavy clay (clay > 50%). Sludge cake without cleaning (5 kg) was mixed with the Bang Len soil (10 kg) at the set ratio of sludge cake: soil by weight was 1:2 by a paddle [11]. Some water (600 mL) was added. The mixture was mixed again by the paddle until it became mushy.

Two machines were used to make the flower pot [11]. They were mixing machine, in which there were two screw parts to mix homogeneously, and shape forming (jigger) machine. The material mixed by the paddle was poured into the mixing machine. The thoroughly mixed material by the mixing machine (clay slurry) was put into the circular plaster mold as a fixed mold. This plaster mold looked like hollow cone (12.0 cm height, inner diameter 15.6 cm at top and 9.6 cm at bottom). The clay slurry was poured into the plaster mold until about half of its volume. The plaster mold of the clay slurry was set under the arm of jigger machine. As the clay slurry came into contact with the vertically rotating head piece, it was pressed and spread until it finally became the shape of plaster mold.

Formations of flower pot (1.5 cm thickness of wall) were dried in the room for 1 day before removing from the plaster mold. They were dried for one more day in a shadow area and drilled (about 15 mm hole in the bottom) for root outlet and water drainage. After that, they were left out of doors for 14 days. The dried weight of flower pot ranged from 507 to 559 g.

Materials for coating the flower pot

Three types of materials, paraffin wax, tapioca starch and modified tapioca starch were selected. Three flower pots were randomly selected for each type of materials.

Paraffin wax

Paraffin wax is a white or colorless soft solid derivable from petroleum. In a pathology laboratory, this is used to impregnate tissue prior to sectioning thin samples of the tissue [12]. This is also used for candle making, for coating for waxed paper or cloth [13], and, as sealant, for jar, cans and bottles [14, 15]. In this study, the paraffin of good quality in white candle (Candles Lucky, Sahasri-utsahagum Company) was used.

Tapioca starch

Starch accumulates in plants as a complex carbohydrate form. There are various types of starch: the common corn/maize starch, tapioca starch, and wheat, rice and potato starch. The starch is insoluble in cold water, but dissolves in warm water. Dissolved starch in warm water makes starch paste [16]. The starch paste is used as thickening, stiffening or gluing agent. In this study, the tapioca starch (Pine Brand, Sitthinan Company Ltd.) from Central Department Store, Rattanathibet, Nonthaburi Province, Thailand was used.

Modified tapioca starch

Starches are modified to enhance their performance in different applications. Modified starch has properties processed using physical, enzymatical, or chemical treatment [17]. This starch is soluble in the cold water and thickens gels without heat. This is practically used as a thickening agent, stabilizer or emulsifier in food products, as disintegrates in pharmaceuticals, and as a binder in coated paper [18, 19]. In this paper, acid-treated modified tapioca starch (KLIC 10, SMS Corporation Company Ltd.) was used. It has the medium viscosity and dissolves in tap water at about 25 °C.

Coating the surface of flower pot

Paraffin wax

Solid paraffin wax in sheet form (about 7.5 kg) was put into the stainless pot. It was heated by a temperature controller of stove until all paraffin completely became liquid. The temperature of the liquid paraffin was kept at 90–95 °C until the coating process was finished. The temperature was checked by a thermometer every 5 min. The flower pot suspended by a rope was immersed into the liquid paraffin for 10 s. The immersed flower pot was left at the room temperature for at least 1 day. The weight of the paraffin used for coating ranged from 15.0 to 17.8 grams (3 samples: average 16.7 g, 3.1 g per pot by 100 g).

Tapioca starch

Four hundred milliliter hot water at 80 °C was poured on 150 g tapioca starch in a pail with stirring it uniformly and rather quickly. This solution of tapioca starch was heated by a heater up to 205 °C. When stirring the solution became tough and rather difficult, 150 mL water was poured. Stirring it was continued until the solution became milky color, namely the solution became starch paste. When the starch paste had high viscosity by the finger testing, the solution was cooled until about 90 °C. Using a clean brush, the starch paste painted all surface of the flower plot. The painted flower pot was dried at the room temperature for at least 1 day. The weight of the starch paste used for coating ranged from 28.1 to 29.8 grams (3 samples: average 29.0 g, 5.4 g per pot by 100 g).

Modified tapioca starch

One hundred milliliter water at 29 °C was poured on 50 g modified tapioca starch in a pail with stirring it uniformly and rather quickly. When stirring the solution became tough and rather difficult, 150 mL water was poured. In total, 250 mL water was warmed by a heater and stirring was continued at 60 °C for 10 min. When the solution became brown jelly and had high viscosity by the finger testing, namely the solution became modified starch paste, stirring was stopped. Using a clean brush, the modified starch paste painted all surface of the flower plot. The painted flower pot was dried at the room temperature for at least 1 day. The weight of the modified starch paste used for coating ranged from 16.2 to 17.8 grams (3 samples average 17.1 g, 3.2 g per pot by 100 g).

Water durability test

Four types of samples were set for the water durability test.

- Paraffin coat: flower pot coated with paraffin wax (3 units).
- Starch coat: flower pot coated with tapioca starch (3 units).
- Modified starch coat: flower pot coated with modified tapioca starch (3 units).
- No coat: flower pot without coating (3 units).

Twelve flower pots were randomly set in the circle of radius 2.5 m with centering a water sprinkler without the margins of the sprinkled area for the purpose of sprinkling the water equally. The sprinkler watered at a flow rate of 924 L h^{-1}, at the maximum volume, at 10 AM for 1 h every day for 12 weeks. The flower pot was checked after 7-day (a week) sprinkling of water whether or not a serious crack occurred on the outside wall or the bottom surface of flower pot. If there was a 1.5 cm long crack on it, the flower pot was judged to be degraded.

Toxicological testing

According to the results of water durability test, and considering easiness in coating the surface of flower pot (using no brush), only paraffin-coated flower pot was used for toxicological testing. Leaching test and seed germination test were used to check negative effect on the ground and the germination.

The transfer factor of the heavy metal into the green leafy vegetable is the order, Cadmium > Copper > Nickel > Lead [20, 21]. In the previous report [11], the researchers measured the concentrations of 9 elements (Arsenic, Cadmium, Chromium, Copper, Lead, Manganese, Mercury, Nickel, Selenium) in the sludge cake. The concentrations of all elements except copper were quite lower than Thai standard [22]. The cadmium concentration was not more than 0.1 mg kg^{-1} [11]. The researchers then estimated that it was unnecessary to measure the cadmium concentration for the toxicity of sludge cake. Consequently, three heavy metals, lead, nickel and copper, were measured in this study. The concentrations of three heavy metals in sludge cake and Bang Len soil (clay) are shown in Table 1.

Leaching test

Procedures to determine the concentrations of heavy metals in the sample were total element analysis (wet digestion) [23]. Five-gram paraffin-coated flower pot was crushed and digested with 5 mL nitric acid for the total element analysis.

Procedure of leaching test was corresponding to USEPASW846, which complied with the standard method of Notification of Ministry of Industry Subject: Disposal of Wastes or Unusable Materials B.E.2005 [24].

Table 1 Concentration of 3 heavy metals in sludge cake, Bang Len soil (clay), Dindeedee soil (plant soil) and crushed material of paraffin-coated flower pot

Sample	Concentration (mg kg^{-1})		
	Lead (Pb)	Nickel (Ni)	Copper (Cu)
Sludge cake	0.588	0.167	10.40
Bang Len soil (Clay)	2.93	2.14	19.20
100% Soil (Dindeedee soil: plant soil)	0.003	0.001	14.50
50% Soil + 50% FP	0.968	0.668	14.61
100% FP	2.388	1.644	15.45
Standard	400[a]	1600[a]	150[b]

Report from Office of Public Health and Environmental Technology Services, Faculty of Public Health, Mahidol University in March 2017

FP flower pot

[a] Soil Quality Standard, Thailand [22]

[b] Criteria for managing contaminated sites in British Columbia, Canada [25]

Table 2 Characteristics of the sludge cake, Bang Len soil (Clay) and Dindeedee soil (plant soil) used in this experiment

Measure	Sludge cake	Bang Len soil (Clay)	Dindeedee soil (Plant soil)
pH	7.87	4.72	8.56
Moisture (%)	62.1	6.4	27.0
Total organic carbon (mg kg^{-1})	1356	216	7560
Total phosphorus (mg kg^{-1})	79	24	292
Total nitrogen (mg kg^{-1})	378	565	3046
Potassium (mg kg^{-1})	120	774	1351

Report from Central Lab, Faculty of Public Health, Mahidol University, in January 2017

Five-gram paraffin-coated flower pot was also crushed and digested with 5 mL nitric acid for the leaching test.

The concentrations of lead, nickel and copper were measured by atomic absorption spectrophotometer (Flameless type). The detection limit of this method is 0.001 mg kg^{-1}. Three samples were used for these testing.

Seed germination test

In the seed germination test, Chinese kale (*Brassica oleracea* var. *alboglabra*) was selected as the seed. A rectangular plastic basket (23 cm wide, 33 cm long, 8 cm deep) was used. It had many rectangular holes at the four-wall sides and bottom. Coconut coir were set about 2 cm depth on the bottom of the plastic basket for good ventilation and good water drainage. The plant soil was used for the seed germination.

The plant soil was brought from Dindeedee, Lopburi Province, Thailand (Dindeedee soil). This includes the fermented leaves of "Samanea saman" and is suitable for flowering plants, ornamental plants, and every vegetable. Their properties are (1) good friability (high porosity) and good ventilation, (2) high ability to keep moisture constant, and (3) good resistant to change to acid property or alkaline property.

The characteristics of the sludge cake, Bang Len soil (clay) and Dindeedee soil (plant soil) used in this experiment are shown in Table 2.

Crushed material of paraffin-coated flower pot was mixed at the appropriate ratio. Three groups were run. They were:

- 100% Soil (control group): 2 kg of the soil.
- 50% Soil and 50% FP: 1 kg of the soil mixed with 1 kg of crushed paraffin-coated flower pot.
- 100% FP: 2 kg of crushed paraffin-coated flower pot.

In each basket, 50 seeds of Chinese kale were planted in 1 cm depth (2 seeds in each hole). Watering was done at 10 AM and 4 PM every day. Observation period for germination test was set 50 days after planting. Three samples in each group were used for this testing.

Statistical analysis

Since the trials were mostly performed three times, only average was shown in this study. Difference in the average was examined by one-way ANOVA. In the comparison of the number of germination of Chinese kale at the same day in the seed germination test, Bonferroni correction was used for multiple comparisons. In the comparison of the number of germination of Chinese kale with that on the 5th day in the seed germination test, Dunnett correction was used for multiple comparisons.

Data handling and statistical analyses were carried out using SPSS version 24.0 (SPSS Japan). The difference was considered statistically significant for $p < 0.05$.

Results

Water durability test

Table 3 shows the number of remaining flower pot at various times of observation period. The durability was different among four types of the coated flower pot. In weeks 1 and 2, all samples in four types were remaining. In week 3, the no coat type began to break. In week 4, the starch coat and the modified starch coat types also began to break. In week 6, the paraffin coat type began to break. On this week, the other type had only 1 remaining set. In week 7,

Table 3 The number of remaining flower pot against the sprinkled water at various time of observation period

Types	Week											
	1	2	3	4	5	6	7	8	9	10	11	12
Paraffin coat	3	3	3	3	3	2	2	2	1	1	0	0
Starch coat	3	3	3	2	2	1	1	0	0	0	0	0
Modified starch coat	3	3	3	2	1	1	0	0	0	0	0	0
No coat	3	3	2	2	2	1	1	1	0	0	0	0

The number shows the remaining flower pot

Table 4 Result of total element analysis and leaching test of three heavy metals

		Flower pot	Thai Std	Canadian Std
TEA (mg kg^{-1})	Lead (Pb)	9.55	400[a]	375[b]
	Nickel (Ni)	1.29	1600[a]	150[b]
	Copper (Cu)	10.38	–[a]	150[b]
LT (μg L^{-1})	Lead (Pb)	0.88	5000[c]	
	Nickel (Ni)	0.12	20,000[c]	
	Copper (Cu)	0.57	25,000[c]	

Figures show the average ($n=3$)

TEA total element analysis, *LT* leaching test, *Std* standard value specified

[a] Soil Quality Standard, Thailand [22]

[b] Criteria for managing contaminated sites in British Columbia, Canada [25]

[c] Notification of the Ministry of Industry, Thailand [26]

the modified starch coat type had no remaining unit. In week 8, the starch coat type had no remaining unit. In week 9, the no coat type had no remaining unit. On this week, the paraffin coat type had still 1 remaining unit.

Judging from the results of the water durability test and easiness in coating the surface of flower pot, the researchers selected paraffin-coated flower pot as the best one. Consequently, it was used for the further experiment in this study.

Toxicological testing

The concentrations of lead and nickel in the total element test were lower than the value of the Soil Quality Standard, Thailand [22]. Copper is not regulated in this act. The concentration of copper was lower than the value of Canadian standard [25] (Table 4).

The results of leaching test of three heavy metals were lower than the values of Notification of the Ministry of Industry, Thailand [26].

The concentrations of heavy metals in sludge cake, Bang Len soil (clay), 100% Soil (Dindeedee soil: plant soil), 50% Soil and 50% FP, and 100% FP are shown in Table 1. The concentrations of three elements except copper were quite lower (less than 10^{-2}) than Thai standard [22]. The concentration of copper was about 10% compared with Canadian standard [25].

Table 5 shows the average number of germination of Chinese kale at various times of observation period. The number of Chinese kale with 100% Soil reached its peak on the 15th day and continued. The ones with 50% Soil and 50% FP, and 100% FP also showed the peak on the 15th day and continued. There was no significant difference in the number of germination among the three groups.

The researchers measured the length of leaf in millimeter when the germination was found. When there were some leaves, the longest value was used. The leaf growth with 100% soil increased continuously (Table 6). The ones with 50% Soil and 50% FP, and 100% FP also had same tendency. The one with 100% Soil was significantly longer than the one with 50% Soil and 50% FP from 10th day and the one with 100% FP from 20th day. The ones in the two

Table 5 Average number of germination of Chinese kale at various time of observation period

	Days after planting									
	5th	10th	15th	20th	25th	30th	35th	40th	45th	50th
100% Soil	32.3	39.3**	43.7**	43.7**	43.7**	43.7**	43.7**	43.7**	43.7**	43.7**
50% Soil + 50% FP	32.7	40.0**	44.3***	44.3***	44.3***	44.3***	44.3***	44.3***	44.3***	44.3***
100% FP	33.7	39.3*	43.3***	43.3***	43.3***	43.3***	43.3***	43.3***	43.3***	43.3***

Figures show the average ($n=3$)

FP flower pot

* $p<0.05$ ** $p<0.01$ *** $p<0.001$ (compared with 5 th day)

Table 6 Average leaf (mm) growth of Chinese kale at various time of observation period

	Days after planting									
	5th	10th	15th	20th	25th	30th	35th	40th	45th	50th
100% Soil	19.4	32.7	41.0	54.7	65.5*	74.0**	81.2***	86.6***	92.4***	95.4***
50% Soil + 50% FP	17.2	25.2***	30.8***, #	37.3***, ##	40.4***, ##	42.4***, ##	43.2***, ##	44.5***, #	45.0***, #	46.1***, #
100% FP	22.2	30.9*	34.6**	40.7***, #	43.7***, ##	44.1***, ##	45.7***, ##	47.4***, #	48.1***, #	49.0***, #

Figures show the average ($n = 3$)

FP flower pot

* $p < 0.05$ ** $p < 0.01$ *** $p < 0.001$ (compared with 5 th day)

# $p < 0.05$ ## $p < 0.01$ (compared with 100% Soil)

groups, 50% Soil and 50% FP, and 100% FP, showed significantly smaller leaf than the one with 100% Soil. There was no significant difference between 50% Soil and 50% FP and 100% FP.

Discussion

Effect of various coating material on water durability of flower pot

The sprinkled water, 924 L h^{-1}, corresponded to 24 mm h^{-1} rain. Against such rain, paraffin coat had all units until 5 week and enough remained until 8 week when the other coats had no remaining unit. In the previous report [11], the flower pot, corresponding to the no coat type in this study, began to break at 15th day against about 26 mm h^{-1} rain. The results of the no coat type showed almost the same situation. Consequently, the paraffin-coated flower pot showed better water resistance. This may show that the paraffin wax strongly protects the sprinkled water, not only at the outside wall in which the water directly contacted but also at the bottom surface facing water on the ground. However, even the paraffin-coated flower pot cannot be used outdoors in the rainy season as the heavy rain would easily collapse it.

Paraffin wax is solid at the room temperature. Its melting point is between about 46 and 68 °C [13], but begins to melt from approximately 37 °C [27]. The research city was Bangkok, Thailand, and it was higher than 37 °C in some days during research period. On such days, paraffin on the surface of flower pot had the possibility to melt and move down to the ground, and then the flower pot seemed to become weak against the sprinkled water. Even such conditions, the flower pot with paraffin coating may be used until 8 weeks.

Effects of paraffin, coated material, on toxicity of sludge cake in the flower pot

The concentrations of three toxicants, lead, nickel and copper, in the paraffin-coated flower pot in the total element analysis and leaching test were almost same as those reported in the previous report [11], but less than those reported earlier [2, 28]. Since there was no difference in the concentrations of heavy metals in the leaching test between the paraffin-coated flower pot (this study) and the no-coated one (the previous report [11]), the paraffin may have no preventive effect for the heavy metals to leach out. However, the concentrations in the leaching test were much lower than the standard of Thailand [26]. Even if the rain would attack the paraffin-coated flower pot, the heavy metals would little meld out into the ground. Paraffin itself is used for coating the surface of fruits [29] and eatable. Even if the paraffin of the paraffin-coated flower pot would melt down to the ground due to a hot environment, the paraffin seems to have no negative effect on the ground.

Sludge cake has the possibility to have some chemicals [11]. Polyacrylamide used in the process to make sludge cake is relatively non-toxic [30]. Aluminum sulfate, however, is toxic to aquatic organism. Because the sludge cake has the possibility to contain it, the flower pot should be used in the area enough far from the water area and not be disposed into the water area after using it.

In the germination test, the groups containing crushed paraffin-coated flower pot showed no decreasing tendency in the number of germination of Chinese kale. Compared with the previous results [8], the average number of germination of Chinese kale in all three groups corresponded to that in 100% soil or 100% flower pot without paraffin coating. These results showed that paraffin had no suppressive effect on the germination.

Three most important nutrients are nitrogen (N), phosphorus (P), and potassium (K). Nitrogen is used by plants for lots of leaf growth and good green color [31]. Phosphorous is used by plants to help form new roots, make seeds, fruit and flowers. It is also used by plants to help fight disease. Potassium helps plants make strong stems and keep growing fast. It is also used to help fight disease. The concentrations of nitrogen and phosphorous in sludge cake and Bang Len soil were about one tenth of the plant soil. The concentration of potassium in them was less than one-half of the plant soil. Lack in these three nutrients in the 50% Soil and 50% FP, and 100% FP groups may play a major role to introduce the slow growth, namely small leaves maybe due to making fewer roots in the leaf growth in the two groups containing crushed paraffin-coated flower pot. Consequently, the paraffin seems to have no suppressive effect on the leaf growth.

Coating flower pot by paraffin has the advantage point. It enables us to use the flower pot longer than no-coated one. The usage of the paraffin-coated one becomes about 2 months from 1 month of no-coated one in the area under protecting rain in an industrial site. Although the paraffin does not prevent the toxicity of the sludge cake in the flower pot, it itself has no health hazard. It is recommended to coating the flower pot with paraffin for the purpose of using it for a long time.

The researchers should mention possible limitation. Researchers were allowed to use the machine for only a limited time and then failed to make enough number of flower pots for durability test. In the leaching test and germination test, it was impossible to show the standard deviations.

Although the sludge from the industry has the possibil ity to content the harmful bacteria [10], even after both chemical and biological treatments, it was not checked. Since the investigators focused on prolonging the lifetime of flower pot on the ground, the economical convenience of the paraffin-coated flower plots was not evaluated. The durability of flower pot in actual use, such as filled with irrigated soil, was also not estimated. However, this study shows that the paraffin-coated flower pot has the possibility to reuse the sludge cake in the horticultural practice of growing the ornamental plant.

Conclusion

Paraffin prolonged the life of flower pot on the ground until 5 weeks at least and may be 8 weeks. It has no preventing effect on the toxicity of the sludge cake in the flower pot, although the sludge cake in the flower pot seems to have no negative effect on the environment at disposing it. The paraffin-coated flower pot seemed to have no suppressive effect on the seed germination or leaf growth. This paraffin-coated flower pot has the possibility to reuse the sludge cake in the horticultural practice of growing the ornamental plant.

Abbreviation

FP: flower pot.

Authors' contributions

UK conceived the study. UK and CP participated in the design of the study, performed the study, and drafted the manuscript. IM helped to make the design and to draft the manuscript, and performed the statistical analysis. All authors read and approved the final manuscript.

Author details

[1] Faculty of Public Health, Bangkokthonburi University, Khet Taweewatana, Bangkok 10170, Thailand. [2] Graduate School of Health and Nursing Science, Wakayama Medical University, Mikazura 580, Wakayama 641-0011, Japan.

Acknowledgements

The authors impress and appreciate associate professor Piangchan Rojanavipart, Faculty of Public Health, Bangkokthonburi University, for her kindness to give us vulnerable advice, associate professor Rawiwan Sangchai, Dean Faculty of Public Health, Bangkokthonburi University for her kindly support in this research, and Miss Supaporn Pilathong, Miss Jiraphon Sookin, Miss Bangon Kwanyeun, the third year students of Faculty of Public Health, Bangkokthonburi University for taking care Chinese kale spouts during the experiment. Many thanks forward to Associate Professor Krisana Teankaprasith, Head of Central Laboratory, Faculty of Public Health, Mahidol University for the chemical analysis. Again the authors appreciate Miss Pramuan Sunpakawe, Research and Development section, Tenma Paper Mill (Thailand) for preparing new sludge cake.

Competing interests

The authors declare that they have no competing interests.

Funding

There was no funding for the research.

References

1. Tchobanoglous G, Burton FL, David Stensel D. Wastewater engineering: treatment and reuse. 4th ed. McGraw Hill Higher Education: Singapore; 2003.
2. Rashid MT, Barry D, Goss M. Paper mill biosolids application to agricultural lands: benefits and environmental concerns with special reference to situation in Canada. Soil Environ. 2006;25(2):85–98.
3. Thompson G, Swain J, Kay M, Forster CF. The treatment of pulp and paper mill effluent: a review. Bioresour Technol. 2001;77(3):275–86.
4. Pokhrel D, Viraraghavan T. Treatment of pulp and paper mill wastewater–a review. Sci Total Environ. 2004;333(1–3):37–58.
5. U.S. Environmental Protection agency, Landfills. Waste - non-hazardous waster - Municipal solid waste; 2016. Landfillshttps://archive.epa.gov/epawaste/nonhaz/municipal/web/html/landfill.html. Accessed 31 Oct 2018.
6. Okuno N, Ishikawa Y, Shimizu A, Yoshida M. Utilization of sludge in building material. Water Sci Technol. 2004;49(10):225–32.
7. Kongmuang U, Kiykaew D, Morioka I. Construction technique of disposable bin from sludge cake and its environmental risk. Environ Health Prev Med. 2015;20:28–35. https://doi.org/10.1007/s12199-014-0420-8 **(Epub 2014 Nov 7.)**.
8. Tiago Natal-da-Luz T, Tidona S, Jesus B, Morais PV, Sousa JP. The use of sewage sludge as soil amendment. The need for an ecotoxicological evaluation. J Soils Sediments. 2009;9:246–60.
9. Soundhirarajan K, Leena R. Utilization and application of sewage sludge as agricultural compost. Int J Adv Eng Res Develop. 2018;5(4):270–9.
10. Department for Environment Food & Rural Affairs. Sewage sludge in agriculture: code of practice for England, Wales and Northern Ireland (Published 23 May 2018). https://www.gov.uk/government/publications/sewage-sludge-in-agriculture-code-of-practice/sewage-sludge-in-agriculture-code-of-practice-for-england-wales-and-northern-ireland. Accessed 3 Oct 2018.
11. Kongmuang U, Sritanaudomchai H, Morioka I. Potential use of sludge cake from paper mill wastewater treatment as degradable flower pot. Environ Health Prev Med. 2016;21(4):258–64. https://doi.org/10.1007/s12199-016-0523-5 **(Epub 2016 Mar 10.)**.
12. Pathology Labs High Density Shelves for Paraffin and Tissue Block Storage. http://www.southwestsolutions.com/healthcare/pathology-lab-paraffin-and-tissue-block-storage-high-density-shelves. Accessed 3 Oct 2018.

13. Paraffin wax. https://en.wikipedia.org/wiki/Paraffin_wax. Accessed 3 Oct 2018.
14. Nunes MA, Ormsby CM, Patel V, Peng T, Boyd A. Paraffin wax as a sealant in sorptivity testing. J Mater Civil Eng. 2015;27(8):04014218. https://doi.org/10.1061/(ASCE)MT.1943-5533.0000508.
15. How to Make Paraffin Wax. https://www.leaf.tv/articles/how-to-make-paraffin-wax/. Accessed 3 Oct 2018.
16. Tapioca. https://en.wikipedia.org/wiki/Tapioca. Accessed 3 Oct 2018.
17. Vaclavik V, Christian WW. Starches in food. In: Vaclavik V, Christian WW, editors. Essentials of food science. 3rd ed. Berlin: Springer; 2007. p. 49–67.
18. Gotlieb KF, Capelle A. Starch derivatization: Fascinating and unique industrial opportunities. Wageningen: Wageningen Academic Publishers; 2005.
19. Modified starch. https://en.wikipedia.org/wiki/Modified_starch. Accessed 3 Oct 2018.
20. Jayadev, Puttaih ET. Assessment of heavy metals uptake in leafy vegetables grown on long term wastewater irrigated soil across Vrishabhavathi River, Bangalore, Karnataka. J Environ Sci Toxicol Food Technol. 2013;7(6):52–5.
21. Intawongse M, Dean JR. Uptake of heavy metals by vegetable plants grown on contaminated soil and their bioavailability in the human gastrointestinal tract. Food Addit Contam. 2006;23(1):36–48.
22. Pollution Control Department, Ministry of Natural Resources and Environment, Thailand. Soil quality standards, http://www.pcd.go.th/info_serv/reg_std_soil01.html. (in Thai) Accessed 3 Oct 2018.
23. Eaton AD, American Public Health Association (APHA), American Water Works Association (AWWA), & Water Environment Federation (WEF). Method 3030D "Digestion for metal" (Total element analysis & leaching test), In: Standard Methods for the examination of Water & Wastewater (21st Ed). Washington, D.C., USA, APHA-AWWA-WEF, 2005.
24. Hazardous Waste Test Methods/SW-846. https://www.epa.gov/hw-sw846. Accessed 3 Oct 2018.
25. Biomedical Waste in a MOE British Colombia Report names needles called sharps as biomedical waste. http://a100.gov.bc.ca/pub/eirs/finishDownloadDocument.do;jsessionid=3wMMZMyZPBD13LLq7KR8yB70mBx6dJLvqFzYs2TGvMnLRmkzZT3x!-952645365?subdocumentId=3732 Accessed 3 Oct 2018.
26. Notification of Ministry of Industry Re: Industrial Waste Disposal B.E. 2548 (2005) http://www.env.go.jp/en/recycle/asian_net/Country_Information/Law_N_Regulation/Thailand/Re_Industrial_Waste_Disposal_BE2548(2005).pdf. Accessed 3 Oct 2018.
27. Freund M, Mozes G. Paraffin products: properties, technologies, applications. Translated by Jakab E. Amsterdam: Elsevier Science Publishers, 1982. p. 121.
28. Legendre BL, Bischoff KP, Gravois KA, Hendrick RD, Arceneaux AE. Sugarcane yields and soil analysis following application of paper-mill sludge. J Am Soc Sugar Cane Tech. 2004;24:60–9.
29. Verma LR, Joshi VK, editors. Postharvest technology of fruits and vegetables: handling, processing, fermentation and waste management. General concepts and principles, vol. 1. New Delhi: Indus Pub. Co; 2000.

13

Increased rate of potassium fertilizer at the time of heading enhances the quality of direct seeded rice

Anjana J. Atapattu[1,2,3], B. D. Rohitha Prasantha[4*], K. S. P. Amaratunga[5] and Buddhi Marambe[6]

Abstract

Background: Potassium (K) is not easily assimilated into organic matter but helps to improve rice quality. Paddy yield and its quality depend on the correct time of fertilization and harvesting (days after flowering) in the field.

Methods: Changes in the grain quality of (*Oryza sativa* L.) were studied in a field experiment over two dry seasons using three rates of muriate of potash (MOP; 60% K_2O) as 12.5, 25 and 37.5 kg/ha applied at the time of heading (7 weeks after planting—WAP). Paddy samples were harvested during 25, 30 (control), 35 and 40 days after 50% flowering (DAFF). Grain yield and physico-chemical characteristics of grain were studied after harvesting.

Results: The impact of seasons and treatments' interactions was not statistically significant ($P > 0.05$) and, hence, data were averaged over two seasons. Length, breadth, true density and bulk density of rice grains were the highest with 37.5 kg MOP/ha applied at heading and harvested at 30–35 DAFF. Crude protein (6.24%) and crude fat (2.61%) contents in grains were the highest when harvested at 40 DAFF and 35–40 DAFF, respectively. Amylose content decreased with increased MOP rates at the time of heading and delayed paddy harvest. The highest average paddy yield (APY; 6.85 t/ha), head rice yield (HRY; 65%) and total rice milling yield (TMY; 67%) were recorded with 37.5 kg MOP/ha applied at heading of rice plant and paddy harvested at 35 DAFF. The APY, HRY and TMY were also 13.8, 7.7 and 5.9% higher, respectively, compared to the control. Applying K fertilizer at a rate 50% more (18.75 kg K/ha) than the recommended rate at the time of heading (7 WAP) and harvesting paddy at optimum maturity (35 DAFF), which is 5 days later than the recommendation, increase the yield and grain quality of direct seeded rice. Harvesting later than 35 DAFF resulted in a 10.5% loss of HRY ($P < 0.05$).

Conclusions: The present study showed that K fertilizer applied at the rate of 37.5 kg MOP/ha at the time of heading 50% higher than the recommended rate is the best among K fertilizer treatments to obtain the highest HRY.

Keywords: Grain quality, Potassium, Time of heading, Harvesting, Milling yield, Paddy, Head rice yield

Background

Paddy (*Oryza sativa* L) is mostly cultivated as a direct seeded crop in Sri Lanka and the annual production is approximately 4.6 million tons resulting in a surplus of rice (de-husked paddy) since 2009 [1]. The annual per capita consumption of rice in Sri Lanka is approximately 116 kg. Expanding cultivation area and adoption of new high yielding varieties are the major causes for increased productivity. Although increasing the achievable yield from new paddy cultivars is critical for sustainable production, processing and cooking quality of rice has now become important as it determines the acceptability of rice for both export and domestic markets. Changing major agronomic practices such as irrigation, fertilization and harvesting at the correct time are important since they are strongly associated with cooking and processing quality of rice [2].

*Correspondence: bdrp@pdn.ac.lk
[4] Department of Food Science & Technology, Faculty of Agriculture, University of Peradeniya, Peradeniya 20400, Sri Lanka
Full list of author information is available at the end of the article

Field crops generally absorb potassium (K) faster than nitrogen (N) or phosphorous (P), where K also plays an important role in ensuring efficient utilization of N [2]. Potassium is not easily assimilated into organic matter as in the case of N and P, but helps in translocation of photosynthetic products and other plant metabolites, thus contributing to improved grain quality [3, 4]. Though K is not a constituent of any plant structure or compound, K input plays a role in many important regulatory processes in the plant such as the grain quality of rice [5]. Rice and wheat require large quantities of K, and a continuous supply of K is necessary up to heading stage after completion of the reproduction stage [2, 6]. It was shown that 200–300 kg K_2O/ha is necessary to obtain 5–10 t/ha of cereal crop yield [7]. Previously studies have also reported that higher K rates result in a stronger aroma, whiter and glassier appearance and lower softness in basmati rice [8]. Potassium fertilization applied as panicle-fertilizer significantly increases the grain yield and quality [9].

Grain yield and its quality depend on the correct judgment on harvesting. Therefore, correct timing of harvesting is important to obtain a higher milling yield with good quality grains [10] fetching higher value for the production in domestic and foreign markets in Europe and Middle East countries. The optimum time for harvesting may be specified by days after flowering of rice panicle [11]. In Sri Lanka, paddy is generally harvested when 80–85% of the panicles turns into yellow-straw color, which is approximately 30–34 days after completion of flowering [12]. Moisture content at this stage is about 20–22%. Allowing the crop to stand in the field after reaching optimum maturity would reduce the percentage of head rice recovery at milling. Postharvest losses of paddy associated with conventional harvesting and handling (before threshing) operations in the field could vary from 1 to 3% and 2 to 7%, respectively [13], amounting to approximately 0.12–0.4 million tons of the total annual production. Hence, determination of the optimum time for harvesting rice varieties is imperative to obtain higher yields and best grain quality. This study was initiated to identify the impact of different rates of K fertilizer application at the time of heading of rice plant and times of harvesting on the final yield and grain quality of rice.

Methods

Crop establishment and growth conditions

Field experiments were carried out during two cultivating seasons (*Yala* season; March to August) in subsequent years, at the seed farm of the CIC Agri Businesses Private Limited (http://www.cic.lk). The experimental site was located in the agro-ecological region "Low Country Dry Zone (LD)" of Sri Lanka (7°30′ N and 80° 50′ E). The minimum night temperature of the site varied from 26 to 28 °C and the maximum day temperature was in the range of 30–33 °C during the cultivating season of both years. The annual rainfall ranged from 1060 mm to 1805 mm. The soil was a low humic glay with an average pH of 5.7 at 0–20 cm depth. The average soil nutrient composition was 14 kg N, 709 kg P_2O_5, 215 kg K_2O, 5.4 kg Zn, and 297 kg Mg per hectare, with approximately 1.4% organic matter. Certified seed paddy (*Oryza sativa* L) variety CIC 300 obtained from "CIC Agri Businesses" was used for this experiment in both seasons. The paddy variety CIC 300, which matures approximately within 3 months, is with low shattering panicles, red pericarp, and long-grain shape known as basmati type. The variety is produced mainly for the export market requirement.

After land preparation, the experimental area was demarcated by plots at a size of 5 m × 4 m using bunds of 0.35 m width and 0.4 m height. The water outlets of plots were prepared ensuring that the drained water will not enter through an inlet of another plot. Pre-germinated seeds of rice were broadcasted as direct seeded rice at the rate of 100 kg/ha while soil moisture was at the field capacity. The total experimental area was about 2000 m^2 (0.2 ha). Irrigation, fertilization and other management practices were done according to the recommendations of the Department of Agriculture (DOA), Sri Lanka (http://www.agridept.gov.lk/), except for potassium fertilizer treatments and time of harvesting.

The crop production package for the paddy variety CIC 300 included the fertilizer recommendation made by the DOA for a rice crop grown in the LD with an average yield of 5 t/ha [14]. The DOA [14] recommended the application of a pre-plant, basal fertilizer mixture comprising urea (46% N): triple super phosphate (TSP; 46% P_2O_5): muriate of potash (MOP; 60% K_2O) in a ratio of 12.5:62.5:37.5 (kg/ha, w/w/w). As the first and second top dressings, urea was applied 2 and 5 weeks after planting (WAP) at 62.5 kg/ha and 100 kg/ha, respectively.

A mixture of 50 kg/ha of urea and 25 kg/ha of MOP was applied 7 WAP (time of heading) as recommended by the DOA. The rate of MOP applied 7 WAP was varied in this experiment as follows: as potassium (K) fertilizer supplement at the time of heading was considered the critical time of heading in rice, three levels of MOP applications at the time of heading were evaluated in this study, namely 12.5 kg MOP/ha (≈6.25 kg K/ha), 25 kg MOP/ha (≈12.5 kg K/ha) and 37.5 kg MOP/ha (≈18.75 kg K/ha). The treatment 25 kg MOP/ha served as the control (recommendation made by the DOA for about 20 years).

The paddy yield was harvested at four different maturity periods (harvesting ages) at 25 days after 50% flowering (DAFF), 30 DAFF, 35 DAFF and 40 DAFF, at each level of MOP application. In the present study, the rice variety reached 50% flowering stage in about 66 days after planting in both seasons (years). Many rice varieties in Sri Lanka complete flowering after 2–3 days from reaching the 50% flowering stage. A previous study had identified the correct time for paddy harvest as 30–34 days of post-completion of flowering [12]. Hence, in the present study, 30 DAFF was considered as the control based on the recommendation of the DOA as the optimum time to harvest paddy in Sri Lanka.

Sampling

Samples were collected using a 2 m × 2 m quadrant placed at the center of plots and the plants were harvested manually. Threshing was done manually by stripping off the grains from the panicle. All paddy samples were sun dried to have approximately 13% moisture (wet basis). The samples were packed in perforated polypropylene sacks and kept at 10 °C until further analysis.

Grain quality and yield measurements

Paddy samples obtained from each treatment site (MOP × DAFF) were weighed to obtain the average paddy yield (APY). The true density of paddy (g/cm^3) and bulk density (g/cm^3) were determined according to the previous method described by Bhattacharya [15]. The gelatinization temperature (GT) value of the samples was measured by the alkali spreading value method [16, 17]. The milled rice samples were classified according to the size and shape classification of rice done by the International Rice Research Institute [18]. Fifty paddy grains were randomly picked from each sample and de-hulled with tweezers. The length (L, major axis) and breadth (B, intermediate axis) of brown rice kernels were measured by a micrometer screw gauge and L/B ratio was calculated [19]. The measurements were repeated 3 times for each sample.

Paddy samples of 150 g were hulled using a laboratory rubber roll Sheller (XXY155-Yanmar, Japan) and milled with a McGill rice miller (McGill mill No. 2 TX, USA). The broken rice grains were removed manually and also passing the milled rice through a sieve (2 mm). The milled rice kernels that had at least 2/3rd of the original kernel length were considered as head rice. The milling percentage of total milled rice yield (TMY) and head rice yield (HRY) recovery was determined as proposed by previous methods [20, 21]. The milled kernels were ground to flour in a laboratory scale grinder and passed through 80-mesh to obtain uniform particle size. Crude protein (CP %) and crude fat (CF %) contents were determined using the standard methods of analysis [17]. Apparent amylose content (AC %) of 100 mg rice flour sample (< 60 μm) was measured according to the AACC 61-03 method [17] using a UV–visible spectrometer (Jenway-6305, UK) at 620 nm.

Experimental design and statistical analysis

The experiment was arranged as a two-factor factorial in a randomized complete block design with three replicates. The MOP fertilizer rates, harvesting at different DAFF and their interactions were considered as fixed effects. The analysis of variance (ANOVA) was conducted using the SAS PROC GLM [22]. Initial tests were carried out to identify the effect of seasons and season x treatment interaction on the measured parameters. Treatment means were compared using Duncan's multiple range test (DMRT) at $P = 0.05$.

Results

The impact of cultivating seasons (two) and season x treatment interactions were not statistically significant ($P > 0.05$) for all the parameters evaluated. Hence, data for each parameter were averaged over two seasons and presented in this section.

Physical quality of rice

The results of the ANOVA carried out for the physical quality parameters of rice are presented in Table 1. The true density (TD) of paddy harvested did not show any significant interaction ($P > 0.05$) between K fertilizer rates and harvesting time. However, the TD significantly varied ($P < 0.05$) among the levels of each treatment (Table 1). The TD of rice grain increased and reached the maximum of 1.52 ± 1.51 g/cm^3 when paddy was harvested between 30 and 35 DAFF (Fig. 1), and decreased to 1.48 ± 0.005 g/cm^3 at 40 DAFF (P < 0.05). Increase in the level K fertilization at the time of heading (7 WAP) also increased the TD of rice ($P < 0.05$). The highest TD of 1.50 ± 0.003 g/cm^3 was observed at 37.5 kg MOP/ha applied at the time of heading. The bulk density (BD) of paddy also increased gradually but significantly ($P < 0.05$) with increased rate of K fertilizer at the time of heading and with delay of harvesting age. It reached its maximum at 35 DAFF (Table 2). The BD increased by about 8–10% with the rate of K fertilizer application higher than 12.5 kg/ha at the time of heading.

There was no significant interaction ($P > 0.05$) between the rate of K fertilizer at the time of heading and different harvesting times on grain dimensions. However, each treatment had a significant impact ($P < 0.05$; Table 1). The grain length (L) and breadth (B) increased significantly with application MOP fertilizer at rates greater than 25 kg/ha at the time of heading and increasing days taken to harvest from 30 DAFF. The highest L and B were detected at 37.5 kg MOP/ha applied at the time of heading and harvesting paddy

Table 1 ANOVA table of potassium fertilizer application at the time of heading and paddy harvest at different days after 50% flowering (DAFF) on the true density, bulk density, length, breadth, crude protein and crude fat of whole grain

Source of variation	df	*P* values*					
		True density	Bulk density	Length	Breadth	Crude protein	Crude fat
Block	2	0.054	0.87	0.82	0.65	0.050	0.08
K	2	0.001	0.001	0.02	0.02	0.01	< 0.0001
DAFF	3	< 0.0001	< 0.0001	< 0.0001	< 0.0001	0.04	< 0.0001
K x DAFF	6	0.78	0.72	0.27	0.81	0.96	0.051

K potassium fertilizer as muriate of potash (MOP), *DAFF* days after 50% flowering

* Significant effects that require means comparison $P < 0.05$

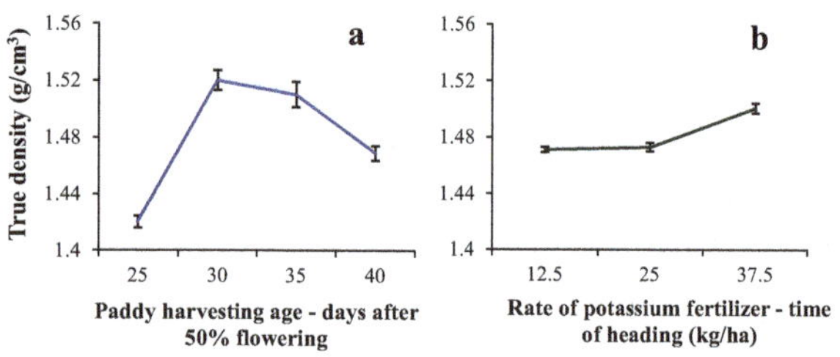

Fig. 1 Average true density of rice grains obtained from two cropping seasons; **a** when paddy was harvested at different days from 50% flowering and **b** application of different rates of muriate of potash at the time of heading. The vertical bars represent the standard error of the mean (30 DAFF and 25 kg MOP/ha were considered as controls)

approximately at 35 DAFF. The L/B ratio of the rice grain in all treatments was greater than 4 (Table 2) and the maximum of 4.43 was observed when K fertilizer was applied at 12.5 kg MOP/ha at the time of heading and paddy was harvested at 25 DAFF. The gelatinization temperature (GT) of the rice variety CIC 300 was 71.7 °C but showed no significant difference ($P > 0.05$) among all treatments combinations.

Proximate composition

The crude protein content (CP %), crude fat content (CF %) and amylose content (AC %) of rice grains did not show a significant interaction ($P > 0.05$) between the two treatments (Tables 1 and 3). The CP content increased significantly ($P < 0.05$) with increasing rate of MOP fertilizer at the time of heading of rice plant and age of paddy at harvesting (Table 4). The highest CP content recorded in this study was 6.24% with the application of 37.5 kg MOP/ha at the time of heading coupled with harvesting at 40 DAFF. The CF content was 3.3% higher when K fertilizer was supplied at the rate of 37.5 kg MOP/ha at the time of heading and 7.8% higher when paddy was harvested at 35–40 DAFF (Table 4) compared to the respective control (application of 25 kg MOP/ha of at the time of heading and harvesting paddy at 30 DAFF). The AC significantly decreased ($P < 0.05$) when the rate of MOP fertilizer applied at the time of heading was increased from 12.5 to 37.5 kg/ha and when harvest was delayed from 25 to 40 DAFF (Table 4).

Average paddy yield

The average paddy yield (APY) was influenced significantly by the interaction effect ($P < 0.05$) between the two treatments (Table 3; Fig. 2). At each K fertilizer level applied, the APY increased with the age of harvesting and recorded the highest yield of 6.85 Mt/ha at 35 DAFF, which was 13.8% higher than that recorded in the control plot (25 kg MOP/ha at the time of heading and harvesting paddy at 30 DAFF). Delaying paddy harvest beyond 35 DAFF showed a marginal decrease in the APY at all MOP levels applied at the time of heading.

Milled rice yield and head rice yield

A significant interaction between treatments was observed for milled rice ($P < 0.05$; Table 3). Figures 3 and 4 illustrate that delaying paddy harvest from 25 to 35 DAFF has resulted in a gradual increase in TMY and HRY across different rates of MOP applied at the time of heading and a significant decrease ($P < 0.05$) thereafter. At the lowest rate of K fertilizer applied at the time of heading (12.5 kg/ha) and early paddy harvest at 25 DAFF, the TMY of rice was the lowest. The highest TMY of 67% was obtained when MOP was applied at the rate of 37.5 kg/ha at the time of heading and the paddy was harvested at 35 DAFF, which was 6% higher than that observed in the control. The highest HRY of 65.3% was recorded from plots where paddy was harvested at 35 DAFF at the highest MOP level (37.5 kg/ha), which was 7.7% higher than the control. However, irrespective of the rate MOP fertilizer application at the time of heading, the HRY sharply decreased when paddy was harvested late at 40 DAFF. The results revealed that harvesting before 30 and after

Table 2 Physical grain quality parameters evaluated using different days of paddy harvest from 50% flowering and potassium fertilizer rates applied at the time of heading

Grain parameters	Harvesting age in days after 50% flowering (DAFF days)[d]				Rate of MOP applied at the time of heading (kg/ha)		
	25	30	35	40	12.5	25	37.5
Length (mm)	6.83 ± 0.04^a	7.23 ± 0.05^b	7.30 ± 0.06^b	7.25 ± 0.05^b	7.10 ± 0.02^a	7.10 ± 0.01^a	7.23 ± 0.02^b
Breadth (mm)	1.54 ± 0.02^a	1.73 ± 0.04^b	1.73 ± 0.03^b	1.74 ± 0.04^b	1.65 ± 0.01^a	1.70 ± 0.01^b	1.75 ± 0.01^c
L/B	4.43 ± 0.03^b	4.18 ± 0.05^a	4.22 ± 0.05^a	4.16 ± 0.04^a	4.30 ± 0.01^a	4.18 ± 0.02^b	4.13 ± 0.02^b
BD (g/cm^3)	0.62 ± 0.01^a	0.68 ± 0.01^b	0.70 ± 0.02^b	0.70 ± 0.01^b	0.64 ± 0.01^a	0.70 ± 0.01^b	0.70 ± 0.01^b

MOP muriate of potash (potassium fertilizer; 60% K_2O), *L/B* length to breadth ratio of the grain, *BD* bulk density of the rice grains

Mean (±SE) followed by the same superscript within a raw is not significantly different ($P > 0.05$) as measured by the DNMRT

[d] Data presented as mean ± SE

Table 3 The ANOVA table of rate of potassium fertilizer and paddy harvest at different days after 50% flowering on the amylose, average paddy yield, total milled rice and head rice yields of whole grain

Source of variation	df	*P* values*			
		Amylose	Average paddy yield	Total milled rice yield	Head rice yield
Block	2	0.96	0.48	0.85	0.20
K	2	0.015	0.014	0.042	0.047
DAFF	3	0.024	< 0.0001	< 0.0001	< 0.0001
K × DAFF	6	0.16	0.44	0.023	0.004

K potassium fertilizer as muriate of potash (MOP), *DAFF* days after 50% flowering

*Significant effects that require means comparison $P < 0.05$

Table 4 Proximate composition of rice grains when paddy was harvested at different days from 50% flowering and potassium fertilizer rates applied at the time of heading

Grain parameters	Harvesting age in days after 50% flowering (DAFF)[d]				Rate of MOP applied at the time of heading (kg/ha)[d]		
	25	30	35	40	12.5	25	37.5
Crude protein (%)	5.86 ± 0.05^a	6.00 ± 0.05^{ab}	6.18 ± 0.03^b	6.24 ± 0.04^c	5.87 ± 0.03^a	6.08 ± 0.05^b	6.24 ± 0.03^c
Crude fat (%)	2.23 ± 0.05^a	2.42 ± 0.06^b	2.60 ± 0.06^c	2.61 ± 0.07^c	2.43 ± 0.04^a	2.44 ± 0.06^a	2.52 ± 0.07^a
Amylose (%)	17.84 ± 0.4^c	17.40 ± 0.3^b	16.10 ± 0.4^a	16.00 ± 0.3^a	17.13 ± 0.4^b	17.6 ± 0.4^b	15.71 ± 0.2^a

MOP muriate of potash (potassium fertilizer; 60% K_2O)

Mean (±SE) followed by the same superscript within a raw is not significantly different ($P > 0.05$) as measured by the DNMRT

[d] Data present as mean ± SE

35 DAFF would result in significantly lower ($P < 0.05$) HRY by 6.2% and 10.5%, respectively.

Discussion

Grain densities

The TD and BD are used to assess the level of grain filling with the maturity states of the rice [23]. The TD indicates the exact material density of the product itself. The increase observed in the TD and BD of rice grains at higher MOP rates applied at 7 WAP with delayed paddy harvest could be due to higher bulking of seeds during grain filling under optimum potassium fertilizer supplement given at the time of heading. Potassium is involved in the transportation of photosynthates from leaves to grain, thereby increasing the dry matter content of the plant and grain. Therefore, changing dry matter or moisture may significantly affect density. Bhattacharya et al. [24] reported that for every 1% increase in moisture or dry matter, the density of the paddy increased by about 7.5 kg/m^3. Thus, application of 37.5 kg MOP/ha at the time of heading and paddy harvested between 30 and 35

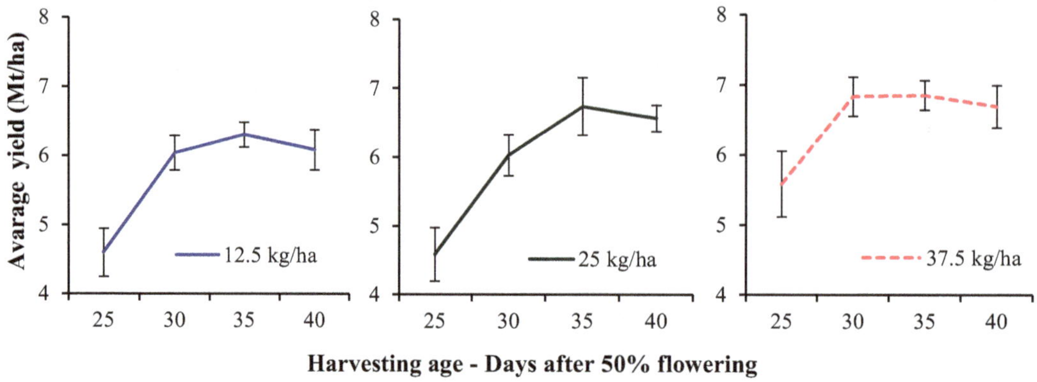

Fig. 2 Average paddy yield harvested at different days from 50% flowering under three rates of MOP fertilizer (kg/ha) applied at the time of heading of two cropping seasons. Vertical bars indicate the standard error of the mean (30 DAFF and 25 kg MOP/ha were the controls)

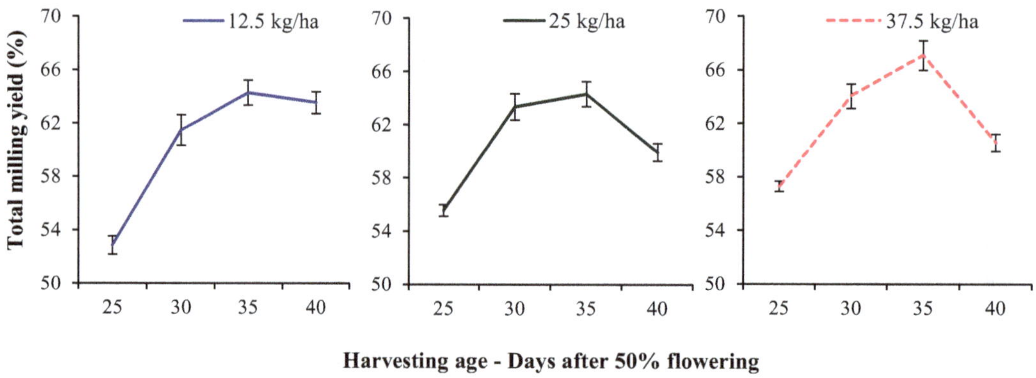

Fig. 3 Total milling yield of rice when paddy was harvested at different days from 50% flowering at three rates of MOP fertilizer applied (kg/ha) at the time of heading. Vertical bars indicate the standard error of the mean (30 DAFF and 25 kg MOP/ha were the controls)

DAFF could have contributed to the increase of dry matter content by about 6.7% and 12.6%, respectively.

Grain dimensions

Quality of rice grain is a complex character composed of many components such as nutritional, appearance, cooking and eating qualities [25]. Thus, the physical properties such as size, shape, uniformity, and general appearance are of utmost importance in determining the final milling quality of rice grains. The head grain dimensions are important when evaluating the rice grain quality as they show clear differences with time due to grain filling [26]. The longer kernel length and other physico-chemical properties were found to be distinctive features of the rice cultivars [27]. The grain L/B ratio of the rice grain in all treatments was greater than 3 (Table 2), which could be classified as 'slender' in shape [19, 27]. Basmati type rice varieties are longer and slender than typical long grain, often with an L/B ratio in the rage of 3.84–5.01 [19].

Gelatinization temperature

The results of the gelatinized temperature (GT) indicated the grains of the rice variety CIC 300 falls under the category high intermediate as described by [18]. This GT value is common for the majority of basmati type long grain rice varieties found in the world. Previous report showed that [28] application of K fertilizer did not affect the gelatinization temperature.

Proximate composition

The crude protein (CP) content of the internationally accepted premium basmati varieties varies between 7 and 9% [27]. Similar to the data obtained in this study. Another study [28] also reported that application of K fertilizer increased the grain protein content and gel consistency. According to previous findings, synthesis of

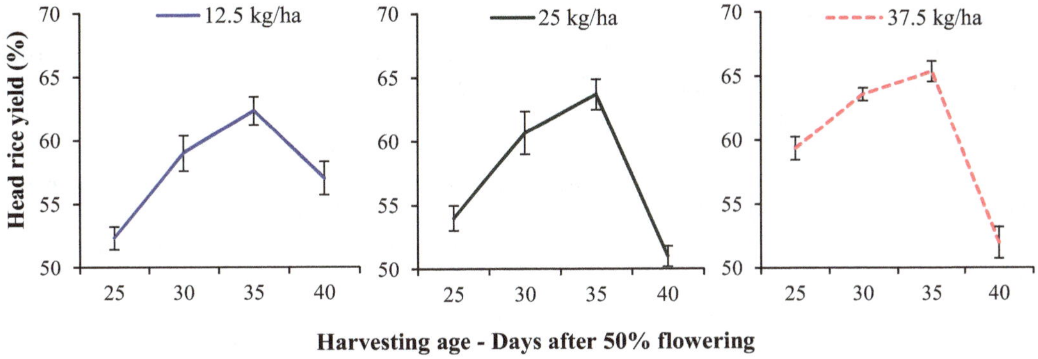

Fig. 4 Average head rice yield when paddy was harvested at different days from 50% flowering at three rates of MOP fertilizer applied (kg/ha) at the time of heading of two cropping seasons. Vertical bars indicate the standard error of the mean (30 DAFF and 25 kg MOP/ha were the controls)

protein is accelerated in the presence of adequate content of K [4, 9]. In the present study, the highest CP content (6.24%) recorded in the export rice variety CIC 300 (harvested at 40 DAFF) was marginally lower than the premium quality requirement for the international market. The maximum protein content in rice grain is accumulated 10 days after flowering [23, 29] and it decreases, however, with grain maturity up to 30–35 days after flowering of non-waxy rice [30]. However, low CP content in the rice grains was observed due to the accumulation of higher levels of amylose at harvesting at 25 DAFF across varying K fertilizer rates at the time of heading of rice plant [2, 9, 33].

Previous studies have reported that [27, 31] the CF content of paddy grains did not significantly differ between *japonica* and *indica* varieties but CF content varied between 2.0 and 3.2% in range. The CF content in the paddy grains in this study became low when paddy was harvested at 25 DAFF and at K fertilizer rate of 12.5 kg MOP/ha at the time of heading. This could be due to that paddy was harvested at an immature stage when the rice plant did not have enough time to translocate fat into the endosperm of seeds [23]. However, Choudhury and Juliano [32] reported that non-starch lipids in developing paddy grain get accumulated up to 16 days after flowering.

Rice with low amylose content (AC) has been reported to affect the eating and cooking qualities of cooked rice [31]. Previous studies have shown that AC increased in the paddy grain up to 20–21 days after flowering due to the activity of starch synthase and then decreased [29, 33]. The lowest AC in rice grains after 25 days flowering reported that AC became constant 30 days after flowering in non-waxy rice [30]. According to Hossain et al. [10], long grain aromatic rice harvested at 30–35 days after flowering was found to be the most suitable for higher head rice out-turn, elongation, volume expansion ratio and AC. It has been shown that the maximum AC can be obtained by application of K fertilizer at 180 kg K_2O/ha and 75 kg K_2O/ha with or without N fertilizer application, respectively [28, 33]. However, the present study found that the increasing rate of K fertilizer application from 12.5 to 37.5 kg MOP/ha at the time of heading reduced the AC significantly ($P<0.05$) by about 8.3%. It is also important to note that soil of the experimental site contained an average of 215 kg of K/ha. Application of high level K fertilizer (MOP) at the time of heading together with N fertilizer at later growth stages (50 kg/ha of urea was added at the time of heading in the present study) or application of K fertilizer at late growth stage of paddy reduces the enzyme activity of starch synthase [33]. Liu et al. [5] showed that the grain filling regulatory enzyme activity is significantly reduced 20 days after heading and under K-deficient conditions. This could be the reason for decreasing amylose content reported with increasing K fertilizer rates applied at the time of heading and with delayed paddy harvest. The results highlight the importance of timing of K fertilizer application to support grain filling of rice plants.

Average paddy yield

It is well known fact that application of K fertilizer significantly increases the grain yield by about 47% with or without organic manure [34, 35]. Potassium deficiency reduces grain size and weight resulting in a direct yield loss if the grain does not respond to K fertilizer applications made after the growth stage of panicle differentiation [6]. According to a nano-material study, application of K as nano-K fertilizer at the rate of 20 kg K_2O/ha showed a significant increase in rice yield and number of seeds per panicle in direct seeded rice [36]. In the present study, the APY may also have decreased when harvested at 40 DAFF due to higher shattering, probably with the lowering of moisture content at the field or due to the source limitation to fulfill the sink demand. The same pattern was also observed with respect to TD of the rice grains. According to the previous reported data, an average agronomic response of 6 kg grain/kg of K for rice and wheat can be obtained by applying K fertilizer at the rate of 37.5 kg/ha [2]. The K uptake requirements for rice varieties that yield between 4 and 8 t/ha have ranged from 17 to 30 kg K/t of grain produced [34].

Total milling rice yield

The TMY is one of the major parameters that determine the quality of rice, especially in a marketing perspective. Two types of milling yield such as HRY and TMY are considered the most important parameters of determining the quality of rice. A variety should possess a high turnout of HRY and total milled rice. Broken rice is generally valued at only 30–50% of the whole grain. The accurate measurement of the amounts and classes of broken grains is important and, hence, the standardized procedures are used for official grading. In the present study, the milling yield was low at early harvests (25 DAFF) as the immature grains did not have enough strength to tolerate the mechanical stresses applied during milling [37]. The results also revealed that the best time for harvesting to achieve a higher rice milling yield of direct seeded rice is between 30 and 35 DAFF similar to the previous study [12]. This pattern of variation was also observed in relation to HRY and true density (Figs. 1a and 2).

Head rice yield

From the consumer's perspective, the HRY is the most important physical quality parameter when determining the grain quality of rice. Harvesting paddy at the optimum crop maturity can give the maximum HRY [10, 38] but late harvesting causes significant reduction of the HRY [39]. Similar to the results of this study, another study had also reported that a higher HRY can be obtained when paddy was harvested between 30 and 35 DAFF [10]. Immature grains and low density of rice (Fig. 1a) may be the probable reason for lower HRY recorded during early stages of harvesting at 25 DAFF. This may also lead to higher grain breakage during milling. Delayed harvesting beyond 35 DAFF would also sharply reduce the true density (Fig. 1a) and, therefore, decrease the grain hardness resulting in higher breakages during milling (>50%). Although the timeliness of harvesting significantly influences the milling yield, harvesting rice at the optimum crop maturity would result in the maximum HRY [38]. Any delay in harvesting causes reduction of the HRY [39] and extended delay in harvesting could lead to significant losses in HRY.

Wang et al. [9] reported that increased application of K fertilizer for paddy can significantly improve the percentages of brown rice, milled rice and HRY while reducing the grain chalkiness and enhancing grain protein content. Structural changes such as ordered arrangement of starch granules inside the seed endosperm are also possible with the increasing K fertilizers [23]. Potassium fertilizer is required for maximum yield and it must be applied during the vegetative growth stage [6]. This was clearly reflected by the results of HRY and TMY in the present study, which was significantly reduced with increasing age of harvesting (DAFF) and high rate of K fertilizer (MOP) application at the time of heading. Application of high doses of K fertilizer could also increase the percentage of rice bran [40]. The results revealed that application of 50% higher rate of MOP fertilizer at the time of heading (37.5 kg MOP/ha; approx. 18.75 kg K/ha) to a low shattering rice variety such as CIC 300 and harvesting paddy 5 days later than the recommended harvesting stage identified by Jayawardena [12] would increase the rice yield (APY, HRY and TMY) and grain quality under direct seeded rice in the dry zone of Sri Lanka.

Conclusion

The results of the present study showed that K fertilizer applied at the rate of 37.5 kg MOP/ha at the time of heading, which is 50% higher than the recommended rate, is the best among K fertilizer treatments to obtain the highest HRY when rice was harvested from 25 to 35 DAFF. The rice variety CIC 300 belongs to the 3 months age class (mature in 90–93 days after sowing) and, thus, the harvesting paddy between 35 and 40 DAFF could be considered as delayed harvesting. Harvesting before 30 or later than 35 DAFF, under different K fertilizer rates applied at the time of heading of the rice plant, would result in heavy losses during milling.

Abbreviations

LD: Low Country Dry Zone; TSP: triple super phosphate; MOP: muriate of potash; WAP: weeks after planting; DAFF: days after 50% flowering; K fertilizer: potassium fertilizer; APY: average paddy yield; TD: true density; BD: bulk density; GT: gelatinization temperature; L: grain length; B: breadth; L/B: length to breadth ratio; TMY: total milled rice yield; HRY: head rice yield; CP: crude protein; CF: crude fat; AC: apparent amylose; ANOVA: analysis of variance; DMRT: Duncan's multiple range test; df: degrees of freedom; SE: standard error.

Authors' contributions

All authors have equally contributed to the conceptualization and designing of the experiments and preparing the manuscript. AJA carried out the field studies, statistical analysis and interpretation. All authors read and approved the final manuscript.

Author details

[1] Agromony Division, Coconut Research Institute, Bandirippuwa Estate, Lunuwila 61150, Sri Lanka. [2] Present Address: Key Laboratory of Tropical Forest Ecology, Xiahuangbanna Tropical Botanical Garden, Chinese Academy of Sciences, Mengla 666303, Yunnan, China. [3] Present Address: University of Chinese Academy of Science, Beijing 100094, China. [4] Department of Food Science &

Technology, Faculty of Agriculture, University of Peradeniya, Peradeniya 20400, Sri Lanka. [5] Department of Agriculture Engineering, Faculty of Agriculture, University of Peradeniya, Peradeniya 20400, Sri Lanka. [6] Department of Crop Science, Faculty of Agriculture, University of Peradeniya, Peradeniya 20400, Sri Lanka.

Acknowledgements
The authors wish to thank Messrs. CIC Agri Business (Pvt) Ltd. for providing the seed material other field facilities to complete this study.

Competing interests
The authors declare that they have no competing interests.

Funding
Not applicable.

References

1. CBSL Annual Report 2016—economic and social statistics of Sri Lanka. Central Bank of Sri Lanka. http://www.cbsl.gov.lk. Accessed 18 Aug 2018.
2. Singh B, Singh Y, Imas P, Jian-chang Z. Potassium nutrition of the rice–wheat cropping system. Adv Agron. 2004;81:203–59.
3. Armengaud P, Breitling R, Amtmann A. The potassium-dependent transcriptome of Arabidopsis reveals a prominent role of jasmonic acid in nutrient signaling. Plant Physiol. 2004;136:2556–76.
4. Armengaud P, Sulpice R, Miller AJ, Stitt M, Amtmann A, Gibon Y. Multi-level analysis of primary metabolism provides new insights into the role of potassium nutrition for glycolysis and nitrogen assimilation in *Arabidopsis thaliana* roots. Plant Physiol. 2009;150:772–85.
5. Liu LJ, Chang EH, Fan MM, Wang ZQ, Yang JC. Effect of potassium and calcium on root exudates and rain quality during grain filling. Acta Agronomica Sinica. 2011;37:661–9.
6. Fageria NK, Slatonb NA, Baligarc VC. Nutrient management for improving low land rice productivity and sustainability. Adv Agron. 2003;80:63–152.
7. Wild A. Russell's—soil condition and plant growth. 12th ed. London: English language book society/Longman, UK; 1988.
8. Srivastava P, Srivastava PC, Singh US, Shrivastava M. Effect of integrated and balanced nutrient application on soil fertility, yield and quality of Basmati rice. Arch Agron Soil Sci. 2008;55:265–84.
9. Wong QS, Zhen RH, Ding YH, Ji ZJ, Cao WX, Huang PS. Effects of potassium fertilizer application on plant potassium accumulation and grain quality of japonica rice. Scientia Agricultura Sinica. 2004;37:1444–50.
10. Hossain MF, Bhuiya MSU, Ahmed M, Mian MH. Effect of harvesting time on the milling and physicochemical properties of aromatic rice. Thai J Agric Sci. 2009;42:91–6.
11. Bose SP, Chattopadhyay P. Harvesting and drying of high moisture paddy. Rice Process Eng Center Rep. 1976;2:33–7.
12. Jayawardena SDG. Effect of grain moisture content, time of harvest and method of drying on milling quality of rice. Trop Agric. 1973;129:103–18.
13. Sartaj IZ, Ekanayake S. Postharvest losses. Trop Agric Res. 1991;3:115–32.
14. DOA. Fertilizer recommendation for rice. 2001. http://www.knowledgebank.irri.org/sriLanka/fert_reco.html. Accessed 18 Aug 2018.
15. Bhattacharya KR. Gelatinization temperature of rice starch and its determination. In: Proceedings of the Workshop on Chemical Aspects of Rice Brain Quality. Los Baños: IRRI; 1979.
16. Bhattacharya KR, Sawbhagya CM, Indhudharaamy YM. Quality profiles of rice. A tentative scheme for classification. J Food Sci. 1982;47:564–9.
17. AACC. Approved Methods of the American Association of Cereal Chemists, American Association of Cereal Chemists Inc., St. Paul, Minnesota, USA; 2000.
18. Cruz ND, Khush GS. Rice grain quality evaluation procedures. In: Singh RK, Singh US, Khush GS, editors. Aromatic Rices. New Delhi: Oxford and IBH Publishing Co Pvt. Ltd; 2000.
19. Kaur S, Panesar PS, Bera MB. Studies on evaluation of grain quality attributes of some basmati and non-basmati rice cultivars. J Food Qual. 2011;34:435–41.
20. Juliano BO. Rice: chemistry and technology. 2nd ed. St. Paul: American Association of Cereal Chemists; 1985.
21. Archer TR, Siebenmorgen TJ. Milling quality as affected by brown rice temperature. Cereal Chem. 1995;72:304–7.
22. SAS Institute. SAS language and procedures, version 6. 1st ed. Cary: SAS Institute; 1990.
23. Yoshida S. Fundamentals of rice crop science. Manila: International Rice Research Institute; 1981.
24. Bhattacharya KR, Sowbhagya CM, Indudhara SYM. Some physical properties of paddy and their interrelation. J Sci Food Agric. 1972;23:171–86.
25. Tan Y-F, Li J-X, Yu S-B, Xing Y-Z, Xu C-G. The three important traits for cooking and eating quality of rice grains are controlled by a single locus in an elite rice hybrid, Shanyou 63. Appl Genet. 1999;99:642–8.
26. Villarreal CP, Juliano BO, Sauphanor B. Grain quality of rice grown in irrigated and upland cultures. Plant Foods Hum Nutr. 1990;40:37–47.
27. Bhattacharjee P, Singhal RS, Kulkarni PR. Basmati rice: a review. Int J Food Sci Technol. 2002;37:1–12.
28. Bahmaniar MA, Ranjbar GA. Response of rice (*Oryza sativa* L.) cooking quality properties to nitrogen and potassium application. Pak J Biol Sci. 2007;10:1880–4.
29. Baun LC, Palmiano EP, Perez CM, Juliano BO. Enzymes of starch metabolism in the developing rice grain. Plant Physiol. 1970;46:429–34.
30. Tamaki M, Ebata M, Tashiro T, Ishikawa M. Physico-ecological studies on quality formation of rice kernel. II. Changes in quality of rice kernel during grain development. Jpn J Crop Sci. 1989;58:659–63.
31. Zhou Z, Robards K, Helliwell S, Blanchard C. Composition and functional properties of rice. Int J Food Sci Technol. 2002;37:849–68.
32. Choudhury NH, Juliano BO. Lipids in developing and mature rice grain. Photochemistry. 1980;19:1063–9.
33. Ling Z, Ha X-Ch, Wei L, Guo-Tao Y. Effect of nitrogen and potassium on starch content and activities of starch syntase in grains of hybrid rice B You 827. Chin J Rice Sci. 2008;22:551–4.
34. Dobermann A, Cruz PC, Cassman KG. Fertilizer inputs, nutrient balance, and soil nutrient-supplying power in intensive, irrigated rice systems. I. Potassium uptake and K balance. Nutr Cycl Agro Ecosys. 1996;46:1–10.
35. Wickramasinhe WMADB, Wijewardena JDH. Soil fertility management and integrated plant nutrition systems in rice cultivation. In: Abeysiriwardena DS de Z, Dissanayake DM N Nugaliyadde L. Rice Congress 2000. Proceedings rice symposium held in Department of Agriculture, Peradeniya, Sri Lanka, 14–15 December 2000, p. 125–140.
36. Sirisena DN, Dissanayake DMN, Somaweera KATN, Karunaratne V, Kottegoda N. Use of nano-K fertilizer as a source of potassium in rice cultivation. Ann Sri Lanka Dep Agric. 2013;15:257–62.
37. Juliano BO, Villareal CP. Grain quality evaluation of world rice. Manila: International Rice Research Institute; 1993.
38. Bal S, Ojha TP. Determination of biological maturity and effect of harvesting and drying conditions on milling quality of paddy. J Agric Eng Res. 1975;20:353–61.
39. Berrio LE, Cuevas-Perez FE. Cultivar differences in milling yields under delayed harvesting of rice. Crop Sci. 1989;29:1510–2.
40. Bahmaniar MA, Ranjbar GA. Effect of nitrogen potassium fertilizer on rice (*Oryza sativa* L.) genotype processing characteristics. Pak J Biol Sci. 2007;10:1829–34.

14

Chemical composition and antibacterial activity of essential oil of *Tetraclinis articulata* (Vahl) Masters branches of eastern Morocco

Fatima Zahra Sadiki[1*], Mostafa El Idrissi[1], Mohammed Sbiti[2], Adrae Lemrhari[3], Adriana Trifan[4], Oana Cioanca[4], Paula Alexandra Postu[5] and Lucian Hritcu[5]

Abstract

Background: Some pathogenic microbial species are becoming less sensitive to antibiotics and developing resistance multiples. The use of essential oils is a serious substitute for treatment with antibiotics in infectious diseases. The present study aims to investigate the chemical composition and to evaluate the antibacterial activity of the essential oil of branches of *Tetraclinis articulata* (Vahl) Masters (Cupressaceae).

Methods: The extraction of the essential oil was realized by hydrodistillation, and the analysis was carried out by gas chromatography coupled with mass spectroscopy (GC–MS/FID). The antibacterial test was carried out using two different methods: the disc diffusion method and the macrodilution method which aims to determine the minimum inhibitory concentration (MIC).

Results: The yield of essential oil was 0.84%. Chromatographic analysis revealed that the major constituents were α-pinene (38.75%), limonene (13.24%), bornyl acetate (8.78%) and camphor (7.68%). The results of the antibacterial activity show that the essential oil exerts a powerful effect to inhibit some strains tested.

Conclusions: The antibacterial activity demonstrated in this study could justify the traditional uses of this plant and could contribute to value its potential in order to develop new bioactive compounds.

Keywords: *Tetraclinis articulata* (Vahl) Masters, Antibacterial activity, MIC, Camphor

Introduction

In the field of anti-infectives, the discovery of new substances has always been the goal of Human, since the main cause of death in the past was infectious diseases [1]. When penicillin therapy was first performed, Human thought he had won the battle against pathogenic bacteria. However, with the appearance of penicillin-resistant bacteria, he had the impression of regression, and with the extraction and synthesis of new powerful antibiotics, the fear dispersed. Nevertheless, like any drug, antibiotics are not totally harmless to the body and can cause many side effects [2]. These antibiotics, which have saved lives and relived the sufferance of millions of people for years to date, are also threatened by the emergence and spread of multi-resistant germs.

In order to overcome the problem of resistance of microorganisms to antibiotics, most of the work is currently oriented towards other antimicrobial agents with a very specific mode of action. Thus, researchers and scientists are trying to find effective and accessible alternatives from natural products extracted from medicinal and aromatic plants, which are known to be endowed with antimicrobial properties [3–11].

Among the aromatic plants, *Tetraclinis articulata* (Vahl) Mast. Synonyms (Cupressaceae), Barbary thuja synonyms of Callitris quadrivalvis Rich and A. Rich., Thuja articulata Vahl, resinous species belonging to the order Pinales, known in Morocco as the "Araar", is characterized by its resistance to destructive agents and the

*Correspondence: fatizahrasad@gmail.com
[1] Laboratory of Molecular Chemistry and Natural Substances, Department of Chemistry, Faculty of Sciences of Meknes, Moulay Ismail University, Bp 11201, Zitoune, Meknes, Morocco
Full list of author information is available at the end of the article

use of its wood in crafts, carpentry and cabinetry. It is endemic in the south western Mediterranean and especially the Maghreb [12]. In Morocco, forests cover about 566.000 ha, and are divided into 6 major biogeographic units at the Rif, oriental, eastern Middle Atlas, valleys of the central plateau and the western plateau, Western Middle Atlas and High Atlas. In traditional medicine, the different parts of the tree, particularly the leaves and branches, are used in the treatment of intestinal and respiratory infections [13].

A research in the scientific literature indicates that there are few reports of studies on the antibacterial properties and chemical composition of essential oil extracted from the branches of the Thuja. In this context, this study was conducted in order to put under scrutiny the chemical composition and to evaluate the antibacterial activity of this essential oil.

Materials and methods

Plant material and extraction

The branches of *Tetraclinis articulata* were harvested in April (2016) in the forest of Tafoughalt-zegzel (Latitude: 34°50′20″N; Longitude: 2°24′49″W), region in East of Morocco. The extraction of essential oil was carried out by hydrodistillation in a Clevenger-type apparatus [14]. Three distillations were performed for 3 h. The essential oil was stored at 4 °C in the dark before proceeding with chromatographic analysis and biological activities.

Chromatographic analysis

The analysis of the essential oil was performed using Agilent 6890 GC–MS system, equipped with a split/splitless injector. GC–MS conditions were line/detector temperature 250 °C, carrier gas (helium, 1 mL/min), split ratio 100:1, and capillary column DB5MS (30 m × 0.25 mm; film thickness 0.25 μm, Agilent, Palo Alto, CA, USA). The temperature program was established to grow with 10 °C/min up to 280 °C. The injected volume was 0.30 μL with a total scan time of 32 min. The GC-FID analysis used an Agilent type 6890 GC connected to a FID detector and all method and analysis parameters were the same as those described above.

Microorganisms studied

Eight clinical bacteria were chosen for their pathogenicity, and they have been isolated from the various samples (Table 1). Isolation and identification of clinical strains were performed at the microbiology laboratory of Military Hospital Moulay Ismail in Meknes.

Tests for antibacterial activity

Determination of the inhibition diameter (solid method)

The diffusion method from a solid disc was used to demonstrate the antibacterial activity. A bacterial suspension was diluted and adjusted to turbidity equal to that of the McFarland standard 0.5 (10^8 CFU/mL). The Mueller–Hinton Agar was poured into Petri dishes 90 mm in diameter. The agar surface was seeded with the microbial suspension. Paper dishes, of 6 mm diameter impregnated with 10 μL of the essential oil, were deposited on the surface of the agar. The dishes were then incubated at 37 °C for 24 h. After incubation, an area or clear halo is present around a disc if the essential oil inhibits bacterial growth. All tests were repeated three times.

Broth dilution method (determination of MIC and MBC)

A cascade dilution was performed in Mueller–Hinton Broth-Tween 80 (0.01%, v/v), so as to obtain a concentration range between 80 μL/mL and 0.3 μL/mL. 13 μL of a bacterial inoculum, equivalent to the standard density of 0.5 McFarland (10^8 UFC/mL), was deposited in each of the tubes of the range. A control of the bacterial growth, for which 13 μL of the standardized inoculum was deposited in MHB-Tween 80 medium (0.01%, v/v), was also carried out. The tubes were incubated at 37 °C for 24 h. The MIC was determined by the lowest concentration of essential oil giving a growth inhibition.

Table 1 Profile of tested bacteria

Bacterial strains	Gram	Profile	Sex	Nature of sampling
Escherichia coli	Gram-negative bacilli	ESBL	Male	Cytobacteriological urine exam
Klebsiella pneumoniae	Gram-negative bacilli	ESBL	Male	Urinary catheter
Proteus mirabilis	Gram-negative bacilli	Penicillinase High level	Male	Cytobacteriological urine exam
Pseudomonas aeruginosa	Gram-negative bacilli	Low level	Female	Coproculture (stool bacteria)
Acinetobacter baumannii	Gram-negative bacilli	Multi-resistant	Male	Cytobacteriological urine exam
Staphylococcus epidermidis	Gram-positive cocci	Multi-resistant	Male	Urethral
Staphylococcus non aureus	Gram-positive cocci	Multi-sensitive	Female	Pus
Staphylococcus aureus	Gram-positive cocci	Multi-sensitive	Male	Pus

ESBL extended-spectrum beta-lactamases

Table 2 Chemical composition of essential oil of *Tetraclinis articulata* branches

Compound	Area	RT	Formula
Tricyclene	0.92	7.92	$C_{10}H_{16}$
α-Pinene	*38.75*	8.24	$C_{10}H_{16}$
Camphene	1.48	8.52	$C_{10}H_{16}$
Sabinene	0.58	8.90	$C_{10}H_{16}$
β-Pinene	1.00	9.01	$C_{10}H_{16}$
Myrcene	2.52	9.19	$C_{10}H_{16}$
Δ-3-Carene	0.22	9.54	$C_{10}H_{16}$
α-Terpinene	0.17	9.69	$C_{10}H_{16}$
Limonene	*13.24*	9.99	$C_{10}H_{16}$
α-Ocimene	0.03	10.18	$C_{10}H_{16}$
Terpinolene	1.17	10.89	$C_{10}H_{16}$
(E)-Ocimene	0.35	11.12	$C_{10}H_{16}$
1,3,8-*p*-Menthatriene	0.27	11.34	$C_{10}H_{14}$
α-Campholenal	0.84	11.59	$C_{10}H_{16}O$
p-Mentha-1,3,8-triene	0.41	11.81	$C_{10}H_{14}$
Camphor	*7.68*	12.03	$C_{10}H_{16}O$
α-Toluenol	0.31	12.20	C_7H_8O
Borneol	2.95	12.39	$C_{10}H_{18}O$
γ-Terpinene	0.74	12.49	$C_{10}H_{16}$
p-Cymenene	0.49	12.58	$C_{10}H_{12}$
Allo-Ocimene	1.59	12.72	$C_{10}H_{16}$
Verbenone	0.68	12.92	$C_{10}H_{14}O$
Carveol	0.88	13.08	$C_{10}H_{16}O$
Methyl carvacrol	0.62	13.28	$C_{11}H_{16}O$
Carvone	0.45	13.43	$C_{10}H_{14}O$
Bornyl acetate	*8.78*	14.07	$C_{12}H_{20}O_2$
Carvacrol	0.94	14.28	$C_{10}H_{14}O$
α-Copaene	0.97	15.37	$C_{15}H_{24}$
β-Elemene	0.48	15.54	$C_{15}H_{24}$
Aromandendrene	1.36	15.78	$C_{15}H_{24}$
β-Caryophyllene	1.11	16.03	$C_{15}H_{24}$
Germacrene D	0.06	16.14	$C_{15}H_{24}$
(Z)-β-Farnesene	0.15	16.39	$C_{15}H_{24}$
α-Humulene	0.54	16.53	$C_{15}H_{24}$
γ-Muurolene	0.10	16.76	$C_{15}H_{24}$
γ-Cadinene	0.61	16.89	$C_{15}H_{24}$
β-Selinene	0.19	17.01	$C_{15}H_{24}$
α-Selinene	0.31	17.10	$C_{15}H_{24}$
γ-Cadinene	0.58	17.40	$C_{15}H_{24}$
(Z)-α-Farnesene	0.04	18.15	$C_{15}H_{24}$
α-Humulene	0.26	18.84	$C_{15}H_{24}$
Cadina-1(2),4-diene	1.23	19.25	$C_{15}H_{24}$

Italic values indicate the percentage of the major compounds of the essential oil

The minimum bactericidal concentration (MBC) was determined from the MIC. The tubes showed no visible growth with the naked eye after incubation and the control tube, were streaked on MHA. The inoculated dishes were incubated for 24 h at 37 °C. The MBC is the lowest concentration that shows no bacterial growth.

Results and discussion

The yield of essential oil

The extraction of essential oil by hydrodistillation from the branches of *Tetraclinis articulata* enables us to obtain a yield of the order of 0.84±0.01%. This rate is relatively higher than that reported by Bourkhis et al. (0.41%) [15]. This shows that certain factors such as the nature of the soil, the climate and quality of the plant material used can influence the secretion of essential oils in a plant.

Identification of chemical composition of essential oil

Forty-two compounds were identified, corresponding to 98.89% of the oil, mainly represented by monoterpenes. The main constituents were α-pinene (38.75%), limonene (13.24%), bornyl acetate (8.78%) and camphor (7.68%). The analysis results are shown in Table 2 and the chromatographic profile is shown in Fig. 1.

Our results are similar to those reported by Bourkhiss et al. [15] who studied the essential oil of branches of *Tetraclinis articulata* (Vahl) Masters of Khemisset region (Northwester Morocco), and have identified thirty-three compounds representing more than 80% of the total chemical composition, the main compounds of which were α-pinene (30.22%), limonene (22.29%), widdrol (5.41%) and bornyl acetate (4.76%).

Similarly, Buhagiar et al. [16] reported that the essential oil of the terminal branches (woody and non-woody) of Thuja from Malta was characterized, respectively, by α-pinene (31.0%; 46.4%), followed by bornyl acetate (19.1%; 7.3%), camphor (18.1%; 7.3%) and limonene (3.8%; 6.2%).

Furthermore, Ait Igri et al. [17] have shown a slightly different composition than ours, bornyl acetate (31.10%), camphor (20.20%) and limonene (7.10%), which were identified as the major compounds of the essential oil of the branches of Thuja collected in the region of Essaouira (south-west Morocco).

This difference was also seen in the work of Tekaya-Karaoui et al. [18], where the main components in the essential oil of the terminal branches (non-woody) of Thuja were muurolene (29.0%) and 4, 6-dimethyl-octane-3,5-dione (22.4%), whereas camphene (43.2%) and (Z)-β-ocimene (11.7%) were the major compounds in the terminal branches (woody).

These differences in chemical composition of essential oil may be attributed to several factors: seasonal and environmental, soil, geographic area, harvest period, extraction method [19–24].

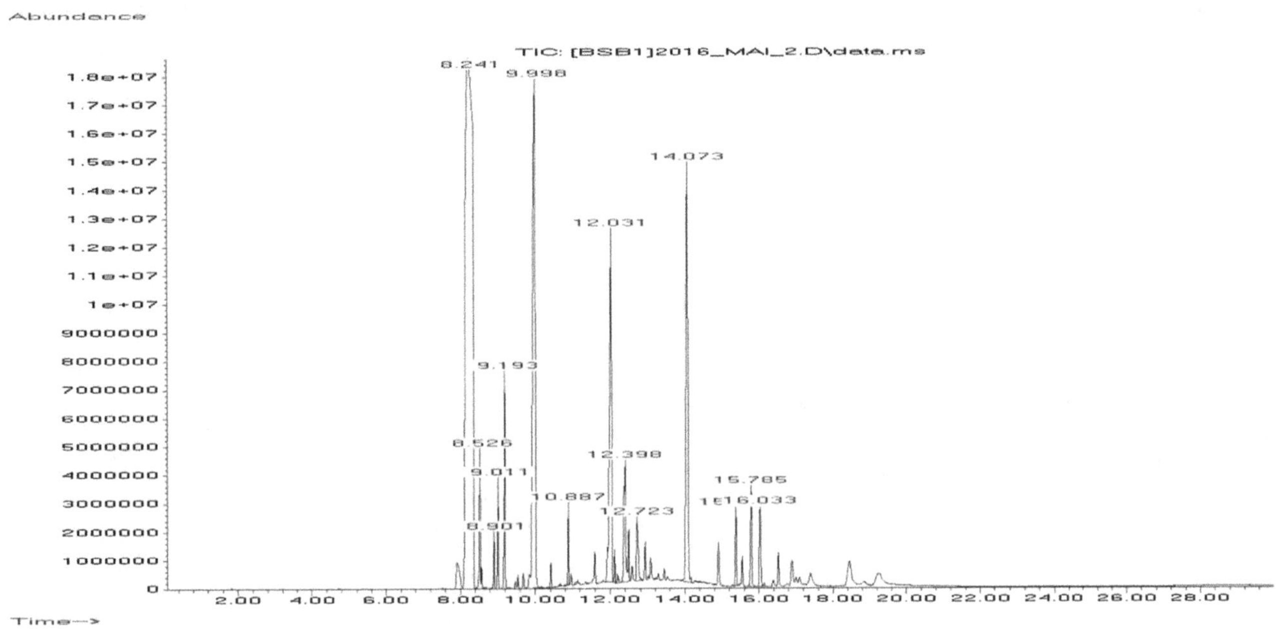

Fig. 1 Chromatographic profile of the essential oil of *Tetraclinis articulata* branches

Table 3 Antibacterial activity of the essential oil tested by disc diffusion assay

Bacterial strains	Zone of inhibition (mm)
Escherichia coli	19.16 ± 0.57
Klebsiella pneumoniae	6.66 ± 0.76
Proteus mirabilis	NA
Pseudomonas aeruginosa	NA
Acinetobacter baumannii	17.16 ± 0.28
Staphylococcus epidermidis	NA
Staphylococcus non aureus	NA
Staphylococcus aureus	9.26 ± 0.30

NA : no activity found

Antibacterial activity

According to the results of the aromatogram shown in Table 3, the essential oil exhibited moderate inhibitory power against *Escherichia coli* and *Acinetobacter baumannii*, which are antibiotic-resistant strains. The essential oil exhibited low activity against *Klebsiella pneumonia* and *Staphylococcus aureus*. The strains *Pseudomonas aeruginosa*, *Proteus mirabilis*, *Staphylococcus epidermidis*, and *Staphylococcus non aureus* showed no sensitivity for the essential oil of branches of *Tetraclinis articulata*.

The MIC values are presented in Table 4. The minimum inhibitory concentration for the two bacteria *Escherichia coli* and *Acinetobacter baumannii* is 2.5 µL/mL. These strains are considered to be the most sensitive by comparison with the other bacteria studied.

The low antimicrobial activity of the essential oil can be explained by its low chemical profile in compounds known for their antimicrobial power, such as certain oxygenated monoterpenes [25, 26].

Our results are in agreement with those obtained by Bourkhiss et al. [15] who tested the same essential oil in vitro against four bacterial and two fungal strains, and only *Staphylococcus aureus* and *Micrococcus luteus* were found sensitive.

The MBC/MIC ratio defines the bacteriostatic or bactericidal character of an essential oil; if this ratio is less than 4, then the oil is considered bactericidal [27].

The MBC/MIC ratio of the essential oil of Thuya branches is between 1 and 2, which confirms that the oil exerts a bactericidal effect (Table 4).

Conclusions

In the present work, we studied the chemical composition and the in vitro antibacterial activity of the essential oil of the branches of *Tetraclinis articulata* (Vahl) Masters of the Eastern Morocco against pathogenic strains of clinical origin. The extraction of the branches by hydrodistillation results in a yield of 0.84 ± 0.01%. For the determination of chemical composition, the essential oil was analysed by gas chromatography coupled with mass spectroscopy. The results show that this oil contains mostly α-pinene (38.75%), limonene (13.24%),

Table 4 Antibacterial parameters (MIC and MBC) of the essential oil of the *Tetraclinis articulata* and their interpretation

Bacterial strains	MIC (μL/mL)	MBC (μL/mL)	MBC/MIC	Interpretation
Escherichia coli	2.5	2.5	1	Bactericidal
Klebsiella pneumoniae	80	80	1	Bactericidal
Acinetobacter baumannii	2.5	5	2	Bactericidal
Staphylococcus aureus	40	80	2	Bactericidal

bornyl acetate (8.78%) and camphor (7.68%). The study of in vitro inhibitor power shows that the essential oil tested has significant antibacterial activity against *Escherichia coli* and *Acinetobacter baumannii,* two strains that are resistant. Generally, the results obtained are promising and open new perspectives in the field of natural applications which can be a viable alternative to synthetic products.

Abbreviations
MIC: minimum inhibitory concentration; MBC: minimum bactericidal concentration; ESBL: extended-spectrum beta-lactamases.

Author's contributions
Not applicable.

Author details
[1] Laboratory of Molecular Chemistry and Natural Substances, Department of Chemistry, Faculty of Sciences of Meknes, Moulay Ismail University, Bp 11201, Zitoune, Meknes, Morocco. [2] Laboratory of Microbiology, Military Hospital Moulay Ismail, Meknes, Morocco. [3] Laboratory of Plant Biotechnology and Molecular Biology, Faculty of Science, Moulay Ismail University, BP 11201, Zitoune Meknes, Morocco. [4] Department of Pharmacognosy, Faculty of Pharmacy, University of Medicine and Pharmacy, "Gr. T. Popa", Iasi, Str. 16University Str., Iasi 700115, Romania. [5] Department of Biology, Faculty of Biology, Alexandru Ioan Cuza University of Iasi, Bv Carol I, No 11, 700505 Iasi, Romania.

Acknowledgements
Not applicable.

Competing interests
The authors declare that they have no competing interests.

Funding
Not applicable.

References

1. Stanier RY, Doudoroff M, Adelberg Ed A. Microbiologie générale. Masson et Cie Editeurs; 1996.
2. Berche P, Gaillard JL, Simone M. Bactériologie: bactéries des infections humaines. Val-de-Meuse: Flammarion Médecine-Sciences; 1989.
3. Djeussi DE, Noumedem JAK, Seukep JA, Fankam AG, Voukeng IK, Tankeo SB, Nkuete AHL, Kuete V. Antibacterial activities of selected edible plants extracts against multidrug-resistant Gram-negative bacteria. BMC Complement Altern Med. 2013. https://doi.org/10.1186/1472-6882-13-164.
4. Djabou N, Lorenzi V, Guinoiseau E, Andreani S, Giuliani MC, Desjobert JM, Bolla JM, Costa J, Berti L, Luciani A, Muselli A. Phytochemical composition of Corsican Teucrium essential oils and antibacterial activity against foodborne or toxi-infectious pathogens. Food Control. 2013. https://doi.org/10.1016/j.foodcont.2012.06.025.
5. Andrade BFTM, Barbosa LN, Probst IS, Júnior AF. Antimicrobial activity of essential oils. J Essent Oil Res. 2014. https://doi.org/10.1080/10412905.2013.860409.
6. Chekol YA, Desta ZY. Determination of antioxidant and antimicrobial activities of leaf extracts of *Otostegia integrifolia*. Chem Cent J. 2018. https://doi.org/10.1186/s13065-018-0433-2.
7. Pombal S, Rodrigues CF, Araújo JP, Rocha PM, Rodilla JM, Diez D, Granja AP, Gomes AC, Silva LA. Antibacterial and antioxidant activity of Portuguese *Lavandula luisieri* (Rozeira) Rivas–Martinez and its relation with their chemical composition. SpringerPlus. 2016. https://doi.org/10.1186/s40064-016-3415-7.
8. Dimitrova L, Zaharieva MM, Popova M, Kostadinova N, Tsvetkova I, Bankova V, Najdenski H. Antimicrobial and antioxidant potential of different solvent extracts of the medicinal plant *Geum urbanum* L. Chem Cent J. 2017. https://doi.org/10.1186/s13065-017-0343-8.
9. Islam MZ, Hossain MT, Paul SC, Hossen F, Mukharjee SK, Sultana N. Evaluation of antioxidant and antibacterial activities of *Crotalaria pallida* stem extract. Clin Phytosci. 2018;1:1. https://doi.org/10.1186/s40816-018-0066-y.
10. Fournomiti M, Kimbaris A, Mantzourani I, Plessas S, Theodoridou I, Papaemmanouil V, Kapsiotis I, Panopoulou M, Stavropoulou E, Bezirtzoglou EE, Alexopoulos A. Antibacterial activity of essential oils of cultivated oregano (*Origanum vulgare*), sage (*Salvia officinalis*), and thyme (*Thymus vulgaris*) against clinical isolates of *Escherichia coli*, *Klebsiella oxytoca*, and *Klebsiella pneumonia*. Microb Ecol Health Dis. 2015. https://doi.org/10.3402/mehd.v26.23289.
11. Asghar N, Naqvi SAR, Hussain Z, Rasool N, Khan ZA, Shahzad SA, Sherazi TA, Janjua MRSA, Nagra SA, Zia-Ul-Haq M, Jaafar HZ. Compositional difference in antioxidant and antibacterial activity of all parts of the *Carica papaya* using different solvents. Chem Cent J. 2016. https://doi.org/10.1186/s13065-016-0149-0.
12. Dakak JE. Relation entre la qualité du bois de thuya de Maghreb (*Tetraclinis articulata*) et ses conditions de développement sur ses principaux sites phytoécologiques de son bloc méridional au Maroc. Nancy: Ecole Nationale du Génie Rural, des Eaux et des Forêts (ENGREEF); 2002.
13. Belkhadar J, Claisse R, Fleurentin J, Younos C. Repertory of standard herbal drugs in the Moroccan pharmacopoeia. J Ethnopharmacol. 1999;35:123–43.
14. Clevenger JF. Apparatus for the determination of volatile oil. J Am Pharm Assoc. 1928;1:1. https://doi.org/10.1002/jps.3080170407.
15. Bourkhiss B, Ouhssine M, Hnach M, Bourkhiss M, Strani B, Farah A. Composition chimique et bioactivité de l'huile essentielle des rameaux de *Tetraclinis Articulata*. Bull Soc Pharm Bordeaux. 2007;146:75–84.
16. Buhagiar J, Podestà MTC, Cioni PL. Essential oil composition of different parts of *Tetraclinis articulata*. Int J Herb Med. 2000. https://doi.org/10.1080/10412905.2000.9712034.
17. Ait Igri M, Holeman M, Ilidrissi A, Berrada M. Contribution à l'étude des huiles essentielles des rameaux et de bois de *Tetraclinis articulata* (Vahl) Masters. Plant Med Phytother. 1990;24:36–43.
18. Tékaya-Karoui A, Ben Jannet H, Mighri Z. Essential oil composition of terminal branches, cones and roots of *Tetraclinis articulata* from Tunisia. Pak J Biol Sci. 2007;10:2495–9.

19. Bertini LM, Pereira AF, Oliveira CLL, Menezes EA, Morais SM, Cunha FA, Cavalcanti ESB. Perfi de sensibilidade de bactérias frente a óleos essenciais de algumas plantas do nordeste do Brasil. Infarma. 2005;17:80–3.
20. Gende LB, Mendiara S, Fernández NJ, Baren CV, Lira LA, Bandoni A, Fritz R, Floris I, Eguaras M. Essentials oils of some *Mentha* spp. and their relation with antimicrobial activity against *Paenibacillus larvae*, the causative agent of American foulbrood in honey bees, by using the bioautography technique. Bull Insect. 2014;67:13–20.
21. Colombo RP, Martíneza AE, Pardo AF, Bidondo LF, Baren VC, Lira LP, Godeas AM. Differential effects of two strains of *Rhizophagus intraradices* on dry biomass and essential oil yield and composition in *Calamintha nepta*. Rev Argent Microbiol. 2013;45:114–8.
22. Kapoor R, Giri B, Mukerji KG. Glomus macrocarpum: a potential bioinoculant to improve essential oil quality and concentration in Dill (*Anethum graveolens* L.) and Carum [*Trachyspermun ammi* (Linn.) Sprague]. World J Microbiol Biotechnol. 2002;1:1. https://doi.org/10.1023/a:1015522100497.
23. Baydar H, Sagdic O, Ozkan G, Karadogan T. Antibacterial activity and composition of essential oils from Origanum, Thymbra and Satureja species with commercial importance in Turkey. Food Control. 2004. https://doi.org/10.1016/S0956-7135(03)00028-8.
24. Koba K, Sanda K, Raynaud C, Nenonene YA, Millet J, Chaumont JP. Activités antimicrobiennes d'huiles essentielles de trois *Cymbopogon sp.* africains vis-à-vis de germes pathogènes d'animaux de compagnie. Ann Méd Vét. 2004;148:202–6.
25. Dorman HJD, Deans SG. Antimicrobial agents from plants: antibacterial activity of plant volatile oils. J Appl Microbiol. 2000;88:308–16.
26. Oussalah M, Caillet S, Saucier L, Lacroix M. Inhibitory effects of selected plant essential oil on the growth of four pathologic bacteria: *E. coli* O157: H7, *Salmonella typhimurium*, *Staphylococcus aureus* and *Listeria monocytogenes*. Food Control. 2007. https://doi.org/10.1016/j.foodcont.2005.11.009.
27. Guinoiseau E. Molécules antibactériennes issues d'huiles essentielles: séparation, identification et mode d'action. PhD thesis. Corse: University of Corse; 2011.

15

Nano and Pheroid technologies for development of foliar iron fertilizers and iron biofortification of soybean grown in South Africa

Jesper T. N. Knijnenburg[1,2], Florentine M. Hilty[1], Janro Oelofse[3], Riaan Buitendag[3], Michael B. Zimmermann[1*], Ismail Cakmak[4] and Anne F. Grobler[5]

Abstract

Background: Foliar iron (Fe) fertilization of crops may increase Fe concentrations in edible portions of plants and improve yield in soils with low available Fe. However, the role of foliar Fe fertilization in increasing seed Fe has not been studied in soybeans (*Glycine max*). In this study, the Pheroid® nutrient delivery technology was combined with $FeSO_4$ or nanostructured $FePO_4$ to develop potential new Fe foliar fertilizers. Eight different treatments including different combinations of $FeSO_4$ and Pheroids were foliarly applied on field-grown soybeans in Northern Cape province in South Africa to investigate their influence on seed nutrient composition and yield.

Results: Confocal and optical microscopy images indicate that $FeSO_4$ or $FePO_4$ was not entrapped in the Pheroids but formed separate precipitates. The average seed Fe of the non-treated plants was 56 ± 3 mg kg^{-1}, and none of the treatments (including the positive controls, $FeSO_4$ and $FeSO_4$ with citrate) significantly increased seed Fe over the control. There was also no significant change in yield or seed Zn, P, protein, or phytic acid. Thus, Pheroids as well as $FeSO_4$ are not suitable as delivery system for Fe to soybean seeds due to Pheroid incompatibility with $FeSO_4$ and poor dispersibility of $FePO_4$.

Conclusions: Because none of the Fe treatments (including positive controls) affected seed Fe concentrations, foliar Fe application may not be effective to increase seed Fe in crops such as soybean that already have high native Fe.

Keywords: Biofortification, Iron, Foliar application, Soybean, Pheroids, Nanotechnology, Fertilizer

Background

Anemia affects close to 2.2 billion people worldwide, with approximately half of the cases being caused by iron (Fe) deficiency (ID) [1]. The high prevalence of ID has major negative impacts on health and economy, and Fe deficiency anemia increases risk for maternal death, low birth weight, and infant mortality [2]. Biofortification is a potentially sustainable and cost-effective approach to reduce micronutrient deficiencies. While selective breeding can provide long-term improvements, agronomic biofortification (i.e., fertilizer application to the soil and/or foliage) can provide a short-term solution [3]. Foliar application has been successful for selenium [4] and particularly for zinc (Zn) to increase grain levels in wheat and rice [5–7]. Since water-soluble ferrous compounds (like ferrous sulfate ($FeSO_4$)) applied to the soil are rapidly converted into unavailable ferric forms [8], it has been suggested that foliar Fe application may be the only way to increase Fe levels in edible plant parts [8], e.g., in rice, wheat, maize, and pea [9–13]. In a few studies, however, no

*Correspondence: michael.zimmermann@hest.ethz.ch
[1] Department of Health Sciences and Technology, Human Nutrition Laboratory, Institute of Food, Nutrition and Health, ETH Zurich, Schmelzbergstrasse 7, 8092 Zurich, Switzerland
Full list of author information is available at the end of the article

clear effect of foliar Fe treatment on grain Fe for barley and oats [14], and wheat [15] was observed.

Legumes, especially soybeans, represent an important part of the diet in many populations [16]. Soybeans are a good source of high-quality protein and non-heme Fe [16, 17]. Due to the wide genetic variety, Fe content of soybean seeds [18] can range from 40 to 90 mg kg^{-1} dry weight (DW) and may, thus, present a good target for agronomic biofortification and selective breeding [19]. Previous studies on foliar or soil treatments focused on correcting Fe deficiency chlorosis in soybean [20]. Yet, soil application of a chelated Fe compound also increased seed Fe concentrations in chlorotic soybeans [21].

The Pheroid® technology (hereafter called Pheroid or Pheroids) is a patented delivery system comprised of essential fatty acids and other ingredients emulsified in water [22]. Proven for active ingredient delivery in the pharmaceutical industry [23], Pheroids also seem to enhance root and foliar uptake and translocation of minerals in plants [24]. Pheroids may, therefore, be a promising vehicle or tool to deliver Fe to the edible plant parts of staple crops.

This study had two main aims: first, the potential of newly developed Pheroid-based nano-Fe foliar fertilizers was studied and compared to positive and negative control treatments. To the best of our knowledge, nanostructured $FePO_4$ has not yet been tested as a source of Fe or P to plants. Second, this study aimed to evaluate whether soybean could be a potentially promising crop for biofortification with Fe through foliar application. Most studies have been carried out on cereal crops [9, 11, 12], while agronomic biofortification of dicotyledonous crops with Fe is relatively unexplored [13].

Two Fe compounds were selected for foliar application onto soybean, namely $FeSO_4$ and nanostructured ferric phosphate ($FePO_4$). The Fe salt $FeSO_4$ is commonly used as an inexpensive foliar fertilizer and has been shown to increase grain Fe in rice [9, 11], wheat [12], and pea [13]. First, Pheroid treatments containing $FeSO_4$ and $FePO_4$ were formulated and characterized. Subsequently, Pheroid-emulsified and bare $FeSO_4$ and $FePO_4$ were applied as foliar sprays to field-grown soybean with the aim to increase Fe concentration in seeds. In addition, seed Zn, P, protein, and phytic acid (PA) were measured, and seed aluminum (Al) was used as indicator for a possible Fe contamination from soil dust or particles during harvest and transport [25].

Methods

Treatment preparation

Eight different treatments were applied (Table 1): distilled water as control treatment (T1), Pheroids (T2), $FeSO_4$ (T3), Pheroids with $FeSO_4$ (T4), $FePO_4$ (T5), Pheroids with $FePO_4$ (T6), $FeSO_4$ with pH adjusted (T7), and $FeSO_4$ with citrate (T8). The Pheroid-containing treatments (T2, T4, and T6) were prepared as a concentrate in the laboratory [24] and diluted to contain 1.2 wt% Pheroids and 0.4 g Fe L^{-1} (7 mM). This Fe concentration is comparable to previous foliar Fe spray studies [26]. The non-Pheroid treatments were all prepared directly before spray application. Treatment T3 was prepared by dissolving $FeSO_4$ (dried, micronized powder, FCC 6, Dr. Paul Lohmann, Emmerthal, Germany) in distilled water to obtain a final concentration of 0.4 g Fe L^{-1} for foliar spray. In the first two spray preparations, some Fe precipitated due to the high pH (around 8) of the local deionized water source. Therefore, two treatment groups T7 and T8 were included for the second and third spray applications as positive (dissolved Fe) control groups. These two Fe treatments were acidified to prevent Fe precipitation. In the case of T7, the required amount of $FeSO_4$ was dissolved in distilled water with pH = 5.5 (adjusted using 0.1 M HCl). For T8, $FeSO_4$ was dissolved in a citric acid solution (Fe:citrate = 1:1 molar ratio) with pH adjusted to 5.5 using NaOH. The resulting pH of T7 and T8 was 4.5 and 4.2, respectively, which is in good agreement with previous studies [27, 28].

For the preparation of T5, $FePO_4$ (specific surface area 28 m^2 g^{-1}, fine powder, FCC 6, Dr. Paul Lohmann) was dispersed in the required amount of distilled water to get 0.4 g Fe L^{-1}, and the sample was shaken vigorously before spray application. No sedimentation was observed between sample preparation and after spray application.

Treatment characterization

The size distribution of the Pheroid vesicles (T2, T4, and T6) was measured in duplicate by light scattering on a Malvern Mastersizer 2000 fitted with a Hydro 2000MU sample dispersion unit (Malvern Instruments, UK). The laser obscuration was kept constant at 12% by diluting the sample in distilled water (stirred at 2000 rpm). A refractive index of 1.481 was used for the Pheroids. Analysis of T5 was done at a lower obscuration of 1–2%, using refractive index 1.7 for $FePO_4$. Zeta potentials of the different treatments were measured on a Malvern Zetasizer Nano ZSP (Malvern Instruments). Samples were diluted in distilled water (1500× dilution for T2, T4, and T6; 100× dilution for T5). Measurements were done twice in duplicate and the average zeta potential was recorded. The infrared spectra of Pheroid vesicles were measured using a Bruker ALPHA FT-IR spectrometer (Bruker Optics, USA). Spectra were collected from 400 to 4000 cm^{-1} with a resolution of 4 cm^{-1} and corrected by subtracting the absorbance spectrum of water.

Optical microscopy analysis of the Pheroid treatments was done on an inverted microscope (Nikon Eclipse

TE2000-E, Nikon Instruments, USA) equipped with a DS-Ri1 Nikon digital camera with real-time imaging. Confocal laser scanning microscopy images were taken with an inverted Nikon D-Eclipse C1 microscope with a 30-mm pinhole and a 60 × 1.40 ApoPlanar oil immersion objective. This microscope was equipped with a diode laser (emission at 409 nm), an Argon Ion laser (emission 488 or 515 nm) and a Helium–Neon polarized laser (emission at 543 nm). Pheroids were labeled with Nile red (1 g L^{-1}) which fluoresced at 650 nm and longer wavelengths. Approximately, 5 μL of sample was placed on a microscope slide, covered with a glass coverslip, and placed with the coverslip face towards the objective.

Trial location

The field trial was carried out in the Northern Cape Province in South Africa (GPS coordinates 28°09′29.1″ S; 024°32′13.7″ E). Soil and fresh leaf analyses were carried out by SGS (Cape Town, South Africa). The soil had a high clay content (40%) with a pH of 6.9 (KCl), with 17.4 mg kg^{-1} available N (1.9 mg kg^{-1} as NH_4^+ and 15.5 mg kg^{-1} as NO_3^-), and 15.4 mg kg^{-1} available P (Bray 1). The available macronutrient concentrations measured by ammonium acetate method were (in mg kg^{-1}): K 446; Na 244; Ca 5116; Mg 1269. The available micronutrient concentrations measured by 0.1 M HCl method were (in mg kg^{-1}): Fe 2.42; Zn 0.41; Mn 1.69; Cu 0.23. Micronutrient concentrations are comparable to the study of Pahlavan-Rad and Pessarakli [12] but lower compared to others [10, 11].

Soybeans (*Glycine max*, cultivar DM4670RR, *Rhizobium* inoculate FDP) were planted with a density of 380,000 plants per hectare, row spacing 300 mm. The plants received 200 kg ha^{-1} monoammonium phosphate. During the study, there was no visual indication of Fe deficiency chlorosis in plants. Leaf samples taken during BBCH growth stage 69 to determine mineral nutritional status of the plants showed that the plants were adequately supplied with mineral nutrients (Additional file 1: Table S1) [29, 30].

Experimental outline

In the field study, eight different treatments were applied, as described in Treatment preparation. The plots (each 10 m^2) for the first 6 treatments were set up in a randomized complete block design (5 by 6) with 5 replicates each, while the plots for treatment 7 and 8 were set up in a chessboard-like pattern (3 by 4) with 6 replicates each. Treatments were applied at the following BBCH growth stages [29] with 2 weeks in between: first spray at stage 69 (end of flowering), second spray at stage 77 (~70% of pods have final length), and third spray at stage 79 (during seed filling). This application sequence was selected based on the recommendations for Zn on wheat [5]. Treatments 1–6 were applied at all three spray occasions, while T7 and T8 were only applied at second and third sprays. Spray application was done in the early mornings (after sunrise) at 500 L ha^{-1} (equivalent to 1.3 mL/plant) using a pressure-regulated knapsack sprayer equipped with a double (spray 1) or single even fan nozzle (spray 2 and 3), each nozzle having spray width 0.5 m. These spray conditions correspond to approximately 20 mg Fe/spray/m^2, similar to previous studies on soybean [31], pea [13] and tomato [32]. Weather conditions were recorded before and after foliar spray; on average, ambient temperature was 21–33 °C, relative humidity was 45–66%, and wind speed was < 2.5 m s^{-1}. There was no rain for at least 36 h after each application.

Plants were manually harvested at full maturity. In each plot, 10 plants were randomly selected and cut off directly above the ground, avoiding the outer rows and at least 0.5 m from each plot edge to avoid side effects. Harvested plants were dried in a drying oven, and dry pods were manually separated from the stems.

Table 1 Composition, pH, and zeta potential of the various applied foliar fertilizer treatments

	Composition	Fe (g L^{-1})	pH (–)	Zeta potential (mV)	Spray application
T1	Control (dH_2O)	0	7.5 ± 0.4		1, 2, 3
T2	Pheroids	0	7.4 ± 0.6	− 25 ± 3	1, 2, 3
T3	$FeSO_4$	0.4	5.3 ± 0.7		1, 2, 3
T4	Pheroids + $FeSO_4$	0.4	3.9 ± 1.2	− 14 ± 2	1, 2, 3
T5	$FePO_4$	0.4	6.7 ± 0.8	− 14 ± 3	1, 2, 3
T6	Pheroids + $FePO_4$	0.4	6.4 ± 0.9	− 24 ± 4	1, 2, 3
T7	$FeSO_4$, pH adjusted	0.4	4.5 ± 0.5		2, 3
T8	$FeSO_4$ + citrate	0.4	4.2 ± 0.1		2, 3

The BBCH growth stages of the spray applications are: 1, stage 69; 2, stage 77; and 3, stage 79

Analysis of seeds

Number of seeds per pod (averaged over 100 pods) and 100-seed weight (averaged over 400 seeds) were determined manually, and seed samples were stored in zip lock bags at room temperature. Before chemical analysis, the seeds were rinsed under flowing nanopure water and dried in a drying oven for 48 h at 50 °C. The dry seeds were ground in a titanium mill (ZM1, Retsch, Germany) with titanium sieve (mesh size 0.25 mm), and ground seed samples were stored in zip lock bags at room temperature until analysis.

The moisture content of the milled samples was measured using a halogen moisture analyzer (HE53, Mettler Toledo, Switzerland). For each measurement, 2 g of sample was dried at 115 °C for 12 min. Averaged over all samples, the moisture content was 50 ± 2 g kg^{-1}. All chemical analyses are reported as normalized per dry weight (DW) of each individual sample.

For mineral analysis, 250 mg of ground sample was digested in triplicate using 65% HNO_3 (sub-boiled, in-house production) in a pressurized microwave digestion system (turboWAVE, MLS, Germany). In each run, a single blank and duplicate reference sample material (rice flour, SRM 1568b, National Institute of Standards & Technology, USA) were used as internal controls. The Fe and Al concentrations were analyzed by graphite furnace atomic absorption spectroscopy (GF-AAS, AA240Z with GTA-120 Graphite Tube Atomizer, Agilent Technologies, USA), and Zn concentrations by flame AAS (FAAS, AA240FS, Agilent Technologies). For each mineral, external calibration was performed using standard Tritisol solutions (Merck, USA).

Seed phytic acid (PA) contents were measured in triplicate by a modified method described by Makower [33]. In each run, a single blank and duplicate in-house reference sample material (hard red wheat bran) were used as controls. Using 500-mg seed powder, the PA was extracted with 12 wt% trichloroacetic acid (≥99%, Sigma-Aldrich, USA) and subsequently precipitated with cerium ($Ce(SO_4)_2 \cdot 4\ H_2O$, ≥98%, Merck). After mineralization for 240 min at 338 °C on a Digest Automat K-438 (Büchi, Switzerland) in concentrated sulfuric acid (95–97%, Sigma-Aldrich), inorganic phosphate was determined according to van Veldhoven and Mannaerts [34] by a microplate spectrophotometer (PowerWave HT, BioTek, USA) and converted into PA concentrations.

Total phosphorus (P) content was measured in triplicate by digesting 250-mg DW ground sample for 90 min at 420 °C on a Digest Automat K-438, using 20-mL concentrated sulfuric acid with a Kjeldahl tablet (Merck). One reference sample material (rice flour, SRM 1568b) and one blank sample were included in each run. The inorganic phosphate concentration was determined as described for the PA analysis.

The total nitrogen (N) content was measured in triplicate according to the Kjeldahl method. To each 250-mg ground sample, 20-mL concentrated sulfuric acid with a Kjeldahl tablet was added, and the sample was digested at 420 °C for 90 min on a Digest Automat K-438. One blank sample was included in each run. The N concentrations were determined by titration on a KjelFlex K-360 (Büchi) connected to a titrator (877 Titrino plus, Metrohm, Switzerland), and 0.1 M HCl (made from Fixanal ampoule, Fluka) titration volumes were converted into seed N content after subtracting the blank. A conversion factor of 6.25 was used to convert N into protein content.

Statistical analysis

Statistical analyses were done using IBM SPSS Statistics (version 23, IBM) and Microsoft Excel (version 2010, Microsoft). Data were subjected to one-way ANOVA, and differences were considered statistically significant for $p \leq 0.05$.

Results

Treatment characterization

Treatment pH and zeta potential

The Pheroid vesicles consist primarily of unsaturated essential fatty acids dispersed in water [35]. Without any Fe, Pheroids (T2) had a neutral pH and zeta potential of −25 mV (Table 1) [35]. The pH of Pheroids with $FeSO_4$ in T4 was significantly lower compared to T2; the presence of dissolved Fe^{2+} most likely catalyzed the fatty acid oxidation that resulted in the formation of acidic oxidation products including carboxylic acids [36] that further reduced the pH to 3.9 in T4. Pheroids are susceptible to oxidation due to their small size, and thus high surface–volume ratio [37]. Any change in Pheroid chemical composition, however, could not be confirmed by infrared spectroscopy analysis (Additional file 1: Figure S1). The pH of $FePO_4$ in water (T5) was 6.7, with zeta potential of −14 mV [38]. Treatment T6 had neutral pH and its zeta potential was the same as that of plain Pheroid (T2) thus $FePO_4$ did not appear to change Pheroid properties. Treatments T7 and T8 had pH of 4.5 and 4.2, respectively, to keep the $FeSO_4$ dissolved and avoid Fe precipitation.

Treatment particle size and morphology

The frequency (Fig. 1a) and cumulative (Fig. 1b) particle size distributions of the different Pheroid-based treatments (T2, T4, and T6) did not show major differences. The bimodally shaped size distributions are typical for Pheroids [39], here with main mode at 2 µm and a smaller mode around 0.2 µm. Characteristic sizes of the size distributions (d_{10}, d_{50}, and d_{90}; inset table in Fig. 1b) are comparable to that of Pheroids analyzed by flow cytometry [23]. The inclusion of $FeSO_4$ (T4) or $FePO_4$ (T6) caused a slight shift towards larger sizes, as is evident from the increase in d_{50} and d_{90}. In T5, the $FePO_4$ agglomerates were larger than the Pheroids, with their main peak at 30 µm, and a smaller, secondary peak around 5 µm.

Optical and confocal microscopy techniques were applied to verify if any Fe entrapment in the Pheroid vesicles had taken place [35]. The Pheroids are spherical and uniform in shape (Fig. 2a), and confocal microscopy clearly shows red-fluorescent Pheroids in contrast to background (Fig. 2d). Addition of $FeSO_4$ to the Pheroids (in T4) did not change Pheroid size or shape (Fig. 2b). Few separate orange–red aggregates/agglomerates with size up to 20 µm are visible (green in the confocal image, Fig. 2e). In both confocal and optical microscopy images, no clear Fe entrapment is visible, and agglomerates are separated from the Pheroids. Most likely, dissolved Fe^{2+} reacted with unsaturated fatty acids in the Pheroids, causing the Fe^{2+} to be oxidized to poorly soluble ferric hydroxides that precipitated. In the presence of Pheroids, the $FePO_4$ agglomerates in T6 (Fig. 2c) were considerably smaller than those in T5 (Fig. 1); possibly the Pheroids inhibited $FePO_4$ re-agglomeration. As for $FeSO_4$, no visual entrapment could be verified (as indicated by the arrows in Fig. 2f).

Yield parameters and soybean composition

Yield analysis

The average number of seeds per pod over all treatments was 2.39 ± 0.10, and average 100-seed weight was 18.2 ± 0.6 g, and there was no statistically significant difference between treatments (Table 2). Leaf analysis during flowering stage indicated that in the current study, there was no Fe deficiency stress in the untreated soybeans (Additional file 1: Table S1). In other studies, there was also no increase in yield after foliar Fe spray [12, 14], probably due to the lack of any visible Fe deficiency chlorosis.

Seed analyses

The average seed Fe concentration over all treatments was 57 ± 4 mg kg^{-1} and foliar application did not increase seed Fe concentrations (Table 2). The average seed Al content was < 4 mg kg^{-1}, confirming that there was no Fe contamination from soil dusts or soil particles [25], as typical Al levels in soybean seeds without soil contamination [40] are 4–12 mg kg^{-1}. Average seed Zn content was 42 ± 2 mg kg^{-1} and not significantly different among the groups. Total P and PA concentrations were also not influenced by the treatments (average concentration of 6.0 ± 0.2 g kg^{-1} and 10.1 ± 1.2 g kg^{-1}, respectively). It was estimated that approximately 54% of seed P is present in the form of PA. Similarly, no significant difference in protein concentration was detected (average concentrations 385 ± 11 g kg^{-1}, Table 2). The nutrient concentrations found in the current study are in the range typical

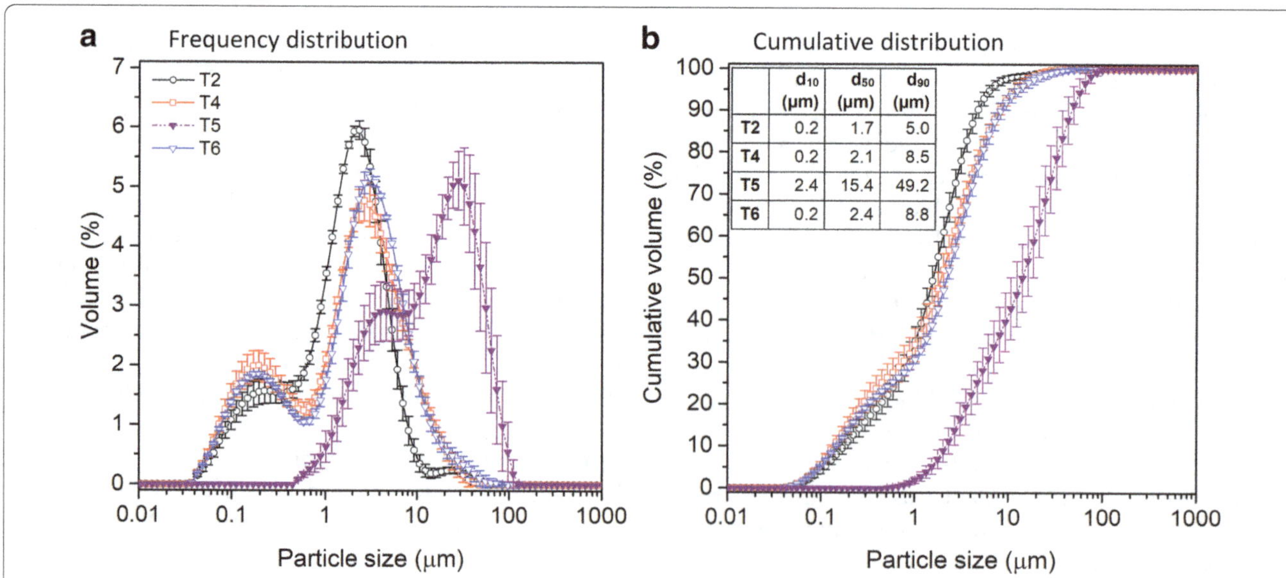

Fig. 1 Average **a** frequency and **b** cumulative particle size distributions of the different foliar treatments, measured after treatment application. Error bars indicate standard deviations between three different treatments. The inset table displays average characteristic sizes (d_{10}, d_{50}, and d_{90}) of each distribution

for soybeans [18, 21, 40, 41]. Using a higher statistical significance level ($p \leq 0.10$) made no difference for any of the outcomes.

Discussion

The main objectives of this study were to combine the Pheroid technology with $FeSO_4$ or nanostructured $FePO_4$ to develop a novel foliar Fe fertilizer, and to test it for enhancement of seed Fe concentrations in soybeans compared with commonly used foliar fertilizers like $FeSO_4$ and $FeSO_4$ with citrate. The results obtained showed that none of the formulations caused any change in soybean seed Fe, Zn, P, PA, and protein concentrations. The cause for this could either come from the formulation of the treatment, or from the chosen crop.

The particle size distribution and surface charge (zeta potential) of the treatment may have an influence on foliar penetration [42, 43]. Plant cuticles have an isoelectric point around 3.0 [44] leading to negatively charged cuticle at pH higher than 3. Since negatively charged Fe compounds enhance foliar penetration [43, 45], our compounds should theoretically have penetrated the leaves (Table 1). Foliar uptake of nanoparticles via stomatal penetration was seen for metal oxides (< 50 nm) in watermelon plants [42] and carboxylate-coated polystyrene nanoparticles (45 nm) in broad bean [46]. Larger particles (1.1 µm) were not taken up [46], since particles with size over 100 nm may block stomata [42]. In our study, the $FePO_4$ agglomerates (in T5) were predominantly larger than 1 µm (Fig. 1) making leaf penetration unlikely [42, 46]. There was no clear evidence for Fe entrapment in the Pheroid vesicles for T4 and T6 (Fig. 2). Ideally, Fe packaging in the Pheroids would have improved foliar penetration and translocation to the seeds. The lack of Fe loading may have resulted in the Pheroids being foliarly absorbed without Fe. In T4, the Fe^{2+} did likely react with the Pheroids causing precipitation of ferric hydroxides outside the Pheroids, which may have partially contributed to the poor soybean response. In one study, ferric chloride showed low penetration compared to chelated

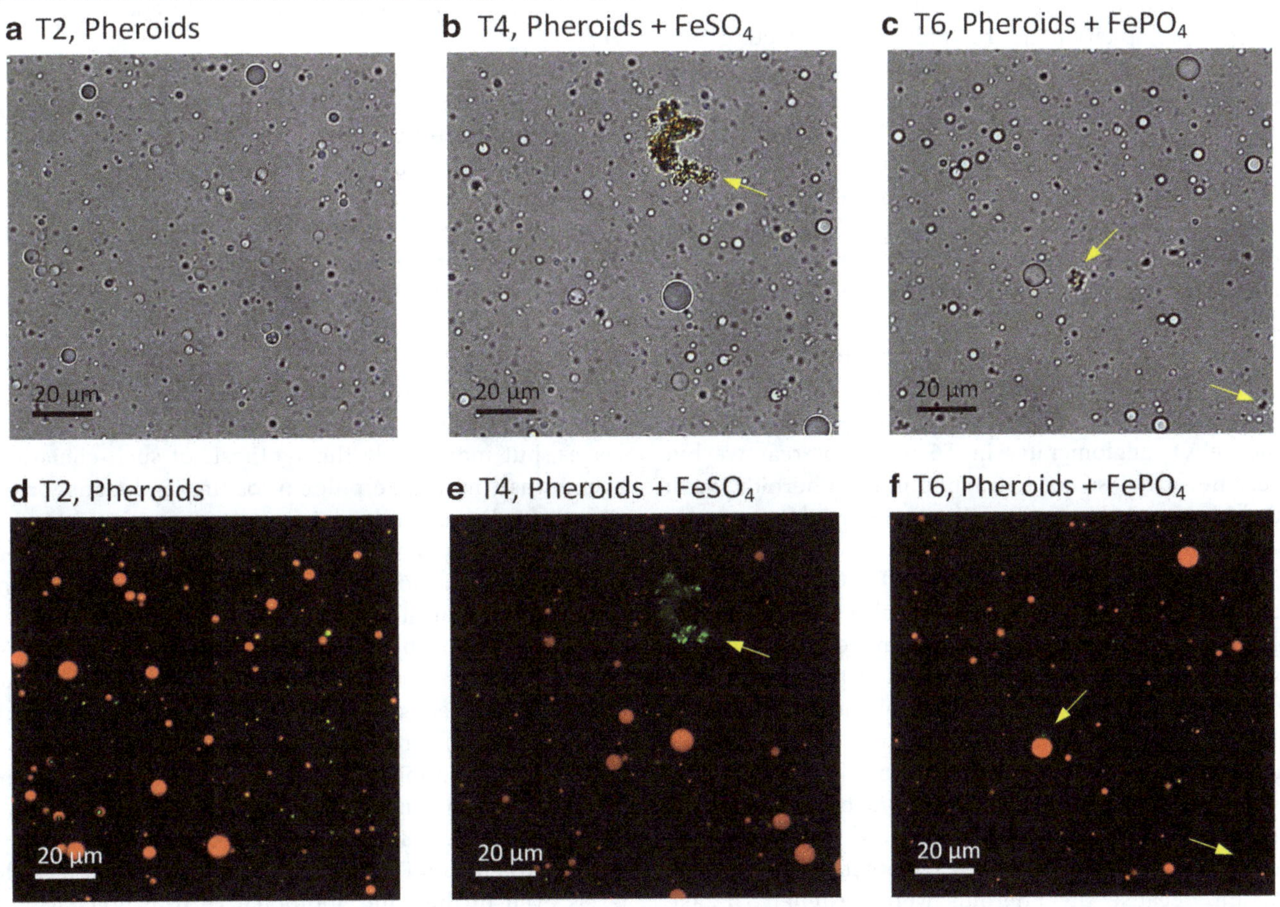

Fig. 2 **a–c** Optical and **d–f** confocal microscopy images of **a**, **d** T2, **b**, **e** T4, and **c**, **f** T6. Scale bars correspond to 20 µm. Confocal images are overlay images of red and green excitation. Pheroids fluoresce red by association with Nile red. Green spots in **d** are most probably an artifact from background noise. Arrows highlight particulate agglomerates, namely ferric precipitates for T4 and $FePO_4$ for T6

Table 2 Number of seeds per pod and weight of 100 seeds, and Fe, Zn, P, and protein concentration of mature soybean seeds after application of various foliar treatments T1–T8

	Seeds/pod (–)	100-seed weight (g)	Fe (mg kg^{-1})	Zn (mg kg^{-1})	P (g kg^{-1})	Protein (g kg^{-1})
T1	2.4 ± 0.1(4)	18.6 ± 0.5(3)	56.0 ± 3.0(5)	42.9 ± 2.3(5)	5.8 ± 0.2(4)	388.1 ± 13.6(4)
T2	2.4 ± 0.1(5)	18.1 ± 0.3(1)	60.1 ± 5.4(9)	42.9 ± 1.3(3)	6.1 ± 0.2(3)	383.8 ± 6.6(2)
T3	2.4 ± 0.1(3)	18.4 ± 0.6(4)	56.3 ± 3.0(5)	41.4 ± 1.0(3)	5.9 ± 0.3(4)	388.4 ± 12.4(3)
T4	2.3 ± 0.1(4)	17.8 ± 0.6(4)	57.2 ± 2.0(4)	41.6 ± 1.4(3)	5.8 ± 0.2(3)	385.6 ± 9.8(3)
T5	2.4 ± 0.1(3)	18.2 ± 0.3(2)	58.1 ± 5.3(9)	42.3 ± 1.7(4)	6.0 ± 0.2(4)	386.1 ± 12.9(3)
T6	2.4 ± 0.1(2)	17.8 ± 0.7(4)	58.0 ± 4.1(7)	42.3 ± 1.8(4)	6.1 ± 0.2(4)	378.0 ± 15.0(4)
T7	2.4 ± 0.1(2)	18.4 ± 0.4(2)	56.0 ± 3.3(6)	43.3 ± 1.4(3)	6.1 ± 0.2(3)	387.3 ± 10.1(3)
T8	2.4 ± 0.1(6)	18.3 ± 0.7(4)	55.8 ± 3.0(5)	42.6 ± 1.9(4)	6.1 ± 0.3(5)	382.8 ± 8.0(2)
	NS	NS	NS	NS	NS	NS

Values are given as mean ± SD, and the coefficient of variation (CV) is given in brackets. The last row shows the statistical significance from one-way ANOVA. NS: no statistically significant difference ($p > 0.05$). For each treatment, $n \geq 5$

Fe compounds caused by precipitation of Fe hydroxides [47]. Also, ferric iron precipitation from inorganic Fe compounds sprayed onto leaves may occur on leaves of plants grown on alkaline soils and, thus, inhibit reduction towards mobile Fe^{2+} [48, 49]. Such reactivity and subsequent precipitation can be limited by protecting Fe^{2+} using chelating agents like EDTA [36], which, however, would increase the cost and decrease attractiveness of such formulations. Combining $FeSO_4$ with citrate (T8), however, also did not increase seed Fe.

The foliar penetration half-life of ferric compounds in poplar leaves was 20–30 h at 100% humidity, with longer times at lower humidity [47]. In common bean, approximately 8% of foliarly applied $FeCl_3$ was absorbed after 48 h [50]. Therefore, the low foliar Fe absorption rate combined with a possible event of rain (even after more than 48 h) may have resulted in the Fe compounds being washed off before foliar absorption. The $FePO_4$ agglomerates in T6 were not reactive but were nevertheless not entrapped in the Pheroids, most likely because the hydrophilic nanoparticles were not compatible with the mainly hydrophobic Pheroid phase [51]. Also, since both Pheroids and $FePO_4$ phase exhibited a negative surface charge (Table 1) they would electrostatically repel each other. Indeed, both hydrophobicity and surface charge are important for Pheroid entrapment: positively charged polymeric nanoparticles could be successfully entrapped in Pheroids, while negatively charged ones were likely not [35].

It is recommended to use a surfactant to improve contact between the treatment and the leaves [14, 43, 52]. In the present study, we opted against a surfactant because the Pheroids were designed to take over this role. The $FeSO_4$ and $FeSO_4$ with citrate were selected as positive control treatments because they have previously shown to be effective foliar fertilizers to increase grain Fe [9–11, 13], even without surfactant [12]. The absence of any response to the positive controls indicates that the crop itself (rather than the formulation) is responsible for the lack of change in seed Fe.

Nitrogen plays a major role in plant mineral accumulation, in both mineral transport and storage. During seed filling, micronutrients such as Fe and Zn are transported to the developing seed via the phloem [53]. If Fe was indeed foliarly absorbed, as would be expected for the positive control treatments (T7 and T8), the remobilization of Fe from leaves to seeds may have been the limiting factor [27]. Seed Fe and Zn concentrations (Table 2) are similar to or lower than leaf Fe and Zn concentrations (Additional file 1: Table S1), respectively, indicating low mineral remobilization efficiency [54]. In pea, it was not the availability of Fe in source leaves, but more likely the synthesis of such chelating agents that limited Fe phloem loading and transportation to seeds [55]. Indeed, improving nutritional N status of plants by soil and/or foliar applications, either with or without foliar Fe fertilizers, enhanced grain Fe concentration in wheat [15]. Under Fe deficiency, a fraction of Fe may be remobilized from senescing leaves [56]. Despite the low available soil Fe in the current study, the soybeans were not Fe deficient (Additional file 1: Table S1), providing no additional incentive for the plants to remobilize Fe from leaf.

Generally, soybean has at least twice as much seed protein as wheat [57, 58]. Because soybean seeds are already rich in protein and Fe, it may be difficult to increase such levels even further, and additional Fe from foliar spray may, thus, have no effect on seed Fe. Analogously, foliar spray of Zn fertilizer resulted in higher increase in grain

Zn of wheat compared to common beans [7]. The smaller increase was ascribed to the high native Zn and protein concentrations in common bean. Moreover, soybean seed mineral concentrations are mostly determined by their genotype and are difficult to overcome using agronomic approaches [59]. Low seed Fe varieties are not likely to increase their seed Fe levels [60] and in general Fe concentrations do not vary more than 10% for a given cultivar [18].

Summarizing, several potential flaws in the formulations have been pointed out, and we conclude that Pheroids may not be suitable for Fe delivery due to their reaction with Fe^{2+}, the difficulty to break up and encapsulate $FePO_4$ agglomerates, and hydrophilic/hydrophobic interactions. The positive controls, however, did not cause any seed response either. It appears that the major obstacle to increase seed Fe concentration is the selection of the crop. Thus, optimization of the formulation may elicit only a limited response because genetics may largely control seed composition limits of soybeans [18, 60].

The HarvestPlus program (http://www.harvestplus.org/) is currently developing more nutritious crops (mainly through plant breeding) with higher amounts of primarily three nutrients: vitamin A, Fe or Zn [61]. The current study demonstrates that Pheroids may not be compatible with Fe. Zinc, however, may be more compatible with Pheroid vesicles due to its redox inactivity and higher stability in solution. While Pheroids may enhance plant uptake of Zn [24], it is not known whether the Zn reaches edible plant parts. In addition, entrapment of vitamin A or provitamin A (carotenoids) in Pheroids could be promising for agronomic biofortification. Given the success of Pheroids in entrapment of pharmaceutical compounds [23] and the fatty acid nature of Pheroids, organic molecules may be easier to entrap than minerals. Their translocation to edible plant parts, however, must still be evaluated.

Conclusions

To our knowledge, this is the first evaluation of nanostructured $FePO_4$ and the Pheroid delivery system for foliar application, and the first attempt to increase Fe content in soybeans by foliar Fe spray application. Novel Pheroids containing $FeSO_4$ and nanosized $FePO_4$, as well as bare $FeSO_4$ and nanosized $FePO_4$, that were foliarly applied on soybeans did not change yield or seed concentrations of Fe, Zn, P, protein, and phytic acid. The lack of soybean response can be partly ascribed to the formulation (reactivity of $FeSO_4$ and no Pheroid entrapment), and the Pheroid technology may not be suitable for foliar Fe fertilization. However, since the positive controls also did not elicit a response, it appears that soybeans are ill-suited for agronomic biofortification due to their inherently high Fe and protein concentrations and tight genetic constraints. Thus, not only formulation optimization, but also careful crop selection is crucial for successful agronomic biofortification.

Abbreviations

DW: dry weight; FAAS: flame atomic absorption spectroscopy; GF-AAS: graphite furnace atomic absorption spectroscopy; ID: iron deficiency; PA: phytic acid; SRM: standard reference material.

Authors' contributions

JTNK designed the study, produced and analyzed the foliar treatments, collected the data, analyzed and interpreted the data, and wrote the manuscript. FMH designed the study, analyzed and interpreted the data, and co-wrote the manuscript. JO designed and carried out the foliar application experiments. RB designed and carried out the foliar application experiments. MBZ designed the study and revised the manuscript. IC designed the study, analyzed and interpreted the data, and co-wrote the manuscript. AFG designed the study and the foliar treatments and revised the manuscript. All authors read and approved the final manuscript.

Author details

[1] Department of Health Sciences and Technology, Human Nutrition Laboratory, Institute of Food, Nutrition and Health, ETH Zurich, Schmelzbergstrasse 7, 8092 Zurich, Switzerland. [2] International College, Khon Kaen University, Khon Kaen 40002, Thailand. [3] Biopher (Pty) Ltd., 5 Ednau Smallholdings, Abrahamskraal Road, Mooiwater, Bloemfontein 9330, South Africa. [4] Faculty of Engineering and Natural Sciences, Sabanci University, 34956 Istanbul, Turkey. [5] DST/NWU Preclinical Drug Development Platform, Faculty of Health Sciences, North-West University, Potchefstroom 2531, South Africa.

Acknowledgements

The authors thank Bennie Basson for generously allowing the use of part of his soybean field for the trial. The authors thank the staff at the GCI-ARC in Potchefstroom for the use of the drying facilities and for their help with soybean processing. The help of Lidija Posavec, Moipone Lebeloane, and Koen van der Vliet with the field trial is kindly acknowledged, and the authors thank Matthew Glyn for his assistance with (confocal) microscopy analysis. The authors thank Nicole Gredig and Christophe Zeder for their assistance with soybean analysis, and Diego Moretti for the stimulating discussions.

Competing interests

The authors declare that they have no competing interests.

Funding

This work was supported by the Swiss National Science Foundation (Doc. Mobility Fellowship P1EZP3_155524) and the Swiss South African Joint Research Programme (project IZLSZ3_149090).

References

1. Kassebaum NJ, Jasrasaria R, Naghavi M, Wulf SK, Johns N, Lozano R, et al. A systematic analysis of global anemia burden from 1990 to 2010. Blood. 2014;123:615–24.
2. Zimmermann MB, Hurrell RF. Nutritional iron deficiency. Lancet. 2007;370:511–20.
3. Cakmak I. Enrichment of cereal grains with zinc: agronomic or genetic biofortification? Plant Soil. 2008;302:1–17.
4. Lyons G, Stangoulis J, Graham R. High-selenium wheat: biofortification for better health. Nutr Res Rev. 2003;16:45–60.
5. Cakmak I, Kalayci M, Kaya Y, Torun AA, Aydin N, Wang Y, et al. Biofortification and localization of zinc in wheat grain. J Agric Food Chem. 2010;58:9092–102.

6. Phattarakul N, Rerkasem B, Li LJ, Wu LH, Zou CQ, Ram H, et al. Biofortification of rice grain with zinc through zinc fertilization in different countries. Plant Soil. 2012;361:131–41.
7. Ram H, Rashid A, Zhang W, Duarte AP, Phattarakul N, Simunji S, et al. Biofortification of wheat, rice and common bean by applying foliar zinc fertilizer along with pesticides in seven countries. Plant Soil. 2016;403:1–12.
8. Rengel Z, Batten GD, Crowley DE. Agronomic approaches for improving the micronutrient density in edible portions of field crops. Field Crop Res. 1999;60:27–40.
9. He WL, Shohag MJI, Wei YY, Feng Y, Yang XE. Iron concentration, bioavailability, and nutritional quality of polished rice affected by different forms of foliar iron fertilizer. Food Chem. 2013;141:4122–6.
10. Zhang YQ, Shi RL, Rezaul KM, Zhang FS, Zou CQ. Iron and zinc concentrations in grain and flour of winter wheat as affected by foliar application. J Agric Food Chem. 2010;58:12268–74.
11. Wei YY, Shohag MJI, Yang XE, Zhang YB. Effects of foliar iron application on iron concentration in polished rice grain and its bioavailability. J Agric Food Chem. 2012;60:11433–9.
12. Pahlavan-Rad MR, Pessarakli M. Response of wheat plants to zinc, iron, and manganese applications and uptake and concentration of zinc, iron, and manganese in wheat grains. Commun Soil Sci Plant Anal. 2009;40:1322–32.
13. Kabir AH, Paltridge N, Stangoulis J. Chlorosis correction and agronomic biofortification in field peas through foliar application of iron fertilizers under Fe deficiency. J Plant Interact. 2016;11:1–4.
14. Gupta UC. Iron status of crops in Prince Edward Island and effect of soil pH on plant iron concentration. Can J Soil Sci. 1991;71:197–202.
15. Aciksoz SB, Yazici A, Ozturk L, Cakmak I. Biofortification of wheat with iron through soil and foliar application of nitrogen and iron fertilizers. Plant Soil. 2011;349:215–25.
16. Messina MJ. Legumes and soybeans: overview of their nutritional profiles and health effects. Am J Clin Nutr. 1999;70:439S–50S.
17. Murray-Kolb LE, Welch R, Theil EC, Beard JL. Women with low iron stores absorb iron from soybeans. Am J Clin Nutr. 2003;77:180–4.
18. Wiersma JV, Moraghan JT. Within-seed distribution of selected mineral elements among soybean genotypes that vary in iron efficiency. Crop Sci. 2013;53:2051–62.
19. Petry N, Boy E, Wirth JP, Hurrell RF. Review: the potential of the common bean (Phaseolus vulgaris) as a vehicle for iron biofortification. Nutrients. 2015;7:1144–73.
20. Rodriguez-Lucena P, Hernandez-Apaolaza L, Lucena JJ. Comparison of iron chelates and complexes supplied as foliar sprays and in nutrient solution to correct iron chlorosis of soybean. J Plant Nutr Soil Sci. 2010;173:120–6.
21. Moraghan JT. Accumulation and within-seed distribution of iron in common bean and soybean. Plant Soil. 2004;264:287–97.
22. Grobler AF, Kotze AF, du Plessis J. The design of a skin-friendly carrier for cosmetic compounds using Pheroid™ technology. In: Wiechers JW, editor. Science and applications of skin delivery systems. Illinois: Allured Publishing Corporation; 2008.
23. du Plessis LH, Lubbe J, Strauss T, Kotze AF. Enhancement of nasal and intestinal calcitonin delivery by the novel Pheroid™ fatty acid based delivery system, and by *N*-trimethyl chitosan chloride. Int J Pharm. 2010;385:181–6.
24. Grobler AF, inventor; North-West University, assignee. Composition in the form of a microemulsion containing free fatty acids and/or free fatty acid derivatives. South Africa patent WO2007096833 2007.
25. Pfeiffer WH, McClafferty B. HarvestPlus: breeding crops for better nutrition. Crop Sci. 2007;47:S88–105.
26. Rios JJ, Carrasco-Gil S, Abadía A, Abadía J. Using perls staining to trace the iron uptake pathway in leaves of a prunus rootstock treated with iron foliar fertilizers. Front Plant Sci. 2016;7:893.
27. El-Jendoubi H, Vázquez S, Calatayud Á, Vavpetič P, Vogel-Mikuš K, Pelicon P, et al. The effects of foliar fertilization with iron sulfate in chlorotic leaves are limited to the treated area. A study with peach trees (*Prunus persica* L. Batsch) grown in the field and sugar beet (*Beta vulgaris* L.) grown in hydroponics. Front Plant Sci. 2014;5:2.
28. Fernandez V, Del Rio V, Abadia J, Abadia A. Foliar iron fertilization of peach (*Prunus persica* (L.) Batsch): effects of iron compounds, surfactants and other adjuvants. Plant Soil. 2006;289:239–52.
29. Munger P, Bleiholder H, Hack H, Hess M, Stauss R, van den Boom T, et al. Phenological growth stages of the soybean plant (*Glycine max* L. Merr.): codification and description according to the BBCH scale. J Agron Crop Sci. 1997;179:209–17.
30. Small HGJ, Ohlrogge AJ. Plant analysis as an aid in fertilizing soybeans and peanuts. In: Walsh LM, Beaton JD, editors. Soil testing and plant analysis. Madison: Soil Science Society of America, Inc.; 1973. p. 315–27.
31. Goos RJ, Johnson BE. A comparison of three methods for reducing iron-deficiency chlorosis in soybean. Agron J. 2000;92:1135–9.
32. Carrasco-Gil S, Rios JJ, Álvarez-Fernández A, Abadía A, García-Mina JM, Abadía J. Effects of individual and combined metal foliar fertilisers on iron- and manganese-deficient *Solanum lycopersicum* plants. Plant Soil. 2016;402:27–45.
33. Makower RU. Extraction and determination of phytic acid in beans (Phaseolus vulgaris). Cereal Chem. 1970;47:288–95.
34. van Veldhoven PP, Mannaerts GP. Inorganic and organic phosphate measurements in the nanomolar range. Anal Biochem. 1987;161:45–8.
35. Chelopo MP, Kalombo L, Wesley-Smith J, Grobler A, Hayeshi R. The fabrication and characterization of a PLGA nanoparticle–Pheroid® combined drug delivery system. J Mater Sci. 2017;52:3133–45.
36. Choe E, Min DB. Mechanisms and factors for edible oil oxidation. Compr Rev Food Sci Food Saf. 2006;5:169–86.
37. Tan CP, Man YBC, Selamat J, Yusoff MSA. Comparative studies of oxidative stability of edible oils by differential scanning calorimetry and oxidative stability index methods. Food Chem. 2002;76:385–9.
38. Yin Y, Zhang H, Wu P, Zhou B, Cai C. Iron phosphate nanostructures synthesized by microwave method and their applications in biosensing. Nanotechnology. 2010;21:425504.
39. Slabbert C, du Plessis LH, Kotze AF. Evaluation of the physical properties and stability of two lipid drug delivery systems containing mefloquine. Int J Pharm. 2011;409:209–15.
40. Spehar CR. Seed quality of soya bean based on mineral composition of seeds of 45 varieties grown in a Brazilian savanna acid soil. Euphytica. 1994;76:127–32.
41. Lolas GM, Palamidis N, Markakis P. Phytic acid total phosphorus relationship in barley, oats, soybeans, and wheat. Cereal Chem. 1976;53:867–71.
42. Wang W-N, Tarafdar JC, Biswas P. Nanoparticle synthesis and delivery by an aerosol route for watermelon plant foliar uptake. J Nanopart Res. 2013;15:1417.
43. Fernandez V, Ebert G. Foliar iron fertilization: a critical review. J Plant Nutr. 2005;28:2113–24.
44. Schonherr J, Huber R. Plant cuticles are polyelectrolytes with isoelectric points around 3. Plant Physiol. 1977;59:145–50.
45. Alidoust D, Isoda A. Effect of γ-Fe_2O_3 nanoparticles on photosynthetic characteristic of soybean (*Glycine max* (L.) Merr.): foliar spray versus soil amendment. Acta Physiol Plant. 2013;35:3365–75.
46. Eichert T, Kurtz A, Steiner U, Goldbach HE. Size exclusion limits and lateral heterogeneity of the stomatal foliar uptake pathway for aqueous solutes and water-suspended nanoparticles. Physiol Plant. 2008;134:151–60.
47. Schonherr J, Fernandez V, Schreiber L. Rates of cuticular penetration of chelated Fe-III: role of humidity, concentration, adjuvants, temperature, and type of chelate. J Agric Food Chem. 2005;53:4484–92.
48. Mengel K, Geurtzen G. Iron chlorosis on calcareous soils—alkaline nutritional conditions as the cause for the chlorosis. J Plant Nutr. 1986;9:161–73.
49. Kosegarten HU, Hoffmann B, Mengel K. Apoplastic pH and Fe^{3+} reduction in intact sunflower leaves. Plant Physiol. 1999;121:1069–79.
50. Bukovac MJ, Wittwer SH. Absorption and mobility of foliar applied nutrients. Plant Physiol. 1957;32:428–35.
51. Berge ND, Ramsburg CA. Oil-in-water emulsions for encapsulated delivery of reactive iron particles. Environ Sci Technol. 2009;43:5060–6.
52. Fernandez V, Sotiropoulos T, Brown P. Foliar fertilization. Scientific principles and field practices. Paris: International Fertilizer Industry Association; 2013.
53. Morrissey J, Guerinot ML. Iron uptake and transport in plants: the good, the bad, and the ionome. Chem Rev. 2009;109:4553–67.
54. Hussain S, Rengel Z, Mohammadi SA, Ebadi-Segherloo A, Maqsood MA. Mapping QTL associated with remobilization of zinc from vegetative tissues into grains of barley (*Hordeum vulgare*). Plant Soil. 2016;399:193–208.

55. Grusak MA. Iron transport to developing ovules of *Pisum sativum*. I. Seed import characteristics and phloem iron-loading capacity of source regions. Plant Physiol. 1994;104:649–55.
56. Zhang CD, Romheld V, Marschner H. Remobilization of iron from primary leaves of bean plants grown at various iron levels. J Plant Nutr. 1996;19:1017–28.
57. Raboy V, Noaman MM, Taylor GA, Pickett SG. Grain phytic acid and protein are highly correlated in winter wheat. Crop Sci. 1991;31:631–5.
58. Rotundo JL, Westgate ME. Meta-analysis of environmental effects on soybean seed composition. Field Crop Res. 2009;110:147–56.
59. Bellaloui N, Smith JR, Gillen AM, Ray JD. Effects of maturity, genotypic background, and temperature on seed mineral composition in near-isogenic soybean lines in the early soybean production system. Crop Sci. 2011;51:1161–71.
60. Wiersma JV. Is seed iron concentration predictive of resistance to iron deficiency in soybean? Crop Sci. 2012;52:471–81.
61. Bouis HE, Saltzman A. Improving nutrition through biofortification: a review of evidence from HarvestPlus, 2003 through 2016. Glob Food Sec. 2017;12:49–58.

16

Effects of climate variability on insect pests of cabbage: adapting alternative planting dates and cropping pattern as control measures

Clovis B. Tanyi*, Christopher Ngosong and Nelson N. Ntonifor

Abstract

Background: Considering the potential impact of climate change on the ecology of insect pests, different planting dates and cropping patterns were investigated as farm-level adaption to control insect pests of cabbage and improve productivity.

Methods: This is a 3 × 4 factorial experiment setup in randomized complete block design including three planting dates (early, normal and late) and four cropping patterns (control—sole cabbage or tomato, tomato intercrop, *Piper* emulsion and insecticide) with four replications each.

Results: Cabbage infestation ranged from 1 to 29 and correlated negatively with planting dates or treatments, which differed ($P < 0.001$) significantly across planting dates, treatments and their interaction, with the highest during early planting. Diamondback moth larvae correlated negatively with planting dates or treatments, ranging from 0 to 13 that differed significantly ($P < 0.001$) across planting dates, treatments and their interaction. Looper larvae correlated negatively with treatments, ranging from 0 to 8 that differed significantly ($P < 0.001$) across planting dates, treatments and their interaction, with highest during normal planting and lowest during late planting. Webworm larvae correlated negatively with planting dates or treatments, ranging from 0 to 13 that differed significantly ($P < 0.001$) across planting dates, treatments and their interaction. The number of sprouted plants ranged from 0 to 6 and differed significantly ($P < 0.001$) across planting dates, treatments and their interaction, with the highest in early planting for control that differed significantly from late planting. Cabbage yield correlated positively with planting dates and ranged from 2.8 to 6.0 tons per hectare that differed significantly ($P < 0.001$) across planting dates, treatments and their interaction, with the highest during normal and late planting dates.

Conclusion: The interaction of planting dates and *Piper* emulsion or intercropping treatments can be effectively used as control measure for insect pests of cabbage leading to greater yield, with late planting as viable farm-level adaptation to climate variability.

Keywords: Climate change, Diamondback moth, Intercropping, Piper botanical, Planting date

Background

Climate variability poses constraints on agricultural production with significant effects on cropping seasons and food security [1]. Sub-Saharan Africa (SSA) is more vulnerable to climate change due to its reliance on agriculture that is highly sensitive to climate variables, and the low capacity for adaptation [2]. Climate change affects most crops including vegetables that are sensitive to climate shocks [3]. Extreme weather events such as heavy rainfall, droughts and heat waves have increased in recent decades, with significant implications on agricultural productivity [4, 5]. Periodic droughts and changes in rainfall frequency or severity can severely influence the ecology of insect pests in arable fields [6–8]. In view of current global climate change scenarios, there is overdue

*Correspondence: tanyi.clovis@yahoo.com
Department of Agronomic and Applied Molecular Sciences, Faculty of Agriculture and Veterinary Medicine, University of Buea, P.O.Box 63, Buea, South West Region, Cameroon

need for sustainable adaptive farm management strategies [9].

Many insect pests hinder cabbage (*Brassica oleracea* L.) cultivation and the most damaging is Diamondback moth—DBM (*Plutella xylostella* L.) (Lepidoptera: Plutellidae) that causes up to 90% yield loss [10]. DBM, Looper *Trichoplusia ni* (Hübner) (Lepidoptera: Noctuidae) and webworm *Hellula undalis* (Fabricius) (Lepidoptera: Pyralidae) can cause sprouting, contamination with frass, or even kill young plants [11]. Reliance on synthetic insecticides to control crop pests usually leads to environmental pollution, pest resistance and high production costs [12–16]. The prevention, avoidance, monitoring and suppression (PAMS) model was successfully used to evaluate pest management strategies [17]. Prevention involves crop rotation and use of resistant cultivars, while avoidance involves alternative planting dates and intercropping [17, 18]. Tomato intercrop [18–20], botanicals and planting dates [21–24] have demonstrated efficacy and cost-effectiveness as control measures for cabbage pests.

Climate variability was projected to cause varying effects on crops including significant yield decrease in eight agricultural regions across Cameroon [25, 26]. Despite limitations in climate data for the study area, recent meteorological data showed significant weather variations with increased annual temperature from 26 °C in 2010 to 27.2 °C in 2012 and 2014, while annual rainfall decreased from 3135 mm in 2013 to 1993 mm in 2014 [27]. These temperature and rainfall fluctuations can affect farming systems by disrupting the normal planting schedule. Consequently, alternative planting and harvesting dates were suggested as potential farm-level adaptations to climate variability in this study area and across Cameroon [28]. Accordingly, Ngondjeb [29] advocated effective agricultural planning for the impact of climate variability in Cameroon. It is therefore important to consider the effects of climate variability on the ecology of insect pests in arable systems, and potential integrated management approach involving alternative planting dates and cropping patterns. Hence, this study was intended to adapt alternative planting dates and cropping pattern as control measures for cabbage pests under climate change scenarios. It was hypothesized that shifting from normal to late planting dates, and the interaction of planting dates and treatments will lead to significant reduction of insect pests and increased cabbage yield.

Methods

Experimental site and setup

This study was conducted at the research and teaching farm of the Faculty of Agriculture and Veterinary Medicine, University of Buea. The site is located in Molyko Buea, situated between latitudes 4°3′N and 4°12′N and longitudes 9°12′E and 9°20′E. The soil is derived from weathered volcanic rocks dominated by silt, clay and sand [30, 31]. The rainfall regime is mono-modal with less pronounced dry season and 85–90% relative humidity, and the dry season starts from November to March [32]. It has a mean annual rainfall of 2800 mm and monthly air temperature ranging from 19 to 30 °C, while soil temperature at 10 cm depths decreases from 25 to 15 °C with increasing elevation from 200 to 2200 m above sea level, respectively [30, 33, 34]. Mean monthly temperature ranged from 17.3 to 19.7 °C and precipitation from 30 to 429 mm in Buea during the experimental period (Table 1).

This is a 3 × 4 factorial experiment setup in randomized complete block design including three planting dates (early—24th January 2016, normal—10th March 2016 and late—15th April 2016) and four cropping patterns (control—sole cabbage or sole tomato, tomato intercrop, *Piper* emulsion and insecticide) with four replications each. The experimental field was cleared manually using a cutlass and partitioned into 16 plots measuring 3 × 4 m each (12 m^2), and manually tilled at about 30 cm depth using a hoe. The sole cabbage plots contained 8 rows and 6 columns of cabbage plants at 50 × 50 cm inter- and intra-row spacing, leading to 48 plants per plot. Cabbage–tomato intercrop plots contained 4 rows and 3 columns of cabbage and tomato plants each, leading to 48 plants per plot (24 plants per crop). These were planted at 50 × 50 cm inter- and intra-row spacing on one alternate row between cabbage and tomato plants. A non-tilled 50-cm buffer zone separated the experimental plots from each other.

Table 1 Mean monthly temperature and precipitation in Buea during the experimental period

Months	Temperature (°C)	Rainfall (mm)
January	19	30
February	19.4	65
March	19.7	161
April	19.6	188
May	19.1	244
June	18.3	264
July	17.3	429
August	17.4	488
September	17.7	482
October	18.3	323
November	18.8	112
December	18.9	29

Source: https://en.climate-data.org/africa/cameroon/southwest/buea-3985/

Plant cultivation

Hybrid cabbage (F1 Green Coronet; STARKE AYRES®, France; that are heat and cold resistant) and tomato (F1 Cobra 26; TECHNISEM®, France; adapted for Sahelian and tropical areas with tolerance to disease and high productivity) seeds were purchased from an agro-shop in Buea Cameroon. The F1 Green Coronet seeds and F1 Cobra 26 seeds were pre-germinated on separate beds at 15×15 cm inter-row spacing on nearby 2.5×1 m nursery beds that were cleared using cutlasses and tilled manually using hoes, and manually irrigated using a watering can. For all planting dates, vigorous cabbage seedlings of similar sizes were transplanted from the nursery, followed by tomato transplant 2 weeks later. After transplanting, all plots were manually irrigated every 2 days to maintain optimum soil moisture for plant growth and performance.

Both cabbage and tomato nurseries were amended with 0.5 kg inorganic fertilizer NPK 20:10:10 + CaO (ADER® Cameroon), and treated with a mixture of synthetic insecticides and fungicides. 35 mL insecticide (K-Optimal; SCPA SIVEX International® France; comprising 15 g/L lambda − cyhalothrine + 20 g/L acetamiprid active ingredients) and 100 g fungicide (Mancozan super; SCPA SIVEX International® France; comprising 640 g/kg mancozeb + 80 g/kg metalaxyl active ingredients) were dissolved in 15 L water and applied using a knapsack sprayer. After transplanting, all experimental plots were manually irrigated every 2 days to maintain optimum soil moisture for plant growth and performance. Weed emergence was monitored regularly on all experimental plots and weeded manually using a hoe.

Fertilizer amendment

All experimental plots were amended with the same type and amount of soil applied organic and inorganic fertilizers. Organic poultry dropping (obtained from a poultry farm in Molyko Buea) was broadcasted on all experimental plots 7 days before the crops were transplanted. Thirty days after transplanting, inorganic NPK 20:10:10 + CaO (ADER® Cameroon) was applied on all experimental plots at 5 g per plant by ringing at 5 cm from plants.

Insecticides and fungicides

The synthetic plots were sprayed with commercial insecticide (K-Optimal; SCPA SIVEX International® France) and fungicide (Mancozan super; SCPA SIVEX International® France). The cabbage–tomato intercrop plots were not sprayed with *Piper* emulsion or synthetic insecticide and the tomato plants served as repellent or attractant of cabbage pests. Neither *Piper* emulsion nor synthetic insecticide was applied in the sole cabbage or sole tomato control plots.

Piper emulsion botanical

The organic *Piper* emulsion botanical comprised West African black pepper (*Piper guineense*) that was harvested from a primary forest at Inokun-Eyumojock in South-western Cameroon and prepared according to Tanyi et al. [35]. Briefly, 250 g crushed sun-dried *Piper* seed powder was dissolved in 1 L vegetable oil (KING'S®, Lagos-Nigeria) and 10 g detergent (SABA®, Douala-Cameroon) was added to produce a sticky emulsion. The mixture was stirred thoroughly and stored in a plastic container at room temperature while laboratory tests were conducted to determine the effective dose for best field results. 50 mL *Piper* emulsion was filtered using a double 169-folded muslin cloth and diluted in 15 L water for field application every 2 weeks, during cold dry early morning periods with minimal drift. *Piper* emulsion was stirred thoroughly to achieve homogeneity and sprayed using a knapsack sprayer on both sides of cabbage leaves for all 192 plants in the respective plots.

Data collection on cabbage pests

Cabbage plants were assessed for pest infestation before and after heading, while the wrapper leaves were monitored regularly for symptoms of pest damage. Five randomly selected plants were tagged on each plot, and visible signs of damage and occurrence of insect pest larvae were assessed and data presented as number of larvae per plant (mean ± SD). Cabbage plants were identified as infested based on observation of diamondback moth, looper and webworm larvae or their damage, and reported as number of infested plants per treatment (mean ± SD). Diamondback moth (*P. xylostella*) is an important cabbage pest with average development time of 25–30 days from egg to pupal stage depending on weather conditions, and a complete cycle of 17–51 days from egg to adult. The annual number of generations varies from four in cold climates to 12 in warm climates. Diamondback moth was assessed on wrapper leaves and identified as small round holes, scratches or skeleton damage on leaves with partially damaged epidermis that gives cabbage leaves a windowpane appearance. An adult female cabbage Looper (*T. ni*) moth lays from 300 to 600 eggs pale yellow and round eggs in 10–12 days adult lifespan that hatch in 3–10 days, with a more rapid pupal stage in warmer temperatures. The occurrence of webworm larvae and plant damage was identified as leaves held together with silk. Each female cabbage webworm (*H. undalis*) lays about 75–250 eggs singly or in groups of 2–3 on the inner tissues of host plants. The hatching larvae bore into stems of growing points, weaving webs into which they place their frass. Webworm pupates in soil within webbed cocoons that include soil particles as a generation develops in about 4–5 weeks at 27 °C, with short-lived adults and the potential to complete 7–8 generations annually.

Data collection on cabbage yield

Sprouted cabbage plants were assessed and recorded as total number of plants with multiple shoots (sprouts) on each plot and reported as number of sprouted cabbage plants per treatment (mean ± SD). Marketable cabbage heads were harvested manually at physical maturity and the yield presented as weight of harvested cabbage heads per treatment (t ha^{-1}, mean ± SD). At harvesting, ten cabbage plants were incised above the soil on each plot and weighed individually using a top loading balance (Brand MK-01, China). Prior to weighing, a cutter was used to remove all damaged leaves.

Statistical analysis

Datasets were assessed for normality using Kolmogorov–Smirnov and Shapiro–Wilk tests before they were subjected to statistical analyses using STATISTICA 9.1 for Windows [36, 37]. Dependent variables (e.g. cabbage infestation or yield, Diamondback moth, Looper and Webworm) were subjected to multivariate analysis of variance (ANOVA, $P<0.05$) to test the effects of planting dates ($n=3$) and treatments ($n=4$) as categorical predictors. Significant data means were compared by post hoc Tukey's HSD test ($P<0.05$), and Spearman rank correlation ($P<0.05$) was performed to determine the degree of association between dependent variables and categorical predictors.

Results

Cabbage infestation

Cabbage infestation ranged from 1 to 29 and correlated negatively with planting dates and treatments, which differed significantly ($P<0.001$) across planting dates (Fig. 1a), treatments (Fig. 1b) and their interaction. The highest infestation occurred during early planting for the control that differed significantly ($P<0.05$) from the normal and late plantings (Fig. 1a). Differences occurred over time for intercropping treatment as the early planting differed significantly ($P<0.05$) from late planting (Fig. 1a). The highest infestation occurred in control that differed significantly ($P<0.05$; Fig. 1b) from the other treatments. Negative correlations ($P<0.05$) occurred between cabbage infestation and planting dates ($r=-0.48$) or treatments ($r=-0.71$).

Diamondback moth larvae

Diamondback moth-DBM larvae correlated negatively planting dates or treatments, ranging from 0 to 13 that differed significantly ($P<0.001$) across planting dates (Fig. 2a), treatments (Fig. 2b) and their interaction. DBM larvae differed significantly ($P<0.05$; Fig. 2a) across planting dates for control, with highest during early planting and lowest during late planting. Significant ($P<0.05$) treatment effects occurred during the early and normal planting, with the highest DBM larvae recorded in the control that differed from the other treatments (Fig. 2b). Treatment effects on DBM larvae were significant ($P<0.05$) during early planting, with highest in control followed by intercropping, *Piper* emulsion and insecticide (Fig. 2b). Negative correlations ($P<0.05$) occurred between DBM larvae and planting dates ($r=-0.33$) or treatments ($r=-0.69$) with correlations during early ($r=-0.89$) and late ($r=-0.87$) plantings.

Looper larvae

Looper larvae correlated negatively with treatments, ranging from 0 to 8 that differed significantly ($P<0.001$) across planting dates (Table 2A), treatments (Table 2B) and their interaction, with highest for control during normal planting and lowest during late planting ($P<0.05$; Table 2A). Negative correlations ($P<0.05$) occurred between looper larvae and treatments for early ($r=-0.54$), normal ($r^*=-0.62$) and late ($r=-0.51$) plantings.

Webworm larvae

Webworm larvae correlated negatively with planting dates and treatments, ranging from 0 to 13 that differed significantly ($P<0.001$) across planting dates (Table 3A), treatments (Table 3B) and their interaction. Webworm larvae differed significantly ($P<0.05$) for the control, with the lowest in late planting and highest in early planting (Table 3B). Negative correlations ($P<0.05$) occurred between webworm larvae and planting dates ($r=-0.54$) or treatments for early ($r=-0.79$), normal ($r^*=-0.75$) and late ($r=-0.51$) plantings.

Cabbage yield

Sprouted plants ranged from 0 to 6 and differed significantly ($P<0.001$) across planting dates (Table 4A), treatments (Table 4B) and their interaction. The highest number of sprouted plants occurred in early planting for control, which differed significantly ($P<0.05$) from late planting (Table 4A). Negative correlations ($P<0.05$) occurred between sprouted plants and planting dates ($r=-0.4$) or treatments for early ($r=-0.9$), normal ($r^*=-0.8$) and late ($r=-0.5$) plantings. Cabbage yield ranged from 2.8 to 6.0 tons per hectare that correlated ($P<0.05$) positively with planting dates ($r=0.6$), and differed significantly ($P<0.001$) across planting dates (Fig. 3a), treatments (Fig. 3b) and their interaction, with highest during normal and late plantings. The late

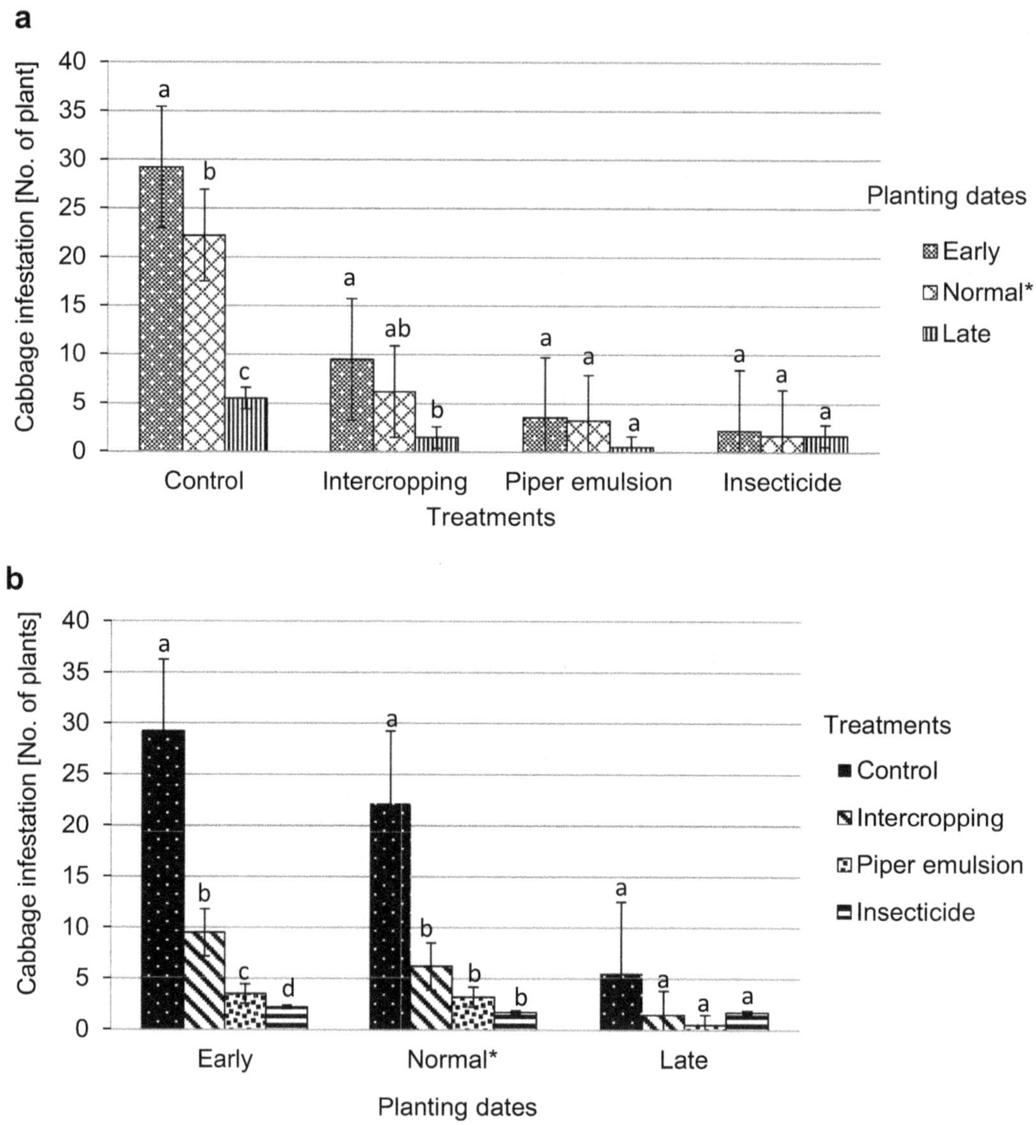

Fig. 1 Effect of planting dates (**a**) (early, normal and late) and treatments (**b**) (control, intercropping, *Piper* emulsion and insecticide) on number of infested cabbage plants (mean ± SD). Values within treatments or planting dates with different letters are significantly different ($P<0.05$). *Data from Tanyi et al. [35]

planting differed significantly ($P<0.05$) from early and normal plantings for the control (Fig. 3).

Discussion

Effect of planting dates on cabbage pests

The results of this study are consistent with current trends in climate dynamics, which require adaptation of cropping seasons to the ecology of insect pests, with particular attention on alternative planting dates and sustainable treatments [6, 38]. These results reflect the overdue need for adapted integrated pest management approaches that incorporate planting dates with local farm management practices under different climate scenarios [39]. The significantly low pest infestation during late planting compared to normal and early plantings can be attributed to current climate dynamics in the study area with decreasing rainfall and increasing temperature [27, 38], and demonstrates the efficacy of alternative planting dates as control measure for cabbage pests [28]. The decrease in diamondback moth larvae from early to late plantings is consistent with other results where diamondback moth occurred in vegetable fields during early and normal planting, but disappeared during late planting [40]. Climate variables play important roles in the ecology of insect pests because of their short generations and high reproductive rates

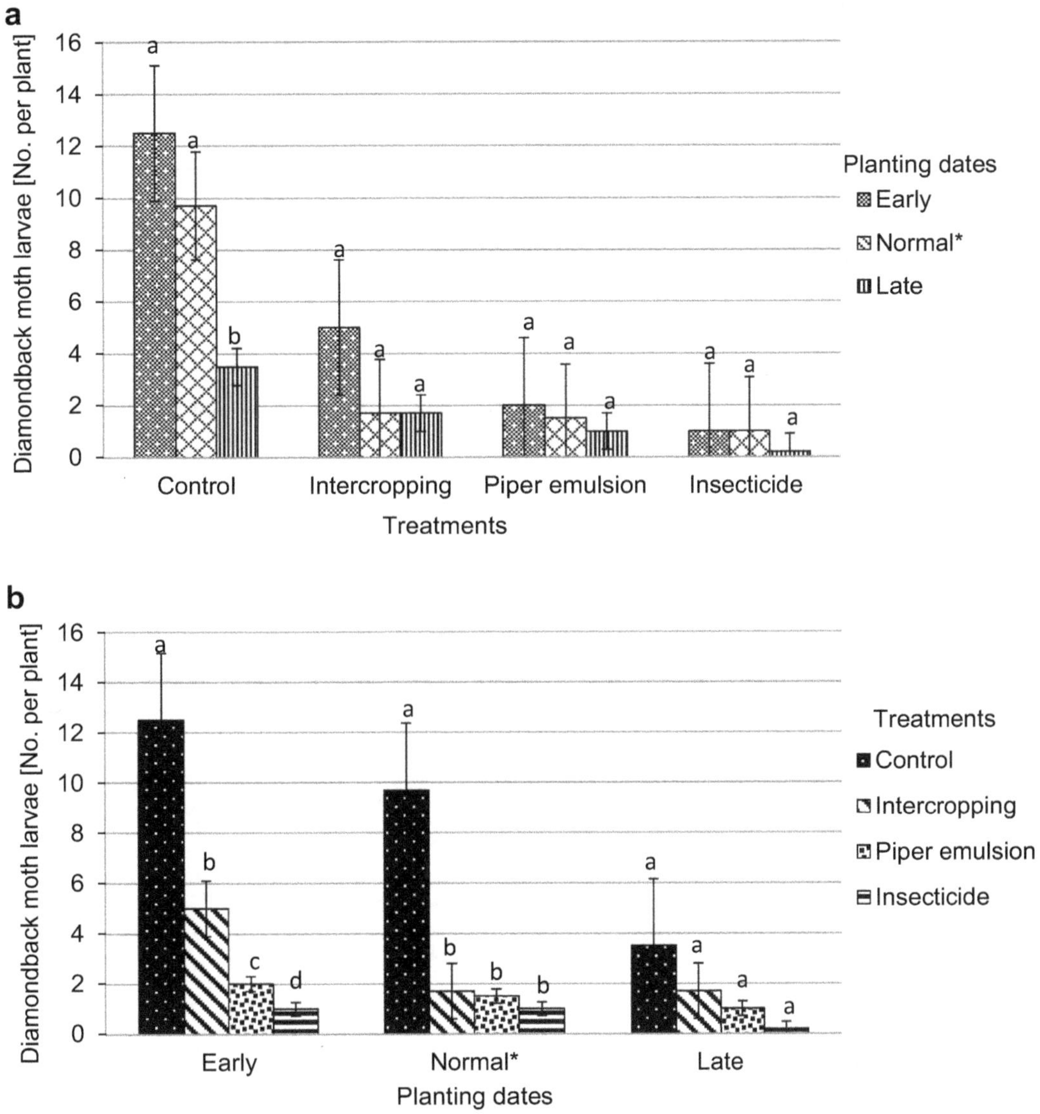

Fig. 2 Effect of planting dates (**a**) (early, normal and late) and treatments (**b**) (control, intercropping, *Piper* emulsion and insecticide) on number of diamondback moth larvae (mean ± SD). Values within treatments or planting dates with different letters are significantly different ($P < 0.05$). *Data from Tanyi et al. [35]

[41, 42], causing significant effects on crop production in Cameroon that require effective planning [29, 43]. In addition, climate variability reduced parasitism of caterpillars leading to variation in the frequency and intensity of herbivory outbreaks [44]. Temperature effects on insect ecology are variable with high temperatures negatively affecting eggs, while low temperatures negatively affect larvae [43, 45]. These results are consistent with the first hypothesis of this study and strongly support a shift to late planting as farm-level adaptation to cabbage pests as compared to the current planting date in the study area.

Besides alternative planting dates, farm management practices demonstrated strong importance as control measure for cabbage pests across the different planting dates. Effectiveness of the evaluated crop protection practices is demonstrated by lack of significant differences in cabbage infestation and pest occurrence between different planting dates for intercropping, *Piper* emulsion and insecticide treatments compared to control [18, 36]. The low insect pest occurrence across treatments during late planting is consistent with the hypothesis of this study, which strongly suggests a shift from the current planting date as farm-level adaptation to climate variability in the study area. This is commensurate with predictions on the

Table 2 Effect of planting dates—A (early, normal and late) and treatments—B (control, intercropping, *Piper* emulsion and insecticide) on number of looper larvae (mean ± SD)

A

Treatments	Planting dates		
	Early	Normal[a]	Late
Control	6.2 ± 2.0a	8.3 ± 1.7a	0.2 ± 0.5b
Intercropping	0.2 ± 0.5a	0.2 ± 0.5a	0.0 ± 0.0a
Piper emulsion	0.0 ± 0.0a	0.5 ± 0.5a	0.0 ± 0.0a
Insecticide	0.7 ± 0.9a	0.7 ± 0.9a	0.5 ± 0.5a

B

Planting dates	Treatments			
	Control	Intercropping	*Piper* emulsion	Insecticide
Early	6.2 ± 2.0a	0.2 ± 0.5a	0.0 ± 0.0a	0.7 ± 0.9a
Normal[a]	8.3 ± 1.7a	0.2 ± 0.5a	0.5 ± 0.5a	0.7 ± 0.9a
Late	0.2 ± 0.5b	0.0 ± 0.0a	0.0 ± 0.0a	0.5 ± 0.5a

Values within rows with different letters are significantly different ($P < 0.05$)

[a] Data taken from Tanyi et al. [35]

Table 3 Effect of planting dates—A (early, normal and late) and treatments—B (control, intercropping, *Piper* emulsion and insecticide) on number of webworm larvae (Mean ± SD)

A

Treatments	Planting dates		
	Early	Normal[a]	Late
Control	12.5 ± 3.1a	5.5 ± 3.1b	1.5 ± 1.7b
Intercropping	4.5 ± 2.6a	4.2 ± 3.0a	0.2 ± 0.5a
Piper emulsion	1.5 ± 2.3a	1.2 ± 0.9a	0.0 ± 0.0a
Insecticide	1.0 ± 0.8a	0.5 ± 1.0a	0.0 ± 0.0a

B

Planting dates	Treatments			
	Control	Intercropping	*Piper* emulsion	Insecticide
Early	12.5 ± 3.1a	4.5 ± 2.6a	1.5 ± 2.3a	1.0 ± 0.8a
Normal[a]	5.5 ± 3.1b	4.2 ± 3.0a	1.2 ± 0.9a	0.5 ± 1.0a
Late	1.5 ± 1.7b	0.2 ± 0.5a	0.0 ± 0.0a	0.0 ± 0.0a

Values within rows with different letters are significantly different ($P < 0.05$)

[a] Data taken from Tanyi et al. [35]

ecology of insect pests under climate change scenarios that combine experimental data and modeling of population dynamics [46]. The positive results for interaction of planting dates and treatments support the second hypothesis of this study, and represent a viable farm-level integrated pest management adaptation to current climate dynamics in the study area.

Effect of treatments on cabbage pests

The ecology of insect pests is highly influenced by temperature, which may either enhance reproduction or decrease mortality of insects, leading to stronger infestations [47, 48]. Climate dynamics can lead to the emergence of new species such as the South American tomato leaf miner *Tuta absoluta* (Meyrick) (Lepidoptera: Gelechiidae) that occurred in Europe [49]. Generally, intercropping, *Piper* emulsion and insecticide demonstrated efficacy as control measures for cabbage pests against the control, irrespective of the planting dates [18]. Tomato intercropping could be considered

Table 4 Effect of planting dates—A (early, normal and late) and treatments—B (control, intercropping, *Piper* emulsion and insecticide) on number of sprouted cabbage plants (mean ± SD)

A

Treatments	Planting dates		
	Early	Normal[a]	Late
Control	6.2 ± 1.7c	5.0 ± 1.1b	0.7 ± 0.9a
Intercropping	0.5 ± 0.5b	1.0 ± 0.0a	0.2 ± 0.5a
Piper emulsion	1.2 ± 0.9b	1.0 ± 0.5a	0.0 ± 0.0a
Insecticide	1.0 ± 0.9b	0.5 ± 0.5a	0.0 ± 0.0a

B

Planting dates	Treatments			
	Control	Intercropping	*Piper* emulsion	Insecticide
Early	6.2 ± 1.7c	0.5 ± 0.5b	1.2 ± 0.9b	1.0 ± 0.9b
Normal	5.0 ± 1.1b	1.0 ± 0.0a	1.0 ± 0.5a	0.5 ± 0.5a
Late	0.7 ± 0.9a	0.2 ± 0.5a	0.0 ± 0.0a	0.0 ± 0.0a

Values within rows with different letters are significantly different ($P < 0.05$)

[a] Data taken from Tanyi et al. [35]

as the most affordable and cost-effective pest management strategy in relation to the other treatments. The impact of tomato intercropping is likely due to the confusing olfactory and visual cues from tomato plants that repelled cabbage pests [18, 50]. *P. guineense* contains isobutyl amide plant secondary metabolites (e.g. piperine, piperiline and natural lipophilic amides) that act as neurotoxins in insects [51–53], which likely resulted in the efficacy of *Piper* emulsion against insect pests across planting dates. This effect of *Piper* emulsion on cabbage infestation and pest occurrence is consistent with reports on the efficacy of *Piper* [54, 55]. These results strongly support the second hypothesis of this study that advocates a combination of alternative planting dates and *Piper* emulsion or intercropping as farm-level adaptation to control cabbage pests under current climate variability in the study area.

Interactive effects of planting dates and treatments on cabbage

Cabbage performance was consistent with the rate of pest infestation, with increased cabbage yield as pest infestation decreased and vice versa. The significantly low number of sprouted plants and high yield during late planting compared to normal and early plantings is consistent with the first hypothesis that suggests a shift in planting dates in the study area. This can be attributed to changes in climate variables such as temperature and rainfall that reduced pest larvae below economic threshold [43, 45]. Cabbage performance is consistent with the second hypothesis of this study that advocates greater cabbage yield due to tomato intercropping as companion crop. Low cabbage yield in the control is consistent with high pest infestation, which corresponds to high leaf and head damage that likely reduced photosynthetic carbon fixation and plant growth [56]. Looper and webworm are considered major yield-reducing cabbage pests, which is consistent with the significant cabbage damage that is comparable to DBM damage, especially during early and late plantings [11]. The increasing cabbage yield from early to late plantings in Cabbage–tomato intercrop compared to insecticide and *Piper* treatments is inconsistent with the recorded trend of pest infestation, which suggests additional factors that improved cabbage yield. The decomposition of dead tomato plants after complete pest damage in the sole tomato and Cabbage–tomato intercrop treatments likely improved soil fertility, plant nutrition and biological processes that favoured cabbage yield [57]. Overall, the relatively high cabbage yield recorded in the late planting and cabbage–tomato intercrop treatment highlights the importance of integrating alternative planting dates and treatments under climate change scenarios.

Conclusion

The interaction of planting dates and *Piper* emulsion or intercropping treatments can be effectively used to control cabbage pests and improve yields, with the late

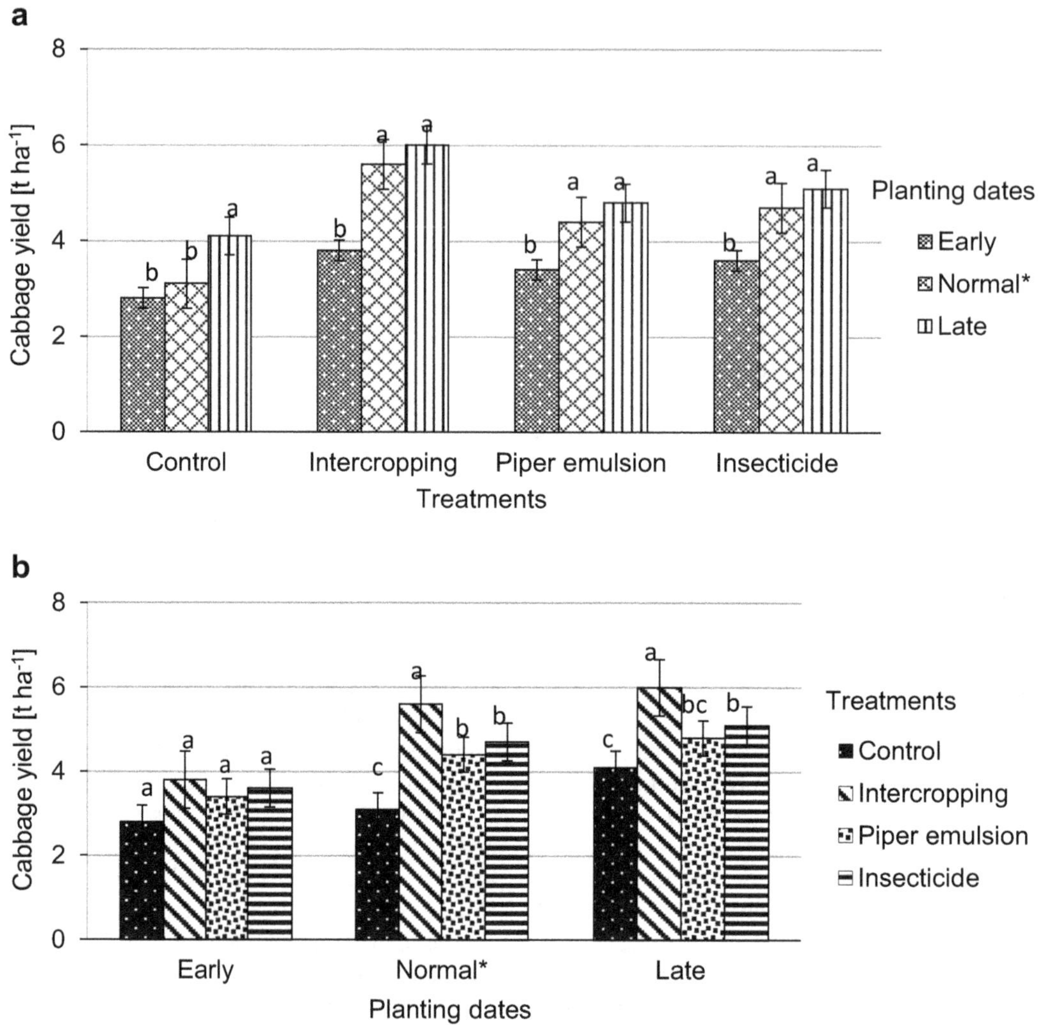

Fig. 3 Impact of planting dates (**a**) (early, normal and late) and treatments (**b**) (control, intercropping, *Piper* emulsion and insecticide) on cabbage yield (t ha^{-1}, mean ± SD); values within treatments or planting dates with different letters are significantly different ($P < 0.05$). *Data from Tanyi et al. [35]

planting date as viable alternative farm-level adaptation to climate variability. These represent cost-effective and environmentally friendly pest management strategy that can be adopted by farmers to control cabbage pests below economic injury threshold and improve yield.

Abbreviations

DBM: diamondback moth; SSA: sub-Saharan Africa; PAMS: prevention, avoidance, monitoring and suppression; SD: standard deviation; ANOVA: analysis of variance; NPK: nitrogen, phosphorus and potassium.

Authors' contributions

This work was carried out in collaboration between all authors. CBT designed, established and managed the experiment, prepared botanical, performed harvest and data collection, processed and analysed data, performed literature searches and wrote the first manuscript draft. CN contributed in experimental design, data processing and statistical analyses, conducted literature searches and coordinated preparation of first manuscript draft. NNN contributed in the experimental design, coordinated the field experimentation and data collection, and supervised manuscript preparation and the overall study. All authors read and approved the final manuscript.

Acknowledgements

We extend gratitude to the Faculty of Agriculture and Veterinary Medicine, University of Buea, and the Ministry of Higher Education–MINESUP Cameroon for Research Grants.

Competing interests

The authors declare that they have no competing interests.

Funding

Not applicable.

References

1. Datta S. Impact of climate change in Indian horticulture—a review. Int J Sci Environ Technol. 2013;2:661–71.
2. Kotir JH. Climate change and variability in sub-Saharan Africa: a review of current and future trends and impacts on agriculture and food security. Environ Dev Sustain. 2011;13:587–605.
3. Kemausuor F, Dwamena E, Bart-Plange A, Kyei-Baffour N. Farmers' perception of climate change in the Ejura-Sekyedumase District of Ghana. J Agric Biol Sci. 2011;6:26–37.
4. Gornall J, Betts R, Burke E, Clark R, Camp J, Willett K, Wiltshire A. Implications of climate change for agricultural productivity in the early twenty-first century. Philos Trans R Soc Br. 2010;365:2973–89.
5. Mirza MMQ. Climate change and extreme weather events: can developing countries adapt? Climate Policy. 2011;3:233–48.
6. Bale J, Masters G, Hodkinson I. Herbivory in global climate change research: direct effects of rising temperature on insect herbivores. J Glob Change Biol. 2002;8:1–16.
7. Das DK, Singh J, Vennila S. Emerging crop pest scenario under the impact of climate change—a brief review. J Agric Phys. 2011;11:13–20.
8. Bagamba F, Bashaasha B, Claessens L, Antle J. Assessing climate change impacts and adaptation strategies for smallholder agricultural systems in Uganda. Afr Crop Sci J. 2012;20:303–16.
9. Lobell DB, Burke MB, Tebaldi C, Mastrandrea MD, Falcon WB, Naylor RL. Prioritizing climate change adaptation needs for food security in 2030. Science. 2008;319:607–10.
10. Mohammad FD, Fauziah I, Mohd RZ, Fairuz K, Abu ZU, Syed A, Rahman SAR, Ismail RM, Hanysyam MN, Norazliza R. Asymmetry effect of intercropping non host crops between cabbage and climatic factor on the population of the diamondback moth (*Plutella Xylostella* L.) and yield. J Agric For Fish. 2014;3:171–7.
11. Mochiah MB, Banful B, Fening KO, Amoabeng BW, Offei K, Bonsu E, Braimah H, Owusu-Akyaw M. Botanicals for the management of insect pests in organic vegetable production. J Entomol Nematol. 2011;3:85–97.
12. Shelton AM, Wyman JA, Cushing NL, Apfelbeck K, Dennehy TJ, Mahr SER, Eigenbrode SD. Insecticide resistance of diamondback moth (Lepidoptera: Plutellidae) in North America. J Econ Entomol. 1993;86:11–9.
13. Shelton AM, Sances FV, Hawley J, Tang JD, Boune M, Jungers D, Collins HL, Farias J. Assessment of insecticide resistance after the outbreak of diamondback moth (Lepidoptera: Plutellidae) in California. J Econ Entomol. 2000;93:931–6.
14. Susila W, Sumiartha K, Nemoto H, Kawai S. The effect of insecticides on populations of diamondback moth, *Plutella xylostella* (Lepidoptera: Yponomeutidae) and its parasitoid, *Diadegma semiclausum* (Hymenoptera: Ichneumonidae) in cabbage. J Int Soc Southeast Asian Agric Sci. 2003;9:132–8.
15. Sarfraz M, Keddie BA. Conserving the efficacy of insecticides against *Plutella xylostella* (L.) (Lepidoptera, Plutellidae). J Appl Entomol. 2005;129:149–57.
16. Xu QC, Xu HL, Qin FF, Tan JY, Liu G, Fujiyama S. Relay-intercropping into tomato decreases cabbage pest incidence. J Food Agric Environ. 2010;8:1037–41.
17. Dudley TS, Marvin KH, Liu TX. Adoption of pest management practices by vegetable growers. J Am Entomol. 2002;48:197–9.
18. Asare-Bediako E, Addo-Quaye AA, Mohammed A. Control of diamondback moth (*Plutella xylostella*) on cabbage (*Brassica oleracea* var capitata) using intercropping with non-host crops. Am J Food Technol. 2010;5:269–74.
19. Endersby NM, Morgan WC. Alternatives to synthetic chemical insecticides for use in crucifer crops. J Biol Agric Hortic. 1991;8:33–52.
20. Said M, Itulya FM. Intercropping and nitrogen management effects of diamondback moth and yield of collards in the highlands of Kenya. J Afr Crop Sci. 2003;2:35–42.
21. Syed TS, Lu YY, Liang GW. Effect of crude extracts from plants on the oviposition behavior of diamondback moth. J South China. 2003;24:87–8.
22. Liu S, Ji M, Zhao L, Wei S, Wang G, Li X, Li L. Preliminary study on bioactivity of two plant extracts against three kinds of pests. Xiandai Nongyao, Shenyang. 2007;6:27–9.
23. Li M, Gao X, Gao Z, Zhao W, Su Z. Insecticidal activity of extracts from forty-eight plants including *Xanthium sibiricum* Patrin. Huanjing Xuebao Jinan. 2008;17:33–7.
24. Sayyed AH, Saeed S, Noor-Ul-Ane M, Crickmore N. Genetic, biochemical, and physiological characterization of spinosad resistance in *Plutella xylostella* (Lepidoptera: Plutellidae). J Econ Entomol. 2008;101:1658–66.
25. Tingem M, Rivington M, Bellocchi G, Azam-Ali S, Colls J. Assessment of the ClimGen stochastic weather generator at Cameroon sites. Afr J Environ Sci Technol. 2008;1:86–92.
26. Tingem M, Rivington M, Azam-Ali S, Colls J. Effects of climate change on crop production in Cameroon. Climate Res. 2007;36:65–77.
27. Fombe L, Amahnui AG. Impacts of climate variability on malaria incidence in the Buea Municipality of Cameroon; implications for malaria control and prevention. Curr J Appl Sci Technol. 2017;25:1–6.
28. Laux P, Jäckel G, Tingem RM, Kunstmann H. Impact of climate change on agricultural productivity under rainfed conditions in Cameroon—a method to improve attainable crop yields by planting date adaptations. Agric For Meteorol. 2010;150:1258–71.
29. Ngondjeb YD. Agriculture and climate change in Cameroon: an assessment of impacts and adaptation options. Afr J Sci Technol Innov. 2013;5:85–94.
30. John P, Ian DE, Robert WP, Laszlo N. Zonation of forest vegetation and soils of Mount Cameroon, West Africa. Plant Ecol. 2007;192:251–69.
31. Ngosong C, Mfombep PM, Njume AC, Tening AS. Comparative advantage of Mucuna and Tithonia residue mulches for improving tropical soil fertility and tomato productivity. Int J Plant Soil Sci. 2016;12:1–13.
32. Fraser PB, Brodie HM, Cheek MD, Healey SJ, Marsden JN, Nning NJ, McRobb A. Plant succession on the 1922 lava flow of Mt. Cameroon. In: Timberlake J, Kativu S, editors. African plants: biodiversity, taxonomy and uses. Richmond: Royal Botanic Garden, Kew; 1999. p. 253–62.
33. Payton RW. Ecology, altitudinal zonation and conservation of tropical rain-forest of Mount Cameroon. Final project-report. R4600, ODA, London; 1993.
34. Fraser PJ, Hall JB. Climate of the Mount Cameroon region, long and medium term rainfall, temperature and sunshine data, SAFS, University of Wales Bangor, MCP–LBG, Limbe; 1998. p. 56.
35. Tanyi CB, Ngosong C, Ntonifor NN. Comparative effects of *Piper guineense* emulsion and Cabbage–tomato intercropping for controlling cabbage pests and improving performance. J Agric Ecol Res Int. 2017;13:1–12.
36. Nana C. Research methods and applied statistics: beginners and advanced learners. Buea: GOOAHEAD; 2012.
37. StatSoft. Statistica 9.1 for windows. StatSoft Inc. Tusla, USA; 2010.
38. Porter JH, Parry ML, Carter TR. The potential effects of climatic change on agricultural insect pests. J Agric For Meteorol. 1991;57:221–40.
39. Ngowi BV, Tonnang HEZ, Mwangi EM, Johansson T, Ambale J, Ndegwa PN. Temperature-dependent phenology of *Plutella xylostella* (Lepidoptera: Plutellidae): Simulation and visualization of current and future distributions along the Eastern Afromontane. Public Libr Sci. 2017;12:1–24.
40. Atwal AS, Dhaliwal GS. Agricultural pests of South Asia and their management. Ludhiana: Kalyani Publisher; 2002. p. 320.
41. Shukla A, Kumar A. The diamond back moth, *Plutella xylostella* a problematic pest of Brassica crop. J Adv Indian Entomol. 2005;1:229–40.
42. Menéndez R. How are insects responding to global warming. J Entomol. 2007;150:355–65.
43. Abril A, Oliveras J, Gomez C. Effect of temperature on the development and survival of the Argentine ant, *Linepithema humile*. J Insect Sci. 2010;10:1–13.
44. Stireman JO, Dyer LA, Janzen DH, Singer MS. Climatic unpredictability and parasitism of caterpillars: implications of global warming. Proc Natl Acad Sci. 2005;102:17384–7.
45. Emily AB, Heather AH, Brent JS, Rebecca HH, Jonathan AN. Winter warming effects on overwinter survival, energy use, and spring emergence of *Cerotoma trifurcata* (Coleoptera: Chrysomelidae). J Agric For Entomol. 2016;19:163–70.
46. Estay SA, Lima M, Labra FA. Predicting insect pest status under climate change scenarios: combining experimental data and population dynamics modeling. J Appl Entomol. 2009;133:491–9.
47. Olesen JE, Bindi M. Consequences of climate change for European agricultural productivity, land use and policy. Eur J Agron. 2002;16:239–62.
48. Jamieson MA, Trowbridge AM, Raffa KF, Lindroth RL. Consequences of climate warming and altered precipitation patterns for plant-insect and multitrophic interactions. Plant Physiol. 2012;160:1719–27.
49. Van Damme V, Berkvens N, Moerkens R, Berckmoes E, Wittemans L, De Vis R, De Clercq P. Overwintering potential of the invasive leafminer *Tuta absoluta* (Meyrick) (Lepidoptera: Gelechiidae) as a pest in greenhouse tomato production in Western Europe. J Pestic Sci. 2014;88:533–41.
50. Krishna MPK, Srinivasan K. Indian mustard as a trap crop of major lepidopterous pests of cabbage. J Trop Pest Manag. 1991;37:26–32.

51. de Paula VF, de Barbosa LCA, Demuner AJ, Pilo-Valeso D, Picanço MC. Synthesis and insecticidal activity of new amide derivatives of piperine. J Pest Manag Sci. 2000;56:168–74.
52. Scott IM. Efficacy of Piper (Piperaceae) extracts for control of common home and garden insect pests. J Econ Entomol. 2004;97:1390–403.
53. Okonkwo CO, Ohaeri OC. Insecticidal potentials of some selected plants. J Chem Pharm Res. 2013;5:370–6.
54. Arong GA, Oku EE, Obhiokhenan AA, Adetunji BA, Mowang DA. Protectant ability of *Xylopia aethiopica* and *Piper guineense* leaves against the cowpea bruchid *Callosobruchus maculatus* (fab.) (Coleoptera: Bruchidae). World J Sci Technol. 2011;1:14–9.
55. Akunne CE, Afonta CN, Mogbo TC, Ononye BU, Ngenegbo UC. Evaluation of the efficacy of mixed powders of *Piper guineense* and *Zingiber officinale* against *Callosobruchus maculatus* (F.) (Coleoptera: Bruchidae). Am J Biol Life Sci. 2014;2:63–7.
56. Pavel IK, Brian F, Christine HF, Robert DH. Plant responses to insect herbivory: interactions between photosynthesis, reactive oxygen species and hormonal signalling pathways. J Plant Cell Environ. 2012;35:441–53.
57. Pane C, Celano G, Piccolo A, Villecco D, Spaccini R, Palese AM, Zaccardelli M. Effects of on-farm composted tomato residues on soil biological activity and yields in a tomato cropping system. J Chem Biol Technol Agric. 2015;2:1–13.

Chemical and rheological characterization of arabinoxylan isolates from rye bran

Denisse Bender[1], Maximilian Schmatz[1], Senad Novalin[1], Renata Nemeth[2], Foteini Chrysanthopoulou[1], Sandor Tömösközi[2], Kitti Török[2], Regine Schoenlechner[1*] and Stefano D'Amico[1]

Abstract

Background: Rye arabinoxylans (AXs) might be used as baking improvers for gluten-free (GF) bread. However, their extraction process still needs to be improved.

Objective and methods: The aim of this study was to simplify AX extraction of rye bran by varying temperature and pH and evaluate its chemical and rheological properties for application in GF bread.

Results and conclusion: The results demonstrated that higher amounts of AX and impurities were extracted with the increasing pH and temperature. AX yield reached values up to 3.3 g AX/100 g bran. Highest ferulic acid (FA) content (117.27 ± 1.46 mg/100 g) was achieved at the mildest extraction conditions (30 °C and 0.17 M NaOH). A/X ratio of isolates ranged between 0.53 and 0.57 and the gluten content between 81.30 and 216.78 ppm. Rheological measurements revealed typical pseudoplastic behavior of isolates. AX extracted at 30 °C and 0.17 M NaOH showed a slightly higher initial viscosity in comparison with the other isolates but was still inferior to carboxylmethylcellulose (CMC) and hydroxypropyl methylcellulose (HPMC).

Keywords: Arabinoxylan, Alkaline extraction, Gelling properties, Rye bran

Background

In wheat bread, the main structure-forming component which gives the dough its viscoelastic properties and its outstanding baking quality is gluten [1]. Therefore, a significant technological challenge arises when this essential structure-building protein has to be replaced. Gluten-free (GF) bread is usually characterized by a poor nutritional and technological quality, generally displaying low volumes and an elevated staling rate [2, 3]. A combination of hydrocolloids, emulsifiers, and proteins has been proven to enhance GF bread's quality as they are able to imitate the three-dimensional gluten network to some extent. Sourdough fermentation has also been used to improve its textural and sensorial properties [4]. Regardless of technological advances, the ability of baking additives to fully match gluten's performance has not been achieved yet [3]. Differences between qualities of gluten-containing and GF breads still remain. A promising alternative to achieve a gluten-like performance in GF bread is by mimicking a rye bread-like structure. This could be accomplished by adding isolated arabinoxylans (AXs) from rye to a GF bread's formulation. It is known that these hydrocolloids can form stable hemicellulose networks by covalent crosslinking between feruloyl-groups with neighborhood AX polymer chains [5]. Moreover, AXs have also the ability to bind proteins by dehydroferulic acid-tyrosine crosslinks. Since strength of the network is highly dependent on intrinsic AX properties, it is believed that by isolating AX with desired characteristics, the potential to imitate a rye dough-like structure could be achieved.

Arabinoxylans are composed of a highly substituted β(1→4)-linked D-xylopyranosyl backbone with single arabinosyl units as main side groups. Ferulic acid is

*Correspondence: regine.schoenlechner@boku.ac.at
[1] Department of Food Science and Technology, BOKU- University of Natural Resources and Life Sciences, Vienna 1180, Austria
Full list of author information is available at the end of the article

additionally attached to arabinosyl units of the polymer chain via ester linkages [6]. Its amount is important for the crosslinking behavior during baking. Depending on their molecular weight, substitution pattern (i.e., A/X ratio), and degree of physical entanglement (e.g., diferulate crosslinking and covalent ester bonds), they are classified into water-soluble AX (WSAX) and water-insoluble AX (WIAX) [7]. Considering that WSAXs have the ability to bind a large amount of water, retain gas in doughs, and slow down staling rate, the importance of AXs in rye bread baking has been proven. Moreover, the molar mass of AXs is determinant to improve bread quality, as chain length and solubility can significantly influence bread properties [8]. It has to be noted that these statements were derived from results conducted with gluten-containing flours. Thus, a similar behavior cannot be transferred directly to GF flours.

Since limited information on the addition of AXs as baking additives is available in GF products, further investigations should be carried out. A recent study showed that the addition of linseed mucilage containing approximately 85% AX to a GF bread formulation led to a significant increase in volume and porosity. The mucilage did not only improve technological quality of GF bread, but proved to be more acceptable in terms of sensory assessment compared with the control [9]. Moreover, Burešová and Kubínek [10] supposed that positive differences among GF doughs made by several GF flours are probably related to the presence and characteristics of AXs.

One of the most common methods to isolate AXs is through water extraction. As most of the AXs in the grain are crosslinked with other cell-wall components to form a structural network, more severe treatments, such as chemical, enzymatic, or mechanically assisted treatment, are usually applied to increase AX solubilization [11]. In addition, the AX extraction yield highly depends on process parameters, such as temperature, pH, extraction time, type of solvent, and raw material-to-solvent ratio. Earlier investigations reported water extraction yields in the range of 1.9–2.1 g AX/100 g rye flour [12, 13], whereas chemical and enzymatic extractions yielded up to 2.5 g AX/100 g rye bran [14] and 1.08–2.1 g AX/100 g rye flour [15], respectively. For rye, no information about mechanically assisted extractions has been reported yet, in comparison with wheat, where low AX yields were achieved [16].

Overall, elevated extraction yields have almost exclusively been achieved by means of severe, as well as time- and cost-intensive treatments, which consequently lead to the degradation of the extracted compounds [17]. A cost-efficient and environmentally friendly alternative to effectively separate AXs from smaller co-extracted molecules is the use of ultrafiltration. During the separation process, a low molecular cutoff membrane allows smaller molecular weight substances to pass through, while retaining larger molecules. High-purity isolates could therefore be obtained.

Based on a previous study carried out by Mansberger et al. [14], the first aim of this study was to simplify the AX extraction procedure and characterize in more detail the effects of a wider range of temperatures and pH values on the chemical composition of isolated rye AXs. The second aim focused on evaluating the rheological properties of the obtained AX isolates to determine their ability to be used as baking improvers in GF bread.

Methods

Materials

Rye bran flour from Good Mills Austria GmbH (Schwechat, Austria) was used as raw material for AX extraction. For enzymatic treatment, subtilisin (Alcalase 2.4 L), alpha-amylase (Termamyl 120 L), amyloglucosidase (AMG 300 L), and bacillolysin (Neutrase 0.8 L), were purchased from Novozymes Ltd. (Bagsvaerd, Denmark). Corolase®LAP, Corolase®PP, Corolase®7089, Papain®, Veron®P, and Veron Amylofresh® were donated by AB Enzymes GmbH (Darmstadt, Germany). For rheological measurements, hydroxypropyl methylcellulose (HPMC, Methocel™ F450; 320–480 cPs) and carboxymethylcellulose (CMC) E466 were donated by Harke GmbH (Mühlheim an der Ruhr, Germany). All the used reagents were of analytic grade and purchased from Sigma-Aldrich (Steinheim, Germany).

Pilot-scale extraction procedure with temperature and pH variations

Based on a previous study [14], the effects of a broader range of temperatures (up to 70 °C) and pH on the extraction of rye bran AXs were tested. For α-amylase inactivation, the bran (10.8 kg) was dry heated at 130 °C for 90 min in an oven (Memmert GmbH &Co. KG, Schwabach, Germany). Sodium hydroxide was used at two concentrations (0.17 and 0.25 M which correspond to pH of 11.83 and 12.80, respectively) and mixed with rye bran in a 1:10 ratio. The extraction was performed at 30, 50, and 70 °C for 100 min with constant agitation. Afterward, the solids were removed using a horizontal centrifuge decanter (Sharples, Waldkraiburg, Germany), and the pH was reduced to pH 8 using phosphoric acid. The supernatant was preserved with 0.15 g/L of potassium sorbate. Proteases (200 mL Alcalase®; 20 g Corolase® PP and 228 mL Corolase LAP®) were added and stirred at 50 °C for 2 h. Then, the pH was decreased again to 6.5 for 16 h for further enzymatic treatment (17 mL Corolase® 7089, 13 mL Papain®; 15 g Veron Amylofresh®; 15 g Veron®P;

200 mL Neutrase®; 150 mL Thermamyl®; 150 mL AMG®). Enzymes were inactivated by heat treatment at 95 °C for 20 min. Subsequently, the extracted solution was ultra/dia-filtrated using a 8-kDa ceramic membrane (TAMI Industries, Hermsdorf, Germany) at 50 °C, and the remaining solution was then portioned. A representative amount of the extract (20 L) was then freeze dried (FreeZone 6, Fa. Labconco, Outside, USA).

Rheological measurements

Rheological characterization was performed using a Kinexus Rheometer pro+ (KNX 2001, Malvern Instruments GmbH, Herrenberg, Germany) at 25 °C. For comparative means and to avoid broad viscosity ranges, especially for HPMC and CMC, a diluted concentration of 0.01% of hydrocolloid was chosen, which was defined in pre-trials (not presented here). Before measurement, 5 mg of each sample was diluted with 50 ml distilled water and immediately poured on a cone–plate geometry (CP1/60). Flow curves were obtained under continuous shearing over a shear rate ranging from 0.01 to 10 s^{-1}. Measurements were carried out in triplicate, and rheological parameters were evaluated using the manufacturer's supplied computer software (rSpace for Kinexus, Malvern Instruments GmbH, Herrenberg, Germany). Rheological behavior of AXs was described by fitting the experimental data to a power law model:

$$\eta = K\dot{\gamma}^{n-1}$$

where η is apparent viscosity (Pa s), K is the consistency coefficient (Pa s^n), $\dot{\gamma}$ is the shear rate (s^{-1}), and n is flow behavior index.

Gluten quantification

Gluten content of AX isolates was quantified by means of a competitive Enzyme Linked Immunosorbent Assay (ELISA) using the R5 antibody (Ridascreen® Gliadin competitive, R-Biopharm, Darmstadt, Germany) according to the manufacturer's instructions [18]. Gliadin concentrations were calculated based on a (4-PL) nonlinear regression curve-fitting model provided by XL-STAT-Base (Addinsoft, Witzenhausen, Germany) and converted into gluten concentration, by multiplying the gliadin concentration by a factor of two [19].

Analytic methods

AX isolates were analyzed by different methods to evaluate the influence of process conditions on their chemical composition. Dry matter was estimated according to ICC-standard method 110/1 [20]. Protein content was performed by the Bradford assay (Roti-Quant®, Carl Roth GmbH, Karlsruhe, Germany). Soluble and insoluble dietary fibers (SDF and IDF) were determined following the standard method of AACC No. 32-07 [21] (Megazyme test kit, Megazyme International Ireland Ltd., Wicklow, Ireland). Total starch was performed according to the standard method of AACC No. 76-13.01 [21] (Megazyme test kit, Megazyme International Ireland Ltd., Wicklow, Ireland). Extraction of free and bound phenolic compounds of rye bran and AX isolates were analyzed as previously reported by Mattila et al. [22]. The measurement of monosaccharide composition was carried out according to the procedure of Sluiter et al. [23] where hydrolyzation into monosaccharides under the process conditions—72% sulfuric acid (30 min at 30 °C), sterilization (60 min at 121 °C), and centrifugation of the sample—was performed. A HPLC equipped with a refractive index detector (Hitachi LaChrom Elite®, Hitachi Europe GmbH, Düsseldorf, Germany) and a Rezek RPM-Monosaccharide Aminex HPX-87P column (300 × 7.80 mm) attached with ionic form H^+/CO_3 deashing guard column (4 × 3.00 mm) (Phenomenex, Aschaffenburg, Germany) were used. All analyses were performed in triplicate, except for monosaccharide composition and gluten content, which were performed in duplicate. The AX content was calculated as the sum of arabinose and xylose fractions.

Statistical evaluation

Statistical analyses were performed using STATGRAPHICS Centurion XVII, version 17.1.04 (Statpoint Technologies, Inc., Warrenton, USA). Statistical significant differences were determined between the samples using ANOVA (analysis of variance; f-test for multiple samples or two samples with $\alpha = 0.05$); and Fishers least significance tests. A p value <0.05 was considered as statistically significant. Significant differences were indicated by different letters in the rows.

Results and discussion

Influence of temperature and pH on pilot scale AX extraction

As seen in rye dough, covalent crosslinks between AX and proteins, as well as other dietary fiber components contribute to form a hemicellulose network [5]. Since AX may not be the only component to play a significant role in GF dough formation, detailed information about chemical composition of the obtained AX isolates at different extraction conditions was provided (Table 1). In such a way, the purity of AX isolates was also defined. Data on Table 1 show that temperature and pH had significant effects on chemical composition and yield. Overall, all AX isolates had a moderate amount of starch and a high concentration of protein, showing that enzymatic treatment was not able to fully purify AX isolates. Increasing the temperature to 70 °C at a pH of 11.83 increased the extractability of SDF to up to 30.30 g out

Table 1 Chemical composition of AX isolates at different extraction conditions

Extraction conditions	Protein (g/100 g)	Starch (g/100 g)	β-Glucan (g/100 g)	IDF (g/100 g)	SDF (g/100 g)
Effect of temperature					
30 °C–0.17 M	19.70 ± 0.13	9.03 ± 0.52[a]	2.59 ± 0.03	1.41 ± 0.24	22.64 ± 0.34[a]
50 °C–0.17 M	20.77 ± 0.68	15.65 ± 0.51[c]	2.54 ± 0.72	0.84 ± 0.11	26.13 ± 0.11[b]
70 °C–0.17 M	22.26 ± 1.03	12.05 ± 0.15[b]	2.35 ± 0.18	1.19 ± 0.03	30.29 ± 0.01[c]
p value[1]	*0.0815*	*0.0014*	*0.8487*	*0.0749*	*0.0001*
Effect of pH[2]					
30 °C–0.25 M	16.68 ± 0.24	14.69 ± 0.01	2.21 ± 0.06	0.75 ± 0.16	25.24 ± 2.40
p value[3]	*0.0041*	*0.0042*	*0.0136*	*0.0835*	*0.2691*

Mean value of triplicate determinations ± standard deviation

IDF Insoluble dietary fiber; *SDF* Soluble dietary fiber

[1] Comparison of isolates extracted at 30, 50, and 70 °C at a fixed pH (0.17 M NaOH: values associated with different lower case letters at the same pH denote significant differences ($p < 0.05$)

[2] 0.17 M NaOH corresponds to a pH of 11.83, while 0.25 M NaOH corresponds to a pH of 12.80

[3] Comparison of isolates extracted at 0.17 and 0.25 M at a fixed temperature (30 °C)

of 100 g rye bran. Due to the strong alkaline solvent, water-insoluble fibers were also extracted. The high pH has been shown to generate oxyanions and improve solubility [24]. Regardless of the extraction conditions, about 2.2–2.6% of β-glucans were recovered.

A similar outcome could be observed in the investigation of Mansberger et al. [14], where only low temperatures of 20 and 30 °C were tested. Nonetheless, the previous investigation was not able to clearly state the individual effects of pH and temperature on yield and physiology of the AX isolates, as both parameters were simultaneously varied (i.e., 20 °C and 0.25 M NaOH; 30 °C and 0.17 M NaOH). Statistical analyses from the current study revealed that an increase of temperature led to significantly higher starch and SDF yields, and starch was further increased with higher pH. On the contrary, the extraction of protein and β-glucan decreased with the increasing pH. IDF content was not affected by either factor.

Monosaccharide composition of isolates is presented in Table 2. As the temperature increased, higher shares of AXs were obtained (not significant). In comparison with Table 1, all the AX contents of the isolates exceeded the amount of total dietary fiber. This discrepancy was attributed to the complexity of the dietary fiber method, in which precise handling skills of more than 30 steps are required. This consequently caused major fiber losses in contrast to the HPLC-RI method. Among the analyzed sugars, only the glucose content was influenced by the altered process conditions. In all processing conditions except for 70 °C, an increase in temperature as well as in pH resulted in significantly higher concentrations of glucose, which was favored by thermal degradation of starch. In contrast, extraction at 70 °C generated

Table 2 Monosaccharide composition of AX isolates

Extraction conditions	Glucose (g/100 g)	Galactose (g/100 g)	Xylose (g/100 g)	Arabinose (g/100 g)	AX content[1] (g/100 g)	A/X ratio
Effect of temperature						
30 °C– 0.17 M	20.45 ± 3.95[a]	5.11 ± 1.45	18.17 ± 3.33	10.36 ± 2.55	28.52	0.57 ± 0.04
50 °C– 0.17 M	32.21 ± 0.10[b]	4.37 ± 0.42	23.46 ± 0.06	12.52 ± 0.56	35.97	0.53 ± 0.02
70 °C– 0.17 M	23.79 ± 0.82[a]	2.68 ± 0.25	22.56 ± 1.05	12.52 ± 0.67	35.07	0.55 ± 0.01
p value[2]	*0.0314*	*0.1440*	*0.1449*	*0.3959*	*0.2200*	*0.5267*
Effect of pH[3]						
30 °C– 0.25 M	33.19 ± 1.28	3.94 ± 0.89	24.05 ± 0.10	13.54 ± 1.29	37.59	0.56 ± 0.04
p value[4]	*0.0492*	*0.4350*	*0.1298*	*0.2565*	*0.1680*	*1.0000*

Mean value of duplicate determinations ± standard deviation

[1] AX was calculated as the sum of arabinose and xylose fractions

[2] Comparison of isolates extracted at 30, 50, and 70 °C at a fixed pH (0.17 M NaOH): values associated with different lower case letters at the same pH denote significant differences ($p < 0.05$)

[3] 0.17 M NaOH corresponds to a pH of 11.83 while 0.25 M NaOH corresponds to a pH of 12.80

[4] Comparison of isolates extracted at 0.17 and 0.25 M at a fixed temperature (30 °C)

isolates with lowest glucose content. It was assumed that a more extensive depolymerization took place, which was produced by higher solubilization and hydrolysis of the starch. Hence, efficient purification of the isolates was achieved. Galactose, xylose, and arabinose contents were not affected by the process parameters. Similarity in A/X ratios for all isolates was visible, lying between 0.53 and 0.57. In comparison, Hell et al. [17] were able to extract around 30% of xylose-based polysaccharides from wheat bran using 1 M sodium hydroxide. They also reported that alkaline (sodium hydroxide) extraction increased the solubilization of solids, especially glucose (~61%), and favored AXs with small ramifications.

FA is the key component of AX structure, due to its ability to connect one AX to another AX-chain or even to other structural polymers [25]. Therefore, it plays an essential role when forming a hemicellulose network. Bound FA and gluten content of the isolates, as well as the extraction yield, are presented in Table 3. The values for yield were calculated and present an approximate estimation of the real AX extraction yield, since not the whole extracted suspension (which amounts were at least 60 L), but only a portion of the isolates were dried due to the limited capacity of the freeze drier. Compared with the investigation of Mansberger et al. [14], the final yields were much lower because the dead volume of ultra/dia-filtration unit was not considered. At higher temperatures, higher amounts of FA crosslinks were broken down, reducing the amount of bound FA and increasing the concentration of free FA, which was removed during ultra/dia-filtration. Determination of gluten content of the AX isolates showed that in parallel to the high protein content of the isolates, the amount of CD toxic gluten remained above 20 ppm, which is the threshold for gluten-free products.

In comparison with the previous study by Mansberger et al. [14] the AX isolates in this present study showed a greater gluten contamination. These differences were mainly attributed to the used quantification method. In contrast to this study, Mansberger, et al. [14] did not use the competitive ELISA test suitable for hydrolyzed samples. Therefore, an underestimation of the gluten content could have been possible. Moreover, the previous study applied longer purification times, in which more extensive gluten depolymerization could be achieved. Gluten concentration was not affected by pH or temperature, except at 70 °C, where it was significantly lower. An explanation for this phenomenon could be that an increase of temperature led to higher solubilization of gluten from rye bran, but simultaneously higher temperatures also increased gluten hydrolysis which was removed at higher rates during purification.

Most appropriate conditions for the extraction of AX were determined according to AX yield and FA content. Even though the lowest temperature showed isolates with the most favorable FA content, a poor extraction yield was achieved. An opposite behavior could be observed at 70 °C. Because the amount of FA has been well correlated with the degree of molecular weight [12], isolates with high FA concentrations were desired. Therefore, intermediate extraction conditions (50 °C and 0.17 M NaOH) were considered as the best compromise. At these conditions, an approximated yield of 3.13 g out of 100 g bran was possible which was even higher compared with several reported studies [12–14]. However, most of the available AXs still remained bound to other polymers in the rye bran matrix or their high molecular weight inhibited solubilization.

The improved isolation procedure has shown superior economic feasibility than traditional alkaline extraction,

Table 3 Influence of process parameters on extraction yield, bound ferulic acid and gluten content of AX-isolates

Extraction conditions	Ferulic acid(mg/100 g), $n = 3$	Gluten content (mg/kg), $n = 4$	Yield[1]
Effect of temperature			
30 °C–0.17 M	117.26 ± 2.06^{c}	216.20 ± 29.01^{b}	1.18
50 °C–0.17 M	38.20 ± 4.30^{b}	216.78 ± 0.77^{b}	3.13
70 °C–0.17 M	15.23 ± 5.72^{a}	81.30 ± 9.42^{a}	3.3
p value[1]	*0.0003*	*0.0049*	
Effect of pH[2]			
30 °C–0.25 M	N.D.	197.88 ± 1.25	1.7
p value[3]		*0.0733*	

Mean value of duplicate or triplicate determinations ± standard deviation

Approximation of yield is expressed as g AX from 100 g rye bran

N.D. Not determined

[1] Comparison of isolates extracted at 30, 50 and 70 °C at a fixed pH (0.17 M NaOH): values associated with different lower case letters at the same pH denote significant differences ($p < 0.05$)

[2] 0.17 M NaOH corresponds to a pH of 11.83 while 0.25 M NaOH corresponds to a pH of 12.80

[3] Comparison of isolates extracted at 0.17 and 0.25 M at a fixed temperature (30 °C)

as costs have been partially reduced. Although industrial scale isolation is somewhat limited by process safety requirements and equipment costs, a sensitive purification procedure has been adapted in order to avoid expensive alcoholic precipitation of AX and to reduce environmental impact. Even with enzymatic purification treatments, the extraction procedure would still cause lower costs compared with traditional purification by precipitation with organic solvents.

Appearance of AX isolates extracted at different pH can be contemplated in Fig. 1. A darkening effect can be seen by increasing the pH, which can mainly be attributed to caramelization and Maillard reaction of co-extracted glucose and probably dextrins.

Viscosity

Knowledge about rheological behavior of hydrocolloids is essential for providing information for appropriate application. Therefore, steady-state shear properties of AX isolates extracted at different process conditions were examined over a specific (0.1–10 s^{-1}) range of shear rates. Steady-state flow curves of AX solutions in comparison with CMC and HPMC are shown in Fig. 2. All samples showed a typical pseudoplastic or shear-thinning behavior, which was also displayed by the flow behavior index (n) < 1 (Table 4). Viscosity drastically decreased with increasing shear rate. This was mainly caused by the faster disruption of polymer network due to external forces, which exceeded the formation rate of new entanglements within the polymer. Cui and Mazza [26] showed similar shear-thinning flow behavior of AX extracted from flaxseed flour in aqueous solutions above 1% (w/w).

HPMC and CMC displayed highest initial viscosities followed by the AX isolated at 30 °C and 0.17 M NaOH. Considering that mildest extraction conditions allowed extraction of AXs with the highest FA content and thus the greatest molecular weight, a slightly higher initial viscosity could be achieved. At this stage of the investigation, viscosity differences between AX isolates were minor, which suggests that shear-rate dependence of flow behavior was not only influenced by molecular weight, but could also be affected by other properties such as fine structure and conformation of the polymer [27]. Also other polymeric impurities such as starch and proteins have to be considered.

Flow curves of hydrocolloids within the linear region were fitted to the Power-Law model in order to describe the apparent viscosity as a function of shear rate (Table 4). All experimental data fitted well with the model, since R^2 values remained above 0.946. Overall, consistency index (K) did not vary at different extraction temperatures and alkalinity. Flow behavior index (n) was estimated to determine pseudoplasticity. This property decreased as pseudoplastic behavior increased [28]. HPMC and AX extracted at 30 °C and 0.17 M displayed the highest pseudoplastic behaviors.

Buksa et al. [29] studied the rheological properties of crosslinked and hydrolyzed AX samples from rye flour. They reported that the highest consistency index and the lowest flow behavior index were found for crosslinked

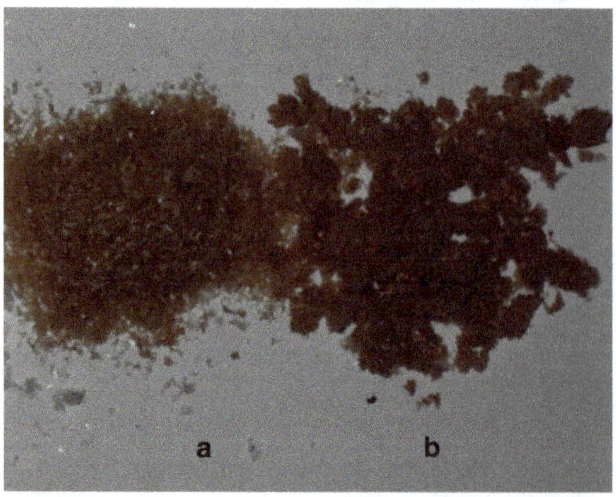

Fig. 1 Appearance of isolated AX at different sodium hydroxide concentrations: **a** 30 °C and 0.17 M NaOH; **b** 30 °C and 0.25 M NaOH

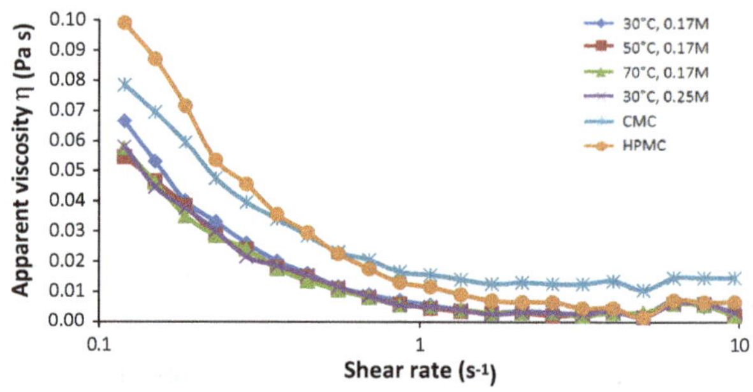

Fig. 2 Shear rate dependence of viscosity for AX isolated at different temperatures and NaOH concentrations in comparison to CMC and HPMC

Table 4 Power law model parameters of AX-solutions prepared at 0.01% in comparison to CMC and HPMC

Hydrocolloid	Model parameters		
	K (Pa s^n)	n	R^2
AX 30 °C, 0.17 M	0.007	0.025	0.996
AX 50 °C, 0.17 M	0.009	0.033	0.977
AX 70 °C, 0.17 M	0.007	0.046	0.995
AX 30 °C, 0.25 M	0.007	0.055	0.946
CMC	0.016	0.266	0.990
HPMC	0.013	0.024	0.991

AX, while the hydrolyzed samples displayed contrary values. These findings would support the results in this investigation. It is believed that AX extracted at the mildest conditions would possess a higher crosslinking ability, since their complex structure would be less damaged. Therefore, further characterization of AX crosslinking ability should be carried out.

One possibility to estimate the degree of crosslinking of AX isolates by rheological methods would be to monitor the viscosity from higher to lower shear rates. The solution would not be able to return to its original viscosity, if crosslinking takes place. Alternatively, gelling ability of AX isolates could also be tested by adding crosslinking enzymes. In this way, functionality between AX isolates would be more visible, as seen by Carvajal-Millan et al. [30] where a dependency between gelling ability and FA content was rheologically confirmed.

Furthermore, future studies should not only focus on optimizing AX extraction yields but also preserve AX functionality. As a next phase, effect of AX isolates in GF bread formulations will be persecuted for improving GF dough stability.

Conclusion

This study could successfully evaluate the influences of various conditions on the AX extraction from rye bran. Among the tested processing conditions, temperature has proven to have a greater influence on the extraction of SDF, which was mainly composed of AXs. On the other hand, the protein solubility was significantly affected by the pH. Overall, high extraction yields could be achieved, even though AX isolates were not fully purified. Since AXs are meant to be used as additives for GF products, gluten detoxification by hydrolysis must be improved before application.

Despite the fact that AX viscosity was still inferior to CMC and HPMC, isolates could have the potential to imitate a rye bread-like structure by the AX crosslinking. The use of higher AX concentrations could provide clearer information about the effect of the extraction conditions on crosslinking properties, although still more research has to be performed in this respect. Hence, knowledge about gelling properties of AX in dough matrices should be the focus of future study.

Abbreviations

A/X ratio: arabinose to xylose ratio; AX: arabinoxylan; CD: celiac disease; CMC: carboxymethylcellulose; FA: ferulic acid; GF: gluten-free; HPMC: hydroxypropyl methylcellulose; WIAX: water-insoluble arabinoxylans; WSAX: water-soluble arabinoxylans.

Authors' contributions

DB performed the statistical analysis of the data and drafted the manuscript. MS carried out AX extractions and chemical characterization of the isolates; SN participated in the AX purification process; RN carried out the rheological characterization of all AX isolates; ST contributed toward the design and coordination of the study; KT performed the monosaccharide compositional analysis and gluten quantification; RS participated in the design of the study and performed the critical revision of the manuscript; SD contributed to the manuscript by way of the interpretation of data and help in drafting the manuscript. All the authors read and approved the final manuscript.

Author details

[1] Department of Food Science and Technology, BOKU- University of Natural Resources and Life Sciences, Vienna 1180, Austria. [2] Department of Applied Biotechnology and Food Science, Budapest University of Technology and Economics, Budapest 1111, Hungary.

Acknowledgements

The authors acknowledge the financial support of the Austrian Scientific Research Fund (FWF) (project 11842-N28) and Hungarian Scientific Research Fund (OTKA) (ANN-114554).

In addition, the authors wish to thank AB Enzymes GmbH and Harke GmbH for supplying the used enzymes and hydrocolloids.

Academic editor: Regine Schoenlechner.

Competing interests

The authors declare that they have no competing interests.

References

1. Sivam AS, Sun-Waterhouse D, Quek S, Perera CO. Properties of bread dough with added fiber polysaccharides and phenolic antioxidants: a review. J Food Sci. 2010;75:R163–74.
2. Matos ME, Rosell CM. Understanding gluten-free dough for reaching breads with physical quality and nutritional balance. J Sci Food Agriculture. 2015;95:653–61.
3. Houben A, Höchstötter A, Becker T. Possibilities to increase the quality in gluten-free bread production: an overview. Eur Food Res Technol. 2012;235:195–208.
4. Moroni AV, Dal Bello F, Arendt EK. Sourdough in gluten-free bread-making: an ancient technology to solve a novel issue? Food Microbiol. 2009;26:676–84.
5. Delcour JA, Poutanen K. Fibre-rich and wholegrain foods: improving quality. New York: Elsevier; 2013. p. 1–459.
6. Rose DJ, Inglett GE. A method for the determination of soluble arabinoxylan released from insoluble substrates by xylanases. Food Anal Methods. 2010;4:66–72.
7. Izydorczyk MS, Biliaderis CG. Cereal arabinoxylans: advances in structure and physicochemical properties. Carbohydr Polym. 1995;28:33–48.
8. Buksa K, Nowotna A, Ziobro R. Application of cross-linked and hydrolyzed arabinoxylans in baking of model rye bread. Food Chem. 2016;192:991–6.
9. Korus J, Witczak T, Ziobro R, Juszczak L. Linseed (*Linum usitatissimum* L.) mucilage as a novel structure forming agent in gluten-free bread. LWT Food Sci Technol. 2015;62:257–64.
10. Burešová I, Kubínek R. The behavior of amaranth, chickpea, millet, corn, quinoa, buckwheat and rice doughs under shear oscillatory and uniaxial elongational tests simulating proving and baking. J Texture Stud. 2016. doi:10.1111/jtxs.12176.
11. Zhang Z, Smith C, Li W. Extraction and modification technology of arabinoxylans from cereal by-products: a critical review. Food Res Int. 2014;65:423–36.
12. Hartmann G, Piber M, Koehler P. Isolation and chemical characterisation of water-extractable arabinoxylans from wheat and rye during bread-making. Eur Food Res Technol. 2005;221:487–92.
13. Buksa K, Ziobro R, Nowotna A, Gambuś H. The influence of native and modified arabinoxylan preparations on baking properties of rye flour. J Cereal Sci. 2013;58:23–30.
14. Mansberger A, D'Amico S, Novalin S, Schmidt J, Tömösközi S, Berghofer E, Schoenlechner R. Pentosan extraction from rye bran on pilot scale for application in gluten-free products. Food Hydrocoll. 2014;35:606–12.

15. Dervilly-Pinel G, Thibault JF, Saulnier L. Experimental evidence for a semiflexible conformation for arabinoxylans. Carbohydr Res. 2001;330:365–72.
16. Jacquemin L, Zeitoun R, Sablayrolles C, Pontalier P, Rigal L. Evaluation of the technical and environmental performances of extraction and purification processes of arabinoxylans from wheat straw and bran. Process Biochem. 2012;47:373–80.
17. Hell J, Donaldson L, Michlmayr H, Kraler M, Kneifel W, Potthast A, Rosenau T, Bohmdorfer S. Effect of pretreatment on arabinoxylan distribution in wheat bran. Carbohydr Polym. 2015;121:18–26.
18. R-Biopharm. Ridascreen® gliadin competitive enzyme immunoassay for the quantitative analysis of peptide fragments of gliadins and corresponding prolamines. 2007.
19. Codex Alimentarius IFS. Codex Standard for foods for special dietary use for persons intolerant to gluten. In: CODEX STAN. 2008; 118–1979.
20. ICC Standard Method No. 110/1. Determination of moisture content of cereals and cereal products (practical method). Vienna: International Association for Cereal Science and Technology; 1976.
21. AACC. Approved methods of the aacc. 10th ed. St. Paul: The Association; 2000.
22. Mattila P, Pihlava J-M, Hellström J. Contents of phenolic acids, alkyl- and alkenylresorcinols, and avenanthramides in commercial grain products. J Agric Food Chem. 2005;53:8290–5.
23. Sluiter A, Hames B, Ruiz R, Scarlata C, Sluiter J, Templeton D, Crocker D. Determination of structural carbohydrates and lignin in biomass, NREL/TP-510-42618. Colorado: National Renewable Energy Laboratory; 2011.
24. Corradini C, Cavazza A, Bignardi C. High-performance anion-exchange chromatography coupled with pulsed electrochemical detection as a powerful tool to evaluate carbohydrates of food interest: principles and applications. Int J Carbohydr Chem. 2012;2012:13.
25. Vinkx C, Delcour J. Rye (*Secale cereale*) arabinoxylans: a critical review. J Cereal Sci. 1996;24:1–14.
26. Cui W, Mazza G. Physicochemical characteristics of flaxseed gum. Food Res Int. 1996;29:397–402.
27. Kale MS, Yadav MP, Hicks KB, Hanah K. Concentration and shear rate dependence of solution viscosity for arabinoxylans from different sources. Food Hydrocoll. 2015;47:178–83.
28. Grigelmo-Miguel N, Ibarz-Ribas A, Martín-Belloso O. Rheology of peach dietary fibre suspensions. J Food Eng. 1999;39:91–9.
29. Buksa K, Ziobro R, Nowotna A, Adamczyk G, Sikora M, Zylewski M. Water binding capacity of rye flours with the addition of native and modified arabinoxylan preparations. J Agric Sci Technol. 2014;16:1083–95.
30. Carvajal-Millan E, Guilbert S, Doublier J-L, Micard V. Arabinoxylan/protein gels: structural, rheological and controlled release properties. Food Hydrocolloids. 2006;20:53–61.

18

NMR-based metabolomics in wine quality control and authentication

Maria Amargianitaki and Apostolos Spyros*

Abstract

A comprehensive summary of research work related to applications of NMR spectroscopy in combination with multivariate statistical analysis techniques for the analysis, quality control, and authentication of wine is presented. NMR spectroscopy is used to obtain the non-volatile metabolic profile and/or phenolic profile of wines, with the help of 2D NMR spectroscopy. Metabolomics is then used as an analytical tool to investigate the variability of the metabolic profile of wines due to a series of different factors involved during wine production, including terroir, pedoclimatic conditions, vintage, vineyard practices, wine-making, barrel maturation, and aging. The discriminating power of the NMR-obtained wine metabolome is utilized as a wholistic analytical approach in efforts to authenticate wine with respect to important economic attributes, such as cultivar, vintage, and geographical origin.

Keywords: Wine, NMR metabolomics, Maturation, Barrel, Vintage, Terroir, Geographical origin

Introduction

In recent years, NMR spectroscopy has found increased application in food science and agriculture, including the quality control [1], authentication [2], and analysis of an immense variety of food products [3–5]. Wine is an important commodity with a high added value, not only from a historical, but also from an economic point of view, especially for wine-producing countries, and thus a great deal of research has been devoted in NMR applications in wine analysis [6–8]. The main advantage of NMR spectroscopy is that it offers a facile and rapid way to obtain the chemical composition of grapes, grape juice, must, and wine, and identify a large variety of minor organic compounds (metabolite profiling), including amino acids [9], organic acids and alcohols [10, 11], sugars, and phenolic compounds [12]. The suitability of NMR analysis as a methodology for the quantitative determination of wine components was recently demonstrated in an international collaborative trial [13]. In fact, the whole NMR spectrum of a wine sample can be considered as a molecular fingerprint, and as such it can be used directly for comparison and identification purposes in metabolite fingerprinting of different wines. This chemical compositional/spectroscopic information constitutes the so-called wine metabolome, which is affected by a multitude of wine-making factors, including agronomic practices and pedoclimatic conditions [14], grape variety [15], fermentation practices [16], and geographical origin [17], and can be used for quality control and authentication purposes. For example, the ability to study geographical origin-related NMR metabolic effects is of extreme importance, since the high added value of wines is obtained from enforcing protected denomination of origin (PDO) and protected geographical indication (PGI) state regulations to control the appellation of wines [2, 5]. Fully automated ^{1}H NMR-based metabolomic products for wine quality control and authentication have been developed and marketed successfully [18].

The rich and complex mole of compositional information revealed by wine metabolomic profiles based on NMR spectra may be unraveled by applying suitable multivariate statistical analyses. In fact, by evaluating a large dataset of NMR spectra, the latter ones are capable of identifying correlations between metabolites and factors (e.g. terroir, vintage, etc.) potentially responsible for a specific metabolomic response as well as classify wines according to these factors [3, 19]. The multivariate statistical analysis models usually applied in wine NMR-based

*Correspondence: aspyros@uoc.gr
NMR Laboratory, Department of Chemistry, University of Crete, P.O. Box 2208, Voutes Campus, 710 03 Heraklion, Crete, Greece

metabolomics are principal component analysis (PCA), partial least squares (PLS), orthogonal projection to latent structures (OPLS), and discriminant analysis (DA). Usually, unsupervised models such as PCA are initially used to obtain an overview of the metabolic variance of a complex NMR dataset, followed by more advanced supervised models (OPLS, OPLS-DA) which are characterized by a stronger discriminatory power.

The ability to follow metabolome changes from the early stages of wine production provides the opportunity to use NMR-based metabolomics as a wholistic approach to obtain information on all aspects of the wine-making procedure, following the product from the vineyard to the bottle. Research efforts aimed toward this direction will be described in detail in the following sections.

Wine metabolites

Although the first application of NMR spectroscopy in wine analysis involved the determination of ethanol content [20], this was soon followed by attempts to discriminate and characterize the minor organic compounds present in wines at smaller concentrations [21]. The most intense peaks in ^{1}H NMR spectra of wines correspond to water and ethanol, being these compounds the main components of wine. NMR spectroscopy can be used effectively for the determination of the ethanol content in wine [22]. Conversely, when the focus is on the minor components of wine, water and ethanol signals must be eliminated spectroscopically, and this can be afforded via a variety of multiple suppression NMR techniques [9, 23]. Wine samples are also usually preconcentrated using freeze drying or an inert gas flow (nitrogen, argon) to physically remove water/ethanol, thus enhancing the detection of weak proton NMR signals of the minor compounds that are present in wine, and facilitating their identification. The merits of different preconcentration methods have been thoroughly discussed by Amaral et al. [24], while Ref. [3] provides an overview of the pros and cons of wine sample preparation methodologies (see Chapter 8). Figure 1 presents some typical ^{1}H NMR spectra of freeze-dried Port wine samples of different age and the more prominent compounds identified in them. The predominant components of wines as determined by NMR spectroscopy were organic acids and diols (tartaric acid, succinic acid, acetic acid, glycerol, 2,3-butanediol, lactic acid), amino acids (proline, alanine, arginine), carbohydrates (glucose, fructose), and polyphenols (gallic acid, 2-phenylethanol, tyrosol). Table 1 presents the major organic compounds most usually identified in wine and their ^{1}H and ^{13}C NMR chemical shifts [23]. Frequently, studies of wine also deal with the analysis of grape juice and must samples, in order to elucidate the biochemical and chemical transformations taking place during wine production. As an example, Fig. 2 presents the ^{1}H NMR spectra of must and wine obtained from cv. Tempranillo (*Vitis vinifera*) grapes from the Rioja region in northern Spain, indicating the differences between must and wine in the organic acid, carbohydrate, and aromatic NMR spectral regions. In particular, it shows that the presence of carbohydrates in wine spectrum was negligible as compared to the must, because of the saccharidic conversion occurred during must fermentation.

Both homonuclear and heteronuclear 2D NMR spectroscopy techniques have been used to assist and verify the assignment of minor organic compounds found in wine [25]. Figure 3 presents the aliphatic region of the homonuclear ^{1}H–^{1}H TOCSY 2D NMR spectrum of a French Cabernet Sauvignon wine [26]. This type of 2D NMR spectrum yields information regarding correlations via ^{1}H through bond spin–spin J couplings and is useful in identifying complete proton spin systems in organic molecules. Heteronuclear ^{1}H–^{13}C HSQC and HMBC 2D NMR spectra have also been used for the assignment of the organic compounds in wine samples [27].

^{13}C NMR spectroscopy has also been used for the identification and quantification of organic compounds in wine [28]. Holland et al. [29] used ^{13}C NMR in order to measure the concentration of amino acids in European wines. Colombo et al. [30] used quantitative ^{13}C nuclear magnetic resonance spectroscopy (qNMR) for the quantification of fructose and ethanol in wines using as an external reference the ^{13}C qNMR spectrum of a sample of known concentration. ^{13}C NMR was also used to examine the metabolic pathway transformations of amino acids during the alcoholic fermentation, since amino acids have an important contribution to the organoleptic characteristics and quality of wines [31].

Solid-state ^{13}C Cross Polarization–Magic Angle Spinning (CP–MAS) NMR spectroscopy has been applied, in combination with 1D and 2D solution-state NMR, for the analysis of the precipitates of Merlot and Cabernet Sauvignon red wines from the Bordeaux region. Figure 4 depicts the ^{13}C CP–MAS NMR spectra of precipitates of red wines after aging in oak barrels. Organic acid carbons appear in the region between 0–50 ppm (aliphatic) and 170–180 ppm (carbonyl groups), polyphenolic carbons between 90 and 140 ppm, and glycerol slightly above 60 ppm, while the 65–90 ppm region is dominated by characteristic carbon contributions from sugar moieties in polysaccharides [32].

Finally, it is worth noting that NMR spectroscopy has also been used to measure the content of acetic acid and detect oxidative wine spoilage by the direct analysis of intact wine bottles, without any sample pretreatment [33, 34].

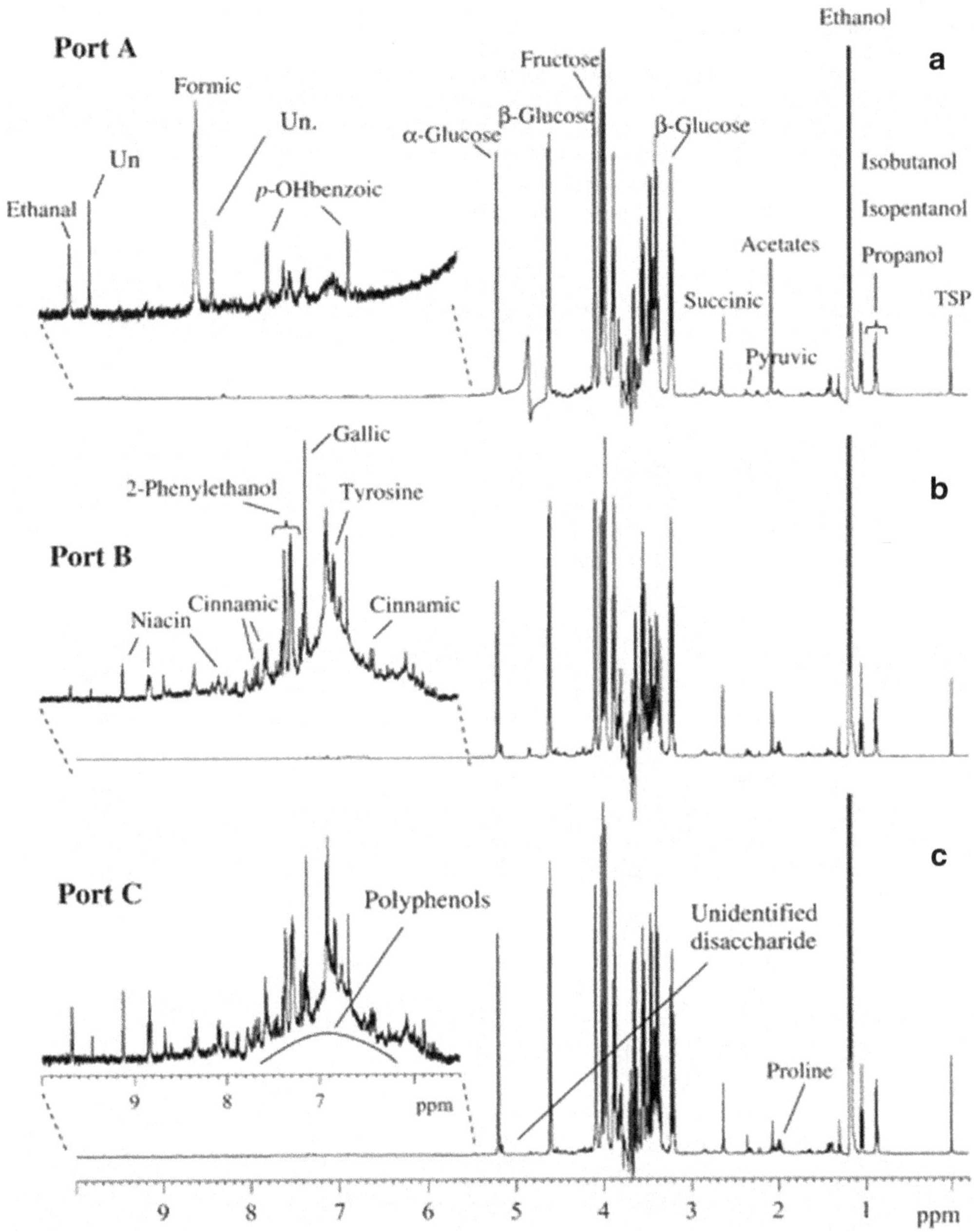

Fig. 1 ^{1}H NMR spectra of Port wines: **a** Port A (20 years old), **b** Port B (3 years old), and **c** Port C (1 year old), and assignment of the main compounds identified. Reprinted with permission from Ref. [23].

Phenolics

There is a great interest in the determination of phenolic compounds found in wine because of their significant role on the organoleptic and sensory characteristics of wines. Polyphenols include phenolic acids, flavanols, stilbenoids, dihydroflavonols, flavanol monomers (catechins), flavanol polymers (proanthocyanidins), and anthocyanins. Since polyphenols are present in very low concentrations in wine, a chromatographic preconcentration step is usually necessary for adequate analysis. ^{1}H NMR and HPLC were used for the determination of the phenolic fraction of Greek wines using adsorption resin XAD-4, in order to concentrate them [12]. The phenolic compounds that were identified were gallic acid, syringic acid, *p*-coumaric acid, *trans*-caffeic acid, (+)-catechin, (−)-epicatechin, ferulic acid, quercetin, kaempferol, and trans-resveratrol. Figure 5 shows the aromatic region of the ^{1}H NMR spectra of

Table 1 ^{1}H and ^{13}C NMR chemical shifts, proton multiplicity, and coupling constants J_{HH} for the organic compounds identified in Port wine, at its natural pH of 3.4–3.7

Compound	δ^1H in ppm (multiplicity, *J* in Hz, assignment)/δ^{13}C in ppm
Acetic acid/acetates	**2.06** (*s*, βCH_3)/23.42
Alanine	**1.47** (*d*, 7.2, βCH_3)/17.17; **3.80** (αCH)/54.79
γ-Aminobutyric acid (GABA)	**1.94** (βCH_2); **2.48** (αCH_2); **3.03** (γCH_2)
Arginine	**1.68** (*m*, γCH_2); **1.90** (*m*, βCH_2); **3.22** (*t*, δCH_2)
Caffeic/*trans*-caftaric acid	**6.42** (*d*, 16.0, αCH); **6.91** (*d*, C5H); **7.12** (*d*, C6H); **7.67** (*d*, 16.0, βCH)
Citric acid	**2.75** (*d*, 16.4, α,γCH)/46.54; **2.95** (*d*, 16.4, α′, γ′CH)
p-Coumaric/CCCC *trans*-coutaric acid	**6.42** (*d*, 16.0, αCH); **6.91** (*d*, C3H/C5H); **7.56** (*d*, C2H/C6H); **7.67** (*d*, 16.0, βCH)
Disaccharide	**5.16** (*d*, 3.9, C1H)/97.02
Ethanal	**2.23** (*d*, 3.0, CH_3); **9.67** (*q*, CH)
Ethanol	**1.17** (*t*, 7.2, CH_3)/19.56; **3.64** (*q*, 7.2, CH_2)/60.19
Ethyl acetate[a]	**1.26** (*t*, 7.2, CH_3); **4.12** (*q*, 7.2, CH_2)
Formic acid	**8.32**(*s*, *HCOOH*)/169.8
α-Fructose	**4.09** (C3H)/84.68
β-Fructose	**4.09** (C3H, C4H)/77.56
Gallic acid	**7.12** (*s*, C2H, C6H, ring)/112.37
α-Glucose	**5.20** (*d*, 3.7, C1H)/95.15
β-Glucose	**4.61** (*d*, 8.0, C1H)/99.10; **3.25** (*dd*, C2H)
p-Hydroxybenzoic acid[a]	**6.67** (*d*, 3.3, C3H, C5H, ring)/113.7; **7.54** (*d*, 3.3, C2H, C6H, ring)/129.9
Isobutanol (2-methyl-1-propanol)	**0.87** (*d*, 6.7, CH_3); **1.73** (*m*, CH); **3.36** (*d*, CH_2OH)
Isopentanol (3-methyl-1-butanol)	**0.88** (*d*, 6.7, CH_3); **1.43** (*q*, CH); **1.64** (CH_2); **3.61** (*t*, 6.7, CH_2OH)
Lactic acid	**1.38** (*d*, 7.0, βCH_3)/22.41; **4.31** (*q*, αCH)
Leucine	**0.95** (*t*, δ, δ′ CH_3); **1.72** (βCH_2, γCH)
Malic acid	**2.71** (*dd*, 6.0; 12.8, βCH)/39.48; **2.82** (*dd*, β′CH); **4.45** (αCH)/69.81
Methanol	**3.35** (*s*, CH_3)/46.15
2-Methyl-1-butanol[a]	**0.87** (*d*, CH_3); **0.88** (*t*, CH_3); **1.10** (*m*, 1/2CH_2); **1.38** (*m*, 1/2CH_2); **1.53** (*m*, CH); **3.38** (*dd*, CH_2OH); **3.47** (*dd*, CH_2OH)
Niacin	**8.0** (*dd*, C5H); **8.82** (*dd*, C4H, C6H); **9.11** (C2H)
2-Phenylethanol	**2.76** (CH_2); **3.74** (CH_2OH); **7.28** (*m*, ring); **7.34** (*m*, ring)
Proline	**1.99** (*m*, γCH_2)/26.40; **2.06** (β′CH)/31.57; **2.33** (*m*, βCH)/31.57; **3.32** (δ′CH)/49.07; **3.42** (δCH)/49.07; **4.11** (αCH)/63.76
1-Propanol	**0.88** (*t*, 7.5, CH_3); **1.53** (*m*, CH_2); **3.55** (*t*, 6.8, CH_2OH)
Pyruvic acid[a]	**2.35** (*s*, βCH_3)
Succinic acid	**2.62** (*s*, α, βCH_2)/31.74
Tartaric acid[a]	**4.41** (*s*, αCH)
Threonine[a]	**1.42** (γCH_3); **4.42** (βCH)
Tyrosine/tyrosol	**6.83** (*d*, C3H, C5H, ring)/118.13; **7.15** (*d*, C2H, C6H, ring)/133.30; **6.87** (*d*, C3H, C5H, ring)/118.46; **7.17** (*d*, C2H, C6H, ring)/133.80
Uridine	**4.17** (C4′H); **4.29** (C3′H); **5.87** (C5H ring, C1′H); **7.88** (C6H ring)

Reprinted with permission from Ref. [23]. American Chemical Society

s singlet, *d* doublet, *t* triplet *q* quartet, *dd* doublet of doublets, *m* multiplet

[a] Tentative identification

Mandilaria and Agiorgitiko wines and the assignment of the phenolic compounds [12]. In another study, two types of German white wines from the Palatinate region were analyzed by extraction with ethyl acetate and addition of Na_2SO_4 to remove the residual water. Using this method, it was possible to identify in wine not only amino acids, carbohydrates, and organic acids, but also phenolic compounds [35].

Anthocyanins, which are responsible for the red color of wines, were analyzed by 1D and 2D NMR spectroscopy in Slovenian wine samples diluted in deuterated methanol, and the anthocyanins that have been identified were peonidin-3-glucoside, petunidin-3-glucoside, cyanidin-3-glucoside, delphinidin-3-glucoside, and malvidin-3-glucoside [36]. Fourier transform near-infrared and ^{1}H NMR spectroscopy techniques combined with

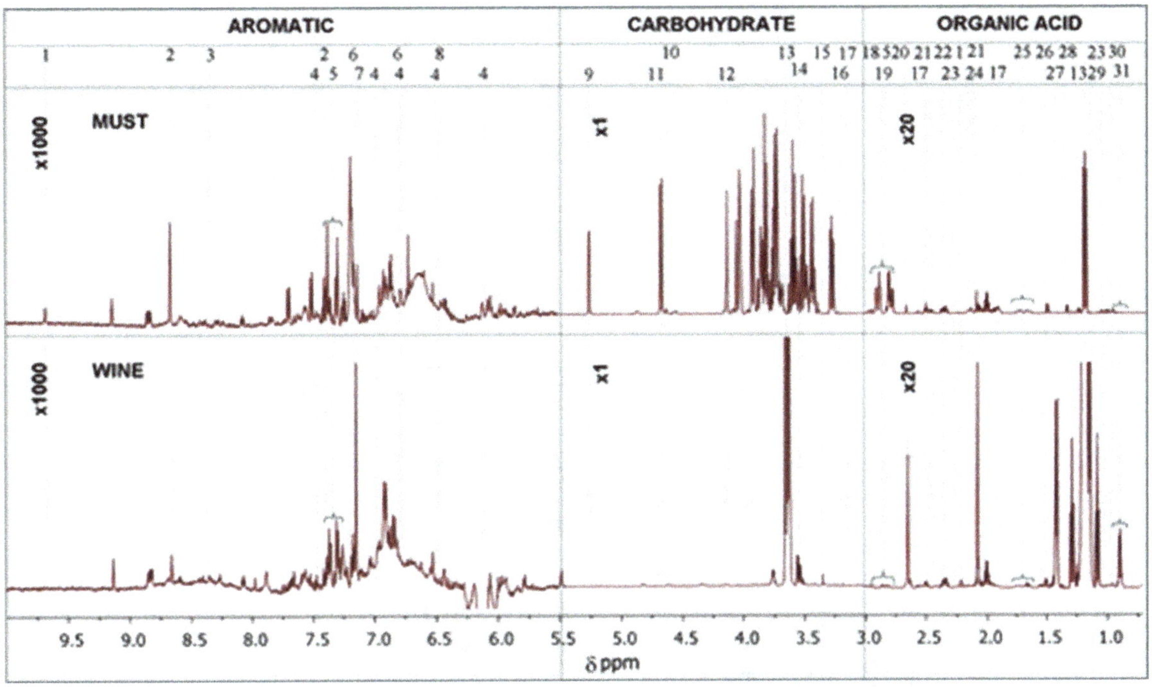

Fig. 2 ^{1}H NMR spectra of must and wine obtained from cv. Tempranillo (*Vitis vinifera*) grapes from the Rioja region in northern Spain. Reprinted with permission from Ref. [50].

classification methods were used for the authentication of anthocyanin content of red wines, in an effort to detect possible adulteration with black rice anthocyanins [37]. By the combination of NMR and LC–MS, the identification of anthocyanins in Pinot Noir, Cabernet Sauvignon, and Merlot red wines produced in Slovenia was obtained [38]. High-performance liquid chromatography coupled with mass spectrometry (LC–MS) and NMR spectrometry (LC–NMR) were used to analyze the composition of anthocyanins of grape berry skins belonging to four different grape varieties [39]. Hyphenated chromatographic/spectroscopic techniques such as LC–NMR/MS were also used for the analysis of the aromatic composition of wine phenolic extracts, with their combination being able to provide more information on the aromatic profile of the wines than standalone methodologies [40]. The chemical structure of tannins, which are considered responsible for the astringency of wines and have a key role in red wine flavors, was studied using chromatographic methods, NMR spectroscopy, and mass spectroscopy [41].

Wine-making

NMR-based metabolomics has been used extensively to study wine fermentation and evaluate the fermentative characteristics of different yeast strains. Son et al. [16] evaluated three *Saccharomyces cerevisiae* yeast strains (RC-212, KIV-1116, and KUBY-501) for their impact on the metabolic changes in must and wine during fermentation. ^{1}H NMR spectroscopy has also been used to investigate the fermentation of Romanian wines with *Saccharomyces cerevisiae* yeast under different conditions: natural fermentation (without additional yeast), induced fermentation using different amounts of selected yeasts, and industrial fermentation [42]. ^{1}H NMR metabolomics was used to successfully differentiate "fiano di Avellino" white wines obtained by fermentation with either a commercial or a selected autochthonous *Saccharomyces cerevisiae* yeast starter [43]. Figure 6 depicts the respective PCA score plot for yeast differentiation, with variance between the two yeasts afforded by higher concentrations of α-glucose and fructose in the autochthonous yeast, while glycerol and succinic acid were positively correlated with the commercial one. Lee et al. [44] used a combination of ^{1}H NMR and gas chromatography metabolite profiling in order to study the fermentative behavior of lactic acid bacteria in grape wines. Quantitative NMR was also used to study the changes in malic and lactic acid concentrations that take place during the alcoholic and malolactic fermentation process [45].

Baiano et al. [46] examined the effect of four different wine-making technologies (traditional white vinification, skin cryomaceration, vinification in a reductive environment, and a combination of the last two procedures) on

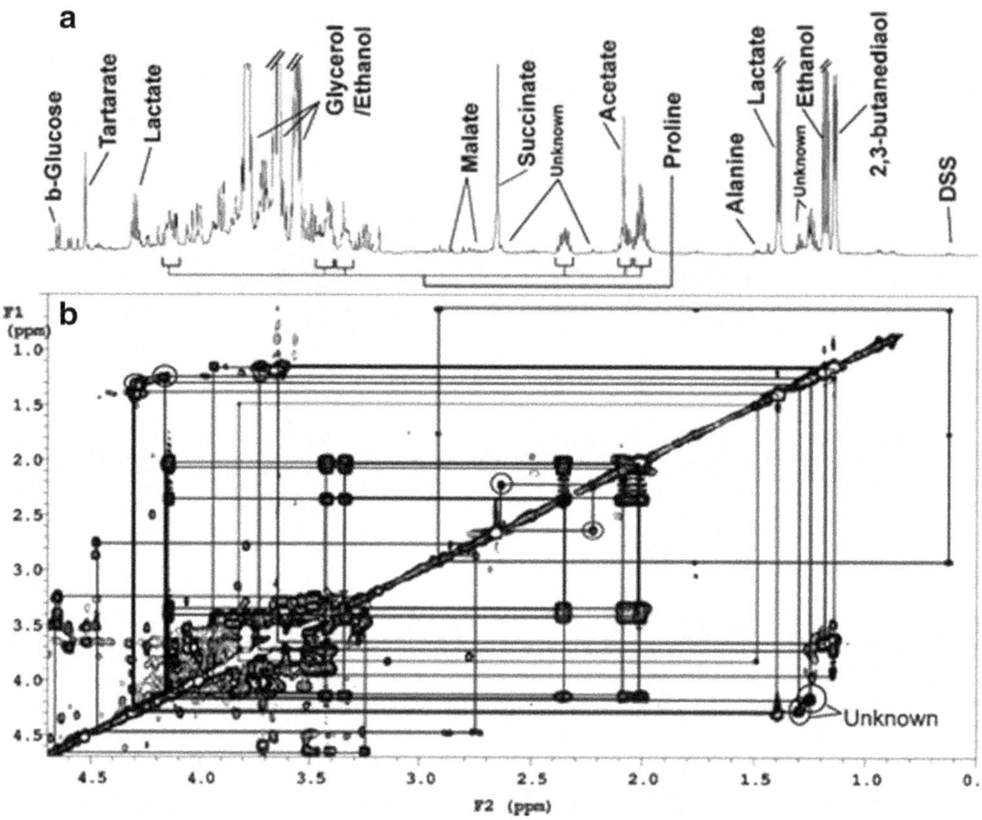

Fig. 3 2D ^{1}H–^{1}H TOCSY NMR spectrum of a French Cabernet Sauvignon wine showing the assignment of major peaks. Reprinted with permission from Ref. [26].

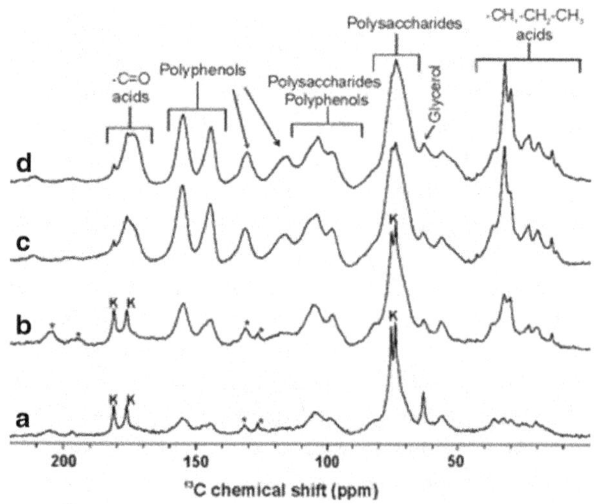

Fig. 4 ^{13}C CP–MAS NMR spectra of wine precipitates from 2012 to 2013 vintages and at different oak barrel aging times *t*: **a** Merlot 2012, $t = 1$ month; **b** Cabernet Sauvignon 2013, $t = 1$ month; **c** Merlot 2013, $t = 4$ months; and **d** Cabernet Sauvignon 2013, $t = 4$ months. A spinning frequency of 10 kHz was used for MAS, residual spinning side bands are marked with (*), and peaks from potassium bitartrate with (K). Reprinted from Ref. [32].

the metabolite profile of Sauvignon blanc wines, reporting strong changes in organic acid concentrations (mainly tartaric acid) and phenolic content. Negroamaro red wines obtained through different wine-making technologies (traditional, ultrasounds, and cryomaceration with dry ice) and two soil management practices (soil tillage and cover crop) were analyzed by ^{1}H NMR metabolomics and were discriminated using both supervised (OPLS-DA) and unsupervised (PCA) multivariate models [47]. ^{1}H NMR spectroscopy with PCA analysis was used to study Cabernet Sauvignon wine samples produced with different cultivation techniques and provided information by connecting technological procedures with the composition changes in wine [48]. Recently, Picone et al. [49] showed that the biodynamic and the organic cultivation managements produce grape berries (*V. vinifera* L. cv. Sangiovese) that can be differentiated using ^{1}H NMR-based metabolomics.

López-Rituerto et al. [50] investigated the La Rioja wine terroir using ^{1}H NMR metabolomics of must and wine samples obtained from different wineries located in this area of central Spain, reporting that the aliphatic alcohols isopentanol and isobutanol were key biomarkers for

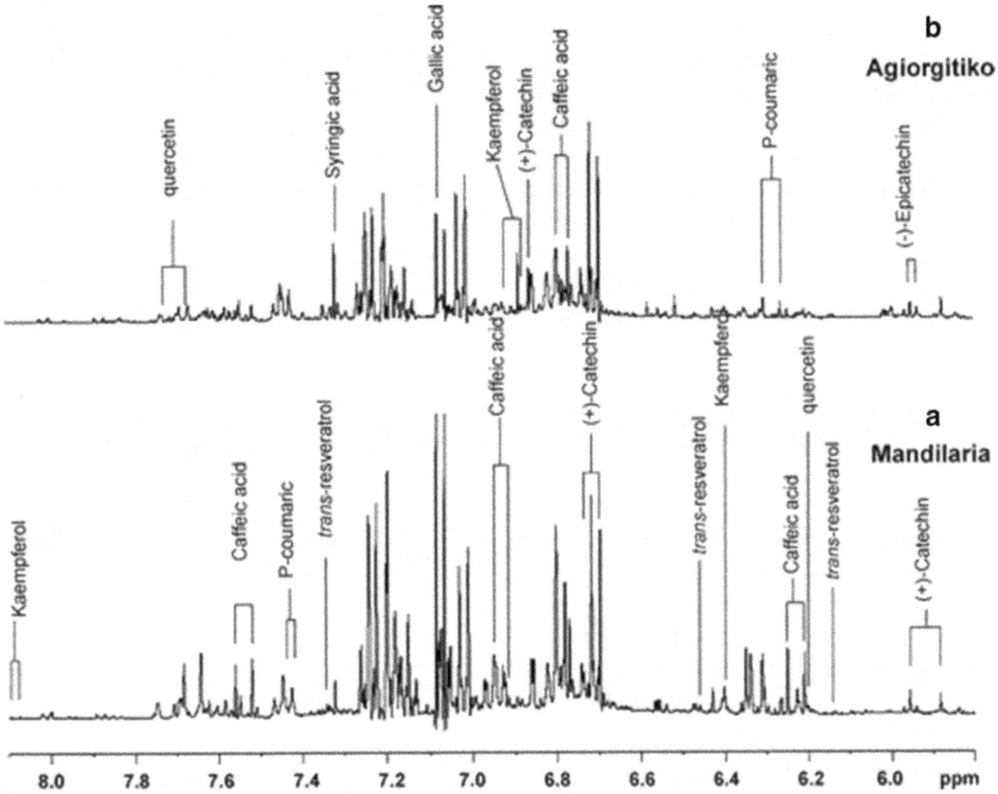

Fig. 5 Representative 400 MHz ^{1}H NMR spectra of wine extracts of **a** Mandilaria and **b** Agiorgitiko wines and assignment of the major polyphenol peaks. Reprinted with permission from Ref. [12].

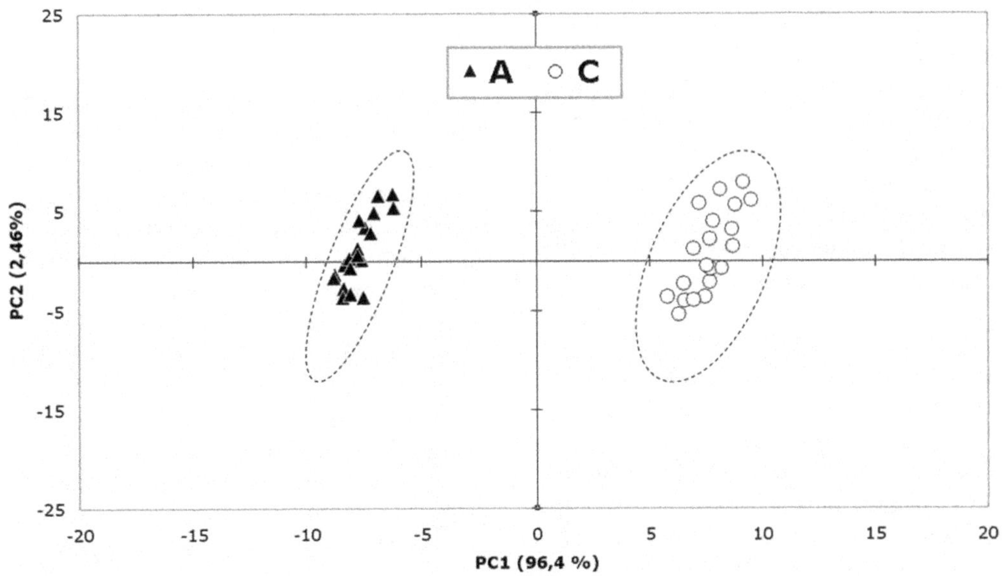

Fig. 6 Principal component analysis (PCA) score plot (PC1 vs PC2) of spectral data obtained from ^{1}H NMR spectra of wines produced with autochthonous (*A*; closed triangles) or commercial (*C*; open circles) yeast starter. The explained variance (%) is reported for each principal component, and dotted ellipses represent the 95% confidence limits for each wine type. Reprinted with permission from Ref. [43].

differentiating wineries located very close geographically. ^{1}H NMR metabolomics has also been used to characterize the terroir effect on the quality of Aglianico red wines from the Campania region of Italy [14]. Multivariate analysis of the NMR metabolite data demonstrated that the differentiation observed among wines produced from three different vineyards was related to microclimate and the carbonate, clay, and organic matter content of soils. In a study of wines produced in the Bordeaux region in France, Pereira et al. [51] showed that the NMR metabolic profile of wines was affected within a given wine cultivar by soil type (gravely, sandy, or clay), but vintage and cultivar effects were more dominant. Da Silva Neto et al. [52] used ^{1}H NMR metabolomics to characterize the metabolic profile of "tropical" wines produced in the region of São Francisco River Valley in northeast Brazil, aiming to compare tropical wines to those produced in more temperate wine-producing zones in Brazil.

Some other agronomical factors that have been studied include the effect of grape maturity and health on wine quality. Chang et al. [53] studied the effect of harvest time on the quality and metabolite composition of wines obtained from grapevine cv. 'Cheongsoo' using ^{1}H NMR metabolomics. It was reported that certain metabolites (proline and arginine) that varied with harvest time were closely associated with the body and balance scores of the wines on sensory tests, and concluded that NMR metabolomics can be used to optimize wine quality by identifying metabolites correlated with positive wine sensory attributes. Hong et al. [54] studied the metabolic effect of Botrytis infection in Champagne base wine using a ^{1}H NMR-based metabolomic approach. PCA analysis of the NMR-obtained metabolite data indicated that Botrytis infection of grape caused fermentative retardation during alcoholic fermentation leading to wines of inferior quality. Figure 7 presents PCA (A) and OPLS-DA (B) score plots of healthy and botrytized base wines indicating the correct classification of healthy wines and wines infected by *Botrytis cinerea* (20 and 40% of infection).

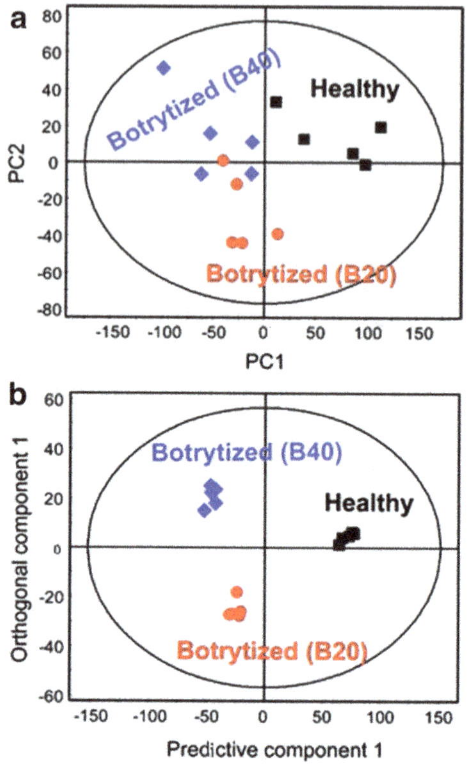

Fig. 7 PCA (**a**) and OPLS-DA (**b**) score plots derived from ^{1}H NMR spectra of healthy and botrytized (B20 and B40) base wines from 2009, indicating metabolic differences among the wines. B20 and B40 represent 20 and 40% *B. cinerea* infection, respectively. Reprinted with permission from Ref. [54].

Cultivar

^{1}H NMR spectroscopy combined with multivariate statistical analysis has been applied successfully for the classification of different wine cultivars, usually involving the study of wines produced from the same geographical region, in order to exclude site-specific effects capable of affecting the wine metabolome. Godelmann et al. [55] analyzed different wine varieties that were produced in Germany using ^{1}H NMR spectroscopy with multivariate data analysis, including PCA, Linear Discrimination Analysis (LDA), and Multivariate analysis of variance (MANOVA). The grape varieties Pinot Noir, Lemberger, Pinot blanc/Pinot gris, Müller-Thurgau, Riesling, and Gewürztraminer were successfully classified, and it was reported that the organic compounds mainly responsible for the observed differentiation of the wine varieties were shikimic acid, caftaric acid, and 2,3-butanediol. In another report, the characterization of Müller-Thurgau and Riesling white cultivars was performed using NMR spectroscopy and orthogonal projections to latent structures (OPLS) statistical analysis methods. Riesling wines were found to have higher levels of catechin, caftarate, valine, proline, malate, and citrate, whereas Müller-Thurgau wines had higher concentrations of quercetin, resveratrol, gallate, leucine, threonine, succinate, and lactate [35]. Wines from the cultivars Agiorgitiko, Mandilaria (red), Moschofilero, and Assyrtiko (white) produced in Greece [12], and Cabernet Sauvignon, Merlot, Feteasca Neagra, Pinot Noir, and Mamaia wines, produced in Romania [56], showed also successful cultivar discrimination using NMR-based metabolomics. Cabernet Sauvignon and Shiraz wines, produced in Australia, showed a clear separation based on their respective metabolite profiles. Cabernet Sauvignon had higher levels of proline, while Shiraz wines had higher levels of sugars (fructose and glucose), succinate, methanol, acetate, and some aliphatic amino acids [57]. Son et al. [15] used ^{1}H NMR spectroscopy coupled with multivariate statistical data analysis (PCA, OPLS-DA) in order to characterize wines from different grape varieties cultivated in South Korea. Figure 8 presents the PCA score plot of the four grape varieties (Muscat Bailey, Campbell Early, Kyoho, and Meoru), demonstrating the clear separation between different wines as a function of the cultivar. Hu et al. [58] examined the influence of grape genotype on the ^{1}H NMR metabolic profile of wines by studying five variously genetically related grapevine (*V. vinifera*) cultivars. It was concluded that genetic differences in grapes are reflected in the composition of related wines which, in turn, can be discriminated by NMR-based metabolomics, and thus the study of single grape cultivars may prove useful in elucidating the genetic lineage of mixed grapevine cultivars. The analysis of wine blends obtained from binary mixtures of a wines using NMR spectroscopy in combination with linear classification models (LDA) and neural networks has also been reported [59].

Geographical origin

NMR-based metabolomics has been extensively used for the classification of wine samples according to their geographical origin, and several studies have shown a clear differentiation among wine samples produced by grapes belonging to the same cultivar but harvested in different regions of the same country. Muscat Bailey variety wines produced in three different regions in South Korea [17], German wines produced in different regions [55], and red and white wine varieties harvested in Apulia (southern Italy) and Slovenia [60] showed a clear differentiation according to the area of production using NMR-based metabolomics. Son et al. [26] classified Cabernet Sauvignon wines harvested in California, France, and Australia using ^{1}H NMR spectroscopy and multivariate statistical analysis. Figure 9 depicts the respective PLS-DA score plot, which shows a clear separation between wine samples originated from the three different countries. Pereira et al. [61] investigated mature grape berries produced in four different regions in Bordeaux (France) and showed that NMR spectroscopy and chemometrics can classify the berries according to their area of production. In another study, commercial wines of Greco Bianco grape variety produced in the Italian regions of Calabria and Campania were investigated using chromatographic methods and NMR spectroscopy. PLS-DA models of the obtained metabolomic data showed good differentiation of the wine samples according to their geographical origin, with total acidity, citric acid, malic acid, succinic acid, lactic acid, total polyphenol index, glucose, and proline/arginine ratio being the main contributors to the classification of the wine samples [62]. For the investigation of the geographical origin of wines, NMR spectroscopy and Isotopic Ratio Mass Spectrometry (IRMS) have also been used, and a deuterium natural abundance quantitative NMR method (SNIF-NMR: Site-specific Natural Isotope Fractionation) to characterize the chemical origin of wine metabolites was developed [63]. Finally, in a recent study, multivariate PCA–DA analyses applied on the combination of NMR and SPME–GC data was used to classify white wine samples from Galicia (NW Spain) according to both varietal and geographical subzones of origin. The results indicated an increased discriminatory extent deriving from the combination of different analytical methodologies in metabolomic studies [64].

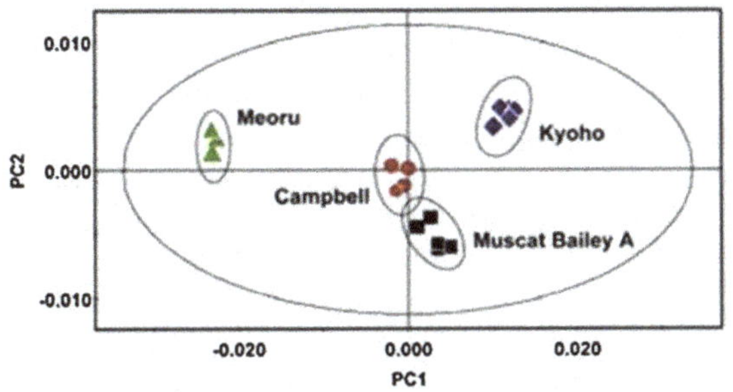

Fig. 8 PCA score plot derived from the ^{1}H NMR spectra of wines vinified from different grape cultivars harvested in October 2008 in South Korea. All wines were fermented with Saccharomyces bayanus PC. Reprinted from Ref. [15.

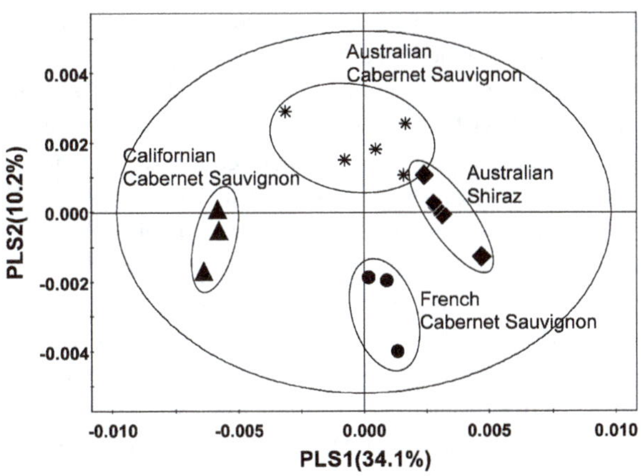

Fig. 9 Partial least squares discriminant analysis (PLS-DA) score plot derived from ^{1}H NMR wine spectra, indicating clear geographical origin separation between Australian, French, and Californian Cabernet Sauvignon wines and Australian Shiraz. Reprinted with permission from Ref. [26].

Vintage

Wine quality is strongly dependent on the vineyard climatic conditions, so vintage is important for the assessment of metabolite profile information obtained from wine samples analyzed by NMR spectroscopy. Macro- and microclimate year-dependent changes can affect the chemical composition of grapes and conversely the metabolite profile of the wines produced. Furthermore, grape cultivars may respond differently to yearly variations in climatic conditions (rainfall, average temperatures, etc.), while small alterations in wine-making practices may also contribute to metabolite profile variations of different vintages. Several publications utilizing multivariate statistical analysis combined with NMR spectroscopy showed differences in the metabolite profiles of wine samples according to the vintage year. Amarone wine produced in Italy showed a clear separation among three different years of harvesting (2005, 2006, 2007) using ^{1}H NMR spectroscopy with PCA and partial least squares discriminant analysis (PLS-DA). The organic compounds that were responsible for the classification of the wine samples according to vintage were mainly amino acids, sugars, and aromatic compounds [11]. Meoru [65] and Chengsoo [53] wines were also classified according to the vintage year. Figure 10 presents the PCA score (A) and loading (B) plots of Meoru wines harvested in 2006 and 2007. There is a clear separation between the two vinification periods and the loading plot (B) shows that this classification is based on differences in lactic acid, γ-aminobutyric acid (GABA), and proline [65]. In another study, white wines produced in 2006 and 2007 showed discrimination using NMR metabolomics, with wines produced in 2006 having higher levels of leucine, phenylalanine, citrate, malate, and phenolics, compared to wines that were produced in 2007, and possessed higher levels of proline, alanine, and succinate [35]. Greek wines Agiorgitiko (red) and Assyrtiko (white) harvested in two different vinification years (2005, 2006) were classified according to vintage by analyzing NMR spectroscopic data with multivariate PLS-DA models [12]. ^{1}H NMR spectroscopy combined with multivariate statistical analysis showed classification of two vintages (2008 and 2009) of wines produced in Germany from different grape varieties, by different enological technologies and from different producing areas [55]. ^{1}H and ^{13}C NMR spectroscopy techniques were also used for the analysis of young and aged Romanian red wines (vintages from 2009 to 2014), demonstrating that NMR metabolite profiles combined with multivariate statistical analysis are able to classify these wines according to the different vintage years [56].

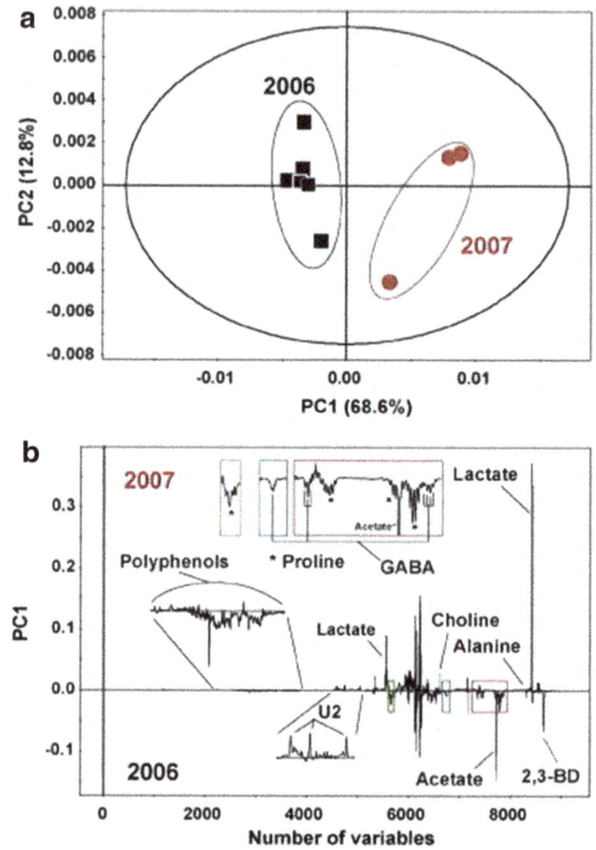

Fig. 10 PCA score (**a**) and loading (**b**) plots derived from the ^{1}H NMR spectra of Meoru wines vinified with *S. cerevisiae* Côte des Blancs (CDB) strain and Meoru grapes harvested in 2006 and 2007. The *boxes* inside panel **b** highlight the changes in GABA and proline levels. *2,3-BD* 2,3-butanediol, *GABA* γ-aminobutyric acid. Reprinted from Ref. [65].

Concluding remarks

The brief summary of the research efforts described in the present article should be convincing evidence of the ability of NMR spectroscopy metabolite profiling, assisted by multivariate statistical analysis modeling, to shed light on the various factors that affect the quality of wine and help establish analytical protocols for the authentication of cultivar, vintage, and geographical origin of wines. Wine quality and authenticity is not only of great economic importance for wine producers, but also significant for building consumer confidence in this high-added-value product. The general picture emerging so far indicates that wine-making procedures and cultivar are factors more strongly reflected in the non-volatile wine NMR metabolome, followed by vintage and geographic origin. The latter factors would benefit from a deeper understanding of the part of the wine metabolome related to minor components (e.g. phenolics, volatiles), which is more strongly affected by ever-changing pedoclimatic conditions contributing to vintage and geographic origin molecular fingerprints.

Further development in the field of NMR-based wine metabolomics will benefit from better systematization of wine studies with respect to sample preparation and analysis protocols and multivariate statistical analysis modeling protocols. This will help the creation of wine metabolic profiling databases that would be continuously updated, and thus could be used in wine quality and authentication applications. Furthermore, the combination of the chemical–analytical profiling capabilities of NMR spectroscopy with other analytical methodologies (LC–MS, GC–MS, etc.), able to sample different parts of the wine's total metabolome (volatiles, phenolics, tannins, aroma compounds, etc.), will undoubtedly become more common in the future. The efficient application of chemometrics to analytical data deriving from advanced techniques will serve as a wholistic method to characterize and investigate on wine production, quality, and authentication, thus permitting to monitor this product from the vineyard to the cellar.

Authors' contributions

The authors contributed equally to this manuscript. Both authors read and approved the final manuscript.

Acknowledgements

Not applicable.

Competing interests

The authors declare that they have no competing interests.

Funding

This research has been co-financed by the European Union (European Social Fund—ESF) and Greek national funds through the Operational Programme "Education and Lifelong Learning" of the National Strategic Reference Framework (NSRF)—Research Funding Programme: THALES (MIS 377213). Investing in knowledge society through the European Social Fund.

References

1. Minoja AP, Napoli C. NMR screening in the quality control of food and nutraceuticals. Food Res Int. 2014;63:126–31.
2. Consonni R, Cagliani LR. Chapter 4—nuclear magnetic resonance and chemometrics to assess geographical origin and quality of traditional food products. In: Steve LT, editor. Advances in food and nutrition research, vol. 59. Cambridge: Academic Press; 2010. p. 87–165.
3. Spyros A, Dais P. NMR spectroscopy in food analysis. Cambridge: RSC; 2012. p. 1–343.
4. Spyros A. Application of NMR in food analysis. In: Ramesh V, editor. Specialist periodical reports: nuclear magnetic resonance, vol. 45. London: RSC; 2016. p. 269–307.
5. Consonni R, Astraka K, Cagliani LR, Nenadis N, Petrakis E, Polissiou M. Authenticity of food. In: Caballero B, Finglas P, Toldrá F. editors. Encyclopedia of food and health. Oxford: Academic Press; 2016. p. 285–93.
6. Hong YS. NMR-based metabolomics in wine science. Magn Reson Chem. 2011;49(SUPPL. 1):S13–21.
7. Fotakis C, Kokkotou K, Zoumpoulakis P, Zervou M. NMR metabolite fingerprinting in grape derived products: an overview. Food Res Int. 2013;54(1):1184–94.
8. Esslinger S, Fauhl-Hassek C, Wittkowski R. Authentication of wine by ^{1}H-NMR spectroscopy: opportunities and challenges. In: Advances in wine research. ACS symposium series, vol. 1203. American Chemical Society; 2015. p. 85–108.
9. Košir IJ, Kidrič J. Identification of amino acids in wines by one-and two-dimensional nuclear magnetic resonance spectroscopy. J Agric Food Chem. 2001;49(1):50–6.
10. Du YY, Bai GY, Zhang X, Liu ML. Classification of wines based on combination of ^{1}H NMR spectroscopy and principal component analysis. Chin J Chem. 2007;25(7):930–6.
11. Consonni R, Cagliani LR, Guantieri V, Simonato B. Identification of metabolic content of selected Amarone wine. Food Chem. 2011;129(2):693–9.
12. Anastasiadi M, Zira A, Magiatis P, Haroutounian SA, Skaltsounis AL, Mikros E. ^{1}H NMR-based metabolomics for the classification of Greek wines according to variety, region, and vintage. Comparison with HPLC data. J Agric Food Chem. 2009;57(23):11067–74.
13. Godelmann R, Kost C, Patz CD, Ristow R, Wachter H. Quantitation of compounds in wine using 1H NMR spectroscopy: description of the method and collaborative study. J AOAC Int. 2016;99(5):1295–304.
14. Mazzei P, Francesca N, Moschetti G, Piccolo A. NMR spectroscopy evaluation of direct relationship between soils and molecular composition of red wines from Aglianico grapes. Anal Chim Acta. 2010;673(2):167–72.
15. Son H-S, Hwang G-S, Ahn H-J, Park W-M, Lee C-H, Hong Y-S. Characterization of wines from grape varieties through multivariate statistical analysis of ^{1}H NMR spectroscopic data. Food Res Int. 2009;42(10):1483–91.
16. Son H-S, Hwang G-S, Kim KM, Kim E-Y, van den Berg F, Park W-M, et al. ^{1}H NMR-based metabolomic approach for understanding the fermentation behaviors of wine yeast strains. Anal Chem. 2009;81(3):1137–45.
17. Son HS, Hwang GS, Kim KM, Ahn HJ, Park WM, Van Den Berg F, et al. Metabolomic studies on geographical grapes and their wines using ^{1}H NMR analysis coupled with multivariate statistics. J Agric Food Chem. 2009;57(4):1481–90.
18. Spraul M, Link M, Schaefer H, Fang F, Schuetz B. Wine analysis to check quality and authenticity by fully-automated ^{1}H-NMR. BIO Web Conf. 2015;5:02022.
19. Spyros A. Nuclear magnetic resonance. In: Nollet LML, Toldrá F, editors. Handbook of food analysis. 3rd ed. Boca Raton: CRC Press; 2015. p. 659.
20. Anders U, Tittgemeier F, Hailer G. ^{1}H-NMR determination of ethanol in wine, winelike beverages, and liquors. Z Lebensm Unters Forsch. 1976;162(1):21–4.
21. Buzaş MC, Chira N, Deleanu C, Roşca S. Identification and quantitative measurement by ^{1}H-NMR spectroscopy of several compounds present in Romanian wines. Rev Chim. 2003;54(10):831–3.
22. Hanke S, Huebner E. A fast method for the determination of the ethanol content in wine using widely available routine nuclear magnetic resonance techniques. J Wine Res. 2014;25(3):137–43.
23. Nilsson M, Duarte IF, Almeida C, Delgadillo I, Goodfellow BJ, Gil AM, et al. High-resolution NMR and diffusion-ordered spectroscopy of port wine. J Agric Food Chem. 2004;52(12):3736–43.
24. Amaral FM, Caro MSB. Investigation of different pre-concentration methods for NMR analyses of Brazilian white wine. Food Chem. 2005;93(3):507–10.
25. Papotti G, Bertelli D, Graziosi R, Silvestri M, Bertacchini L, Durante C, et al. Application of one- and two-dimensional NMR spectroscopy for the characterization of protected designation of Origin Lambrusco wines of Modena. J Agric Food Chem. 2013;61(8):1741–6.
26. Son HS, Ki MK, Van Den Berg F, Hwang GS, Park WM, Lee CH, et al. ^{1}H nuclear magnetic resonance-based metabolomic characterization of wines by grape varieties and production areas. J Agric Food Chem. 2008;56(17):8007–16.
27. Viggiani L, Morelli MAC. Characterization of wines by nuclear magnetic resonance: a work study on wines from the Basilicata region in Italy. J Agric Food Chem. 2008;56(18):8273–9.
28. Gil AM, Duarte IF. High-resolution nuclear magnetic resonance spectroscopy of wine, beer, and spirits. In: Webb GA, editor. Modern magnetic resonance. Dordrecht: Springer; 2006. p. 1689–94.
29. Holland MV, Bernreuther A, Reniero F. The use of amino acids as a fingerprint for the monitoring of European wines. In: Belton P, Delgadillo I, Gil A, Webb G. editors. Magnetic Resonance in Food Science. Cambridge: RSC; 1995. p. 136–45.
30. Colombo C, Aupic C, Lewis AR, Pinto BM. In situ determination of fructose isomer concentrations in wine using (13)C quantitative nuclear magnetic resonance spectroscopy. J Agric Food Chem. 2015;63(38):8551–9.

31. López-Rituerto E, Avenoza A, Busto JH, Peregrina JM. Evidence of metabolic transformations of amino acids into higher alcohols through ^{13}C NMR studies of wine alcoholic fermentation. J Agric Food Chem. 2010;58(8):4923–7.
32. Prakash S, Iturmendi N, Grelard A, Moine V, Dufourc E. Quantitative analysis of Bordeaux red wine precipitates by solid-state NMR: role of tartrates and polyphenols. Food Chem. 2016;199:229–37.
33. Weekley AJ, Bruins P, Sisto M, Augustine MP. Using NMR to study full intact wine bottles. J Magn Reson. 2003;161(1):91–8.
34. Sobieski DN, Mulvihill G, Broz JS, Augustine MP. Towards rapid throughput NMR studies of full wine bottles. Solid State Nucl Magn Reson. 2006;29(1–3):191–8.
35. Ali K, Maltese F, Toepfer R, Choi YH, Verpoorte R. Metabolic characterization of Palatinate German white wines according to sensory attributes, varieties, and vintages using NMR spectroscopy and multivariate data analyses. J Biomol NMR. 2011;49(3–4):255–66.
36. Košir IJ, Kidrič J. Use of modern nuclear magnetic resonance spectroscopy in wine analysis: determination of minor compounds. Anal Chim Acta. 2002;458(1):77–84.
37. Ferrari E, Foca G, Vignali M, Tassi L, Ulrici A. Adulteration of the anthocyanin content of red wines: perspectives for authentication by Fourier Transform-Near InfraRed and ^{1}H NMR spectroscopies. Anal Chim Acta. 2011;701(2):139–51.
38. Košir IJ, Lapornik B, Andrenšek S, Wondra AG, Vrhovšek U, Kidrič J. Identification of anthocyanins in wines by liquid chromatography, liquid chromatography-mass spectrometry and nuclear magnetic resonance. Anal Chim Acta. 2004;513(1):277–82.
39. De la Cruz AA, Hilbert G, Rivière C, Mengin V, Ollat N, Bordenave L, et al. Anthocyanin identification and composition of wild *Vitis* spp. accessions by using LC-MS and LC-NMR. Anal Chim Acta. 2012;732:145–52.
40. Gil AM, Duarte IF, Godejohann M, Braumann U, Maraschin M, Spraul M. Characterization of the aromatic composition of some liquid foods by nuclear magnetic resonance spectrometry and liquid chromatography with nuclear magnetic resonance and mass spectrometric detection. Anal Chim Acta. 2003;488(1):35–51.
41. Absalon C, Fabre S, Tarascou I, Fouquet E, Pianet I. New strategies to study the chemical nature of wine oligomeric procyanidins. Anal Bioanal Chem. 2011;401(5):1485–95.
42. Hanganu A, Todasca MC, Chira NA, Rosca S. Influence of common and selected yeasts on wine composition studied using ^{1}H-NMR spectroscopy. Rev Chim. 2011;62(7):689–92.
43. Mazzei P, Spaccini R, Francesca N, Moschetti G, Piccolo A. Metabolomic by ^{1}H NMR spectroscopy differentiates "Fiano Di Avellino" white wines obtained with different yeast strains. J Agric Food Chem. 2013;61(45):10816–22.
44. Lee JE, Hong YS, Lee CH. Characterization of fermentative behaviors of lactic acid bacteria in grape wines through 1H NMR- and GC-based metabolic profiling. J Agric Food Chem. 2009;57(11):4810–7.
45. Avenoza A, Busto JH, Canal N, Peregrina JM. Time course of the evolution of malic and lactic acids in the alcoholic and malolactic fermentation of grape must by quantitative ^{1}H NMR (qHNMR) spectroscopy. J Agric Food Chem. 2006;54(13):4715–20.
46. Baiano A, Terracone C, Longobardi F, Ventrella A, Agostiano A, Del Nobile MA. Effects of different vinification technologies on physical and chemical characteristics of Sauvignon blanc wines. Food Chem. 2012;135(4):2694–701.
47. De Pascali SA, Coletta A, Del Coco L, Basile T, Gambacorta G, Fanizzi FP. Viticultural practice and winemaking effects on metabolic profile of Negroamaro. Food Chem. 2014;161:112–9.
48. Todasca MC, Fotescu L, Chira NA, Deleanu C. Composition changes in wines produced by different growing techniques examined through ^{1}H-NMR spectroscopy. Rev Chim. 2011;62(2):131–4.
49. Picone G, Trimigno A, Tessarin P, Donnini S, Rombolà AD, Capozzi F. ^{1}H NMR foodomics reveals that the biodynamic and the organic cultivation managements produce different grape berries (*Vitis vinifera* L. cv. Sangiovese). Food Chem. 2016;213:187–95.
50. Lopez-Rituerto E, Savorani F, Avenoza A, Busto JH, Peregrina JM, Engelsen SB. Investigations of La Rioja terroir for wine production using ^{1}H NMR metabolomics. J Agric Food Chem. 2012;60(13):3452–61.
51. Pereira GE, Gaudillère J-P, van Leeuwen C, Hilbert G, Maucourt M, Deborde C, et al. ^{1}H-NMR metabolic profiling of wines from three cultivars, three soil types and two contrasting vintages. J Int Sci Vigne Vin. 2007;41(2):7.
52. Da Silva Neto HG, Da Silva JBP, Pereira GE, Hallwass F. Determination of metabolite profiles in tropical wines by ^{1}H NMR spectroscopy and chemometrics. Magn Reson Chem. 2009;47(SUPPL. 1):S127–9.
53. Chang EH, Jung SM, Park SJ, Noh JH, Hur YY, Nam JC, et al. Wine quality of grapevine 'Cheongsoo' and the related metabolites on proton nuclear magnetic resonance (NMR) spectroscopy at the different harvest times. Plant OMICS. 2014;7(2):80–6.
54. Hong YS, Cilindre C, Liger-Belair G, Jeandet P, Hertkorn N, Schmitt-Kopplin P. Metabolic influence of *Botrytis cinerea* infection in champagne base wine. J Agric Food Chem. 2011;59(13):7237–45.
55. Godelmann R, Fang F, Humpfer E, Schütz B, Bansbach M, Schäfer H, et al. Targeted and nontargeted wine analysis by ^{1}H NMR spectroscopy combined with multivariate statistical analysis. Differentiation of important parameters: grape variety, geographical origin, year of vintage. J Agric Food Chem. 2013;61(23):5610–9.
56. Geana EI, Popescu R, Costinel D, Dinca OR, Ionete RE, Stefanescu I, et al. Classification of red wines using suitable markers coupled with multivariate statistic analysis. Food Chem. 2016;192:1015–24.
57. Rochfort S, Ezernieks V, Bastian SEP, Downey MO. Sensory attributes of wine influenced by variety and berry shading discriminated by NMR metabolomics. Food Chem. 2010;121(4):1296–304.
58. Hu B, Yue Y, Zhu Y, Wen W, Zhang F, Hardie JW. Proton nuclear magnetic resonance-spectroscopic discrimination of wines reflects Genetic homology of several different grape (*V. vinifera* L.) cultivars. PLoS ONE. 2015;10(12):e0142840.
59. Imparato G, Di Paolo E, Braca A, Lamanna R. Nuclear magnetic resonance profiling of wine blends. J Agric Food Chem. 2011;59(9):4429–34.
60. Brescia MA, Košir IJ, Caldarola V, Kidrič J, Sacco A. Chemometric classification of Apulian and Slovenian wines using ^{1}H NMR and ICP-OES together with HPICE data. J Agric Food Chem. 2003;51(1):21–6.
61. Pereira GE, Gaudillere JP, Van Leeuwen C, Hilbert G, Lavialle O, Maucourt M, et al. ^{1}H NMR and chemometrics to characterize mature grape berries in four wine-growing areas in Bordeaux, France. J Agric Food Chem. 2005;53(16):6382–9.
62. Caruso M, Galgano F, Castiglione Morelli MA, Viggiani L, Lencioni L, Giussani B, et al. Chemical profile of white wines produced from 'Greco bianco' grape variety in different Italian areas by nuclear magnetic resonance (NMR) and conventional physicochemical analyses. J Agric Food Chem. 2012;60(1):7–15.
63. Pîrnău A, Bogdan M, Măgdaş DA, Stătescu D. Isotopic analysis of some Romanian wines by ^{2}H NMR and IRMS. Food Biophys. 2013;8(1):24–8.
64. Martin-Pastor M, Guitian E, Riguera R. Joint NMR and solid-phase microextraction–gas chromatography chemometric approach for very complex mixtures: grape and zone identification in wines. Anal Chem. 2016;88(12):6239–46.
65. Lee JE, Hwang GS, Van Den Berg F, Lee CH, Hong YS. Evidence of vintage effects on grape wines using ^{1}H NMR-based metabolomic study. Anal Chim Acta. 2009;648(1):71–6.

19

Multiple benefits of legumes for agriculture sustainability

Fabio Stagnari[1*], Albino Maggio[2], Angelica Galieni[1] and Michele Pisante[1]

Abstract

Food security, lowering the risk of climate change and meeting the increasing demand for energy will increasingly be critical challenges in the years to come. Producing sustainably is therefore becoming central in agriculture and food systems. Legume crops could play an important role in this context by delivering multiple services in line with sustainability principles. In addition to serving as fundamental, worldwide source of high-quality food and feed, legumes contribute to reduce the emission of greenhouse gases, as they release 5–7 times less GHG per unit area compared with other crops; allow the sequestration of carbon in soils with values estimated from 7.21 g kg^{-1} DM, 23.6 versus 21.8 g C kg^{-1} year; and induce a saving of fossil energy inputs in the system thanks to N fertilizer reduction, corresponding to 277 kg ha^{-1} of CO_2 per year. Legumes could also be competitive crops and, due to their environmental and socioeconomic benefits, could be introduced in modern cropping systems to increase crop diversity and reduce use of external inputs. They also perform well in conservation systems, intercropping systems, which are very important in developing countries as well as in low-input and low-yield farming systems. Legumes fix the atmospheric nitrogen, release in the soil high-quality organic matter and facilitate soil nutrients' circulation and water retention. Based on these multiple functions, legume crops have high potential for conservation agriculture, being functional either as growing crop or as crop residue.

Keywords: Soil fertility, Conservation agriculture, Sustainable agricultural systems, Food security, Climate change, Greenhouse gas, Energy

Introduction

Global population will hit 9.6 billion people by 2050 [108] and will face global challenges among which achieving food security, lowering the risk of climate change by reducing the net release of greenhouse gases into the atmosphere and meeting the increasing demand for energy are the most critical ones. In particular, the impact of climate change and associated biotic and abiotic stresses to which crop systems will be increasingly exposed pose serious implications for global food production [119].

To meet these challenges, a policy framework needs to be developed in which the *sustainability* of production/consumption patterns becomes central. In this context, food legumes and legume-inclusive production systems can play important roles by delivering multiple services in line with sustainability principles. Indeed, legumes play central roles [112]: (1) at food-system level, both for human and animal consumption, as a source of plant proteins and with an increasingly importance in improving humans health [106]; (2) at production-system level, due to the capacity to fix atmospheric nitrogen making them potentially highly suitable for inclusion in low-input cropping systems, and due to their role in mitigating greenhouse gases emissions [53]; and (3) at cropping-system levels, as diversification crops in agroecosystems based on few major species, breaking the cycles of pests and diseases and contributing to balance the deficit in plant protein production in many areas of the world, including Europe [43, 48, 72, 78, 116].

Leguminosae family comprises 800 genera and 20,000 species [54] and represents the third largest family of flowering plants. Some legumes are considered weeds of

*Correspondence: fstagnari@unite.it
[1] Faculty of Biosciences and Technologies for Agriculture Food and Environment, University of Teramo, Coste Sant'Agostino, Teramo, TE, Italy
Full list of author information is available at the end of the article

Table 1 Trends for word acreage (million ha) and yield (t ha^{-1}) for legume crops included in FAOSTAT classification starting from 1974 to 2014 [23]; the major three cereal crops are also reported, for comparison

	Harvested area (Million ha)					Yield (t ha^{-1})				
	1974	1984	1994	2004	2014	1974	1984	1994	2004	2014
Legume crops										
Bambara bean	0.05	0.05	0.09	0.12	0.37	0.67	0.66	0.64	0.65	0.77
Dry bean	23.9	26.3	26.7	27.3	30.14	0.53	0.6	0.65	0.67	0.83
Faba bean	3.98	3.32	2.48	2.65	2.37	1.07	1.29	1.45	1.62	1.82
Chickpea	10.6	9.85	9.96	10.5	14.8	0.56	0.67	0.71	0.8	0.96
Cowpea	4.7	3.66	7.35	9.18	12.52	0.35	0.31	0.38	0.45	0.45
Groundnut	19.9	18.2	22	23.7	25.68	0.94	1.1	1.3	1.54	1.65
Lentil	2.03	2.56	3.43	3.85	4.52	0.61	0.68	0.81	0.93	1.08
Lupin	0.76	1.06	1.56	1.05	0.76	0.84	1.05	0.78	1.18	1.3
Pea	8.13	8.91	7.65	6.34	6.87	1.22	1.3	1.88	1.85	1.65
Pigeon pea	3.04	3.61	4.24	4.72	6.67	0.54	0.78	0.74	0.7	0.73
Soybean	37.4	52.9	62.5	91.6	117.72	1.41	1.71	2.18	2.24	2.62
French bean	0.22	0.17	0.22	0.23	0.20[a]	5.76	6.93	7.44	9.04	9.32[a]
Vetch	1.52	1.29	0.93	0.89	0.52	1.24	1.21	1.12	1.43	1.71
Pulses, nes	5.67	5.65	5.33	4.27	6.1	0.54	0.58	0.65	0.82	0.84
Vegetables, leguminous nes	0.14	0.18	0.18	0.25	0.24[a]	5.41	5.09	5.18	6.54	6.86[a]
Major cereal crops										
Wheat	222.12	230.77	215.12	216.57	221.62	1.62	2.22	2.45	2.92	3.29
Maize	119.86	127.76	137.99	147.45	183.32	2.56	3.53	4.12	4.94	5.66
Rice (paddy)	136.89	144.24	147.29	150.58	163.25	2.43	3.23	3.66	4.03	4.54

In Table 2, for each legume crop, item name and code as well as FAO definitions are reported

[a] Data are referred to year 2013 (2014 data not available)

cereal crops, while others are major grain crops; these latter species are known as grain legumes, or pulses,[1] and represent the focus of this review. For some of these species, the trends for word acreage and yield are available, as reported in Table 1.

Despite the growing trends observed during the 50-year period between 1974 and 2014 for some warm-season legumes (e.g. soybean, cowpea, dry bean, groundnut, and pigeon pea), the acreages of several temperate legumes (e.g. pea, faba bean, lupin, french bean and vetch) have declined worldwide with differences between world Regions (Table 3). In any case, food legumes occupy a minimal part of arable land, mostly dominated by cereal crops [99]; soybean represents the most important and cultivated legume, acreage of which reached 117.72 million ha in 2014 (steadily increased over years, see also Table 3), which is about that of the other grain legumes, but still far below the major cereals (e.g. rice, wheat, maize). Such trend is mainly associated to the expansion of more specialized and intensive production systems [82].

[1] Soybean and groundnuts are not defined by FAO as 'pulse crops'.

Market forces stimulating specialization of cropping systems as non-marketable benefits of diversification, like cultivation/introduction of legumes in the farming system, do not deliver immediate and/or apparent profits [82]. This is, however, not equally perceived throughout the globe, and there is indeed a remarkable diversity in grain legumes' production trends across the world (Table 3).

The European decline in grain legume's production is not mirrored by other regions of the world such as Canada or Australia, where legume's cultivation has been increasing over the last few decades. In these areas, monoculture of cereals, which relies on frequent summer-fallowing and use of mechanical tillage, has been replaced by extended and diversified crop rotations together with the use of conservation tillage [122]. Furthermore, supply chains and markets are inadequately developed for most legume crops (see also [66], for France) with the exception of soybean, for which the global market is well developed [85]. Nevertheless, soybean areas in Europe are constrained by climatic factors although there is considerable potential to develop new varieties suitable to flourish under cool growing conditions [123].

Table 2 Definition of legume crops focused in Table 1 and corresponding item name in FAOSTAT

Legume crop	Scientific name	Corresponding FAO item name and code	FAO definition
Bambara bean	*Voandzeia subterranea*	Bambara bean [203]	Bambara groundnut, earth pea. These beans are grown underground in a similar way to groundnuts
Dry bean	–	Beans, dry [176]	*Phaseolus* spp.: kidney, haricot bean (*Ph. vulgaris*); lima, butter bean (*Ph. lunatus*); adzuki bean (*Ph. angularis*); mungo bean, golden, green gram (*Ph. aureus*); black gram, urd (*Ph. mungo*); scarlet runner bean (*Ph. coccineus*); rice bean (*Ph. calcaratus*); moth bean (*Ph. aconitifolius*); tepary bean (*Ph. acutifolius*). Several countries also include some types of beans commonly classified as *Vigna* (*angularis, mungo, radiata, aconitifolia*)
Faba bean	*Vicia faba*	Broad beans, horse beans, dry [181]	*Vicia faba*: horse-bean (var. *equina*); broad bean (var. *major*); field bean (var. *minor*)
Chickpea	*Cicer arietinum*	Chick peas [191]	Chickpea, Bengal gram, garbanzos (*Cicer arietinum*).
Cowpea	*Vigna ungiculanta*	Cow peas, dry [195]	Cowpea, blackeye pea/bean (*Vigna sinensis*; *Dolichos sinensis*)
Groundnut	*Arachis hypogaea*	Groundnuts, with shell [242]	*Arachis hypogaea*. For trade data, groundnuts in shell are converted at 70% and reported on a shelled basis
Lentil	*Lens esculenta*	Lentils [201]	*Lens esculenta; Ervum lens*
Lupin	–	Lupins [210]	*Lupinus spp.* Used primarily for feed, though in some parts of Africa and in Latin America some varieties are cultivated for human food
Pea	–	Peas, dry [187]	Garden pea *(Pisum sativum)*; field pea *(P. arvense)*
Pigeon pea	*Cajanus cajan*	Pigeon peas [197]	Pigeon pea, cajan pea, Congo bean *(Cajanus cajan)*
Soybean	*Glycine max*	Soybeans [236]	*Glycine soja*
French bean	–	String beans [423]	*Phaseolus vulgaris*; *Vigna* spp. Not for shelling
Vetches	*Vicia sativa*	Vetches [205]	Spring/common vetch (*Vicia sativa*). Used mainly for animal feed
Pulses, nes	–	Pulses, nes [211]	Including inter alia: lablab or hyacinth bean (*Dolichos* spp.); jack or sword bean (*Canavalia* spp.); winged bean (*Psophocarpus tetragonolobus*); guar bean (*Cyamopsis tetragonoloba*); velvet bean (*Stizolobium* spp.); yam bean (*Pachyrrhizus erosus*); *Vigna* spp. other than those included in 176 and 195; other pulses that are not identified separately because of their minor relevance at the international level. Because of their limited local importance, some countries report pulses under this heading that are classified individually by FAO
Vegetables, leguminous nes	–	Vegetables, leguminous nes [420]	*Vicia faba*. For shelling

The low diffusion of legumes' cultivation is also due to reduced and unstable yields and susceptibility to biotic and abiotic stress conditions; the average yields for unit area have increased (soybean +86%, lentil +77%, groundnut +75%, chickpea +70%) less than cereal crops (+104%, on average) (Table 1). Moreover, legume cultivation depends not only on the effect of farmers' choices, although they play a central role for such decision, but also on policymakers who have the responsibility to provide effective strategies to support the integration of legumes into cropping systems. This aspect is particularly relevant if the overall objective for future agricultural systems is to promote sustainability, improve resource use efficiency and preserve the environment [82].

Grain legumes impacts on atmosphere and soil quality

Among the many important benefits that legumes deliver to society, their role in contributing to climate change mitigation has been rarely addressed. Legumes can (1) lower the emission of greenhouse gases (GHG) such as carbon dioxide (CO_2) and nitrous oxide (N_2O) compared

Table 3 Trend for Region Δ acreage (%) during the 50-year period starting from 1974 to 2014 for legume crops included in FAOSTAT classification [23]; the major three cereal crops are also reported, for comparison

	Δ harvested area 1974–2014 (%)					
	Africa	**Northern America**	**South America**	**Asia**	**Europe**	**Oceania**
Legume crops						
Bambara bean	+612	–	–	–	–	–
Dry bean	+207	+16	−20	+25	−84	+1778
Faba bean	+7	Disappeared	−53	−59	−54	+75,085
Chickpea	+30	Appeared	+1	+37	−35	–
Cowpea	+168	Appeared	Appeared	+402	+153	–
Groundnut	+69	−10	−22	+6	+16	−39
Lentil	−20	+3376	−75	+72	−45	Appeared
Lupin	−82	–	+577	−89	−64	+315
Pea	+49	+1119	+7	−21	−63	+578
Pigeon pea	+226	–	−83	+108	–	–
Soybean	+642	+71	+882	+116	+291	−10
French bean	Appeared[a]	−39[a]	+129[a]	+66[a]	−18[a]	+122[a]
Vetch	+109	–	–	−73	−80	+4757
Pulses, nes	+20	–	−69	−15	+73	+7648
Vegetables, leguminous nes	+180[a]	Appeared[a]	+118[a]	+23[a]	−31[a]	−52[a]
Major cereal crops						
Wheat	+11	−20	+16	+39	−33	+51
Maize	+98	+29	+45	+76	+21	+31
Rice (paddy)	+185	+15	−16	+16	−28	+2

In Table 2, for each legume crop, item name and code as well as FAO definitions are reported

[a] Data are referred to year 2013 (2014 data not available)

with agricultural systems based on mineral N fertilization, (2) have an important role in the sequestration of carbon in soils, and (3) reduce the overall fossil energy inputs in the system.

Greenhouse gas emissions

The introduction of legumes into agricultural rotations help in reducing the use of fertilizers and energy in arable systems and consequently lowering the GHG emissions [52]. N fertilizer savings across Europe [51], in rotations including leguminous crops, range around 277 kg ha^{-1} of CO_2 per year (1 kg $N = 3.15$ kg CO_2, [42]. It has been reported that half of the CO_2 generated during NH3 production would be reused if the NH3 was converted to urea. This is, however, only a time shift of CO_2 release in the atmosphere since, once the urea is applied to the soil, the hydrolyzation activity by urease will release CO_2 originally captured during urea production [39]. Considering an efficiency of 2.6–3.7 kg CO_2 generated per kilogram of N synthesized, the annual global fertilizer leads to a release of 300 Tg of CO_2 into the atmosphere each year [42]. Some studies indicate that at global scale, the amount of CO_2 respired from the root systems of N_2-fixing legumes could be higher than the CO_2 generated during N-fertilizer production [42]. However, it is important to emphasize that the CO_2 respired from nodulated roots of legumes comes from the atmosphere through the photosynthesis activity. Conversely, all the CO_2 released during the process of N-fertilizer synthesis derives from fossil energy, thus determining a net contribution to atmospheric amount of CO_2 [42].

N_2O represents about 5–6% of the total atmospheric GHG, but it is much more active[2] than CO_2 [21]. Agriculture represents the main source of anthropogenic N_2O emissions (about 60%; [84], due to both animal and crop production [38]). A majority of these emissions result from the application of nitrogen fertilizers [84]: every 100 kg of N fertilizer about 1.0 kg of N is emitted as N_2O [42], although different amounts depend on several factors including N application rate, soil organic C content, soil pH, and texture [78, 88]. Denitrification processes are the most important source of N_2O in most cropping and pasture systems [76, 88, 102].

In the recent years, several studies have focalized on the role of legumes in the reduction of GHG emissions. Jeuffroy et al. [44] demonstrated that legume crops

[2] N_2O absorbs approximately 292 times as much infra-red radiation per kilogram as CO_2.

emit around 5–7 times less GHG per unit area compared with other crops. Measuring N_2O fluxes, they showed that peas emitted 69 kg N_2O ha^{-1}, far less than winter wheat (368 kg N_2O ha^{-1}) and rape (534 kg N_2O ha^{-1}). Clune et al. [19] reviewed different life cycle-assessment (LCA) studies on GHG emissions carried out from 2000 to 2015 around the world (despite the used literature was predominately European centric) highlighting that pulses have a very low Global Warming Potential (GWP) values (0.50–0.51 kg CO_2 eq kg^{-1} produce or bone-free meat[3]). In a comparison between vetch and barley under Mediterranean environments and alkaline soil, N_2O emissions were higher for barley than vetch; furthermore, the N_2O fluxes derived from the synthetic fertilizers added to the crops were 2.5 times higher in barley compared with vetch [29]. In two field experiments conducted in a black Vertosol in sub-tropical Australia, Schwenke et al. [95] demonstrated that the cumulative N_2O emissions from N-fertilized canola greatly exceeded those from chickpea, faba bean and field pea (385 vs. 166, 166 and 135 g N_2O-N ha^{-1}, respectively). The same authors highlighted that grain legumes significantly reduced their emission factors suggesting that legume-fixed N is a less-emissive form of N input to the soil than fertilizer N.

Nevertheless, it is important to highlight that the influence of legumes in reducing GHG depends also on the management of agro-ecosystems in which they are included. When faba bean was grown as mono cropping, it led to threefold higher cumulative N_2O emissions than that of unfertilized wheat (441 vs. 152 g N_2O ha^{-1}, respectively); conversely, when faba bean was mixed with wheat (intercropping system), cumulative N_2O emissions fluxes were 31% lower than that of N-fertilized wheat [96]. Anyway, the benefits derived from the introduction of legumes in crop rotations become significant when commercially relevant rates of N fertilizer are applied [42].

The mitigation in terms of GHG emissions is also obtained by adopting sustainable agricultural systems, such as conservation tillage and conservation agriculture systems, which are suitable for the cultivation of both grain and green-manure legumes (see "Grain legumes and conservation agriculture" section).

In conclusion, it is noteworthy to underline that field tests and experimental analyses on GHG emissions, and in particular on N_2O, provided quite different results [89] due to the influences of differences of several variables, including climatic, soil and management conditions [45, 78, 88].

In general, N_2O losses from soils covered with legumes are certainly lower than those from both N_2O fertilized grasslands and non-legume crops, as also indicated by Jensen et al. [42] who report a mean of 3.22 kg N_2O-N ha^{-1}, calculated from 67 site years of data. In addition, there is no direct association between N_2O emissions and biological nitrogen fixation [42], since organic N from legume residues is decomposed, mineralized and rapidly immobilized by microorganisms [78]. Emissions of N_2O could occur either during nitrification or due to denitrification, being affected by timing of mineralized N supply [20]: the asynchrony between N supply and utilization from the following crops enhances N loss, especially in winter/early spring in cold wet soils [64].

Soil properties

Cultivation and cropping may cause significant SOC losses through decomposition of humus [18]. Shifting from pasture to cropping systems may result in loss of soil C stocks between 25 and 43% [101].

Legume-based systems improve several aspects of soil fertility, such as SOC and humus content, N and P availability [42]. With respect to SOC, grain legumes can increase it in several ways, by supplying biomass, organic C, and N [27, 53], as well as releasing hydrogen gas as by-product of BNF, which promotes bacterial legume nodules' development in the rhizosphere [49].

In sandy soils, the beneficial effect of grain legumes after three years of study was registered in terms of higher content of SOC compared with soils with oats (7.21 g kg^{-1} DM, on average). Specifically, cultivation of pea exerted the most positive action to organic carbon content (7.58 g kg^{-1}, after harvest, on average), whereas narrow-leaved lupin had the least effect (7.23 g kg^{-1}, on average) [30]. In southern America (Argentina), the intercropping of soybean with maize at different rates favoured a SOC accumulation of 23.6 g C kg^{-1} versus 21.8 g C kg^{-1} of the sole maize; the greatest potential for enhancing SOC stocks occurred in the 2:3 (maize:soybean) intercrop configuration [11]. Furthermore, just only amending the soil with soybean residues allows to obtain an increase of 38.5% in SOC [11].

Thanks to BNF, legumes also affect significantly soil N availability; by using legumes as winter crops in rice–bean and rice–vetch combination, rice residue N content is enhanced by 9.7–20.5%, with values ranging from 1.87 to 1.93 g N kg^{-1} soil [120]. It needs to be underlines that a majority of studies on the role of legumes for soil

[3] In the study of Clune et al. [19], each GWP value recorded from the literature data was converted into a common functional unit and system boundary in kg CO_2 eq kg^{-1} bone-free meat (BFM), using the conversion ratios identified in the literature.

N fertility have investigated the shoot N content. In this regard, Carranca et al. [15] found that 7–11% of total legume N was associated with root and nodules and an allocation of 11–14 kg N fixed t^{-1} belowground dry matter, representing half the amount of total aboveground plant.

In intercropping cowpea–maize, Latati et al. [50] found an increase in P availability at rhizosphere level associated with significant acidification (−0.73 U) than in sole cropping. Wang et al. [115], assessing properties related to N and P cycling in the rhizosphere of wheat and grain legumes (faba bean and white lupin) grown in monoculture or in wheat/legume mixtures, found that the less-labile organic P pools (i.e. NaOH-extractable P pools and acid-extractable P pools) significantly accumulated in the rhizosphere of legumes. However, the P uptake and the changes in rhizosphere soil P pools seem to depend also on legume species. Compared with the unplanted soil, the depletion of labile P pools (resin P and $NaHCO_3$-P inorganic) was the greatest in the rhizosphere of faba bean (54 and 39%) with respect to chickpea, white lupin, yellow lupin and narrow-leafed lupin [31]. Of the less-labile P pools, NaOH-P inorganic was depleted in the rhizosphere of faba bean, while NaOH-P organic and residual P were most strongly depleted in the rhizosphere of white lupin [31].

Also in North Rift, Kenya Region, in well-drained, extremely deep, friable clay, acid humic top soil, the effects of cultivation and incorporation of lupine and garden pea were significant in terms of soil-available P with respect to fallow, with lupine showing higher P availability than pea (from 20.3 to 31.0% higher).

Although there is a general agreement on the influence of grain legumes on rhizosphere properties in terms of N supply, SOC and P availability, the magnitude of the impact varied across legume species, soil properties and climatic conditions. Among these, soil type represents the major factor determining plant growth, rhizosphere nutrient dynamics and microbial community structure. The pattern of depletion and accumulation of some macro- and micronutrients differed also between cropping systems (i.e. monoculture, mixed culture, narrow crop rotations) as well as among soil management strategies (i.e. tillage, no-tillage).

Role of grain legumes in cropping systems

Legumes could be competitive crops, in terms of environmental and socioeconomic benefits, with potential to be introduced in modern cropping systems, which are characterized by a decreasing crop diversity [24, 80] and an excessive use of external inputs (i.e. fertilizers and agrochemicals).

Grain legumes into crop-sequences

In the recent years, many studies have focused on the sustainable re-introduction of grain legumes into crop rotations,[4] based on their positive effects on yield and quality characteristics on subsequent crops [46, 82, 103]. However, assessment of the rotational advantages/disadvantages should be based on a pairwise comparison between legume and non-legume pre-crops [82]. Some experimental designs involving multi-year and multispecies rotations do not provide information on yield benefits to the subsequent species in the rotation sequence. Therefore, it is difficult to formulate adequate conclusions [2].

The agronomic pre-crop benefits of grain legumes can be divided into a 'nitrogen effect' component and 'break crop effect' component. The 'nitrogen effect' component is a result of the N provision from BNF [77], which is highest in situations of low N fertilization to subsequent crop cycles [82]. The second one (break crop effect) includes non-legume-specific benefits, such as improvements of soil organic matter and structure [34], phosphorus mobilization [98], soil water retention and availability [2], and reduced pressure from diseases and weeds [87]. In this case, benefits are highest in cereal-dominated rotations [82].

Several authors have reviewed the yield benefits of legumes for subsequent cereal crops.

In Australia, Angus et al. [2] reported higher yield of wheat after legumes (field peas, lupins, faba beans, chickpeas and lentils) than those of wheat after wheat. In particular for a wheat–wheat yield of 4.0 t ha^{-1}, the mean grain legume-wheat yield was 5.2 t ha^{-1} (+30% on average). Other studies from Australia quantified yield benefits compared to pure cereal crop sequences at 40–50% for low N levels and 10–17% for high N levels [3].

In Europe yields benefits of grain legumes have been shown to strongly depend on climatic factors which affect N dynamics in soils [52]. In temperate environments, cereals yield is on average 17 and 21% higher in grain-legume based systems than wheat monocropping, under standard and moderate fertilization levels, respectively [40]. Conversely, yield benefits are lower in Mediterranean climates where water availability is the limiting factor to cereal yields [46, 61, 62].

The yield advantage to subsequent cereal crops provided by legumes depends also on the species and amounts of fixed N [114, 121]. Field pea and faba bean

[4] According to Angus et al. [2], crop-sequences experiments can be classified into rotation experiments and break crop experiments. Rotation strictly defined, refers to a recurring sequence of crops, forages and fallows, or more loosely defined, to a cropping sequence that contains fallows, or crops and forages in addition to the locally dominant species. A break crop generally refers to a single alternative crop followed by the dominant species.

accumulate about 130 and 153 kg N ha^{-1} in their aboveground biomass, respectively [77] and significant quantities (30–60% of the accumulated total N) may also be stored in belowground biomass [77]. Differences in BNF patterns are also found between the same species. For example, Mokgehle et al. [69] compared 25 groundnut varieties for plant BNF at three differing agro-ecologies in South Africa, highlighting N-fixed range between 76 and 188 kg ha^{-1}, depending also on soil and environmental conditions as well as on N-uptake. Other factors influencing BNF include salinity and sodicity (alkalinity) of soils, as observed in chickpea [83], common bean [22] and faba bean [109].

It is, however, rather difficult to quantify the legume dependent increase in N uptake in subsequent crops, versus other sources of N [46, 77]. In temperate environments of Australia, measurements of the additional N-nitrate available to wheat crops following legumes instead of cereals, averaged around 37 kg N ha^{-1} [17]. In Denmark, nitrogen uptake in crops that follow legume crops has been reported to increase by 23–59% after field pea and narrow-leafed lupin on different soil types [40], but only 14–15% for durum wheat following vetch in a semi-arid Mediterranean environment [28]. Increased N uptake of crops after grain legumes reached up to 61% or 36 kg ha^{-1} for a vetch-barley rotation in Cyprus [74]. Further, some legume residues have beneficial effects on some quality aspects of the subsequent crops in southern Italy [104].

Among other beneficial effects brought about by legumes, the production of hydrogen gas (H_2) as a by-product of BNF greatly affects the composition of the soil microbial population, further favouring the development of plant growth-promoting bacteria [2].

Some grain legumes, including chickpea, pigeon pea and white lupin can mobilize fixed forms of soil P through the secretion of organic acids such as citrate and malate and other P mobilizing compounds from their roots [36]. Among grain legumes, white lupin most strongly solubilize P, a function that can be facilitated by its proteoid roots that may englobe small portions of soil [2]. Glasshouse experiments using a highly P-fixing soil showed better wheat growth following white lupin than soybean [37], suggesting that the cereal was able to access P made available by the previous white lupin break crop. 'Break crop' effects also include increased soil water content, since the break-crop stubble can affect retention of soil water and infiltration and retention of rain water [47]. A species-specific response has also been documented. Soil profiles after pea field can be wetter than after a wheat crop [2]. In Saskatchewan, Canada, Miller et al. [68] reported that post-harvest soil water status up to 122 cm-depth was 31 and 49 mm greater for all legumes (field pea, lentil and chick pea) with respect of wheat under loam and clay soils, respectively. This was primarily due to increased plant water use efficiency. Lentil in rotation with cereals has been shown to increase total grain production by increasing residual soil water in dry areas of Saskatchewan [25].

In general, grain legumes are not susceptible to the same pests and diseases as the main cereal crops (non-host), resulting suitable as break crops in wheat-based rotations [121]. Grain legumes as break crops can also contribute to weed control [97] by contrasting their specialization and helping stabilizing the agricultural crop weed community composition [7].

Despite the described beneficial effects, there are still concerns on the introduction of grain legumes into cropping sequences. Cropping systems that include legume crops in farm rotations must be supported by best crop-management practices (e.g. N fertilization rates and timing, soil management, weeding, irrigation), which often do not match standard techniques normally applied by farmers. For example, some possible risks in terms of nitrate leaching associated to grain legumes cultivation can be counteracted by including cover crops in the system [33, 81]. Additional reasons may explain why grain legumes are not very common in high-input cropping systems. These include (1) their low and unstable yields [16, 86]; (2) inadequate policy support [14]; (3) lack of proper quantification (and recognition) of long-term benefits of legumes within cropping systems [82]. However, other efforts could be addressed, for example, to breeding programs for improved crop cultivars, to better sustain livelihood and increase the economic return to farmers. Indeed, during last years significant progresses in breeding for quality traits for food [110] and feed uses [79], as well as for resistances to biotic [91] and abiotic stresses [4] are being achieved, but several others, many of which are controlled quantitatively by multiple genes, have been more difficult to achieve.

Grain legumes in intercropping

Intercropping systems consist in simultaneous growth of two or more crop species on the same area and at the same time [13]. Intercropping is widely used in developing countries or in low-input and low-yield farming systems [73]. Despite several recognized beneficial aspects of intercropping such as better pest control [60], competitive yields with reduced inputs [70, 107], pollution mitigation [63], more stable aggregate food or forage yields per unit area [100], there are a number of constrains that make intercropping not common in modern agriculture, such as example the request of a single and standardized product and the suitability for mechanization or use of other inputs as a prerogative in intensive farming system [13]. It is therefore necessary to optimize intercropping systems to enhance resource-use efficiency and crop yield simultaneously [55], while also promoting multiple ecosystem services (see also [13]). Most recent research has focalized on the potential of intercropping in sustainable productions and in particular on grain legumes that can fix N_2 through biological mechanisms (BNF). Indeed, legumes are pivotal in many intercropping systems, and of the top 10 most frequently used intercrop species listed by Hauggaard-Nielsen and Jensen [32], seven are legumes One of the basic spatial arrangements used in intercropping is strip intercropping, in which two or more crops grow together in strips wide enough to permit separate crop production using inputs but close enough for the crops to interact. The current challenge is how to determine an optimal intercropping width to maximise the resources use efficiency and, consequently, the crop productivity. In a maize-bean strip intercropping, Mahallati et al. [65] suggested that strip width of 2 and 3 rows was superior compared with monoculture and other strip intercropping combinations in terms of radiation absorption, radiation use efficiency and biological yields of both species, also allowing to an improve of total land productivity and land equivalent ratio (1.39 and 1.37). Gao et al. [26] showed a total yield increase of 65 and 71% in a system of 1 and 2 rows of maize (planted at a higher density in intercropping) alternated with 3 rows of soybean compared with both crops grown as monoculture. However, Liu et al. [59] showed a reduction in the photosynthetically active radiation and R:FR ratio at the top of soybean canopy intercropped with maize - under two intercropping patterns: 1 row of maize with 1 row of soybean; 2 rows of maize with 2 rows of soybean - leading to increased internode lengths, plant height and specific leaf area (SLA), but reduced branching of soybean plants. In order to gain sufficient light in the most shaded border rows of the neighbouring, shorter crops, efforts could be addressed to (i) the selection of highly productive maize cultivars with reduced canopy height and LAI; (ii) the increase of the strip width under a higher fraction of direct PAR; (iii) the selection of crops and cultivars suitable under the shade levels that likely occur in strip-intercropping systems with maize [71].

The increase in N availability in intercrops hosting legumes occurs because the competition for soil N from legumes is weaker than from other plants. Moreover, non-legumes obtain additional N from that released by legumes into the soil [56, 117] or via mycorrhizal fungi [113]. Legumes can contribute up to 15% of the N in an intercropped cereal [57], thus increasing biomass production and carry-over effects [75], reducing synthetic mineral N-fertilizer use and mitigating N_2O fluxes [9, 96]. However, the adoption of grain legume intercropping systems should benefit from the identification of suitable legumes that are less susceptible to N fertilizer-induced inhibition of BNF—that is, legumes that sustain higher %BNF in the presence of increasing soil mineral N. To this purpose, Rose et al. [90] indicated that faba bean is more suitable as intercrop than chickpea when supplementary N fertilizer additions are required, with about 40%BNF and 29%BNF maintained in faba bean and chickpea, respectively, supplying both crops with 150 kg N ha^{-1}.

BNF represents the most common plant growth stimulating factor that can also improve crop competition with respect to weeds in both organic and sustainable farming systems [10]. Grain legumes are weak suppressors of weeds, but mixing species in the same cropping system could represent a valid way to improve the ability of the crop itself to suppress weeds [41, 94]. In a wheat-chickpea intercropping system (20 cm spacing without weeding treatment) it was observed a 69.7% reduction in weed biomass and 70% in weed population as compared to un-weeded monocrop wheat at 20 cm spacing [6]. Similar results on weed smothering have been obtained by Midya et al. [67] in rice-blackgram (20 cm) intercropping system although the deferred seeding of blackgram in rice field (30 cm) with one weeding may be recommended for both better yield and weed suppression.

Direct mutual benefits in cereal-legumes intercropping involve below-ground processes in which cereals while benefiting of legumes-fixed N, increase Fe and Zn bioavailability to the companion legumes [118].

Physiology, agronomy and ecology can simultaneously contribute to the improvement of intercropping systems, allowing to enhance crop productivity and resource-use efficiency, so making intercropping a viable approach for sustainable intensification, particularly in regions with impoverished soils and economies where measured benefits have been greatest [93]. But to realize these goals, major efforts in research programs still remain. For example: (1) breeding for intercrops; (2) better understanding of the interactions between plants and other organisms in crop systems, focusing on the roles of above- and below-ground interactions of plants with other organisms; (3) improving agricultural engineering and management, i.e. developing new machinery that can till, weed and harvest at small spatial scales and in complex configurations to encourage the uptake of intercropping without greater demands for labour [58]; (4) adoption of a wider 'systems thinking' through the enactment of schemes, including payment for ecosystem services [105].

Grain legumes and conservation agriculture

Legumes have some characteristics particularly suitable for sustainable cropping systems and conservation agriculture, and making them functional either as growing crop or as crop residue. Conservation agriculture is based on minimal soil disturbance and permanent soil cover combined with rotations [35]. As previously described, major advantages of legumes include the amount of nitrogen fixed into the soil and the high quality of the organic matter released to the soil in term of C/N ratio. Some legume species have also deep root systems, which facilitate nutrients solubilization by root exudates and their uptake/recycling as well as water infiltration in deeper soil layers.

Many countries already rely on conservation agriculture. Brazil has implemented conservation agriculture systems using soybean as legume crop. Grain legumes like lentil, chickpea, pea and faba bean play a major role in conservation agriculture in North America, Australia, and Turkey. In Australia, some advantages of minimum tillage for grain legumes have been quantified for water-limited environments. Some studies indicate that the majority of grain-legumes producers use direct seeding after a legume pre-crop [1]. This change from conventional tillage (CT) to reduced or no tillage (NT) systems (with at least 30% of the soil surface covered) would lead to significant positive impacts on SOC [18]. In contrast, other results indicate that such positive effects are limited to the first 20 cm depth, while little or no difference between CT and NT in total SOC can be seen lower down the soil profile [5, 111]. Such findings suggest that C stock changes in the soil are mainly dependent to the net N-balance in the system. With high N harvest index legumes, SOC stocks are not preserved due to the high amount of N taken off from the field into the grain [42]. Conversely, the effect of legumes on soil carbon sequestration is more detectable for forage, green-manures and cover-crops which return to the soil large amounts of organic C and N [52]. Boddey et al. [12] indicate that vetch under no tillage may increase SOC stocks under NT (0–100 cm) at a rate between 0.48 and 1.53 Mg C ha^{-1} per year [42].

The implementation of practices of conservation tillage could significantly reduce the GWP, especially when a grain legumes is added to the rotation. In Mediterranean agro-ecosystems, Guardia et al. [29] compared three tillage treatments (i.e. no tillage: NT, minimum tillage: MT, conventional tillage: CT) and two crops (i.e. vetch, barley) and recorded the emission of N_2O, CH_4 and CO_2 during one year. Authors found a significant 'tillage × crop' interaction on cumulative N_2O emissions with vetch releasing higher N_2O amount than barley only in CT and MT, whereas similar fluxes were observed under NT. This was attributable to the soil water-filled pore space, dissolved organic carbon content and denitrification losses, in spite of the presumable predominance of nitrification. In any case, the most sustainable crop and tillage treatments in terms of GWP were represented by the non-fertilized vetch and NT, due to higher carbon sequestration, lower fuel consumption and the absence of mineral N fertilizers [29]. In subtropical Ultisol, under legume cover crops, NT soil exhibited increased N_2O emissions with respect to CT soil (531 vs. 217 kg CO_2 eq ha^{-1} $year^{-1}$); however, emissions of this gas from NT soil were fully offset by CO_2 retention in soil organic matter (−2063 to −3940 kg CO_2 ha^{-1} $year^{-1}$) [8]. Moreover, NT soil under legume cover crops behaved as a net sink for GHG (GWP ranged from −971 to −2818 kg CO_2 eq ha^{-1} $year^{-1}$) [8].

The expansion of ecological-based approaches like conservation agriculture opens opportunities to food legumes to be profitably included in sustainable cropping systems. There are still major challenges for conservation agriculture that need to be overcome, including the development of effective methods for weed control (see also [92]) that can avoid the use of herbicides or tillage. Overall conservation agriculture is an environmentally sustainable production system that may boost the incorporation of grain legumes within large and small-scale farming.

Conclusion

The roles and importance of grain legumes in a context of sustainability in agriculture could be enhanced by the emerging research opportunities for the major topics discussed above.

A major task in the future will be the selection of legume species and cultivars which could be effectively introduced across different cropping systems. An important point concerns balancing yield, which gives economic return, with the environmental and agronomic benefits.

Some priority areas seem emerge. Nitrogen fixation activity of grain legumes should be evaluated in relation with soil, climatic, plant characteristics and management conditions to find the suitable approach to achieve the best improvements. With this respect, the ability of the host plant to store fixed nitrogen appears to be a major component of increasing nitrogen fixation input. A particular focus should be paid also to the study of abiotic stress limitations and in particular water deficit, salinity and thermal shocks require extensive investigation.

Legumes that can recover unavailable forms of soil phosphorus could be major assets in future cropping systems. Consequently, those legumes which are able to accumulate phosphorus from forms normally unavailable need to be further studied, since phosphorus represents an expensive and limiting resource in several cropping systems.

Because of the growing request for plant products, i.e. protein and oils, and to the increased economic and environmental pressures on agro-eco systems, it emerges that grain legumes would play a major role in future cropping systems.

Abbreviations

GHG: greenhouse gases; CO_2: carbon dioxide; N_2O: nitrous oxide; LCA: life cycle assessment; GWP: global warming potential; BNF: biological nitrogen fixation; NT: no tillage; CT: conventional tillage.

Authors' contributions

FS participated in the topic literature view and selection, and in drafting of the manuscript. AM drafted the manuscript and revised it critically. AG participated in the topic literature view and selection, and in the drafting of the manuscript. MP drafted the manuscript and revised it critically. All authors read and approved the final manuscript.

Author details

[1] Faculty of Biosciences and Technologies for Agriculture Food and Environment, University of Teramo, Coste Sant'Agostino, Teramo, TE, Italy. [2] Department of Agricultural Science, University of Naples Federico II, Via Università 100, Portici, Italy.

Authors information

FS: Associate Professor of Agronomy and Crop Sciences at University of Teramo (ITALY). AM: Associate Professor of Agronomy and Crop Sciences at University Federico II°, Naples (ITALY). AG: Researcher in Agronomy and Crop Sciences at University of Teramo (ITALY). MP: Full Professor of Agronomy and Crop Sciences at University of Teramo (ITALY).

Competing interests

The authors declare that they have no competing interests.

Funding

No funds were received for this work.

References

1. Alpmann D, Braun J, Schäfer BC. Analyse einer Befragung unter erfolgreichen Körnerleguminosen anbauern im konventionellen Landbau. Erste Ergebnisse aus dem Forschungsprojekt LeguAN. In: Wintertagung DLG, Im Fokus: Heimische Körnerleguminosen vom Anbau bis zur Nutzung. Berlin; 2013.
2. Angus JF, Kirkegaard JA, Hunt JR, Ryan MH, Ohlander L, Peoples MB. Break crops and rotations for wheat. Crop Pasture Sci. 2015;66:523–52.
3. Angus JF, van Herwaarden AF, Howe GN. Productivity and break crop effects of winter-growing oilseeds. Animal Prod Sci. 1991;31:669–77.
4. Araújo SS, Beebe S, Crespi M, Delbreil B, González EM, Gruber V, et al. Abiotic stress responses in legumes: strategies used to cope with environmental challenges. Crit Rev Plant Sci. 2014;34:237–80.
5. Baker JM, Ochsner TE, Venterea RT, Griffis TJ. Tillage and carbon sequestration-what do we really know? Agric Ecosyst Environ. 2007;118:1–5.
6. Banik P, Midya A, Sarkar BK, Ghose SS. Wheat and chickpea intercropping systems in an additive series experiment: advantages and weed smothering. Eur J Agron. 2006;24:325–32.
7. Barbery P. Weed management in organic agriculture: are we addressing the right issues. Weed Res. 2002;42:177–93.
8. Bayer C, Gomes J, Zanatta JA, Vieira FCB, Dieckow J. Mitigating greenhouse gas emissions from a subtropical Ultisol by using long-term no-tillage in combination with legume cover crops. Soil Tillage Res. 2016;161:86–94.
9. Beaudette C, Bradley RL, Whalen JK, McVetty PBE, Vessey K, Smith DL. Tree-based intercropping does not compromise canola (*Brassica napus* L.) seed oil yield and reduces soil nitrous oxide emissions. Agric Ecosyst Environ. 2010;139:33–9.
10. Berry PM, Sylvester-Bradley R, Philipps L, Hatch SP, Cuttle FW, Gosling P. Is the productivity of organic farms restricted by the supply of available nitrogen? Soil Use Manag. 2002;18:248–55.
11. Bichel A, Oelbermann M, Voroney P, Echarte L. Sequestration of native soil organic carbon and residue carbon in complex agroecosystems. Carbon Manag. 2016;7:1–10.
12. Boddey RM, Jantalia CP, Zanatta JA, Conceição PC, Bayer C, Mielniczuk J, et al. Carbon accumulation at depth in Ferralsols under zero-till subtropical agriculture in southern Brazil. Global Change Biol. 2010;16:784–95.
13. Brooker RW, Bennett AE, Cong W-F, Daniell TJ, George TS, Hallett PD, et al. Improving intercropping: a synthesis of research in agronomy, plant physiology and ecology. New Phytol. 2015;206:107–17.
14. Bues A, Preißel S, Reckling M, Zander P, Kuhlmann T, Topp K, et al. The Environmental Role of Protein Crops in the New Common Agricultural Policy. European Parliament, Directorate General for Internal Policies, Policy Department B: Structural and Cohesion Policies, Agricultural and Rural DevelopmentIP/B/AGRI/IC/2012-067; 2013. Access www.europarl.europa.eu/studies.
15. Carranca C, Torres MO, Madeira M. Underestimated role of legume roots for soil N fertility. Agron Sustain Dev. 2015;35:1095–102.
16. Cernay C, Ben-Ari T, Pelzer E, Meynard J-M, Makowski D. Estimating variability in grain legume yields across Europe and the Americas. Sci Rep. 2015;5:11171.
17. Chalk PM. Dynamics of biologically fixed N in legume-cereal rotations: a review. Aust J Agric Res. 1998;49:303–16.
18. Christopher SF, Lal R. Nitrogen management affects carbon sequestration in North American cropland soils. Crit Rev Plant Sci. 2007;26:45–64.
19. Clune S, Crossin E, Verghese K. Systematic review of greenhouse gas emissions for different fresh food categories. J Clean Prod. 2017;140:766–83.
20. Crews TE, Peoples MB. Can the synchrony of nitrogen supply and crop demand be improved in legume and fertilizer-based agroecosystems? A Review. Nutr Cycl Agroecosyst. 2005;72:101–20.
21. Crutzen PJ, Mosier AR, Smith KA, Winiwarter W. N_2O release from agro-biofuel production negates global warming reduction by replacing fossil fuels. Atmos Chem Phys Discuss. 2007;7:11191–205.
22. Faghire M, Mohamed F, Taoufiq K, Faghire R, Bargaz A, Mandri B, et al. Genotypic variation of nodules'enzymatic activities in symbiotic nitrogen fixation among common bean (*Phaseolus vulgaris* L.) genotypes grown under salinity constraint. Symbiosis. 2013;60:115–22.
23. FAOSTAT. www.faostat.fao.org (visited 18 November 2016).
24. FAO. The State of the World's Land and Water Resources for Food and Agriculture (SOLAW)—Managing Systems at Risk. Food and Agriculture Organization of the United Nations, Rome and Earthscan, London; 2011.
25. Gan Y, Hamel C, Kutcher HR, Poppy L. Lentil enhances agroecosystem productivity with increased residual soil water and nitrogen. Renew Agr Food Syst 2016:1–12. doi:10.1017/S1742170516000223.
26. Gao Y, Duan A, Qiu X, Liu Z, Sun J, Zhang J, Wang H. Distribution of roots and root length density in a maize/soybean strip intercropping system. Agric Water Manag. 2010;98:199–212.
27. Garrigues E, Corson MS, Walter C, Angers DA, van der Werf H. Soil-quality indicators in LCA: method presentation with a case study. In: Corson MS, van der Werf HMG, editors. Proceedings of the 8th international conference on life cycle assessment in the agri-food sector, 1–4 October 2012, INRA, Saint Malo; 2012. p 163–68.

28. Giambalvo D, Stringi L, Durante G, Amato G, Frenda AS. Nitrogen efficiency component analysis in wheat under rainfed. Mediterranean conditions: effects of crop rotation and nitrogen fertilization. In: Cantero-Martínez C, Gabiña D, editors. Mediterranean rainfed agriculture: strategies for sustainability. Mediterranean Agronomic Institute of Zaragoza, Zaragoza; 2004. p. 169–73.
29. Guardia G, Tellez-Rio A, García-Marco S, Martin-Lammerding D, Tenorio JL, Ibáñez MÁ, Vallejo A. Effect of tillage and crop (cereal versus legume) on greenhouse gas emissions and Global Warming Potential in a non-irrigated Mediterranean field. Agric Ecosyst Environ. 2016;221:187–97.
30. Hajduk E, Właśniewski S, Szpunar-Krok E. Influence of legume crops on content of organic carbon in sandy soil. Soil Sci Ann. 2015;66:52–6.
31. Hassan HM, Hasbullah H, Marschner P. Growth and rhizosphere P pools of legume–wheat rotations at low P supply. Biol Fertil Soils. 2013;49:41–9.
32. Hauggaard-Nielsen H, Jensen ES. Facilitative root interactions in intercrops. Plant Soil. 2005;274:237–50.
33. Hauggaard-Nielsen H, Mundus S, Jensen ES. Nitrogen dynamics following grain legumes and catch crops and the effects on succeeding cereal crops. Nutr Cycl Agroecosyst. 2009;84:281–91.
34. Hernanz JL, Sanchez-Giron V, Navarrete L. Soil carbon sequestration and stratification in a cereal/leguminous crop rotation with three tillage systems in semiarid conditions. Agric Ecosyst Environ. 2009;133:114–22.
35. Hobbs PR, Sayre K, Gupta R. The role of conservation agriculture in sustainable agriculture. Philos Trans R Soc Lond, Ser B. 2008;363:543–55.
36. Hocking PJ. Organic acids exuded from roots in phosphorus uptake and aluminum tolerance of plants in acid soils. Adv Agron. 2001;74:63–97.
37. Hocking PJ, Randall PJ. Better growth and phosphorus nutrition of sorghum and wheat following organic acid secreting crops. In: Horst WJ, et al., editors. Proceedings of the 14th international plant nutrition colloquium Germany. Dordrecht: Kluwer Academic Publishers; 2001. p. 548–9.
38. IPCC. Climate change 2007: Synthesis report. summary for policymakers. Intergovernmental panel on climate change (IPCC); 2007.
39. Jenkinson DS. The impact of humans on the nitrogen cycle, with focus on temperate agriculture. Plant Soil. 2001;228:3–15.
40. Jensen CR, Joernsgaard B, Andersen MN, Christiansen JL, Mogensen VO, Friis P, Petersen CT. The effect of lupins as compared with peas and oats on the yield of the subsequent winter barley crop. Eur J Agron. 2004;20:405–18.
41. Jensen ES, Ambus P, Bellostas N, Boisen S, Brisson N, Corre-Hellou G, et al. Intercropping of cereals and grain legumes for increased production, weed control, im-proved product quality and prevention of N–losses in European organic farming systems. In: International conferences: joint organic congress - Theme 4: crop systems and soils, 9 May 2006.
42. Jensen ES, Peoples MB, Boddey RM, Gresshoff PM, Hauggaard-Nielsen H, Alves BJ, Morrison MJ. Legumes for mitigation of climate change and the provision of feedstock for biofuels and biorefineries. A review. Agron Sustain Dev. 2012;32:329–64.
43. Jensen ES, Peoples MB, Hauggaard-Nielsen H. Faba bean in cropping systems. Field Crops Res. 2010;115:203–16.
44. Jeuffroy MH, Baranger E, Carrouée B, Chezelles ED, Gosme M, Hénault C. Nitrous oxide emissions from crop rotations including wheat, oilseed rape and dry peas. Biogeosciences. 2013;10:1787–97.
45. Jones SK, Rees RM, Skiba UM, Ball BC. Influence of organic and mineral N fertiliser on N_2O fluxes from a temperate grassland. Agric Ecosyst Environ. 2007;121:74–83.
46. Kirkegaard JA, Christen O, Krupinsky J, Layzell DB. Break crop benefits in temperate wheat production. Field Crop Res. 2008;107:185–95.
47. Kirkegaard JA, Ryan MH. Magnitude and mechanisms of persistent crop sequence effects on wheat. Field Crops Res. 2014;164:154–65.
48. Köpke U, Nemecek T. Ecological services of faba bean. Field Crop Res. 2010;115:217–33.
49. La Favre JS, Focht DD. Conservation in soil of H_2 liberated from N_2 fixation by H up-nodules. Appl Environ Microb. 1983;46:304–11.
50. Latati M, Bargaz A, Belarbi B, Lazali M, Benlahrech S, Tellah S. The intercropping common bean with maize improves the rhizobial efficiency, resource use and grain yield under low phosphorus availability. Eur J Agron. 2016;72:80–90.
51. Legume Futures Report 4.2. Reckling M, Schläfke N, Hecker J-M, Bachinger J, Zander P, Bergkvist G, et al. Generation and evaluation of legume-supported crop rotations in five case study regions across Europe; 2014. Available from www.legumefutures.de.
52. Legume Futures Report 1.6. Reckling M, Preissel S, Zander P, Topp CFE, Watson CA, Murphy-Bokern D, Stoddard FL. Effects of legume cropping on farming and food systems; 2014. Available from www.legumefutures.de.
53. Lemke RL, Zhong Z, Campbell CA, Zentner RP. Can pulse crops play a role in mitigating greenhouse gases from North American agriculture? Agron J. 2007;99:1719–25.
54. Lewis G, Schrire B, Mackinder B, Lock M. Legumes of the World. Kew: Royal Botanic Gardens; 2005.
55. Li L, Tilman D, Lambers H, Zhang F-S. Biodiversity and over yielding: insights from belowground facilitation of intercropping in agriculture. New Phytol. 2014;203:63–9.
56. Li L, Zhang L-Z, Zhang F-Z. Crop mixtures and the mechanisms of overyielding. In: Levin SA, editor. Encyclopedia of biodiversity, vol. 2. 2nd ed. Waltham: Academic Press; 2013. p. 382–95.
57. Li YF, Ran W, Zhang RP, Sun SB, Xu GH. Facilitated legume nodulation, phosphate uptake and nitrogen transfer by arbuscular inoculation in an upland rice and mung bean intercropping system. Plant Soil. 2009;315:285–96.
58. Lithourgidis AS, Dordas CA, Damalas CA, Vlachostergios D. Annual intercrops: an alternative pathway for sustainable agriculture. Aust J Crop Sci. 2011;5:396.
59. Liu X, Rahman T, Song C, Su B, Yang F, Yong T, et al. Changes in light environment, morphology, growth and yield of soybean in maize-soybean intercropping systems. Field Crop Res. 2017;200:38–46.
60. Lopes T, Hatt S, Xu Q, Chen J, Liu Y, Francis F. Wheat (*Triticum aestivum* L.)-based intercropping systems for biological pest control: a review. Pest Manag Sci. 2016;72:2193–202.
61. López-Bellido L, Munoz-Romero V, Benítez-Vega J, Fernández-García P, Redondo R, López-Bellido RJ. Wheat response to nitrogen splitting applied to a Vertisols in different tillage systems and crop-ping rotations under typical Mediterranean climatic conditions. Eur J Agron. 2012;43:24–32.
62. López-Bellido L, Munoz-Romero V, López-Bellido RJ. Nitrate accumulation in the soil profile: long-term effects of tillage, rotation and N rate in a Mediterranean Vertisol. Soil Tillage Res. 2013;130:18–23.
63. Luo S, Yu L, Liu Y, Zhang Y, Yang W, Li Z, Wang J. Effects of reduced nitrogen input on productivity and N_2O emissions in a sugarcane/soybean intercropping system. Eur J Agron. 2016;81:78–85.
64. Magid J, Henriksen O, Thorup-Kristensen K, Mueller T. Disproportionately high N-mineralisation rates from green manures at low temperatures-implications for modelling and management in cool temperate agro-ecosystems. Plant Soil. 2001;228:73–82.
65. Mahallati MN, Koocheki A, Mondani F, Feizi H, Amirmoradi S. Determination of optimal strip width in strip intercropping of maize (*Zea mays* L.) and bean (*Phaseolus vulgaris* L.) in Northeast Iran. J Clean Prod. 2015;106:343–50.
66. Meynard JM, Messéan A, Charlier A, Charrier F, Farès M, Le Bail M, et al. Crop diversification: obstacles and levers. Study of farms and supply chains. Synopsis of the study report. INRA, Paris. 2013.
67. Midya A, Bhattacharjee K, Ghose SS, Banik P. Deferred seeding of blackgram (*Phaseolus mungo* L.) in rice (*Oryza sativa* L.) field on yield advantages and smothering of weeds. J Agron Crop Sci. 2005;191:195–201.
68. Miller PR, Gan Y, McConkey BG, McDonald CL. Pulse crops for the Northern Great Plains: I. Grain productivity and residual effects on soil water and nitrogen. Agron J. 2003;95:972–9.
69. Mokgehle SN, Dakora FD, Mathews C. Variation in N_2 fixation and N contribution by 25 groundnut (*Arachis hypogaea* L.) varieties grown in different agro-ecologies, measured using 15 N natural abundance. Agric Ecosyst Environ. 2014;195:161–72.
70. Monti M, Pellicanò A, Santonoceto C, Preiti G, Pristeri A. Yield components and nitrogen use in cereal-pea intercrops in Mediterranean environment. Field Crop Res. 2016;196:379–88.
71. Munz S, Graeff-Hönninger S, Lizaso JI, Chen Q, Claupein W. Modeling light availability for a subordinate crop within a strip–intercropping system. Field Crop Res. 2014;155:77–89.

72. Nemecek T, von Richthofen J-S, Dubois G, Casta P, Charles R, Pahl H. Environmental impacts of introducing grain legumes into European crop rotations. Eur J Agron. 2008;28:380–93.
73. Ngwira AR, Aune JB, Mkwinda S. On-farm evaluation of yield and economic benefit of short term maize legume intercropping systems under conservation agriculture in Malawi. Field Crop Res. 2012;132:149–57.
74. Papastylianou I. Effect of rotation system and N fertiliser on barley and vetch grown in various crop combinations and cycle lengths. J Agric Sci. 2004;142:41–8.
75. Pappa VA, Rees RM, Walker RL, Baddeley JA, Watson CA. Legumes intercropped with spring barley contribute to increased biomass production and carry-over effects. J Agric Sci. 2012;150:584–94.
76. Peoples MB, Boyer EW, Goulding KWT, Heffer P, Ochwoh VA, Vanlauwe B, et al. Pathways of nitrogen loss and their impacts on human health and the environment. In: Mosier AR, Syers KJ, Freney JR, editors. Agriculture and the nitrogen cycle, the Scientific Committee on Problems of the Environment (SCOPE). Covelo: Island Press; 2004. p. 53–69.
77. Peoples MB, Brockwell J, Herridge DF, Rochester IJ, Alves BJR, Urquiaga S, et al. The contributions of nitrogen-fixing crop legumes to the productivity of agricultural systems. Symbiosis. 2009;48:1–17.
78. Peoples MB, Hauggaard-Nielsen H, Jensen ES. The potential environmental benefits and risks derived from legumes in rotations. In: Emerich DW, Krishnan HB, editors. Nitrogen fixation in crop production. Madison: American Society of Agronomy, Crop Science Society of America, Soil Science Society of America; 2009. p. 349–85.
79. Phelan P, Moloney AP, McGeough EJ, Humphreys J, Bertilsson J, O'Riordan E, O'Kiely P. Forage legumes for grazing and conserving in ruminant production systems. Crit Rev Plant Sci. 2015;34:281–326.
80. Plaza-Bonilla D, Nolot J-M, Raffaillac D, Justes E. Innovative cropping systems to reduce N inputs and maintain wheat yields by inserting grain legumes and cover crops in southwestern France. Eur J Agron. 2016. doi:10.1016/j.eja.2016.05.010.
81. Plaza-Bonilla D, Nolot JM, Raffaillac D, Justes E. Cover crops mitigate nitrate leaching in cropping systems including grain legumes: field evidence and model simulations. Agric Ecosyst Environ. 2015;212:1–12.
82. Preissel S, Reckling M, Schläfke N, Zander P. Magnitude and farm-economic value of grain legume pre-crop benefits in Europe: a review. Field Crop Res. 2015;175:64–79.
83. Rao DLN, Giller KE, Yeo AR, Flowers TJ. The effects of salinity and sodicity upon nodulation and nitrogen fixation in chickpea (*Cicer arietinum*). Ann Bot Lond. 2002;89:563–70.
84. Reay DS, Davidson EA, Smith KA, Smith P, Melillo JM, et al. Global agriculture and nitrous oxide emissions. Nat Clim Change. 2012;2:410–6.
85. Reckling M, Hecker J-M, Bergkvist G, Watson CA, Zander P, Schläfke N, et al. A cropping system assessment framework—evaluating effects of introducing legumes into crop rotations. Eur J Agron. 2016;76:186–97.
86. Reckling M, Döring T, Stein-Bachinger K, Bloch R, Bachinger J. Yield stability of grain legumes in an organically managed monitoring experiment. Aspects Appl Biol. 2015;128:57–62.
87. Robson MC, Fowler SM, Lampkin NH, Leifert C, Leitch M, Robinson D, et al. The agronomic and economic potential of breakcrops for ley/arable rotations in temperate organic agriculture. Adv Agron. 2002;77:369–427.
88. Rochester IJ. Estimating nitrous oxide emissions from flood irrigated alkaline grey clays. Aust J Soil Res. 2003;41:197–206.
89. Rochette P, Janzen HH. Towards a revised coefficient for estimating N_2O emissions from legumes. Nutr Cycl Agroecosyst. 2005;73:171–9.
90. Rose TJ, Julia CC, Shepherd M, Rose MT, Van Zwieten L. Faba bean is less susceptible to fertiliser N impacts on biological N_2 fixation than chickpea in monoculture and intercropping systems. Biol Fert Soils. 2016;52:271–6.
91. Rubiales D, Fondevilla S, Chen W, Gentzbittel L, Higgins TJV, Castillejo MA, et al. Achievements and challenges in legume breeding for pest and disease resistance. Crit Rev Plant Sci. 2014;34:195–236.
92. Rühlemann L, Schmidtke K. Evaluation of monocropped and intercropped grain legumes for cover cropping in no-tillage and reduced tillage organic agriculture. Eur J Agron. 2015;65:83–94.
93. Rusinamhodzi L, Corbeels M, Nyamangara J, Giller KE. Maize–grain legume intercropping is an attractive option for ecological intensification that reduces climatic risk for smallholder farmers in central Mozambique. Field CropRes. 2012;136:12–22.
94. Šarūnaitė L, Deveikytė I, Kadžiulienė Ž. Intercropping spring wheat with grain legume for increased production in an organic crop rotation. Žemdirbystė = Agric. 2010;97:51–8.
95. Schwenke GD, Herridge DF, Scheer C, Rowlings DW, Haigh BM, McMullen KG. Soil N_2O emissions under N_2-fixing legumes and N-fertilised canola: a reappraisal of emissions factor calculations. Agric Ecosyst Environ. 2015;202:232–42.
96. Senbayram M, Wenthe C, Lingner A, Isselstein J, Steinmann H, Kaya C, Köbke S. Legume-based mixed intercropping systems may lower agricultural born N_2O emissions. Energy Sustain Soc. 2016;6:2.
97. Seymour M, Kirkegaard JA, Peoples MB, White PF, French RJ. Break-crop benefits to wheat in Western Australia-insights from over three decades of research. Crop Pasture Sci. 2012;63:1–16.
98. Shen J, Yuan L, Zhang J, Li H, Bai Z, Chen X, et al. Phosphorus dynamics: from soil to plant. Plant Physiol. 2011;156:997–1005.
99. Siddique KH, Johansen C, Turner NC, Jeuffroy MH, Hashem A, Sakar D, Gan Y, Alghamdi SS. Innovations in agronomy for food legumes: a review. Agron Sustain Dev. 2012;32:45–64.
100. Smith J, Pearce BD, Wolfe M, Martin S. Reconciling productivity with protection of the environment: Is temperate agroforestry the answer? Renew Agric Food Syst. 2013;28:80–92.
101. Soussana JF, Loiseau P, Vuichard N, Ceschia E, Balesdent J, Chevallier T, Arrouays D. Carbon cycling and sequestration opportunities in temperate grasslands. Soil Use Manag. 2004;20:219–30.
102. Soussana JF, Tallec T, Blanfort V. Mitigating the greenhouse gas balance of ruminant production systems through carbon sequestration in grasslands. Animal. 2010;4:334–50.
103. St. Luce M, Grant CA, Zebarth BJ, Ziadi N, O'Donovan JT, Blackshaw RE, et al. Legumes can reduce economic optimum nitrogen rates and increase yields in a wheat-canola cropping sequence in western Canada. Field Crop Res. 2015;179:12–25.
104. Stagnari F, Pisante M. Managing faba bean residues to enhance the fruit quality of the melon (*Cucumis melo* L.) Crop. Sci. Hort. 2010;126:317–23.
105. Swinton SM, Lupi F, Robertson GP, Hamilton SK. Ecosystem services and agriculture: cultivating agricultural ecosystems for diverse benefits. Ecol Econ. 2007;64:245–52.
106. Tharanathan RN, Mahadevamma S. Grain legumes—a boon to human nutrition. Trends Food Sci Tech. 2003;14:507–18.
107. Tosti G, Guiducci M. Durum wheat–faba bean temporary intercropping: effects on nitrogen supply and wheat quality. Eur J Agron. 2010;33:157–65.
108. United Nations: World population prospects: The 2012 revision, key findings and advance tables. Working paper no. ESA/P/WP.227; 2013 (United Nations, Department of Economic and Social Affairs, Population Division, New York).
109. Van Zwieten L, Rose T, Herridge D, Kimber S, Rust J, Cowie A, Morris S. Enhanced biological N_2 fixation and yield of faba bean (*Vicia faba* L.) in an acid soil following biochar addition: dissection of causal mechanisms. Plant Soil. 2015;395:7–20.
110. Vaz Patto MC, Amarowicz R, Aryee AN, Boye JI, Chung HJ, Martín-Cabrejas MA, Domoney C. Achievements and challenges in improving the nutritional quality of food legumes. Crit Rev Plant Sci. 2015;34:105–43.
111. VandenBygaart AJ, Gregorich EG, Angers DA. Influence of agricultural management on soil organic carbon: a compendium and assessment of Canadian studies. Can J Soil Sci. 2003;83:363–80.
112. Voisin AS, Guéguen J, Huyghe C, Jeuffroy MH, Magrini MB, Meynard JM, et al. Legumes for feed, food, biomaterials and bioenergy in Europe: a review. Agron Sustain Dev. 2014;34:361–80.
113. Wahbi S, Maghraoui T, Hafidi M, Sanguin H, Oufdou K, Prin Y, et al. Enhanced transfer of biologically fixed N from faba bean to intercropped wheat through mycorrhizal symbiosis. Appl Soil Ecol. 2016;107:91–8.
114. Walley FL, Clayton GW, Miller PR, Carr PM, Lafond GP. Nitrogen economy of pulse crop production in the Northern Great Plains. Agron J. 2007;99:1710–8.
115. Wang Y, Marschner P, Zhang F. Phosphorus pools and other soil properties in the rhizosphere of wheat and legumes growing in three soils in monoculture or as a mixture of wheat and legume. Plant Soil. 2012;354:283–98.

116. Westhoek H, Rood T, van den Berg M, Janse J, Nijdam D, Reudink M, Stehfest E. The protein puzzle. The consumption and production of meat, dairy and fish in the European Union. Netherlands Environmental Assessment Agency (PBL); 2011.
117. White PJ, George TS, Gregory PJ, Bengough AG, Hallett PD, McKenzie BM. Matching roots to their environment. Ann Bot Lond. 2013;112:207–22.
118. Xue Y, Xia H, Christie P, Zhang Z, Li L, Tang C. Crop acquisition of phosphorus, iron and zinc from soil in cereal/legume intercropping systems: a critical review. Ann Bot Lond 2016;117:363–77. doi:10.1093/aob/mcv182..
119. Yadav SS, Hunter D, Redden B, Nang M, Yadava DK, Habibi AB. Impact of climate change on agriculture production, food, and nutritional security. In: Redden R, Yadav SS, Maxted N, Dulloo MS, Guarino L, Smith P, editors. Crop wild relatives and climate change. New Jersey, USA: Wiley; 2015. p. 1–23.
120. Yu Y, Xue L, Yang L. Winter legumes in rice crop rotations reduces nitrogen loss, and improves rice yield and soil nitrogen supply. Agron Sustain Dev. 2014;34:633–40.
121. Zander P, Amjath-Babu TS, Preissel S, Reckling M, Bues A, Schläfke N, et al. Grain legume decline and potential recovery in European agriculture: a review. Agron Sustain Dev. 2016;36:26.
122. Zentner RP, Wall DD, Nagy CN, Smith EG, Young DL, Miller PR, et al. Economics of crop diversification and soil tillage opportunities in the Canadian prairies. Agron J. 2002;94:216–30.
123. Zimmer S, Messmer M, Haase T, Piepho HP, Mindermann A, Schulz H, et al. Effects of soybean variety and *Bradyrhizobium* strains on yield, protein content and biological nitrogen fixation under cool growing conditions in Germany. Eur J Agron. 2016;72:38–46.

20

Plant growth promotion using microbial IAA producers in conjunction with azolla: a novel approach

V. Raut[1,5], I. Shaikh[1,5], B. Naphade[2,5], K. Prashar[3,5] and N. Adhapure[4,5*]

Abstract

Studies were conducted to find efficient Indole acetic acid producing strains and assessing the effectiveness of combined dose of azolla (dried) and IAA producers on plant growth. Out of ten isolates, only two isolates, viz, F1 and G2 were found to be potent IAA producers and were used for further experiments. It was found that the isolate F1 and G2 produced maximum amount of IAA at 0.9% tryptophan and incubation for 72 h at room temperature. The optimum pH for IAA production for isolates F1 and G2 was found to be at pH 9. The isolates were also assessed for the presence of other plant growth promoting activity, such as hydrogen cyanide production, oxalate solubilization, phosphate solubilization and zinc solubilization. Effect of IAA producers on seed germination was done by pot experiments carried out into sets. Isolates F1 and G2 were found efficient for plant growth promotion in desired aspects. Effect of combined dose of azolla and IAA producers on seed germination was assessed by pot experiments. For this, dried powder of azolla was added during sowing of culture treated seeds. No further culture supply was given to the seeds. Azolla has high protein content. Slow protein releasing property of azolla can be useful to provide constant supply of the precursor (tryptophan) for IAA synthesis by isolate. In addition to the plant exudates, azolla may also give constant supply of tryptophan. From all the assessed samples, the isolates FI and G2 showed efficient growth of plants in all aspects. G2 was found to be more potent than F1. So as to assess the compatibility of isolate in natural conditions, field studies were done. Four plots designed having individual size 9 × 5 m, were used. It was found that plots 3 (only culture) and 4 (culture with azolla) showed better plant growth as compared to others. On the basis of IAA producing ability of G2, pot experiments and field studies, it was concluded that isolate G2 can be used alone or in conjunction with azolla as a biofertilizer to get better plant growth.

Background

Rhizosphere supports large and active microbial population capable of exerting beneficial, neutral and detrimental effect on the plants. As plant roots grow through soil, they release water-soluble compounds such as amino acids, sugars and organic acids that supply food for the microorganisms. In return the microorganisms provide nutrients for the plants. All this activity makes the rhizosphere, the most dynamic environment in soil [1]. The plant–microbe interaction in the rhizosphere is responsible for increasing plant health and soil fertility [2]. Plant growth promoting rhizobacteria are a group of free living soil bacteria, which have the ability to promote growth and yield of crop plant by direct and indirect mechanism [3]. Phytohormones play an important role as signals and regulators of growth and development in plants. Auxins, among them in particular, indole-3-acetic acid (IAA), are the most studied plant growth regulators, and this includes physiological, biochemical and genetic aspects [4].

Indole acetic acid (IAA) is a natural auxin which is also synthesized in many species of non-seeded plants, many bacteria, fungi and algae. The amino acid tryptophan is commonly regarded as the precursor for the biosynthesis of auxin in plants [5]. Therefore, screening of bacteria for their, in vitro potential of auxin production could provide

*Correspondence: adhapurenn@gmail.com

[4] Department of Biotechnology and Microbiology, Vivekanand Arts Sardar Dalipsingh Commerce and Science College, Aurangabad, MS, India

Full list of author information is available at the end of the article

a reliable base for selection of effective plant growth promoting bacteria.

The production and concentration of IAA could also have influenced by other factors besides species or strain of rhizobacteria such as PSB and medium conditions, growth stage and availability of substrates [6]. In the presence of Trp microbes release greater quantities of IAA [7, 8, 9]), and other related compounds like gibberellins and cytokinins [9, 10].

Azolla is a genus of heterosporous aquatic fern which contains within the leaf cavities the N2-fixing cyanobacterium *Anabaena azollae*. The host *Azolla* synthesizes and provides photosynthates essential for the nitrogenase activity of the cyanobiont, while *A. azollae* supplies fixed nitrogen to the fern [11]. The cyanobacterial strains belonging to the genera *Nostoc* and *Anabaena* comprised 80% of the rhizosphere isolates, which were also efficient in enhancing the germination and growth of wheat seeds and exhibited significantly high protein accumulation and IAA production [12]. In addition to its influence on soils physical properties, azolla is important in the cycling of nutrients. When the azolla is incorporated into the soil and humus is formed, these nutrients are slowly released into the soil as decomposition progresses. Even though azolla appears to be a rather delicate plant that would rapidly decompose, it actually takes six weeks or more for most of the nutrients to be released because the plant has rather high lignin content. Slow decomposition gives a natural slow release effect that is ideal for efficient absorption of the nutrients released. Another factor in azolla effectiveness as a green manure is its low carbon to nitrogen ratio of about 10:1. This high ratio ensures that azolla nitrogen will not be tied up by bacteria that are involved in decomposition of an overabundance of carbonaceous plant residues [13]. Nitrogen compounds released include small amounts of amide, peptides, and free amino acids, with most as nitrate and ammonium. Released N is readily taken up by surrounding organisms, including vascular plants, fungi, actinomycetes, and bacteria [14].

There are three primary methods for applying *Azolla* to crops. First, it may be grown in the field as a monocrop during the fallow season and then incorporated into the soil before planting the target crop. Second, *Azolla* may be grown as an intercrop among the target crop. Third, natural or deliberately cultured growths of *Azolla* may be harvested from ponds, swamps, or flooded fields and applied to a variety of target crops, either by incorporating it into the soft before planting the crop or by applying it as much on top of the soil around the bases of crop plants. Often, a combination of these methods of application is utilized [15].

As we know, IAA should be supplied in optimum amount to plants. Using only IAA producers at the time of sowing cannot guarantee the production of IAA as it is tryptophan limiting. Natural source for tryptophan is the exudates from plants. The carbonaceous compounds released through root exudates includes sugars, polysaccharides, amino acids, aromatic acids, aliphatic acids, fatty acids, sterols, phenolics, enzymes, proteins, plant growth regulators and secondary metabolites. Secondary metabolites such as flavonoids of plants are released during seed germination [16]. In this study, azolla has been applied in conjunction with IAA producers, considering that slow nutrient release property of azolla could be a source of controlled tryptophan supply to IAA producers.

Today, biofertilizers have emerged as a highly potent alternative to chemical fertilizers due to their eco-friendly, easy to apply, non-toxic and cost effective nature. Also, they make nutrients that are naturally abundant in soil or atmosphere, usable for plants and act as supplements to agrochemicals. Biofertilizer is a substance which contains living microorganisms which when applied to seed, plant surfaces, or soil colonizes the rhizosphere or the interior of the plant and promotes growth by increasing the availability of primary nutrients to the host plant [17]. Assuming all the above parameters a better approach of using Azolla and IAA producing microorganisms as biofertilizer was proposed and the studies were carried out accordingly.

Methods

Collection of soil sample

Soil samples were collected from three locations, viz. Garden soil (rhizospheric soil sample) GS, Farm soil (non-rhizospheric soil sample) FS and College campus soil (non rhizospheric soil sample) CS. Non rhizospheric soil sample was basically a soil from a ploughed farm. This was taken for getting the natural/robust microbial flora from a farm soil. Rhizospheric soil samples were actually taken from the rhizosphere of garden plants. This was done for getting maximum IAA producers as they are present in such environment. The purpose of taking rhizospheric and non-rhizospheric soil sample was to assess the diverse microflora for the said IAA producing activity. All samples were collected in sterile containers. The soil was then crushed in sterile mortar and pestle aseptically to get a fine powder and this powdered soil was then used for further experiments.

Enrichment of IAA producing bacteria

Five gram of each soil sample was inoculated in 100 ml conical flask containing sterile nutrient broth (NB) supplemented with 0.1% of tryptophan [18]. The flaks were labeled as, GS, FS and CS for garden soil, farm soil and college campus soil, respectively.

The flasks were kept for enrichment by incubating at room temperature for 24 h. IAA was determined in vitro by the method of Salkowski [18]. From each of the enriched flasks, 1.5 ml of culture was taken in centrifuge tube and centrifuged for 5 min at 8000 rpm, 1 ml of supernatant was taken and mixed with 3 ml of Salkowski's reagent and incubated for 30 min in dark for development of pink color. Spectrophotometric analysis of each sample was done for IAA detection. The soil samples, which showed higher absorbance at 520 nm were considered as efficient IAA producers and chosen for further experimental procedure.

Isolation of bacterial isolates for IAA production

From the selected flasks (FS and GS), loopful of cultures were streaked on two sterile (N.A + 0.1% tryptophan) plates separately, and incubated for 24 h at room temperature for obtaining isolated colonies.

Screening for IAA producing microorganisms

After incubation, ten different colonies were selected (five colonies from each plate) and were labeled according to their source, viz. For FS sample isolates: F1, F2, F3, F4, and F5 and for GS sample isolates: G1, G2, G3, G4, and G5. These isolated colonies were inoculated in sterile 10 ml (NB + 0.1% tryptophan) tubes separately and incubated at room temperature for 24 h.

Simultaneously, 1 ml of culture from each flask (FS and GS) was added in flask containing 25 ml sterile (NB + 0.1% tryptophan) medium separately and were labeled as consortiums, viz. FC (Farm soil consortium) GC (Garden soil consortium). These flasks were then incubated at room temperature for 24 h.

After incubation, 1.5 ml broth from each tube and flask were taken in respective centrifuge tubes and centrifuged at 8000 rpm for 5 min. 1 ml of supernatant from each tube was mixed with 3 ml of Salkowski's reagent and the tubes were incubated at room temperature in dark for 30 min and observed for pink color development [19]. Optical densities of all these samples were taken at 520 nm and their IAA concentrations were measured according to standard graph values. The isolates, which showed maximum IAA concentration were selected for further experiments.

Effect of different factors on IAA production

Effect of tryptophan concentration

Two hundred milliliter of sterile NB was distributed in 20 sterile test tubes. The tubes were supplemented with varying concentration of tryptophan, viz. 0.1, 0.3, 0.5, 0.7 and 0.9% (four tubes each with same tryptophan concentration). Then the tubes were inoculated with isolates (F1 and G2) and consortiums (FC and GC) for different incubation periods, i.e., 24, 48, 72, and 96 h, respectively, and the tubes were assessed for IAA concentration by Salkowski's method after each 24 h.

Effect of pH

One hundred and sixty milliliter of sterile (NB + 0.1% tryptophan) was prepared and distributed in 16 sterile test tubes with pH 5, 7, 9 and 11 (4 tubes each with same pH), respectively. The tubes prepared according to respective pH were inoculated with isolates and consortium and labeled according to culture inoculated and kept for incubation at room temperature for 24, 48, 72, and 96 h. All the incubated samples were assessed for measuring IAA concentration by Salkowski's method after each 24 h.

Assessment of other plant growth promoting activities of the isolates

Hydrogen cyanide production

Isolates (F1 and G2) and consortiums (FC and GC) were tested for the production of HCN by spreading them on sterile N.A plate amended with 0.44% of glycine. A circular disk of Whatman's filter paper No. 1 of 10 cm diameter was soaked in 2% sodium carbonate in 0.5% picric acid solution and placed in the lid of the plates [20]. One plate was uninoculated and kept as control. Plates were sealed with para film and incubated at room temperature for 24 h. After incubation the plates were observed for the development of yellow to light brown, moderate brown or strong brown color on filter paper by referring the control plate.

Oxalate solubilization

The agar plates containing oxalic acid degradation medium were prepared according to the method of Dickman and Mitra [21], all the samples (isolates and consortiums) were spotted on the respective plates and one plate was kept uninoculated which was referred as control. The plates were incubated at room temperature for 7 days. After incubation the plates were observed for formation of clear halo around the bacterial spots.

Phosphate solubilization

Evaluation of tricalcium phosphate solubilization of isolates (F1 and G2) and consortiums (FC and GC) was assessed using Pikovskaya's agar. Each bacterial culture was spot-inoculated in the centre of the plate [22]. After incubation for 48 h at room temperature, each plate was observed for zone of clearance, for the indication of inorganic phosphate solubilization [23].

Zinc solubilization

Zinc solubilization ability of the isolates (F1 and G2) and consortiums (FC and GC) were detected by spotting the log phase culture of bacterial strains on Tris-minimal medium plates having zinc phosphate, and zinc carbonate as source of insoluble inorganic zinc [20]. The inoculated plates were then incubated at room temperature for 3 days, and observed for the clearing zone around the colonies (due to the solubilization of inorganic zinc by bacteria).

Biochemical characterization

Biochemical studies were carried out for the identification of isolates (F1 and G2) as follows: Catalase test, Methyl-Red Test, Voges–Proskauer Test, Citrate Utilization Test, Phenylalanine Deamination Test, Urea Hydrolysis Test, Starch Hydrolysis Test, Casein Hydrolysis Test, Triple Sugar Iron (TSI) test, catalase, oxidase, urease and sugar fermentation tests.

Pot experiment 1: effect of IAA producer on plant growth

Ten pots of equal size and volume were taken. The pots were equally distributed in two batches (5 pots in each batch). The pots were filled with sterile soil and wetted with sterile distilled water. The seeds of soyabean were selected for sowing in pots of both the batches. Prior to sowing, the seeds were surface sterilized by washing under tap water and placed in beaker containing distilled water. A Pinch of SDS was added and seeds were shaken gently for 1–2 min. The water was discarded and 70% ethanol was added for 1–2 min. Ethanol was discarded and seeds were placed in sterile distilled water, a pinch of $HgCl_2$ was added and kept for 15 min. Further 3–4 washes of distilled water were given and all the water was discarded. After surface sterilization the seeds were treated with cultures, F1, G2 and FC, GC respectively.

The pots were labeled according to the cultures by which they were treated. One pot was sown with culture untreated seeds and kept as control. For batch 2, similar procedure was followed as of batch 1, but pots were supplied continuously with their respective cultures at interval of 24 h. Both the batches were supplied with sterile distilled water at every 24 h till 1 month and observed for growth of plantlets. After 1 month the plantlets were gently removed from pots by immersing them in water as to remove the soil. The plants of both the batches were assessed for root length, shoot length, primary–secondary leaves and chlorophyll content.

Root length measurement

All the plantlets of batch 1 and batch 2 were removed gently from pots and each root of each plantlet was measured manually using scale. The average root measurement for each pot of both the batches was calculated.

Shoot length measurement

The shoots of all the plantlets of both the batches were excised and measured manually using scale.

Primary and secondary leaves counting

The leaves of each plantlet were counted manually and noted according to their nature (i.e., primary and secondary).

Measuring chlorophyll content

From both the batches, one plantlet from each pot was taken as a sample for assessing the chlorophyll content. From each of the selected plant 1 gm of finely cut well mixed leaf tissue was taken in clean mortar. Tissue was ground to fine pulp with the addition of 20 ml of 80% acetone. Centrifuged (5000 rpm for 5 min) and transferred supernatant to volumetric flask. Residual tissue was ground again with 20 ml of 80% acetone, centrifuged and the supernatant was transferred to a 100 ml volumetric flask. This procedure was repeated until the residue became colorless. Mortar and pestle was thoroughly washed with 80% acetone and the clear washings were collected in 100 ml volumetric flask. Volume was made up to 100 ml with 80% acetone. Absorbance of solutions was read at 645 nm, 663 nm against the solvent (80% acetone) blank. Following formula was used for determination of chlorophyll A and B.

$$\text{Mg chlorophyll A/g tissue} = 12.7(A_{663}) - 2.69(A_{645}) \times V/1000 \times W$$

$$\text{Mg chlorophyll B/g tissue} = 22.9(A_{645}) - 4.68(A_{663}) \times V/1000 \times W$$

where, A = absorbance at specific wavelength, V = final volume of chlorophyll extract in 80% of acetone, W = fresh weight of tissue extracted

Cultivation of Azolla

For cultivating azolla, a pit of size 7 × 4 feet and of depth 20 cm was created by arranging bricks and placing plastic mat over the bricks for water holding. The pit was built under shade. A pit can also be done on ground by digging and placing plastic mat in it. The pit was filled with water up to the edge of the pit and soil mixed with cattle manure which was added to the pit as per requirement. The azolla culture was brought form Krishi Vidnyan Kendra, Kharpudi, Jalna. The azolla culture was inoculated in the prepared pit and kept for growth. After sufficient amount of growth, azolla was taken from the pit, dried and powdered, which was used for pot experiment 2 and field studies.

Pot experiment 2: effect of co inoculation of IAA producer and azolla on plant growth

For this experiment, six pots of equal size and volume were taken. The pots were filled with sterile soil and wetted with sterile distilled water. Prior to sowing, the seeds were surface sterilized as described in pot experiment 1 and treated with our sample cultures, respectively. At the time of sowing a pinch of azolla powder was added with each seed. In the remaining two pots, Seeds in one pot were sown only with azolla powder without treating it by sample cultures and the other pot was kept blank with no treatment of azolla powder and sample cultures which was kept as control. All the samples were labeled according to their sample cultures (i.e., F1, G2 and FC, GC, respectively). Both the batches were supplied with sterile distilled water at every 24 h till one month and observed for growth of plantlets. After one month the plantlets were gently removed from pots by immersing them in water as to remove the soil. The plants were assessed for root length, shoot length, primary–secondary leaves and chlorophyll content. This was done using the procedure followed in pot experiment 1.

Field study: effect of co inoculation of IAA producer and azolla on plant growth

From the results of pot experiments 1 and 2, the isolate G2 found to be most efficient among all the samples and chosen for this experiment. In the selected field, four plots of equal size (9 × 5 m) were selected for the experiment. In each plot, 15 rows were created manually for sowing of seeds and draining water from the spaces of adjacent rows. Seeds were sown near the bottom of rows to get access of water draining through it. Soyabean seeds were selected for sowing. The four plots were labeled as 1, 2, 3 and 4. In plot 1, seeds were sown without treating with culture and azolla powder and kept as control. Seeds of plot 2 were sown only with azolla powder without treating with culture. Plot 3 consisted with seeds sown by treating only with culture. In plot 4, seeds were sown with culture treatment and supplement of azolla powder. Supply of water was given to each plot, at the interval of every 2–3 days for one month and observed for plant growth. After a month, one plant from each plot was selected for assessing root length; shoot length, primary–secondary leaves and chlorophyll content. The selection of plant, for the assessment of above parameters was selected on the majority of maximum shoot length among all the plants of each plot. The assessment of above parameters was done using the same procedures as followed in pot experiments 1 and 2.

Results

Enrichment of IAA producing bacteria

After assessing using Salkowski's reagent, it was found that among all three enriched soil samples, two samples, viz. FS and GS showed formation of pink color with high intensity than CS. So, for screening of IAA producers only GS and FS were used and not the CS.

Screening of bacterial isolates for IAA production

The assessment of all the ten isolates (F1, F2, F3, F4, F5 and G1, G2, G3, G4, G5) by Salkowski's method, it was found that, among all the samples, F1 and G2 showed maximum absorbance for IAA at 520 nm. GC and FC also showed good IAA production.

Effect of different factors on IAA production

Effect of tryptophan concentration on IAA production

The samples inoculated with varying concentration of tryptophan, viz. 0.1, 0.3, 0.5, 0.7 and 0.9% in NB medium, showed gradual increase in production of IAA with the increase of tryptophan concentration and incubation period. The samples were assessed by Salkowski's method. The isolates (F1 and G2) showed maximum IAA production at 0.9% concentration of tryptophan and incubation of 72 h, whereas maximum IAA productivity of consortiums (FC and GC) was also observed at 0.9% concentration of tryptophan but by incubating for 96 h. The OD values and concentration of IAA produced by all the samples with respect to varying concentration of tryptophan and incubation periods, is represented in Table 1 and Figs. 1 and 2.

Effect of pH on IAA production

All the samples inoculated in the (NB + 0.1% tryptophan) medium containing varying pH, viz. 5, 7, 9 and 11 showed different concentration of IAA production at different incubation periods. After assessing the samples by Salkowski's method, it was observed that F1 and G2 showed maximum IAA production at pH 9 but at different incubation periods, i.e., 24 and 72 h. The consortium FC showed maximum IAA productivity at pH 9 and incubation at 48 h, whereas GC showed maximum IAA production at pH 11 and incubation at 72 h. The OD values and concentration of IAA produced by all the samples with respect to varying pH and incubation periods, is represented in Table 2 and Fig. 3.

IAA production abilities of all four cultures at pH 9.0 was assessed using one way ANOVA and it was found that at 0.05 level the means are significantly different ($F = 120.52152, p = 5.3568$)

Table 1 Effects of tryptophan concentration on IAA production

Culture samples	Tryptophan concentration (%)	Concentration of IAA (μg/ml)			
		24 h	48 h	72 h	96 h
F1	0.1	9.5 (±0.4)	9 (±0.4)	11.6 (0.8)	10.1 (±0.4)
	0.3	10.4 (±0.4)	17.8 (0.8)	18.1 (±0.8)	16 (0.8)
	0.5	16.4 (±0.4)	21 (±1.2)	25 (±1.6)	22.6 (±1.2)
	0.7	10.4 (±0.4)	30.1 (±1.5)	30.2 (±1.6)	28 (±1.6)
	0.9	21.5 (±1.2)	27 (±1.6)	32.5 (±1.5)	30.1 (±1.2)
G2	0.1	10.1 (±0.4)	16 (0.8)	19 (±0.8)	17.3 (0.8)
	0.3	15.1 (±0.8)	27.5 (±1.4)	27.7 (±1.8)	30 (±1.6)
	0.5	19 (±0.8)	30.1 (±1.6)	40 (±2.2)	39 (±2.1)
	0.7	25.5 (±1.6)	38.5 (±2.0)	37.7 (±2.1)	43 (±2.1)
	0.9	23 (±1.2)	39.5 (±2.1)	47 (2.3)	34 (1.6)
FC	0.1	5.5 (±0.4)	10 (±0.4)	10.4 (±0.4)	9.2 (±0.4)
	0.3	9.5 (±0.4)	5.7 (±0.4)	25.5 (±1.6)	25.6 (±1.6)
	0.5	10.9 (±0.4)	21.5 (±1.2)	30 (±1.6)	32 (±1.4)
	0.7	14.5 (±0.5)	25.5 (±1.6)	40 (±2.1)	43.5 (±2.1)
	0.9	14.2 (±0.5)	31 (1.7)	42.5 (2)	47.5 (2.0)
GC	0.1	2.5 (±0.4)	1.6 (±0.4)	6.5 (±0.4)	6.2 (±0.4)
	0.3	3 (± 0.4)	6 (±0.5)	13.6 (0.8)	13.2 (0.8)
	0.5	2.4 (±0.4)	6.6 (±0.5)	17.7 (±0.8)	19.5 (±0.8)
	0.7	4.9 (±0.4)	13.5 (0.8)	20 (±1.0)	22 (±1.2)
	0.9	4.1 (0.6)	16.9 (0.8)	22.5 (1.4)	26 (1.0)

The IAA producing ability of all four cultures at 72 h with 0.9% tryptophan concentration was assessed by One way ANOVA and found that at 0.05 level the means are significantly different ($F = 106.60588$, $p = 8.64269$)

Experiments were done in triplicates ± SD

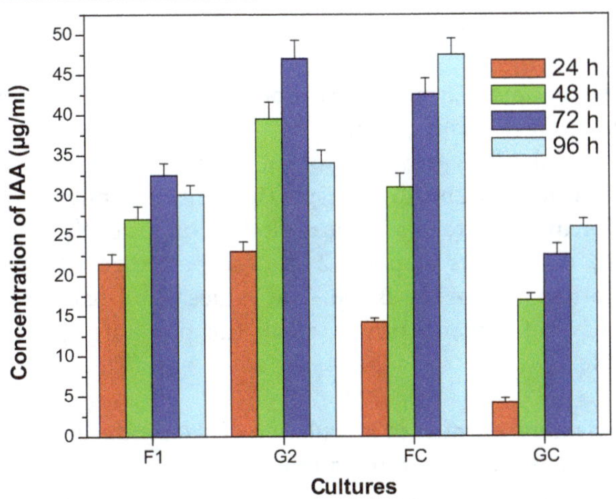

Fig. 1 Efficiency of cultures to produce IAA using 0.9% tryptophan

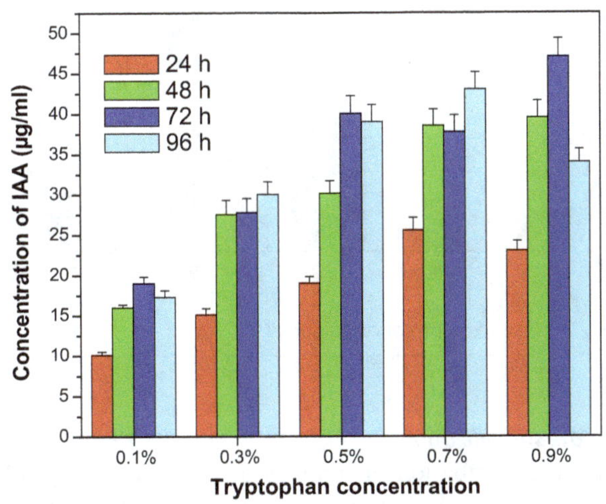

Fig. 2 Effect of tryptophan concentration on IAA production by G 2

Assessment of other plant growth promoting activities of the isolates

Hydrogen cyanide production

On assessment of all the samples (F1 and G2) and (FC and GC) by Bakkers and Schippers method for detection of HCN production it was found that among the isolates, strong brown color was observed in plate of F1 and no brown color was observed in G2 plate, whereas in consortiums, light brown color was observed in G2 plate. The difference of brown color formation of all the sample plates can be seen in figure given below.

Table 2 Effects of pH on IAA production

Culture samples	pH range	Concentration of IAA (in µg/ml)			
		24 h	48 h	72 h	96 h
F1	5	0.9 (±0.1)	3.6 (±0.4)	1.4 (±0.2)	1.1 (±0.1)
	7	7.3 (0.4)	10.1 (±0.5)	0.9 (±0.1)	8.1 (±0.4)
	9	15.2 (±0.8)	13.1 (±0.5)	12.6 (±0.5)	0.9 (±0.1)
	11	0.5 (±0.1)	5 (±0.4)	0.9 (±0.1)	1 (±0.1)
G2	5	0.6 (±0.1)	1.1 (±0.1)	1.1 (±0.1)	1.1 (±0.1)
	7	8.6 (±0.4)	12 (±0.5)	14.8 (±0.6)	14.6 (±0.6)
	9	10.5 (±0.5)	17.1 (±0.6)	19.5 (±0.8)	18.2 (±0.8)
	11	1.3 (±0.1)	5.3 (±0.4)	17.5 (±0.8)	14.8 (±0.6)
FC	5	0.7 (±0.1)	1.9 (±0.2)	6.6 (±0.4)	6.7 (±0.4)
	7	4.6 (±0.4)	7.9 (±0.4)	7.9 (±0.4)	6.8 (±0.4)
	9	9.6 (±0.4)	15.9 (±0.8)	13.5 (±0.8)	8.7 (±0.4)
	11	0.9 (±0.1)	8.2 (±0.4)	15.1 (±0.8)	14.5 (±0.6)
GC	5	1.1 (±0.1)	0.8 (±0.1)	1.5 (±0.1)	3.6 (±0.4)
	7	1 (±0.1)	3.4 (±0.4)	7 (±0.4)	6.7 (±0.4)
	9	6.7 (±0.4)	9.9 (±0.4)	9.4 (±0.4)	6.2 (±0.4)
	11	1.1 (±0.1)	6.5 (±0.4)	13.1 (±0.8)	12.3 (±0.8)

Experiments were done in triplicates ± SD

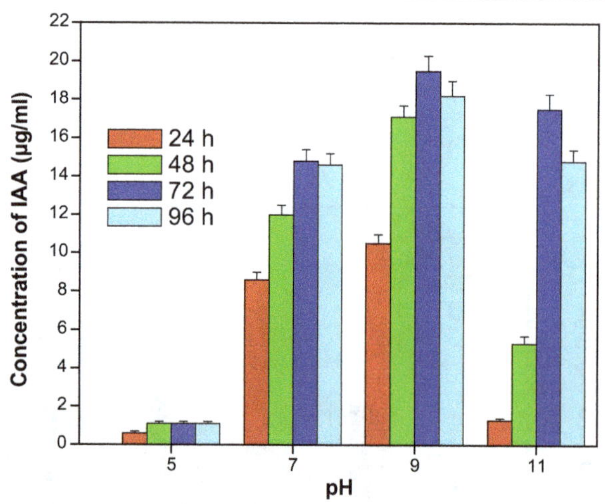

Fig. 3 Effect of pH on concentration of IAA produced by G2

Oxalate solubilization

After the incubation of samples on oxalic acid degradation selective medium for 7 days, no clear halos were observed around the bacterial spots inoculated on the plates. This indicated that none of the sample produced oxalate oxidase.

Phosphate solubilization

After the incubation of samples on Pikovskaya's agar for 2 days, zone of clearance was not observed around any of the bacterial spots inoculated on the plates. This indicated that none of the sample had phosphate solubilizing activity.

Zinc solubilization

The samples when incubated on Tris minimal media for 3 days, it was found that none of the samples were able to grow on the medium, which nullified the ability of zinc solubilization of the samples.

Biochemical characteristics of the isolates

On the basis of cultural, morphological and biochemical characteristics according to Bergey's Manual of Determinative Bacteriology (9th edition) it was found that F1 and G2 belongs to *Micrococcus* sp. and *Pseudomonas* sp. respectively.

Pot experiment 1: effect of IAA producer on plant growth

All the plantlets grown in pots of both the batches supplied with their respective cultures showed varied growth pattern after a period of one month. It was seen that plants of Batch 1 showed better growth than Batch 2. Among all the culture supplied pots, the pots supplied with isolates F1 and G2 showed better plant growth. The values of all parameters assessed for plant growth are given in Table 3.

Cultivation of azolla

On inoculation of 500 g azolla culture in pit a double yield was obtained, i.e., 1000 gm in 15 days. The dried powder of azolla was taken for pot experiment 2 and field studies.

Pot experiment 2: effect of co inoculation of IAA producer and azolla on plant growth

All the plantlets grown in pots of this batch supplied with azolla powder and their respective cultures showed varied growth pattern after a period of one month. It was seen that plants of pot G2 showed better growth than all other pots of the batch. The values of all parameters assessed for plant growth of all the pots, are given in Table 4.

Field study: effect of co inoculation of IAA producer and azolla on Seed Germination

After one month, the plants grown in all four plots, viz Plot 1, Plot 2, Plot 3, and Plot 4 showed varied growth on assessment by different parameters given in Table 5. It was found that, plants belong to Plot 1 and Plot 2 showed better plant growth.

Discussion

Soil is considered as the source of finding diverse kind of bacterial species, so the soil was considered as our priority for isolating IAA producers. Many researchers have

Table 3 Growth assessment of pot experiment 1

Growth assessment (±SD)	Batch 1					One way ANOVA	Batch 2					One way ANOVA
	Control	F1	G2	FC	GC		Control	F1	G2	FC	GC	
Root length (cm) (±0.01)	5.98	6.02	10.22	4.70	8.72	$F = 239.2$ $p = 7.03$ significantly different at 0.05 level	6.25	3.33	6.59	9.04	5.76	$F = 427.6$ $p = 3.95$ significantly different at 0.05 level
Shoot length (cm) (±0.12)	20.86	22.69	25.42	20.25	21.59	$F = 147.5$ $p = 7.60$ significantly different at 0.05 level	28.22	22.23	17.82	28.63	14.80	$F = 3124.2$ $p = 1.99$ significantly different at 0.05 level
Pri-leaves (cm) (±0.01)	4.33	4.33	4.33	4.66	4.33	$F = 0.2$ $p = 0.93$ NOT significantly different at 0.05 level	4.0	5.0	4.0	4.0	4.0	$F = 0.2$ $p = 0.93$ NOT significantly different at 0.05 level
Sec-leaves (cm) (±0.01)	4.66	4.66	4.66	4.66	4.66	$F = 0$ $p = 1$ NOT significantly different at 0.05 level	4.6	3.0	3.3	4.6	3.0	$F = 11.16$ $p = 0.00$ significantly different at 0.05 level
Chlorophyll—A (mg) (±0.12)	14.92	12.26	14.70	14.81	17.39	$F = 129.4$ $p = 1.439$ significantly different at 0.05 level	18.73	15.37	17.77	20.37	17.0	$F = 406.0$ $p = 5.12$ significantly different at 0.05 level
Chlorophyll—B (mg) (±0.12)	10.47	9.60	11.83	10.29	12.49	$F = 67.04$ $p = 3.492$ significantly different at 0.05 level	13.93	12.79	14.10	16.47	14.43	$F = 240.7$ $p = 6.818$ significantly different at 0.05 level

Experiments were done in triplicates ± SD

Table 4 Growth assessment of pot experiment 2

Growth assessment	Sample cultures + Azolla						One way ANOVA
	Control	Azolla	F1	G2	FC	GC	
Root length (cm) (±0.12)	3.23	6.74	5.25	4.66	6.38	5.56	$F = 153.33$ $p = 1.925$ significantly different at 0.05 level
Shoot length (cm) (±0.12)	28.51	31.02	31.5	36.32	27.18	36.48	$F = 2492.28$ $p = 0$ significantly different at 0.05 level
Pri-leaves (cm) (±0.01)	7.33	8.66	8.66	10.66	6.33	7.33	$F = 20.7$ $p = 1.60$ significantly different at 0.05 level
Sec-leaves (cm) (±0.01)	5.33	5.33	5	5.66	7.66	5.33	$F = 10.28$ $p = 5.17$ significantly different at 0.05 level
Chlorophyll—A(mg) (±0.12)	30.47	20.15	21.3	24.12	27.42	19.28	$F = 3837.6984$ $p = 0$ significantly different at 0.05 level
Chlorophyll—B(mg) (±0.12)	22.19	15.18	16.24	19.23	22.65	17.63	$F = 3730.4$ $p = 0$ significantly different at 0.05 level

Experiments were done in triplicates ± SD

Table 5 Field studies: effect of co inoculation of IAA producer and Azolla on seed germination

Plant growth assessment	Plot 1 (control)	Plot 2 (only azolla)	Plot 3 (only culture)	Plot 4 (culture + Azolla)
Root length (cm)	5.42	5.58	3.95	3.05
Shoot length (cm)	14	14	15	15
Primary leaves (cm)	15	7	7	9
Sec-leaves (cm)	10	10	10	9
Chlorophyll—A (mg)	18.78	10.27	12.18	19.04
Chlorophyll—B (mg)	12.35	7.31	8.69	13.04

used already isolated cultures or standard cultures for assessing the IAA production ability of it [24]. The traditional procedure of first isolating the rhizobial cultures and then screening them for IAA production was time consuming as, Sahasrabudhe [1] have isolated 60 cultures from rhizosphere of groundnut among them 40 cultures showed IAA production. Harikrishnan et al. [25] isolated 95 cultures of actinomycetes, out of 95, 65 isolates were found to be producing IAA. Taking in consideration the ability of soil samples to consist IAA producers, we have used soil samples for enrichment IAA producing micro-organisms. Rather going for isolating the rhizospheric bacteria and then assessing the IAA producing ability of it, we have used a different way. We have enriched the soil samples directly in tryptophan containing medium and observed for pink coloration through Salkowski's test. This has reduced the long traditional procedure of isolating each bacteria and subsequent assessment of each one for IAA production. By this method we have found that CS does not have any bacteria with a potent IAA producing ability, whereas GS and FS soil samples contained maximum IAA producers (or bacteria with potent IAA producing ability).

The isolates which were obtained from GS (G1,G2,G3,G4,G5) and FS (FS1,FS2,FS3,FS4,FS5) were separately assessed for their IAA producing ability. In addition to this a natural consortium of bacteria present in GS and FS was also maintained and named as GC and FC, respectively, and both these consortiums were also assessed for IAA producing ability. GC and FC were also used for further experiments along with F1 and G2. The reason behind using the isolates (F1 and G2) and consortiums (GC and FC) separately, was to use to different approaches, viz. "top down" and "bottom up" for getting the better IAA producer. In top down approach the natural consortium was used as it is for getting the better IAA producer and in bottom up approach each isolate was tested for getting better IAA producer. If the difference of IAA production between isolate and the consortiums would not have been statistically significant, we would have suggested to use the consortium rather than the isolates. This is because the natural consortium would be more stable in field conditions as compared to the isolates. But if we look at the results, the difference of IAA production between isolates and consortiums was statistically significant and it was found that G2 was better among them.

The cultures were assessed for different parameters such as, effect of tryptophan concentration and pH on yield of IAA production and it was found that at 0.9% concentration of tryptophan the isolates F1 and G2 showed maximum production of IAA, i.e., 32.5 µg/ml and 47 µg/ml on incubation at 72 h. The consortiums, FC and GC also performed well in IAA production at 0.9% concentration of tryptophan, i.e., 47.5 µg/ml and 26 µg/ml, but took 96 h.

On different pH range, fluctuations in IAA productivity of all samples were observed, among the isolates F1 showed higher IAA production at pH 9, i.e., 15.2 µg/ml in 24 h and G2 produced maximum, i.e., 19.5 µg/ml IAA at pH 9 in 72 h, whereas in case of consortiums, FC showed maximum IAA synthesis, i.e., 15.9 µg/ml at pH 9 and GC showed maximum IAA, i.e., 13.1 µg/ml. Thokal et al. [26] have also assessed the effect of pH values 4, 5, 6, 7 and 8 and they found pH 7 as optimum for their cultures. *Pantoea agglomerans* produced maximum IAA production at pH 7 [27].

On assessing all the samples for parameters such as, oxalate solubilization, zinc solubilization, HCN (hydrogen cyanide) production [20] and phosphate solubilization [22], it was found that in HCN production only the isolate F1 showed positive result, whereas all the samples were incapable of producing HCN. For rest of the parameters, none of the sample culture was found to be potent. Ajay et al. [28] worked on the isolating, screening and characterization of bacteria from rhizospheric soils for different plant growth promoting activities, IAA production, phosphate solubilization, HCN production, siderophore production, ACC deaminase activity and catalase.

Though the IAA production ability of cultures was tested invitro, it was necessary to assess the same in practical conditions; hence the pot experiments were performed. The cultures were supplied at the vicinity of roots. This step was carried by assuming that there will be the dominance of microbes, present in supplied culture, in the competing environment of rhizosphere for nutrition.

Considering this, studies were carried in pot experiment 1 and it was found that, among the two sets, Set 1 consisting of sowing culture treated seeds showed better plant growth than Set 2, which consists of sowing untreated seeds with regular culture supply. The assessment of different growth parameters of plants from both the sets showed that, G2 is the superior isolate for plant growth. Though the other samples outperformed G2 in some parameters, G2 was found better for all the plant growth promoting abilities when taken as average. By observing this scenario, we can say that the reason of poor growth of set 2 may be due to the higher production of IAA which has posed a negative effect on plant growth. So further experiments were carried out by sowing pretreated seeds.

Pot experiment 2 was carried out with the aid of azolla. As azolla decomposes slowly in soil, it may provide a consistent limited supply of proteins and nutrients for the growth of rhizospheric bacteria [13]. Azolla could be a consistent provider of tryptophan (a precursor of IAA) for IAA producing microbe in rhizosphere. By assessing all growth parameters in pot experiment 2 it was found that G2 as an efficient plant growth promoting microorganism.

As we know many cultures which show good results in laboratory conditions or in controlled pot experimental conditions fails substantially when applied to natural conditions. The reason is compatibility and adaptability of cultures in such natural field conditions. So, considering this, it was decided to use a combination of G2 and Azolla on fields with suitable positive and negative controls. Among the four plots allotted for this experiment, on an average plot 2 (only azolla) and plot 4 (culture + azolla) showed efficient results in terms of plant growth assessment.

Conclusions

From the overall study, we can say that the isolate G2 is a potent IAA producer and an efficient plant growth promoter. Combined dose of G2 and dried azolla can be used as a good biofertilizer.

Abbreviations

IAA: indole acetic acid; FC: farm soil consortium; GC: garden soil consortium; NB: nutrient broth; NA: nutrient agar; HCN: hydrogen cyanide; RT: room temperature.

Authors' contributions

RV and SI developed the pot experiment and most of the laboratory analysis, interpretation of the results and manuscript writing. NB and PK offered advice during the development of the experiment, interpretation of the results. NA provided general conception and interpretation of results and manuscript writing. All authors read and approved the final manuscript.

Author details

[1] Department of Biotechnology, Badrinarayan Barwale College, Jalna, MS, India. [2] Department of Microbiology, Badrinarayan Barwale College, Jalna, MS, India. [3] Department of Chemistry, Badrinarayan Barwale College, Jalna, MS, India. [4] Department of Biotechnology and Microbiology, Vivekanand Arts Sardar Dalipsingh Commerce and Science College, Aurangabad, MS, India. [5] Dr. Babasaheb Ambedkar Marathwada University, Aurangabad, MS, India.

Acknowledgements

We are thankful to Mr. Sajid Shaikh Assistant professor New Arts Commerce and Science College Ahmednagar (M.S.) India for useful technical discussions related to work.

Competing interests

The authors declare that they have no competing interests.

Funding

No funding from any funding agency was received for carrying out this research work.

References

1. Sahasrabudhe Madhuri M. Screening of rhizobia for indole acetic acid production. Ann Biol Res. 2011;2(4):460–8.
2. Ahmad F, Ahmad I, Khan MS. Screening of free-living rhizospheric bacteria for their multiple plant growth promoting activities. Microbiol Res. 2008;163(2):173–181. doi:10.1016/j.micres.2006.04.001.
3. Verma JP, Yadav J, Tiwari KN, Singh L, Singh V. Impact of plant growth promoting rhizobacteria on crop production. Int J Agri Res. 2010;5(11):954–83.
4. Elena S, Anton L, Birgitta B. Evidence for production of the phytohormone indole-3-acetic acid by cyanobacteria. Planta. 2002;215:229–38. doi:10.1007/s00425-002-0749-x.
5. Varalakshmi P, Malliga P. Evidence for production of Indole-3-acetic acid from a fresh water cyanobacteria (*Oscillatoria annae*) on the growth of *H. annus*. Int J Sci Res Publ. 2012;2(3):2250–3153.
6. Frankenberger WT Jr, Arshad M. Phytohormones in soils: microbial production and function. New York: Marcel Dekker, Inc; 1995. p. 5–40.
7. Arshad M. Frankenberger WT. Microbial production of plant hormones, Plant Soil. 1991;133:1. doi:10.1007/BF00011893.
8. Brown ME. Plant growth substances produced by microorganisms of soil and rhizosphere. J Appl Bacteriol. 1972;43:443–451.
9. Sarwar M, Frankenberger WT. Tryptophan dependent biosynthesis of auxins in soil. Plant Soil. 1994;160:97–104.
10. Barea JM, Brown ME. Effects on plant growth by *Azotobacter paspali* related to synthesis of plant growth regulating substances. J Appl Bacteriol. 1974;37:583–93.
11. Forni C, Riov J, GrillicCaiola M, Tel-Or E. Indole-3-acetic acid (IAA) production by *Arthrobacter* species isolated from Azolla. J Gen Microbiol. 1991;138:377–81.
12. Prasanna R, Jaiswal P, Nayak S, Sood A, Kaushik BD. Cyanobacterial diversity in the rhizosphere of rice and its ecological significance. Indian J Microbiol. 2008;49:89–97. doi:10.1007/s12088-009-0009-x.
13. Lumpkin TA, Plucknett DL. Azolla as a green manure: use and management in crop production. Westview Press, Boulder, Colorado; 1982.
14. Belnap J. 19 Factors influencing nitrogen fixation and nitrogen release in biological soil crusts. Ecol Stud, vol 150; 2001.
15. Wagner GM. Azolla: a review of its biology and utilisation. Bot Rev. 1997;63(1):1–26.
16. Kamble KD, Galerao DK. Indole acetic acid production from *Pseudomonas* species isolated from rhizosphere of garden plants in Amravati. IJAPBC. 2015;4(1):2277–4688.
17. Ahsan ML, Ali A, Ahmed I. Bio-fertilizer a highly potent alternative to chemical fertilizers: uses and future prospects. J Chem Eng Biol Sci. 2012;6(4):10–23.
18. Shahab Sadaf, Ahmed Nuzhat, Khan Nasreen S. Indole acetic acid production and enhanced plant growth promotion by indigenous PSBs. Afr J Agri Res. 2009;4(11):1312–6.
19. Khamna S, Yokota A, Lumyong S. Actinomycetes isolated from medicinal plant rhizosphere soils: diversity and screening of antifungal compounds, indole-3-acetic acid and siderophore production. World J Microbiol Biotechnol. 2008;25:649–55. doi:10.1007/s11274-008-9933-x.
20. Kumar P, Dubey RC, Maheshwari DK. Bacillus strains isolated from rhizosphere showed plant growth promoting and antagonistic activity against phytopathogens. Microbiol Res. 2012;167(8):493–9.
21. Dickman MB, Mitra A. *Arabidopsis thaliana* as a model for studying *Sclerotinia sclerotiorum* pathogenesis. Physiol Mol Plant Pathol. 1992;41:255–63.
22. Laslo E, Gyorgy E, Gyongyver M, Tamas E, Beata A, Szabolcs L. Screening of plant growth promoting rhizobacteria as potential microbial inoculants. Crop Prot. 2012;40:43–8.
23. Malboobi M, Owlia P, Behbahani M, Sarokhani E, Moradi S, Yakhchali B, Deljou A, Heravi KM. Solubilization of organic and inorganic phosphates by three highly efficient soil bacterial isolates. World J Microbiol Biotechnol. 2009;25:1471–7.
24. Sobral JK, Araújo WL, Mendes R, Geraldi IO, Pizzirani-Kleiner AA, Azevedo JL. Isolation and characterization of soybean-associated bacteria and their potential for plant growth promotion. Environ Microbiol. 2004;6(12):1244–51.
25. Harikrishnan H, Shanmugaiah V, Balasubramanian N. Optimization for production of Indole acetic acid (IAA) by plant growth promoting *Streptomyces* sp VSMGT1014 isolated from rice rhizosphere. Int J Curr Microbiol Appl Sci. 2014;3(8):158–71.
26. Thokal PJ, Shelar BL, Shaikh SH, Adhapure NN. Microbial optimized production of indole acetic acid and assessment of other plant growth promoting activities. Int J Sci Nat. 2013;4(4):627–32.
27. Apine OA, Jadhav JP. Optimization of medium for indole-3-acetic acid production using *Pantoea agglomerans* strain PVM. J Appl Microbiol. 2011;110(5):1235.
28. Ajay Kumar, Kumar Amit, Devi Shikha, Patil Sandip, Payal Chandani, Negi Sushila. Isolation, screening and characterization of bacteria from rhizospheric soils for different plant growth promotion (PGP) activities; an in vitro study. Recent Res Sci Technol. 2012;4(1):01–5.

Indigenous rhizobia population influences the effectiveness of *Rhizobium* inoculation and need of inorganic N for common bean (*Phaseolus vulgaris* L.) production in eastern Ethiopia

Anteneh Argaw[1*] and Angaw Tsigie[2]

Abstract

Background: Supplement with inorganic N application is essential to improve the common bean production in sub-Saharan Africa. However, the influence of indigenous rhizobial population on the inorganic N requirement with *Rhizobium* inoculation to secure sustainable way of common bean production system is not well known. The effect of different rates of N application either alone or in combination with *Rhizobium* inoculation on the nodulation, yield and yield traits of common bean cultivated in soils with different rhizobial population were conducted.

Methods: Twelve treatments were produced by factorially combined six levels of N fertilizer (0, 20, 40, 60, 80 and 100 kg N ha^{-1}) and two *Rhizobium* inoculation treatments (inoculated and uninoculated). The treatments were laid out in randomized completely block design and all treatments were replicated three times.

Result: Regardless of soil types, nodule number and nodule dry weight decreased with increasing rates of N application. 20 kg N ha^{-1} both alone and in combination with *Rhizobium* inoculation resulted in the largest nodulation in all soil types. The largest nodulation were induced in soil with large rhizobial population. *Rhizobium* inoculation significantly ($P < 0.05$) improved yield and yield traits of common bean. Moreover, our result revealed that the largest values of investigated traits were observed in inoculated treatment, as compared to the corresponding N rates of uninoculated treatments. The 20, 100 and 40 kg N ha^{-1} treatments resulted in significantly greater plant total tissue N at soil types with small, medium and large rhizobial population, respectively, as compared to unfertilized control. The highest total biomass yield (TBY) and grain yield (GY) at soil types with small and medium rhizobial population were obtained by the 100 kg N ha^{-1} treatment in combination with *Rhizobium* inoculation, while 20 and 40 kg N ha^{-1} applications produced the greatest TBY and GY, respectively, in soil with large rhizobial population.

Conclusion: These results indicate that N requirement is varied based on rhizobial population and effectiveness of native rhizobia in N_2 fixation.

Keywords: Common bean (*Phaseolus vulgaris* L.), Ethiopia, Indigenous rhizobia, *Rhizobium leguminosarum* bv. Phaseoli

*Correspondence: antenehargaw@gmail.com
[1] College of Agriculture and Environmental Sciences, School of Natural Resources Management and Environmental Sciences, Haramaya University, Dire Dawa, Ethiopia
Full list of author information is available at the end of the article

Background

Common bean (*Phaseolus vulgaris* L.) is one of most important food legume cultivated on greater than 4 million ha, providing the main source of dietary protein and carbohydrate for eastern and central African peoples, and containing about 20–25 % of the protein [1]. In Ethiopia, common bean is one of the major grain legumes, with its production centered in small farmers' fields where the soil fertility is depleted. Furthermore, the use of nitrogen (N) fertilizer is limited and average yields are low, usually less than 1 ton ha^{-1} [2]. In contrast, some studies indicated that up to 4600 kg ha^{-1} seed yield of common bean were obtained from research managed experimental plots in Ethiopia [3–5]. Low soil fertility, especially low N content, is the most important limiting nutrient status for common bean production in the tropics, including Ethiopia [6, 7].

Due to high price of N mineral fertilizers, their use by subsistent farmers in sub-Saharan Africa (SSA) to increase crop production has been limited. This condition has therefore necessitated an approach to crop production that emphasizes biological N_2 fixation (BNF). However, common bean is considered as a poor nitrogen fixing plant in comparison to other grain legumes due to its promiscuous nature, i.e., it forms symbiosis with many rhizobia species [8]. On top of this, several findings revealed that common bean generally responds poorly to *Rhizobium* inoculation under field conditions [9–11]. The reason for the failure to respond inoculation is believed to be due to a high and inefficient population of native common bean nodulating rhizobia in soil [9, 12, 13].

It is indicated that N_2 fixation in common bean can be increased through highly efficient *Rhizobium* inoculation [14]. Moreover, elite *Rhizobium* isolates improved the productivity of common bean, although in soil with high rhizobial nodulating population [15–17]. Asad et al. [15] found that the efficient *Rhizobium* inoculation did not fulfill the N needs of common bean. In most cases, common bean is able to fix up to 50 kg N ha^{-1} [18], which is less than 50 % of the plant N requirement [19]. Therefore, common bean requires mineral N application to achieve substantial yields under the current cropping system in SSA. However, N mineral recommended rates >40 kg N ha^{-1} suppress nodulation and N_2 fixation [10, 20, 21]. On the other hand, the application of a small amount of fertilizer N (<30 kg N ha^{-1}) enhanced nodulation, but grain yield improvement was not satisfactory [10, 22]. Higher stimulation of plant growth, N_2 fixation and grain yield of common bean have been recorded at low levels of N fertilizer applied with *Rhizobium* inoculation [17, 23–25]. Therefore, inoculation trials must emphasize not only the benefits of common bean inoculation, but also the combination of that practice with N fertilization, in order to achieve a decrease in mineral N input, whilst still obtaining maximum yields. However, the amount of N required in conjunction with elite *Rhizobium* inoculation to get maximum yield of common bean in soil with different rhizobial population is unknown. Hence, the objective of this study was to investigate the effect of naturalized common bean rhizobia population on the N rates of applied when alone or in combination with *Rhizobium* inoculation on nodulation, yield and yield traits of common bean in major growing areas of eastern Ethiopia.

Methods

Study areas

Field experiments were conducted on four locations of Eastern Ethiopia having different indigenous rhizobia nodulating common bean in 2012 cropping season. The experimental sites were located in the Hirna [N09°13.157″ and E041°06.488″ at an altitude of 1808 m above sea level (m.a.s.l.)], Fedis (N09°06.941″ and E042°04.835″ at an altitude of 1669 m.a.s.l.) Babillae (N09°13.234″ and E042°19.407″ 1669 m.a.s.l.) and Haramaya (N09°24.954″ and E042°02.037″ at an altitude of 2020 m.a.s.l.) agricultural research centers. The soils had not been inoculated before with rhizobia isolates nodulating common bean. Before sowing, soil samples were taken from 0–20 cm depth to determine baseline soil properties. Soil samples were air-dried, crushed, and passed through a 2-mm sieve prior to physical and chemical analysis. Details of physical and chemical characteristics of the soil of experimental sites are given in Table 1.

Sources of seeds and *Rhizobium* strain

A common bean var. Dursitu was supplied by Lowland Pulses Research Project, Haramaya University, Ethiopia. Variety was selected based on their yield, their maturity time, and its better performance in eastern Ethiopia. Strain of *Rhizobium leguminosarum* bv. Phaseoli (HU*PvR*-16) was obtained from Biofertilizer Research and Production Project, Haramaya University (Haramaya, Ethiopia). This strain was selected because it was previously found efficient while tested in this region on two improved varieties of common bean under laboratory and greenhouse conditions [26].

Inoculums preparation

Agar slope of HU*PvR*-16 strain was obtained from Soil Microbiology Research Laboratory, Haramaya University, Ethiopia. For purification, this isolate was preliminarily cultured in YEMA medium (10 g mannitol, 1 g yeast-extract, 1 g KH_2PO_4, 0.1 g NaCl, and 0.2 g $MgSO_4·7H_2O$ per liter, pH 6.8) and incubated at 28 °C for 5 days. The pure colony of the isolate was later transferred to YEM broth

Table 1 Soil analysis of experimental sites before sowing

Soil properties	Hirna soil	Babillae soil	Haramaya soil	Fedis soil
pH in H_2O	7.25	6.66	7.84	7.76
EC (mS/cm)	0.06	0.04	0.14	0.06
Organic carbon (%)	1.65	0.56	1.96	1.32
Total nitrogen (%)	0.16	0.06	0.12	0.12
Available P (mg kg^{-1})	27.11	2.22	9.94	1.78
Ca (cmol(+)kg^{-1})	39.88	4.18	31	23.12
Mg (cmol(+)kg^{-1})	9.00	3.5	8.7	12.87
Na (cmol(+)kg^{-1})	0.14	0.15	0.33	0.12
K (cmol(+)kg^{-1})	0.80	0.34	0.14	1.09
CEC (cmol(+)kg^{-1})	40.03	6.59	25.98	32.22
Zn (mg kg^{-1})	0.95	0.26	0.11	0.10
B (mg kg^{-1})	0.83	ND	0.15	0.75
NH_4-N (mg kg^{-1})	33.77	25.57	–	20.10
NO_3-N (mg kg^{-1})	33.74	27.98	–	27.75
Clay (g kg^{-1})	49	18	33	36
Silt (g kg^{-1})	39	6	18	45
Sand (g kg^{-1})	12	79	49	19
Textural class	Clay	Sandy loam	Sandy clay loam	Silty clay loam
Number of indigenous rhizobia of common bean g^{-1} soil	1.1×10^4	<10	2.8×10^3	2.5×10^2

medium and incubated at 28 °C for 5 days with gentle shaking at 120 rpm in shaker incubator. By this procedure, cell density in the culture was estimated by measuring optical density (540 nm) to determine whether the *Rhizobium* culture reached the middle or late logarithmic phase. *Rhizobium* inoculant was prepared by mixing 30 g of sterilized decomposed filter-mud with 15 ml of broth culture containing HU*PvR*-16 strain in polyethylene bags. After incubating the inoculated filter-mud for 2 weeks at 28 °C, the count of the *Rhizobium* was reached 1×10^9 g^{-1} of inoculant. Populations of rhizobia in the inoculants were determined by duplicate plate counts (Vincent, 1970).

Enumeration of indigenous rhizobia nodulating common bean

The initial indigenous rhizobia population was determined by the plant infection technique, using inoculation of serially diluted soil on germinated common bean seedling for nodulation assessment following the method of Brockwell et al. [27]. This experiment was conducted under controlled condition in growth chamber. The most probable number (MPN) was calculated from the most likely number, using the MPN tables of Vincent [28]. The rhizobial population that nodulated common bean in all study sites are indicated in Table 1.

Experimental design

Field trials on three soil types which had different rhizobial population nodulating common bean were conducted in order to investigate the effect of indigenous rhizobial population on the effect of N rates when applied alone or in combination with *Rhizobium* inoculation. The treatment effects were evaluated via determination of nodulation, yield and yield traits of common bean. The experimental design was a split plot in randomized complete block design (RCBD) with three replications. Main plot treatments consisted of six levels of inorganic N: 0, 20, 40, 60, 80 and 100 kg N ha^{-1}. Two *Rhizobium* inoculations (inoculated and uninoculated) were assigned as subplot treatments. Nitrogen fertilizer in each level was divided into two equal parts: (1) the first part of the N (20 kg N ha^{-1}) was applied along the furrow by hand and incorporated before planting time, and (2) the remaining parts were applied at flowering stages (R_3-stage).

The area was moldboard-plowed and disked before planting. The sizes of the main and subplot were 3×5 m^2 and 3×2 m^2, respectively. There were five rows per subplot and the spacing was 40 cm between rows, 10 cm between plants, 1 m between subplots and 1.5 between main plots. Disinfected seeds of common bean were sown after they were moistened with a 20 % solution of sucrose and then inoculated (7 g inoculant per kg seed) with *Rhizobium.* Inoculated seeds were hand planted on July 7, 2012. Phosphorus (P) was uniformly applied at planting at rate of 20 kg P ha^{-1} as triple superphosphate. Two seeds were sown per hill. After germination, the plants were thinned to one seedling per hill to obtain about 30 plants per row.

Weeds were controlled over the growth period with hand hoeing. At late flowering and early pod setting stage (R_3 stage), five plants from central rows were randomly chosen and harvested to record number of nodule plant^{-1} (NN), nodule dry weight plant^{-1} (NDW) and shoot dry weight plant^{-1} (SDW). Shoots of the plants were dried and later ground to pass a 0.5 cm sieve. Total N determinations were done by the Kjeldahl method of Bremner [29]. At physiological maturity stage on October 30, 2012, yield and yield traits of common bean were recorded. Number of pods plant^{-1} (NPP), number of seeds plant^{-1} (NSP), 100 seed weight, total biomass yield (TBY) and grain yield (at 13 % moisture content) (GY) were determined.

Data analysis

Data were submitted to analysis of variance using SAS version 9.1. Statistically significant differences between means were also determined by the LSD test. The bar graphs were constructed using Microsoft excel version 10.

Result and discussion

Nodulation

The soil samples from four experimental sites were aseptically collected to enumerate the rhizobia population that nodulated common bean by using the plant infection method. The result of this experiment revealed that the rhizobia of common bean varied from <10 to 10^5 g^{-1} of soil. Based on this rhizobial population, the experimental sites were grouped into three soil types. Accordingly, Babillae soil had <100 rhizobia of common bean g^{-1} soil. Rhizobia population >1000 g^{-1} of soil was found in Haramaya and Hirna soils. While the rhizobial population in Fedis soil was between 100 and 1000 g^{-1} soil. Therefore, Babillae, Fedis, and Haramaya and Hirna soils were categorized into low, medium and high rhizobia containing soil types, as it has been previously described by Howieson and Ballard [30]. None of common bean cultivating history in Babillae and Fedis sites could be attributed to lower rhizobial population nodulating common bean [31]. Continuous cultivation of the host plant could increase rhizobial population at Haramaya and Hirna soils [32].

Different rates of N application either alone or in combination with *Rhizobium leguminosarum* bv. Phaseoli inoculation significantly ($P < 0.05$) affected common bean nodulation in all study sites (Table 2). In all soil types, the NN and NDW were significantly decreased with increasing rates of N application either alone or in combination with inoculation. Other authors have found similar trends of nodulation along N rates with different

Table 2 Nodulation status and shoot dry weight of common bean var. Dursitu along different rates of N application with and without inoculation of *Rhizobium leguminosarum* bv. *Phaseoli* at selected areas of eastern Ethiopia

Treatments	NN			NDW			SDW		
	Soil type 1	Soil type 2	Soil type 3	Soil type 1	Soil type 2	Soil type 3	Soil type 1	Soil type 2	Soil type 3
Control	28.33d	71.67bcd	163.00abcd	0.1217cd	0.1920bc	0.4407b	36.23cd	31.37e	49.92b
20 kg N ha^{-1}	88.67c	146.67a	206.00a	0.1880b	0.4097a	0.3399bc	35.30cd	38.03cde	58.13ab
40 kg N ha^{-1}	65.00c	106.67b	149.67abcde	0.1263cd	0.3963a	0.2299bc	41.17bc	48.57bcd	60.77ab
60 kg N ha^{-1}	18.67d	45.00def	91.33cde	0.0437d	0.1727bc	0.1300c	49.97ab	53.23abc	63.60ab
80 kg N ha^{-1}	31.00d	45.00def	93.00cde	0.1033cd	0.1610bcd	0.1621c	47.53ab	55.70ab	63.67ab
100 kg N ha^{-1}	22.33d	20.00f	68.00e	0.0193d	0.0160e	0.2127bc	44.83abc	56.33ab	76.47a
Rhizobium sp.	82.33c	90.67bc	186.67ab	0.3063ab	0.2533b	0.7184a	30.77d	39.00cde	59.07ab
Rhizobium sp. + 20 kg N ha^{-1}	161.67a	64.33cde	178.67abc	0.3900a	0.2010bc	0.7111a	42.63bc	36.40de	70.50ab
Rhizobium sp. + 40 kg N ha^{-1}	124.33b	46.67def	111.50bcde	0.3207ab	0.0897cde	0.3337bc	52.93a	64.90a	78.03a
Rhizobium sp. + 60 kg N ha^{-1}	26.00d	27.33ef	71.50de	0.1530cd	0.0516de	0.1596c	47.47ab	52.83abc	75.68a
Rhizobium sp. + 80 kg N ha^{-1}	17.33d	20.00f	170.00abc	0.0287d	0.0340e	0.3238bc	50.77ab	51.70abcd	76.38a
Rhizobium sp. + 100 kg N ha^{-1}	19.67d	22.33f	63.17e	0.0210d	0.0279e	0.2124bc	43.93abc	51.97abcd	66.92ab
Mean	57.11	58.86	129.38	0.1518	0.1671	0.3312	43.63	48.35	66.59
F value	66.17***	29.26***	7.05***	21.55***	33.36***	12.78***	11.46***	9.80***	3.71**
LSD	30.06	37.15	93.59	0.1389	0.119	0.27	10.22	16.20	22.32
CV (%)	17.88	21.44	36.85	31.07	24.19	41.54	7.95	11.38	17.07

Means in the same column followed by the same letter are not significantly different at the 5 % probability level by Tukey's test

NS non-significant, *NN* nodule number, *NDW* nodule dry weight, *SDW* shoot dry weight

* Significant at 0.05; ** significant at 0.01; *** significant at 0.001

legume pulses [21, 33, 34]. In soil with low rhizobial population, *Rhizobium* inoculation alone produced significantly higher NN and NDW than the uninoculated treatments. In contrast, the data indicated the non-significant increases of NN and NDW due to inoculation in soil types with medium and high rhizobial population. Similarly, the effect of inoculation on nodulation of common bean varied due to different indigenous rhizobial population, as previously observed by Asad et al. [15]. In soil having low rhizobial population, *Rhizobium* inoculated with 20 and 40 kg N ha^{-1} resulted in significantly higher NN than the remaining N treatments. Moreover, *Rhizobium* inoculated with 20 kg N ha^{-1} produced the highest NDW. The overall effect of inoculation on NN and NDW in this soil type was higher than for the corresponding N rates of applied without inoculation (Fig. 1a, b), thus showing the competitive advantage of inoculated *Rhizobium* over the indigenous rhizobial population. Similarly, low rate of N with inoculation in soil with low indigenous rhizobial population resulted in an enhancement of nodulation [24]. In contrast, N application as low as 15 kg N ha^{-1} and applied at sowing, suppressed nodule dry weight of common bean [25]. Furthermore, N applied at planting had beneficial effect on nodulation at late growth stage [35].

The soil type with medium rhizobial population responded to the 20 kg N ha^{-1} with significantly higher NN and NDW than the other treatments, followed in the order by that obtained from 40 kg N ha^{-1}. On top of this, it was also observed a slight reduction of NN and NDW due to inoculation as compared to the uninoculated treatment (Fig. 1a, b). Similar result was previously observed by Msumali and Kipe-Nolt, [36], when indigenous rhizobia may have been more potent than inoculated *Rhizobium* strain. On the other hand, for soil having high rhizobial population, the 20 kg N ha^{-1} treatment alone gave the highest NN, although without significant difference from that obtained with either *Rhizobium* inoculation alone or 40 kg N ha^{-1} alone or inoculation applied with 20 kg N ha^{-1}. Moreover, the highest NDW were obtained from *Rhizobium* inoculation applied either alone or in combination with 20 kg N ha^{-1}. In general, it was observed a slight increment of NN and a remarkable improvement of NDW due to inoculation, as compared to the uninoculated treatments with corresponding rates of N (Fig. 1a, b). Similar to this finding, a significant improvement of nodulation due to inoculation in soil with >1000 rhizobial population was observed by Hungria et al. [17]. Inoculation together with a 20 kg N ha^{-1} treatment resulted in the highest NDW at soil with low and high rhizobial population. While the 20 kg N ha^{-1} treatment alone gave the highest NDW in soil with medium rhizobial population. The highest NN produced in soil with low, medium and high rhizobial population were 161.67, 146.67 and 206.00, respectively. These various NN in different soil types could be attributed to different number of rhizobia present in these three soil types. Similarly, Patrick and Lowther [37] found that the size of rhizobial population affects nodulation. Previous work confirmed that the largest the soil native rhizobial population, the greater is the nodulation [38]. The lowest NN and NDW produced in soil having relatively lower rhizobial population was previously observed by Chemining'wa and Vessey [21].

Shoot dry weight

The experimental treatments (different rates of N application solely and in combination with *Rhizobium* inoculation) resulted in significant variation of SDW (Table 2). In all soil types, a significant improvement of SDW was observed with increasing rates of N applied either alone or in combination with *Rhizobium* inoculation. In soil with low and medium rhizobial population, *Rhizobium* inoculation together with the 40 kg N ha^{-1} treatment, in which the highest nodulation was produced, resulted in significantly higher SDW than those produced at N rates lower than 40 kg N ha^{-1} in both inoculation treatments. This indicates the importance of higher nodulation for SDW production of common bean. In contrast, the high N rate application decreased nodulation, but an increase of above ground biomass production was observed [21]. In soil having high rhizobial population, the non-significant difference in SDW among N treatments, excluding the control, was observed. This implies that the sole N reserves in soil with lower N rate of application could have been sufficient for maximum shoot biomass production of common bean at R_3 stage. The highest SDW obtained from soils having low, medium and high rhizobial population were 52.93, 64.90 and 78.03 g, respectively. All these shoot biomass were obtained from *Rhizobium* inoculated with 40 kg N ha^{-1}. It has been shown that number of native rhizobia had a detrimental impact on productivity of above ground dry biomass [39]. Denton et al. [40] found increased shoot biomass production with increased *Rhizobium* inoculation rate. In all soil types, the lowest SDW were produced for the control treatments (uninoculated and unfertilized). This suggests that the N is the major limiting nutrient for common bean production in all study sites.

Number of pods per plant

The treatments of this experiment affected significantly ($P < 0.05$) the NPP in all soil types (Table 3). In soil having low rhizobial population, 20 kg N ha^{-1} alone gave the highest NPP. Inoculation with 60 kg N ha^{-1} resulted in significantly higher NPP than those produced

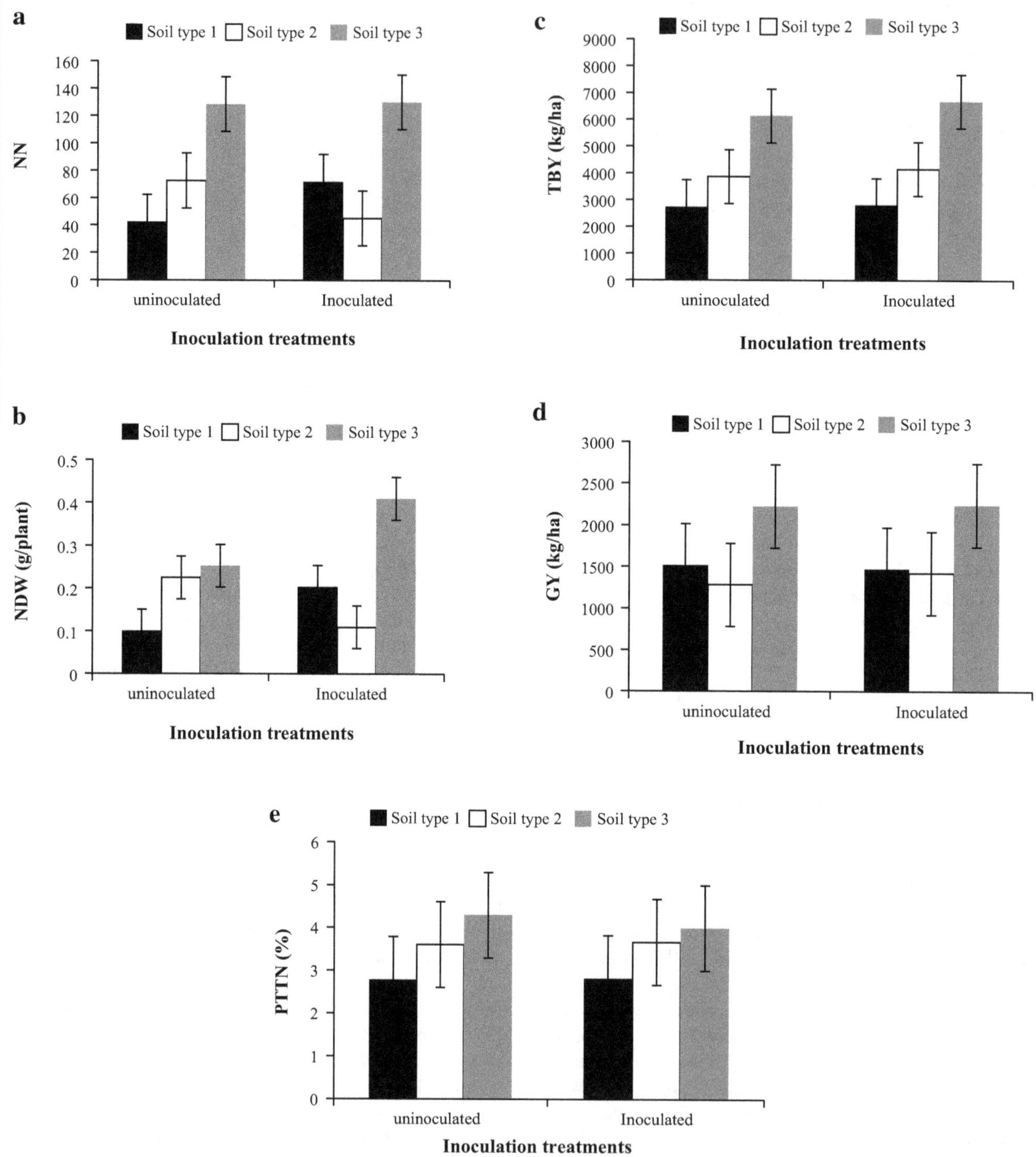

Fig. 1 The effect of indigenous rhizobial population on the effectiveness of inoculation on **a** nodule number (NN), **b** nodule dry weight (NDW), **c** grain yield (GY), **d** total biomass yield (TBY), **e** plant total tissue N (PTTN). Soil type 1—soil having <100 rhizobial population, soil type 2—soil having rhizobial population between 100 and 1000 and soil type 3—soil having rhizobial population >1000

for the control and inoculation alone, in soils having medium and high rhizobial population. Similarly, a significant improvement of NPP ranging from 20.2 at the control to 24.15 at N treated plants was obtained from faba bean [41]. The highest NPP produced in soils having low, medium and high were 18.63, 24.99 and 31.05,

Table 3 Number of pods per plant, number of seeds per pod and 100 seed weight of common bean var. Dursitu along different rates of N application with and without inoculation of *Rhizobium leguminosarum* bv. *Phaseoli* at selected areas of eastern Ethiopia

Treatments	NPP			NSP			100 seed weight		
	Soil type 1	Soil type 2	Soil type 3	Soil type 1	Soil type 2	Soil type 3	Soil type 1	Soil type 2	Soil type 3
Control	12.55bc	10.33d	20.89c	4.61b	5.40bc	5.82b	19.77c	20.90cd	19.33a
20 kg N ha^{-1}	18.63d	18.66ab	20.83c	6.78a	6.07abc	6.62ab	21.30ab	22.80abc	19.18a
40 kg N ha^{-1}	17.11ab	23.22ab	23.00bc	6.55a	6.30ab	6.70ab	21.37ab	22.07bc	18.92a
60 kg N ha^{-1}	16.00abc	18.00abc	25.22abc	6.44a	6.97a	6.92ab	21.17abc	22.13abc	19.42a
80 kg N ha^{-1}	14.89abc	22.00ab	25.83abc	6.99a	6.40ab	7.13a	21.20abc	22.90ab	19.42a
100 kg N ha^{-1}	16.44ab	22.22ab	25.78abc	6.66a	6.63a	6.70ab	21.57a	22.90ab	19.02a
Rhizobium sp.	10.78c	11.22 cd	23.66bc	5.33ab	5.50bc	6.50ab	20.00bc	19.90d	19.28a
Rhizobium sp. + 20 kg N ha^{-1}	13.66abc	16.44bcd	29.33ab	6.89a	5.20c	6.65ab	21.07abc	22.57abc	18.87a
Rhizobium sp. + 40 kg N ha^{-1}	16.00abc	22.00ab	29.11ab	6.66a	6.30ab	6.85ab	21.40ab	23.03ab	18.97a
Rhizobium sp. + 60 kg N ha^{-1}	16.22abc	24.99a	31.05a	6.22ab	6.30ab	6.35ab	22.17a	22.20abc	18.83a
Rhizobium sp. + 80 kg N ha^{-1}	16.44ab	22.11ab	28.89ab	6.99a	6.40ab	6.72ab	21.47ab	21.37bcd	18.87a
Rhizobium sp. + 100 kg N ha^{-1}	15.22abc	22.22ab	27.05abc	6.89a	6.00abc	6.38ab	20.97abc	24.07a	19.10a
Mean	15.33	19.45	25.89	6.42	6.12	6.61	21.12	22.24	19.10
F value	3.93**	11.06***	5.86***	4.68***	6.87***	2.11*	5.09***	8.41***	0.21 ns
LSD	5.48	7.20	6.66	1.72	1.02	1.10	1.49	1.94	2.30
CV (%)	12.13	12.58	13.10	9.11	5.64	8.48	2.38	2.96	6.12

Means in the same column followed by the same letter are not significantly different at the 5 % probability level by Tukey's test

NS non-significant, *NPP* number of pods per plant, *NSP* number of seeds per pod

* Significant at 0.05; ** significant at 0.01; *** significant at 0.001

respectively, confirming the positive effect of indigenous rhizobial population on NPP.

Number of seeds per pod

The data revealed significant variation of NSP due to the treatments (Table 3). In all soil types, NSP increased with increasing rates of N application solely and in combination with *Rhizobium* inoculation. Previous study reported that N nutrition increased the seeds per pod [42, 43]. In soil having low rhizobial population, significantly higher NSP (6.99) was recorded at 80 kg N ha^{-1} and inoculation in conjunction with 80 kg N ha^{-1} as compared to the control. In soils having medium rhizobial population, 60 kg N ha^{-1} resulted in significantly higher NSP than those produced at N rates below 20 kg N ha^{-1} in both inoculation treatments. 80 kg N ha^{-1} resulted in the highest NSP in soil having high rhizobial population. Non-significant differences in NSP obtained from different N rates, excluding control were observed for both inoculation treatments, in soils with low and high rhizobial population.

100 seed weight

The 100 seed weight of common bean was significantly ($P < 0.05$) varied due to the treatments, except those observed in soil having high rhizobial population, in which this trait exhibited a non-significant difference (Table 3). The non-significant difference in common bean seed size was previously observed by Mulas et al. [44]. Similar to this, 100 seed weight of soybean was not improved by either inoculation or N fertilizer application [45]. In soil having low rhizobial population, significantly higher 100 seed weight was obtained from 100 kg N ha^{-1} and *Rhizobium* inoculated with 60 kg N ha^{-1}, as compared to the control treatment. The sole 100 kg N ha^{-1} addition and in combination with *Rhizobium* inoculation gave significantly higher 100 seed weight compared to the control, in soil having medium rhizobial population. This result is in agreement with that obtained previously by El Hardi and Elsheikh [46] who found that inoculation and N application significantly improved 100 seed weight of chickpea over the control. N application increased 100 seed weight by 7.4 % over those produced for the uninoculated treatment [41].

Total biomass yield

The significant variation of TBY due to treatments was observed at $P \leq 0.05$ (Table 4). In all soil types, *Rhizobium* inoculated with 100 kg N ha^{-1} resulted in significantly higher TBY than those produced in the control treatment and inoculation alone. This result supports the findings of Mulas et al. [44] that inoculation and

Table 4 Total biomass yield, grain yield and total plant tissue N of common bean var. Dursitu along different rates of N application with and without inoculation of *Rhizobium leguminosarum* bv. *Phaseoli* at selected areas of eastern Ethiopia

Treatments	TBY			GY			PTTN		
	Soil type 1	Soil type 2	Soil type 3	Soil type 1	Soil type 2	Soil type 3	Soil type 1	Soil type 2	Soil type 3
Control	2055.6ef	2077.8c	4917.6c	1025.65f	1082.13cd	1946.05bc	2.3800de	2.5200c	4.0983ab
20 kg N ha^{-1}	2277.8def	4018.5ab	5609.5bc	1254.35e	1270.19bc	2121.48abc	3.2433a	3.5533b	4.5300ab
40 kg N ha^{-1}	2631.5cd	4333.3a	6351.4abc	1568.52c	1307.13bc	2262.27ab	2.8533abcd	3.7800ab	4.6533a
60 kg N ha^{-1}	3131.5abc	4055.6a	6481.5abc	1727.63bc	1289.44bc	2207.04abc	2.5567bcde	3.6933ab	4.0767ab
80 kg N ha^{-1}	3082.4bc	4388.9a	6721.8ab	1674.26bc	1280.46bc	2380.60a	2.5167cde	3.8600ab	4.2900ab
100 kg N ha^{-1}	3260.8ab	4374.1a	6742.8ab	1830.09b	1454.26ab	2416.44a	3.1567abc	4.2000a	4.1200ab
Rhizobium sp.	1787.0f	3133.3b	5472.6bc	996.39f	1005.37d	1941.67c	2.1600e	3.3833b	4.0517ab
Rhizobium sp. + 20 kg N ha^{-1}	2148.1def	4087.0a	7405.6a	1302.22de	1258.70bc	2169.17abc	3.1800ab	3.3633b	4.1867ab
Rhizobium sp. + 40 kg N ha^{-1}	2498.1de	4249.2a	6668.1ab	1211.48ef	1569.72a	2381.16a	2.6567abcde	3.6800ab	4.0800ab
Rhizobium sp. + 60 kg N ha^{-1}	3375.4ab	4277.8a	6851.6ab	1697.59bc	1461.85ab	2366.06a	2.7000abcde	3.8200ab	4.0033ab
Rhizobium sp. + 80 kg N ha^{-1}	3421.4ab	4462.2a	6486.1abc	1511.57cd	1571.57a	2240.34abc	3.2000ab	3.9200ab	3.9467ab
Rhizobium sp. + 100 kg N ha^{-1}	3648.1a	4740.7a	7222.2a	2089.54a	1653.89a	2329.40a	3.0133abcd	3.8767ab	3.7583b
Mean	2776.48	4016.54	6410.89	1490.78	1350.39	2230.14	2.8014	3.6375	4.1496
F value	32.12***	16.30***	4.95***	57.11***	19.75***	5.99***	8.10***	14.64***	1.73 ns
LSD	557.86	913.75	15,380.6	226.81	226.53	318.67	0.6467	0.5621	0.8943
CV (%)	6.82	7.73	12.56	5.17	5.70	7.28	7.84	5.25	10.98

Means in the same column followed by the same letter are not significantly different at the 5 % probability level by Tukey's test

NS non-significant, *TBY* total biomass yield, GY grain yield, *PTTN* plant total tissue N

* Significant at 0.05; ** significant at 0.01; *** significant at 0.001

inorganic N application enhanced the common bean production in all soil type regardless of the indigenous rhizobial population. Significant enhancement of above ground biomass production by 22 % due to *Rhizobium* inoculated with inorganic N over the uninoculated treatment was also observed by Chemining'wa and Vessey [21]. In soil having medium rhizobial population, it was observed a non-significant difference in TBY produced at 20 kg N ha^{-1} and beyond rates of N, with both inoculation treatments. This could have confirmed the presence of effective common bean–rhizobia symbiosis at low inorganic N [47] and thus satisfy the N need for boost the biomass production. The control treatment, inoculation alone, and 20 kg N ha^{-1} alone gave significantly lower TBY than those produced in other treatments in soil having high rhizobial population. In soil having low rhizobial population, statistically lower TBY was produced for control treatment, and for those with 20 and 40 kg N ha^{-1} either alone or in combination with *Rhizobium* inoculation, as compared to those produced in other treatments. This may indicate that the native rhizobia in this site are capable but not effective to fix N_2 to satisfy the N requirement of common bean. The presence of higher rhizobial population is not an indicator of the symbiotic effectiveness between rhizobia and N derived from the atmosphere [31]. In all soil types, the present study revealed that inoculation slightly increased the TBY as compared to those obtained from the corresponding N treatments without *Rhizobium* inoculation (Fig. 1c). The highest TBY produced in soils having low, medium and high rhizobial population were 3648.1, 4740.7 and 7222.2 kg ha^{-1}, respectively. The highest biomass in soil having high rhizobial population was previously confirmed by Furseth et al. [48]. These authors found that yield of soybean positively correlated with soil indigenous rhizobial population across environments.

Grain yield

The GY of common bean revealed significant variation due to treatments at $P \leq 0.05$ (Table 4). In soil having low rhizobial population, increasing rates of N application increased GY production, though the highest nodulation and PTTN were recorded at 20 kg N ha^{-1}. The 100 kg N ha^{-1} addition in conjunction with inoculation also gave significantly higher GY than those produced in other treatments. Similarly, inhibition effect of higher N application on nodulation and nitrogenase enzyme without affecting grain yield was previously determined by Rai [23]. Our result may be confirm that the previous findings, although N mineral reduced the nodulation and N_2 fixation and yields of common bean can be improved by increasing N availability [49]. da Silveira et al. [47] also found that common bean responded well up to 200 kg N ha^{-1}. Asad et al. [15] observed that nodulation improvement due to inoculation had not significantly enhanced plant biomass production. This present study also confirms that the common bean–rhizobia symbiosis was not satisfactory to fulfill the N needs as it has been previously observed by Fesonko et al. [50]. Contrary to this, Denton et al. [40] showed that improvement of shoot N increased the grain yield of faba bean by 1 Mg ha^{-1} in soil having low rhizobial population. This indicates that biological N_2 fixation alone is not a sufficient N requirement of common bean to get the local attainable yield in the prevailing environmental condition.

Inorganic N at 100 kg N ha^{-1} with *Rhizobium* inoculation gave the highest GY at soil having medium rhizobial population as it has been observed in PTTN. This may indicate that N derived from indigenous and inoculated rhizobia alone did not satisfy the N requirement of common bean. Voisin et al. [51] observed that high N requirement at the seed setting stage was supported by both N_2 fixation and mineral N supply. Although N fertilizer enhanced the plant N uptake, the common bean production did not improve [25]. The present study indicates a non-significant difference in GY obtained from inoculation when either coupled to 40, 60, 80 and 100 kg N ha^{-1} treatment, or sole application of 100 kg N ha^{-1} addition. This may indicate the inoculated *Rhizobium* satisfies the N requirement of common bean with relatively low inorganic N application.

In soil having high rhizobial population, the 40 kg N ha^{-1} applied in conjunction with *Rhizobium* resulted in significantly higher GY than those produced at the control and inoculation alone. Similarly, a previous finding recommended N application to maximize common bean yield in soil having high indigenous rhizobial population [31]. In this soil type, it was recorded a non-significant difference in GY at different N rates of application, excluding control. Besides this, treatment with low N rate applied together with both inoculation treatments, produced statistically similar GY with those found at higher N rates application. The synergetic effect between low rates of N application and *Rhizobium* inoculation on common bean production in soil having rhizobial population >1000 was previously confirmed by Hungria et al. [17]. These authors found the highest seed yield from treatments with *Rhizobium* inoculated together with 15 kg N ha^{-1} applied at planting and further 15 kg N ha^{-1} addition at early flowering stages. Due to the fact that biological N_2 fixation is not active at early stage of common bean, a starter dose of N application is required to enhance plant growth and eventually improve the grain yield production [35, 52, 53]. Nitrogen application in addition to starter N reduced nodulation and failed to increase the common bean yield [24]. Vargas et al. [31] found that increasing rates of N decreased the number of nodules from inoculated cells, whereas it increased that by indigenous rhizobia.

The overall effect of inoculation on GY was slightly decreased in soil having low rhizobial population but slightly improved in soil having medium and high rhizobial population, as compared to that obtained at corresponding rates of N application without inoculation (Fig. 1d). This result supports the finding of Hungria et al. [54] who showed that inoculation increased the yield of common bean as compared to the control treatment. A previous study also reported that an inoculation response was observed in soil having rhizobial population between 300 and 1000 g^{-1} soil [55, 56]. The highest GY produced for soil types having low, medium and high rhizobial population were 2089.54, 1653.89 and 2381.16 kg ha^{-1}, respectively. These GY are comparable with those previously produced at 100 kg N ha^{-1} N, for in which 2000 kg ha^{-1} was reported [47]. At least 29 % faba bean yield advantage was provided for soil having higher rhizobial population over that yield obtained from low rhizobial population, as determined by Sorwle and Mytton [41].

Plant total tissue N

The PTTN was significantly varied due to different treatments at $P \leq 0.05$ (Table 4). In soil having low and high rhizobial population, N application beyond 20 kg N ha^{-1} with both inoculation treatments resulted in a slight decrease of PTTN. 20 kg N ha^{-1} application resulted in the highest PTTN in soil having low rhizobial population. The lower PTTN at higher N application could be attributed to the negative effect of inorganic N on fixed N through the inhibition of the nitrogenase activity [20, 57].

In soil having medium rhizobial population, progressively larger rates of N application increased the PTTN, though the highest NN and NDW was produced at 40 kg N ha^{-1}. The 100 kg N ha^{-1} treatment gave significantly higher PTTN than those obtained from either the control treatment or that with 20 kg N ha^{-1} applied alone or in combination with *Rhizobium* inoculation. This could be due to low effectiveness of indigenous rhizobia of common bean in N_2 fixation. Asad et al. [15] found the significant improvement of plant total N accumulation of common bean was due to N application rather than to inoculation. This was attributed to the fact that common bean rarely derives more than 50 % of its N from symbiotic N_2 fixation [58].

In soil type having high rhizobial population, 40 kg N ha^{-1} alone resulted in significantly higher PTTN than that obtained from inoculation with the 100 kg N ha^{-1} addition. Shutsrirung et al. [59] demonstrated that inoculation may not be effective in soil having high effective native rhizobial population. In top of this, PTTN was also decreased with increasing rates of N application following both inoculation treatments. This indicates that the native rhizobia in this site might have been more effective in N_2 fixation than the inoculated stain. Furthermore, inoculation slightly increased the PTTN in soil having low and medium rhizobial population whereas it showed a decreasing trend in soil with high rhizobial population, in comparison to PTTN found for corresponding N rates without inoculation (Fig. 1e). The highest PTTN was observed at low, medium and high rhizobial population and reached 3.24, 4.20 and 4.65 %, respectively. The highest PTTN could be attributed to a higher nodulation induced by the high rhizobial population and, thus, an enhanced N_2 fixation [38, 48]. Due to synergetic effect of a starter N application on symbiotic N_2 fixation, the highest accumulated N in plants was recorded for the 20 kg N ha^{-1} application in soil having high rhizobial population [24].

Correlation analysis

The significant correlation among investigated traits of common bean was observed in all soil types at $P < 0.05$ (Tables 5, 6, 7). The amount of NN produced at soil with low rhizobial nodulating population inversely correlated with TBY ($r = -0.6754$; $P \leq 0.05$) and with GY ($r = -0.5629$; $P \leq 0.05$) (Table 5). In this soil type, a negative and strong correlation between NDW with TBY ($r = -0.7563$; $P \leq 0.01$), GY ($r = -0.6953$; $P \leq 0.05$) was observed. NN had an inverse and strong correlation with both SDW ($r = -0.5958$; $P \leq 0.05$) and GY ($r = -0.6090$; $P \leq 0.05$) in soil with medium rhizobial population (Table 6). Conversely, NDW and GY revealed an inverse and significant correlation ($r = -0.6356$; $P \leq 0.05$). In soil with high rhizobial population, the correlation analyses between the NN and both TBY ($r = 0.5599$; $P \leq 0.05$) and GY ($r = 0.7538$; $P \leq 0.01$) were positive (Table 7), whereas NDW had an inverse and significant correlation with GY ($r = -0.6917$; $P \leq 0.05$). In general, this finding indicates that N mineral fertilizer rather than symbiotic N_2 fixation had a determinant effect on common bean yield. This result supports the results of the other authors [60], in which NDW and N fixation showed negative correlation with yield related parameters. La Favre and Eaglesham [61] also observed an inverse and significant relationship between the starting N application and both NN and NDW. In contrast, Aggarwal [62], who conducted experiment in Malawi, found positive and significant correlations between NN and GY for common bean under inorganic N treatment. Mothapo et al. [38] also reported that plant biomass production correlated positively with NDW when inoculation alone was applied.

In all soil types, also SDW of common bean had a significant correlation with NPP, NSP, 100 seed weight, TBY and GY. However, NPP and NSP had the non-significant association with SDW with low and high rhizobial population (Tables 5, 6, 7). In particular, in soil with low rhizobial population, all listed growth traits had positive and significant correlation with SDW at $P \leq 0.05$ (Table 5). SDW had also a positive and significant ($P \leq 0.05$) correlation with 100 seed weight ($r = 0.5087$) and TBY ($r = 0.6996$), and even highly significant ($P \leq 0.01$) with NPP ($r = 0.7719$), NSP ($r = 0.7751$) and GY ($r = 0.7270$). In the same soil, a positive and strong correlation was also observed between SDW and either NPP ($r = 0.8726$; $P \leq 0.001$), or TBY ($r = 7378$; $P \leq 0.01$) or GY ($r = 0.7496$; $P \leq 0.01$). Highly significant relationships among SDW, shoot N content and seed yield are previously shown by Pereira and Bliss [63].

Table 5 Correlation among the investigated traits of common bean treated different rates of N with and without inoculation in soil having rhizobial population nodulating common bean <100 g^{-1} soil

Traits	NN	NDW	SDW	NPP	NSP	100 SW	TBY	GY	HI
NN									
NDW	0.93***								
SDW	−0.19 ns	−0.28 ns							
NPP	−0.15 ns	−0.39 ns	0.44 ns						
NSP	0.11 ns	−0.14 ns	0.59*	0.66*					
100 SW	−0.08 ns	−0.21 ns	0.67*	0.78**	0.74**				
TBY	−0.68*	−0.76**	0.69*	0.50*	0.56*	0.65*			
GY	−0.56*	−0.70*	0.52*	0.46 ns	0.59*	0.58*	0.91***		
HI	0.13 ns	−0.04 ns	0.08 ns	−0.05 ns	0.30 ns	0.08 ns	0.09 ns	−0.00 ns	
PTTN	0.13 ns	−0.18 ns	0.23 ns	0.66*	0.72*	0.56*	0.36 ns	0.38 ns	0.52*

Ns non-significant, *NN-Nodule* number per plant, *NDW* nodule dry weight per plant (g $plant^{-1}$), *SDW* shoot dry weight (g $plant^{-1}$), *NPP* number of pods per plant, *NSP* number of seeds per pod, *100 SW* 100 seed weight (g), *TBY* total biomass yield (kg ha^{-1}), *GY* grain yield (kg ha^{-1}), *HI* harvest index

* Significant at 0.05; ** highly significant at 0.01; *** very highly significant at 0.001

Table 6 Correlation among the investigated traits of common bean treated different rates of N with and without inoculation in soil having rhizobial population nodulating common bean between 100 and 1000 g^{-1} soil

Traits	NN	NDW	SDW	NPP	NSP	SW	TBY	GY	HI
NN									
NDW	0.96***								
SDW	−0.60*	−0.55*							
NPP	−0.40 ns	−0.34 ns	0.77**						
NSP	−0.35 ns	−0.26 ns	0.78**	0.66*					
SW	−0.26 ns	−0.26 ns	0.51*	0.66*	0.34 ns				
TBY	−0.36 ns	−0.29 ns	0.70*	0.88***	0.57*	0.71**			
GY	−0.61*	−0.64*	0.73**	0.81**	0.49 ns	0.69*	0.77**		
HI	−0.25 ns	−0.29 ns	0.19 ns	0.26 ns	0.16 ns	0.46 ns	0.17 ns	0.50*	
PTTN	−0.42 ns	−0.36 ns	0.77**	0.84**	0.72**	0.52*	0.91***	0.69*	0.03 ns

ns non-significant, *NN-Nodule* number per plant, *NDW* nodule dry weight per plant (g $plant^{-1}$), *SDW* shoot dry weight (g $plant^{-1}$), *NPP* number of pods per plant, *NSP* number of seeds per pod, *100 SW* 100 seed weight (g), *TBY* total biomass yield (kg ha^{-1}), GY grain yield (kg ha^{-1}), *HI* harvest index

* Significant at 0.05; ** highly significant at 0.01; *** very highly significant at 0.001

Table 7 Correlation among the investigated traits of common bean treated different rates of N with and without inoculation in soil having rhizobial population nodulating common bean >1000 g^{-1} soil

Traits	NN	NDW	SDW	NPP	NSP	SW	TBY	GY	HI
NN									
NDW	0.72**								
SDW	0.45 ns	−0.23 ns							
NPP	0.45 ns	−0.10 ns	0.87***						
NSP	0.18 ns	−0.29 ns	0.41 ns	0.27 ns					
SW	0.00 ns	−0.07 ns	−0.68*	−0.62*	0.05 ns				
TBY	0.56*	−0.27 ns	0.74**	0.79**	0.49 ns	−0.50 ns			
GY	0.75**	−0.69*	0.75**	0.61*	0.55*	−0.39 ns	0.77**		
HI	0.33 ns	0.34 ns	0.08 ns	−0.15 ns	−0.28 ns	0.06 ns	−0.49 ns	−0.25 ns	
PTTN	0.44 ns	−0.02 ns	0.38 ns	−0.53*	0.27 ns	0.03 ns	−0.26 ns	−0.09 ns	0.11 ns

Ns non-significant, *NN-Nodule* number per plant, *NDW* nodule dry weight per plant (g $plant^{-1}$), *SDW* shoot dry weight (g $plant^{-1}$), *NPP* number of pods per plant, *NSP* number of seeds per pod, *100 SW* 100 seed weight (g), *TBY* total biomass yield (kg ha^{-1}), *GY* grain yield (kg ha^{-1}), *HI* harvest index

* Significant at 0.05; ** highly significant at 0.01; *** very highly significant at 0.001

In sites with soil having a low rhizobial population, GY of common bean correlated positively with SDW ($r = 0.5152$; $P \leq 0.05$), NSP ($r = 0.5907$; $P \leq 0.05$), 100 seed weight ($r = 0.5791$; $P \leq 0.05$) and TBY ($r = 0.9121$; $P \leq 0.001$) (Table 5). Similarly, strong and positive correlation was observed between GY with SDW ($r = 0.7271$; $P \leq 0.01$), NPP ($r = 0.8098$; $P \leq 0.01$), 100 seed weight ($r = 0.6939$; $P \leq 0.05$) and TBY ($r = 0.7713$; $P \leq 0.01$) in experimental soil with medium rhizobial population (Table 6). In soil showing a high rhizobial population, we noted a positive correlation between GY and SDW ($r = 0.7496$; $P \leq 0.01$), NPP ($r = 0.6067$; $P \leq 0.050$), NSP ($r = 0.5474$; $P \leq 0.05$) and TBY ($r = 0.7712$; $P \leq 0.01$) (Table 7). Similarly, Bayuelo-Jiménez et al. [64] had previously indicated a positive and significant correlation among yield and yield components of common bean.

The correlation analysis indicated a positive relationship between PTTN and NPP ($r = 0.6644$; $P \leq 0.05$), NSP ($r = 0.7195$; $P \leq 0.05$) and 100 seed weight ($r = 0.5562$; $P < 0.05$) in experimental sites showing soil with low rhizobial population (Table 5). In those with medium rhizobial population, a positive and strong correlation between PTTN and SDW ($r = 0.7665$; $P \leq 0.01$), NPP ($r = 8386$; $P \leq 0.01$), NSP ($r = 0.7154$; $P \leq 0.01$), 100 seed weight ($r = 0.5206$; $P \leq 0.05$), TBY ($r = 0.9114$; $P \leq 0.001$) and GY ($r = 0.6941$; $P \leq 0.05$) (Table 6). These results agree with the previous studies, which reported the significant correlation between plant N accumulation with seed yield, seed weight and total biomass [45, 50]. Ruiz-Díez et al. [45] also demonstrated that plant N accumulation was the most suitable trait for the selection of highly effective and highly competitive *Rhizobium* isolates. In contrast, in soil with high rhizobial population, PTTN was only significantly but inversely correlated with NPP ($r = -0.5296$; $P \leq 0.05$), thus, indicating the negative impact of mineral N treatment on soil productivity, despite the plant N accumulation was noted (Table 7). The present study also indicated a non-significant relationship between PTTN and both NN and NDW in all soil types. Similar result was previously observed by da Silva et al. [24] who reported that the correlation between fixed N and nodulation was decreased with increasing rates of N application. In contrast to this, Tsia et al. [25] found a positive and significant relationship between both NN and NDW and N in shoots at 45 days after emergency.

Conclusion

The results of this study indicate the significant effect of N treatments on the nodulation, yield and yield traits of common bean in the major growing areas of eastern Ethiopia. Our experiments indicated the need of inoculation of *Rhizobium* beside that of a starting N dose in soil having <100 rhizobial population g^{-1} of soil. An N requirement to reach the highest yield of common bean was not affected by the native rhizobial population. The effect of *Rhizobium* inoculation on nodulation of common bean appears to depend on the type of soil. Our data also indicated that the amount of mineral N required for maximum seed yield is dependent on the original soil rhizobial population. Moreover, the indigenous rhizobial population that nodulates common bean also affects the bean production. Further research on effectiveness of the combination between indigenous rhizobia and inoculated rhizobial population would be recommended.

Authors' contributions

Both of us participated equally starting from the development of the research idea, writing proposal and competing research grant and development of this manuscript. But I participated more in the management and collection of data from the field experiment, which is why I am the first author of this manuscript. Both authors read and approved the final manuscript.

Author details

[1] College of Agriculture and Environmental Sciences, School of Natural Resources Management and Environmental Sciences, Haramaya University, Dire Dawa, Ethiopia. [2] Ethiopian Institute of Agricultural Research, Holleta Agricultural Research Center, Holleta, Ethiopia.

Acknowledgements

This work was carried out with the support of Ethiopia Institute of Agricultural Research under National Biofertilizer Development Project hosted at Holleta Agricultural Research Center, Soil microbiology research group. The author also thanks Berhanu Mengistu, Dejene Ayenew and Girmaye Mekonnen for managing the field experiments.

Competing interests

The authors declare that they have no competing interests.

References

1. Broughton WJ, Hernandez G, Blair M, Beebe S, Gepts P, Vanderleyden J. Beans (*Phaseolus* spp.)—model food legumes. Plant Soil. 2003;252:55–128.
2. Abebe G, Mazengia W, Nikus O. Soil fertility and crop management research of lowland food legumes in the rift valley. In Proceedings of the workshop on food and forage legumes. Edited by Ali K, Ahmed S, Beniwal S, Kenneni G, Malhotra RS, Makkouk K, Halila MH. International Center for Agriculture Research in the Dry Areas (ICARDA), Aleppo, Syria, 2006. pp. 157–166.
3. IAR. Melkassa Agricultural Research Center Progress Report for the period 1995/96. Melkassa: Institute of Agricultural Research (IAR); 1997.
4. IAR. Melkassa Agrictural Research Center Progress Report for the period 1997/98. Melkassa: Institute of Agricultural Research (IAR); 1998.
5. Assefa T, Assefa H, Kimani P: Development of improved Haricot bean germplasm for the Mid-and Low-altitude sub-humid agro-ecologies of Ethiopia. In Proceedings of the Workshop on food and forage legumes. Edited by Ali K, Ahmed S, Beniwal S, Kenneni G, Malhotra RS, Makkouk K, Halila MH. International Center for Agriculture Research in the Dry Areas (ICARDA), Aleppo, Syria; 2006. pp. 87–94.
6. Jansa J, Bationo A, Frossard E, Rao IM. Options for improving plant nutrition to increase common bean productivity in Africa. In: Bationo A, editor. Fighting poverty in Sub-Saharan Africa: the multiple roles of legumes in integrated soil fertility management. Dordrecht: Springer Science + Business Media B.V.; 2011. p. 201–40.
7. Lunze L, Abang MM, Buruchara R, Ugen MA, Nabahungu NL, Rachier GO, Ngongo M, Rao I. Integrated Soil Fertility Management in Bean Based Cropping Systems of Eastern, Central and Southern Africa. In: Soil fertility improvement and integrated nutrient management—a global perspective. Edited by Whalen J. InTech; 2012. pp. 239–272.
8. Hardarson G, Bliss FA, Cigales-Riveri MR, Henson RA, Kipe-Nolt JA, Longeri L, Manrique A, Pena-Cabriales JJ, Pereira PAA, Sanabria CA, Tsai SM. Genotypic variation in biological nitrogen fixation by common bean. Plant Soil. 1993;152:59–70.
9. Graham PH, Apolitano C, Ferrera R, Halliday J, Lepiz R, Menéndez D, Ríos R, Saito SMT, Viteri S. The International Bean Inoculation Trail (IBIT). Results for the 1978–1979 Trial. In: Graham PH, Harris S, editors. Biological Nitrogen Fixation Technology for Tropical Agriculture. CIAT, Cali; 1981.
10. Graham PH. Some problems of nodulation and symbiotic fixation in *Phaseolus vulgaris* L.: a review. Field Crops Res. 1981;4:93–112.
11. Buttery BR, Park SJ, Findlay WI. Growth and yield of white bean (*Phaseolus vulgaris* L.) in response to nitrogen, phosphorus and potassium fertilizer and to inoculation with *Rhizobium. Can.* J Plant Sci. 1987;67:425–32.
12. Rodriguez-Navarro DN, Buendia AM, Camacho M, Lukas MM, Santamatria C. Characterization of *Rhizobium* spp. bean isolates from South West Spain. Soil Biol Biochem. 2000;32:1601–13.
13. Brockwell J, Bottomley PJ. Recent advances in inoculant technology and prospects for the future. Soil Biol Biochem. 1995;27:683–97.
14. Giller KE, Cadisch G. Future benefits from biological nitrogen fixation: an ecological approach to agriculture. Plant Soil. 1995;174:255–77.
15. Asad S, Malik KA, Hafeez FY. Competition between inoculated and indigenous *Rhizobium/Bradyrhizobium* spp. strains for nodulation of grain and fodder legumes in Pakistan. Biol Fertil Soils. 1991;12:107–11.
16. Mostasso L, Mostasso FL, Vargas MAT, Hungria M. Selection of bean (*Phaseolus vulgaris*) rhizobial strains for the Brazilian Cerrados. Field Crops Res. 2002;73:121–32.
17. Hungria M, Campo RJ, Mendes IC. Benefits of inoculation of the common bean (*Phaseolus vulgaris*) crop with efficient and competitive *Rhizobium tropici* strains. Biol Fertil Soils. 2003;39:88–93.
18. Bliss FA. Utilizing the potential for increased nitrogen fixation in common bean. Plant Soil. 1993;152:157–60.
19. van Kessel C, Hartely C. Agricultural management of grain legumes: has it led to an increase in nitrogen fixation. Field Crops Res. 2000;65:165–81.
20. Saxena AK, Rathi SK, Tilak KVBR. Selection and evaluation of nitrate-tolerant strains of *Rhizobium leguminosarum* biovar *viceae* specific to the lentil. Biol Fertil Soils. 1996;22:126–30.
21. Chemingwa GN, Vessey JK. The abundance and efficacy of *Rhizobium leguminosarum* bv'. *viciae* in cultivated soils of the eastern Canadian prairie. Soil Biol Biochem. 2006;38:294–302.

22. Santalla M, Amurrio JM, Rodiño AP, de Ron AM. Variation in traits affecting nodulation of common bean under intercropping with maize and sole cropping. Euphytica. 2001;122:243–55.
23. Rai R. Effect of nitrogen levels and Rhizobium strains on symbiotic N_2 fixation and grain yield of *Phaseolus vulgaris* L. genotypes in normal and saline-sodic soils. Biol Fertil Soils. 1992;14:293–9.
24. da Silva PM, Tsai SM, Bonetti R. Response to inoculation and N fertilization for increased yield and biological nitrogen fixation of common bean (*Phaseolus vulgaris* L.). Plant Soil. 1993;152:123–30.
25. Tsai SM, DA Silva PM, Cabezas WL, Bonetti R. Variability in nitrogen fixation of common bean (*Phaseolus vulgaris* L) intercropped with maize. Plant Soil. 1993;152:93–101.
26. Argaw A: Symbiotic and phenotypic characterization of rhizobia nodulating common bean (Phaseolus vulgaris L.) from Eastern Ethiopia. *MSC thesis*, Addis Ababa University, Addis Ababa, Ethiopia; 2007.
27. Brockwell J. Accuracy of a plant-infection technique for counting populations of *Rhizobium trifolii*. Appl Microbiol. 1963;2:377–83.
28. Vincent JM: *A manual for the practical study of root-nodule bacteria*. (IBP handbook no 15) Blackwell, Oxford; 1970.
29. Bremner JM: Inorganic forms of nitrogen. In: *Methods of soil analysis*. Edited by Black C.A. et al. Part 2, 2nd ed., agron. Monogr. 9. ASA and SSSA, Madison, WI; 1965.
30. Howieson JG, Ballard RA. Optimising the symbiosis in stressful and competitive environments in southern Australia—some contemporary thoughts. Soil Biol Biochem. 2004;36:1261–73.
31. Vargas MAT, Mendes IC, Hungria M. Response of field-grown bean (*Phaseolus vulgaris* L.) to Rhizobium inoculation and nitrogen fertilization in two Cerrados soils. Biol Fertil Soils. 2000;32:228–33.
32. Vlassak K, Vanderleyden J, Franco AA. Competition and persistence of *Rhizobium etli* in tropical soil during successive bean (*Phaseolus vulgaris* L.) cultures. Biol Fertil Soils. 1996;21:61–8.
33. Leidi EO, Rodriguez Navarro DN. Nitrogen and phosphorus availability limit N fixation in bean. New Phytol. 2000;147:337–46.
34. Li YY, Yu CB, Cheng X, Li CJ, Sun HJ, Zhang FS, Lambers H, Li L. Intercropping alleviates the inhibitory effect of N fertilization on nodulation and symbiotic N fixation of faba bean. Plant Soil. 2009;323:295–308.
35. Eaglesham AR, Hassouna S, Seegers R. Nitrogen fixation by cowpea and soybean. Agron J. 1983;75:61–6.
36. Msumali GP, Kipe-Nolt JA. The usefulness of bean nodulation data relative to other symbiotic parameters in judging success of *Rhizobium* inoculation. Biol Agric Hortic. 2002;19:355–64.
37. Patrick HN, Lowther WL. Influence of the number of rhizobia on the nodulation and establishment of *Trifolium ambiguum*. Soil Biol Biochem. 1995;27:717–20.
38. Mothapo NV, Grossman JM, Sooksa-nguan T, Maul J, Bräuer SL, Shi W. Cropping history affects nodulation and symbiotic efficiency of distinct hairy vetch (*Vicia villosa* Roth.) genotypes with resident soil rhizobia. Biol Fertil Soils. 2013;49:871–9.
39. Ballard RA, Charman N, McInnes A, Davidson JA. Size, symbiotic effectiveness and genetic diversity of field pea rhizobia (*Rhizobium leguminosarum* bv. *viciae*) populations in South Australian soils. Soil Biol Biochem. 2004;36:1347–55.
40. Denton MD, Pearce DJ, Peoples MB. Nitrogen contributions from faba bean (*Vicia faba* L.) reliant on soil rhizobia or inoculation. Plant Soil. 2013;365:363–74.
41. Sorwli FK, Mytton LR. Nitrogen limitations to field bean productivity: a comparison of combined nitrogen applications with *Rhizobium* inoculation. Plant Soil. 1986;94:267–75.
42. Deshwal VK, Dubey RC, Maheshwari DK. Isolation of plant growth-promoting Bradyrhizobium (Arachis) sp. with biocontrol potential against Macrophomina phaseolina causing charcoal rot of peanut. Curr Sci. 2003;84:443–8.
43. Mazen MM, El-Batanony NH, Abd El-Monium MM, Massoud ON. Cultural filtrate of *Rhizobium* spp. and arbuscular mycorrhiza are potential biological control agents against root rot fungal diseases of faba bean. Glob J Biotechnol Biochem. 2008;3:32–41.
44. Mulas D, García-Fraile P, Carro L, Ramírez-Bahena MH, Velázquez E, González-Andrés F. Distribution and efficiency of *Rhizobium leguminosarum* strains nodulating *Phaseolus vulgaris* in Northern Spanish soils: selection of native strains that replace conventional N fertilization. Soil Biol Biochem. 2011;43:2283–93.
45. Ruiz-Díez B, Fajardo S, Fernández-Pascual M. Selection of rhizobia from agronomic legumes grown in semiarid soils to be employed as bioinoculants. Agron J. 2012;104:550–9.
46. El Hardi EA, Elsheikh EAE. Effect of *Rhizobium* inoculation and nitrogen fertilization on yield and protein content of six chickpea (*Cicer arietinum* L.) cultivars in marginal soils under irrigation. Nutr Cycl Agroecosys. 1999;54:57–63.
47. da Silveira PM, Braz AJBP, Kliemann HJ, et al. Nitrogen fertilization of common bean grown under no-tillage system after several cover crops. Pesqui Agropecu Bras. 2005;40:377–81.
48. Furseth BJ, Conley SP, Ané JM. Soybean response to soil rhizobia and seed-applied rhizobia inoculants in Wisconsin. Crop Sci. 2012;52:339–44.
49. Herridge DF. Contributions of fixed nitrogen and soil nitrate to the nitrogen economy of irrigated soybean. Soil Biol Biochem. 1988;20:711–7.
50. Fesenko AN, Provorov NA, Orlova IE, Orlov VP, Simarov BV. Selection of *Rhizobium leguminosarum* bv. *viceae* strains for inoculation of *Pisum sativum* L. cultivars: analysis of symbiotic efficiency and nodulation competitiveness. Plant Soil. 1995;172:189–98.
51. Voisin A, Salon C, Munier-Jolain NG, Ney B. Quantitative effects of soil nitrate, growth potential and phenology on symbiotic nitrogen fixation of pea (*Pisum sativum* L.). Plant Soil. 2002;243:31–42.
52. Sanginga N, Thottappilly G, Dashiell K. Effectiveness of rhizobia nodulating recent promiscuous soyabean selections in the moist savanna of Nigeria. Soil Biol Biochem. 2000;32:127–33.
53. Gan YT, Warkentin TD, McDonald CL, Zentner RP, Vandenberg A. Seed yield and yield stability of chickpea in response to cropping systems and soil fertility in Northern Latitudes. Agron J. 2009;101:1113–22.
54. Hungria M, Andrade DS, Chueire LMO, Probanza A, GuttierrezMaero FJ, Megias M. Isolation and characterization of new efficient and competitive bean (*Phaseolus vulgaris* L.) rhizobia from Brazil. Soil Biol Biochem. 2000;32:1515–28.
55. Herridge DF. Inoculation technology for legumes. In: Dilworth MJ, James EK, Sprent JI, Newton WE (eds). Leguminous nitrogen-fixing symbioses. Dordrecht: Kluwer; 2008.
56. Thies JE, Singleton PW, Bohlool B. Influence of the size of indigenous rhizobial populations on establishment and symbiotic performance of introduced rhizobia on field grown legumes. Appl Environ Microb. 1991;57:19–28.
57. Sanginga N, Wirkom LE, Okogun A, Akobundu IO, Carsky RJ, Tian G. Nodulation and estimation of symbiotic nitrogen fixation by herbaceous and shrub legumes in Guinea savanna in Nigeria. Biol Fertil Soils. 1996;23:441–8.
58. Wortmann CS. Nutrient dynamics in a climbing bean and sorghum crop rotation in the Central Africa highlands. Nutr Cycl Agroecosys. 2001;61:267–72.
59. Shutsrirung A, Sutigoolabud P, Santasup C, Senoo K, Tajima S, Hisamatsu M, Bhromsiri A. Symbiotic efficiency and compatibility of native rhizobia in northern Thailand with different soybean cultivars. Soil Sci Plant Nutr. 2002;48:491–9.
60. Nleya T, Walley FL, Vandenberg A. Response of determinate and indeterminate common bean genotypes to *rhizobium* inoculant in a short season rainfed production system in the canadian prairie. J Plant Nutr. 2009;32:44–57.
61. LA Favre AK, Eaglesham ARJ. The effects of a high level of N, applied at planting, on nodulation of soybean (*Glycine max* (L.) Merr.) by diverse strains of *Bradyrhizobium*. Plant Soil. 1987;102:267–70.
62. Aggarwal VD. An overview of SADC/CIAT bean activities in Malawi. Bean Res. 1994;8:15–31.
63. Pereira PAA, Bliss FA. Selection of common bean (*Phaseolus vulgaris* L.) for N_2 fixation at different levels of available phosphorus under field and environmentally-controlled conditions. Plant Soil. 1989;115:75–82.
64. Bayuelo-Jiménez JS, Peña-Valdivia CB, Rogelio Aguirre RJ. Yield components of samples of two wild Mexican common bean (*Phaseolus vulgaris* L.) populations grown under cultivation. S Afr J Plant Soil. 1999;16:197–203.

22

Hygienic characteristics of radishes grown in soil contaminated with *Stenotrophomonas maltophilia*

Alessandro Miceli, Alessandra Martorana, Giancarlo Moschetti and Luca Settanni*

Abstract

Background: *Stenotrophomonas maltophilia* is a plant growth-promoter. This bacterium is also implicated in human diseases. Thus, after the use of this bacterium in agriculture, the safety of the final products has to be verified. Due to the ubiquitous presence of *S. maltophilia* in soil, in this study a massive contamination was simulated to evaluate the growth and safety of *Raphanus sativus* L..

Results: Different inoculums and soil treatment conditions were tested. Soils were analysed weekly and the radishes at harvest for their microbial loads and presence/persistence of *S. maltophilia* LMG 6606. The concentration of the bacterium added in the different trials decreased during the first week, but increased thereafter and determined a significant increase of growth parameters of radishes.

Conclusions: The addition of *S. maltophilia* LMG 6606 to non-autoclaved soil enhanced the productivity of radishes. The bacterium did not internalize in the hypocotyls, but colonized the external surface ensuring the safety of the products. Thus, a sanitizing bath of hypocotyls before consumption is necessary.

Keywords: Hygienic safety, Microbial internalization, Plant growth, *Raphanus sativus* L., *Stenotrophomonas maltophilia*

Background

Soil fertility is a complex concept that involves many interacting parameters. Cultivated plants may suffer nutritional stresses when the amount or availability of soil nutrients is lower than that required for sustaining metabolic processes in each growth stage [1]. Thus, restoring of nutrients and enhancing their availability by improving soil characteristics and efficiency of plants, are the main objectives of the modern agriculture. Due to the increasing sensitivity to environmental and economic issues, researchers and consumers are more and more aware of the impact of agriculture on the environment. Lowering the use of chemical inputs and the search for alternative ways to improve a more sustainable agriculture is a current challenge [2].

Sustainable agricultural production systems may be obtained by enhancing the uptake efficiency of nutrient of plants, that might be achieved through a better comprehension of the role of plant–microbe–soil interaction and the association of soil microorganisms with roots in the rhizosphere [3]. Within soil microbiota, plant growth-promoting (PGP) microorganisms constitute a heterogeneous group of bacteria and fungi that have gained particular importance for their stimulating effects. They are able to promote directly plant metabolism by nutritional and/or hormonal ways [4–9] as well as indirectly through the production of antimicrobial compounds, the reduction of iron available to phytopathogens, the synthesis of fungal cell wall-lysing enzymes, the competition with detrimental microorganisms for colonization sites on the roots, and the induced systemic resistance [5, 10].

The most studied PGP rhizobacteria belong to Gram-negative genera [11]. Among these, *Stenotrophomonas maltophilia* is an important species and represents one of the four dominant bacterial species in the rhizosphere of European cereal cultivations [12]. *S. maltophilia* is a Gammaproteobacterium distributed worldwide [13] and

*Correspondence: luca.settanni@unipa.it
Department of Agricultural and Forest Science, University of Palermo, Viale delle Scienze 4, 90128 Palermo, Italy

typically found in soil where it plays a defining role in the nitrogen and sulphur cycles [14–16] and it often dominates the microbial communities detected outside and inside the plants [13]. However, *S. maltophilia* is a known human pathogen [17]. It may be responsible for respiratory tract infections, bacteremia, biliary sepsis, infections of the bones and joints, urinary tract and soft tissues, endophthalmitis, eye infections, endocarditis and meningitis [18].

In the last years, the change in lifestyle and the consequent need of consuming ready-to-eat foods determined the increase of fresh-cut vegetable use. Since these products do not undergo any treatment before consumption, their hygienic safety is of paramount importance. Among fresh vegetables, the request of radish is on the increase [19] due to its positive effect on the consumer's health [20–22].

Radish (*Raphanus sativus* L.) is an important vegetable of the *Brassicaceae* family, grown and consumed all over the world, due to its wide adaptation, high yield, and high nutritional content. It is normally consumed in salads and comes in a variety of forms and skin colors. Most known varieties are round and red-skinned. The most popular part for eating is the napiform hypocotyl, although the entire plant is edible and the tops can be used as a leaf vegetable. Radishes are eaten raw or cooked or processed by pickling, canning or drying [23]. Radish is not only a vegetable crop but also an important source of medicinal compounds. In fact, it is used by people with different gastrointestinal, biliary, hepatic, urinary and respiratory disorders, and in cardiovascular diseases such as hypertension [24]. An important quality characteristic of radish is the total antioxidant activity due to molecules, such as ascorbic acid and phenol, mainly phenolic acids, with free radical scavenging activity [25, 26]. The protective role of these molecules in the prevention of human degenerative diseases has been widely demonstrated by many studies [27–29]. Moreover, as other Crucifers, radish contains many other compounds as glucosinolates that are associated with cancer protection [30].

The interaction between *S. maltophilia* and *R. sativus* has been recently approached in a floating cultivation system [31]. However, their reciprocal effects in soil and the safety aspects related to the safety of the radishes have not been evaluated yet. In the present study high levels of *S. maltophilia* were added to soil, simulating a massive environmental contamination, to: investigate the survival of *S. maltophilia* in soil during the whole crop cycle of *R. sativus*; evaluate the effects of this bacterium on plants; monitor its transfer to the radishes; and determine their viability in the radishes ready for consumption.

Methods

Microbial strain, plant seeds and experimental plan

Stenotrophomonas maltophilia LMG 6606, a strain originating from rhyzosphere as reported in the strain details provided by the Belgian Co-ordinated Collection of Micro-organisms (BCCM/LMG), was propagated in Nutrient Broth (NB) (Oxoid, Milan, Italy) at 28 °C for 24 h. Seeds of radish (*R. sativus* L.) cultivar Saxa three were purchased from Blumen (Piacenza, Italy).

The experimental plan included eight different conditions for the growth of radish plants: ASS, autoclaved soil inoculated with *S. maltophilia* LMG 6606; AS, autoclaved soil added with Ringer's solution (Sigma-Aldrich, Milan, Italy); nASS, non-autoclaved soil inoculated with *S. maltophilia* LMG 6606; nAS, non-autoclaved soil added with Ringer's solution; ASSwS, autoclaved soil weekly inoculated with *S. maltophilia* LMG 6606; ASwR, autoclaved soil weekly added with Ringer's solution; nASSwS, non-autoclaved soil weekly inoculated with *S. maltophilia* LMG 6606; nASwR, non-autoclaved soil weekly added with Ringer's solution. Four replicate pots were produced for each trial. Two independent experiments were performed in two consecutive weeks.

Inoculation of soil and plant development

The pots used in this work were 13.5 cm × 13.5 cm × 16 cm and were filled with 2.5 L of commercial soil SER CA-V7 (Vigorplant Italia srl, Piacenza, Italy). This soil is a mixture of slightly or fully decomposed raised bog peat (pH 6.0) fertilized with 800 g m^{-3} of a mineral fertilizer (NPK 12-11-18). The pots were previously treated with a NaClO solution (5 % *v*/*v*) for 24 h. A part of the bulk soil, placed in autoclave bags, was autoclaved twice (in two consecutive days) for 70 min at 120 °C [32]. The fresh inoculums of *S. maltophilia* LMG 6606 was prepared after overnight development as reported by Settanni et al. [33]. The cell suspension was added to the autoclaved and non-autoclaved soil in the ratio 1:10 (*v*/*v*), vigorously mixed with sterile spoons to obtain the homogenous distribution of the bacterial inocula and transferred into the pots (trials ASS, nASS, ASSwS and nASSwS). The trials AS, nAS, ASwR and nASwR were prepared with Ringer's solution. Seeds of radish were sown in five dibblings for each pot. Pots were watered from below (sub-irrigation) with sterile water, in order to avoid further microbial contamination to the soil and kept at 25 °C in a climatic chamber till seed germination. From the third day, the trials ASSwS and nASSwS were weekly added with 300 mL of *S. maltophilia* LMG 6606 cell suspension (concentration at about 10^8 CFU mL^{-1}), while the trials ASwR and nASwR were added with the same volume of Ringer's solution.

After seed emergence, only five plants per pot were left to grow (274 plants m^{-2}). The pots were transferred in an unheated plastic greenhouse and received the same volume of water that varied daily according to environmental conditions and plant needs until harvesting.

Microbiological analyses

The sampling for microbiological analysis included soil during plant growth, aseptically collected as described by Settanni et al. [33] at T_0 and at 7-day intervals, and hypocotyls at harvest, aseptically collected as reported by Settanni et al. [31]: four hypocotyls were collected from each replicate of all trials, two hypocotyls for the direct microbial count and two for the internal *S. maltophilia* LMG 6606 detection.

Soil samples (10 g) were diluted (1:10) with sodium pyrophosphate (0.16 % *w*/*v*) solution in sterile flasks under agitation (10 min at 150 rpm). Radishes (approximately 10 g) were first subjected to the removal of the soil adhering to the surface as described by Brandl et al. [34] and then homogenised in Ringer's solution by a stomacher (BagMixer® 400, Interscience, Saint Nom, France) at the maximum speed for 2 min. The decimal serial dilutions of both soil and radishes continued in Ringer's solution. Total mesophilic count (TMC) were determined on Plate Count Agar (PCA) (Oxoid), incubated aerobically at 30 °C for 72 h; presumptive *S. maltophilia* were enumerated on vancomycin–imipenem–amphotericin B (VIA) agar [35], incubated aerobically at 30 °C for 48 h. Plate counts were performed in duplicate.

Data from bacterial counts were averaged and converted to log CFU g^{-1} dry weight (dw) for soil samples and to log CFU g^{-1} for radishes. Moisture of soil (5 g) was obtained after drying (24 h at 105 ± 1 °C) in an oven and weighting the residual.

Recognition of *S. maltophilia* LMG 6606 and evaluation of the internalization

Approximately ten colonies from the two highest dilutions of sample suspensions at each analysis were picked up from VIA agar plates based on their morphology (colour, edge, surface and elevation) and cultured in NB overnight at 30 °C. The cultures were sub-cultivated onto NA and stored in glycerol (20 %, *v*/*v*) stocks at −80 °C. The isolates and *S. maltophilia* LMG 6606 were analysed by randomly amplified polymorphic DNA-PCR (RAPD-PCR) as described by Settanni et al. [31].

The internal presence of *S. maltophilia* LMG 6606 in radishes was investigated as reported by Settanni et al. [31]. Briefly, radishes of each trials were collected, superficially sterilized and then transferred in stomacher bags, added with VIA broth (final ratio 1:10) and homogenized as reported above.

Analyses of plants

Plants were harvested 4 weeks after sowing and washed accurately with tap water. After air drying, leaves, hypocotyls and roots were separated and the number of leaves and their area, the root elongation and the radish diameter were recorded on four plants for each replicate of each trial. Leaf area of each plant was calculated by digital image analysis. Leaves were scanned (Epson Perfection 4180 Photo, Seiko Epson Corp. Japan) with 350 dpi of resolution and the images were saved in TIFF format. The images were analysed with the ImageJ 1.46r software (National Institutes Health, Bethesda, MD, USA).

Fresh and dry biomass of the different plant parts were calculated by weighting before and after they were oven dried to a constant weight at 80 °C.

Color of radishes was measured on two points of four hypocotyls from each replicates of all trials, using a colorimeter (Chroma Meter CR-400C, Minolta, Osaka, Japan). The Hunter scale parameters were determined: *L* (lightness, ranging from 0 to 100, from black to white), *a* (positive values indicating redness and negative values, greenness) and *b* (positive values indicate yellowness and negative values, blueness).

The firmness of radishes was determined using a digital penetrometer (mod. 53205, TR Snc. Italy) equipped with a flat 6 mm diameter stainless steel cylinder probe. Four hypocotyls were punched and the mean peak force was calculated in Newton.

Radishes (50 g) were homogenized in 50 mL of distilled water and homogenate centrifuged (5000 rpm, 15 min); the supernatant was taken for analysis of soluble solids content (SSC), ascorbic acid and $N–NO_3^-$. SSC was measured using a digital refractometer (MTD-045nD, Three-In-One Enterprises Co. Ltd. Taiwan). Ascorbic acid and nitrate content were measured by the Reflectoquant test strips and a RQflex hand-held reflectometer (Merck, Darmstadt, Germany) [procedures described in Art. 1.16971.0001 and 1.16981.0001 by Merck (http://www.merckmillipore.com/chemicals)].

Water use efficiency (WUE) was calculated as the ratio between total dry weight of plant and total amount of water supplied.

Statistical analyses

The study was carried out in a completely randomized design. To determine the effects of microorganisms and time on soil microbial load, a two-way ANOVA was carried out. A one way ANOVA was performed for other data. When a significant F value was detected, Tukey–Kramer's multiple range test was used to determine differences among microbial populations and plant parameters of the different trials (significance level $P < 0.05$).

Principal components analysis was employed to investigate any underlying relationship among the different trials based on the agronomic and quality parameters of radish plants at harvest. The input matrix for the analysis consisted of leaf number, leaf fresh weight, leaf dry matter, leaf area, root length, root fresh weight, root dry matter, radish fresh weight, radish dry matter, radish diameter, firmness, L^*, a^*, b^*, SSC, nitrate, ascorbic acid, plant WUE. For the selection of the optimum number of principal components (PCs), factors with eigenvalues greater than 1.0 were retained. In addition the plot of the PCs enabled the investigation of correlations between the variables of the input data set. To this end, the initial variables were projected into the subspace defined by the reduced number of PCs (first and second components) and correlated variables were identified. In the current approach, the Principal components analysis was implemented with SPSS version 14.0 (SPSS Inc. Chicago, IL, USA).

Results

Microbiological analyses

The microbiological counts of soil during the growth of radishes are reported in Table 1. The interaction microorganisms x time resulted significant for both PCA and VIA media ($P < 0.001$). The autoclaved soil un-inoculated showed a TMC of 3.53 Log CFU gdw^{-1} and, surprisingly, the bacterial load estimated on VIA was at almost 2 Log CFU gdw^{-1}, whereas the non-autoclaved un-inoculated soil showed levels of ca. 7 Log CFU gdw^{-1} on both PCA and VIA media. Soon after inoculation, the concentration of TMC and *S. maltophilia* LMG 6606 or stenotrophomonads were above 10^8 CFU gdw^{-1}. In general, TMC was at higher levels than the microbial developments detected on VIA, but the trials nASS, nASSwS, nAS and nASwR showed an opposite behaviour at 7 day. During radish growth, the highest microbial concentrations were observed for the trial AS and ASwR, whose TMC, at the second week, were 9.58 and 9.57 Log CFU gdw^{-1}, respectively. Lower concentrations were estimated for the trials nASS, nASSwS, nAS and nASwR, for which TMC was below 8 Log CFU gdw^{-1} for the entire period of observation. The concentrations on VIA evaluated for the trials weekly added with *S. maltophilia* LMG 6606 (ASSwS and nASSwS) showed trends almost comparable to those displayed by the corresponding trials not subjected to the weekly additions (ASS and nASS), but the levels estimated for the second were slightly lower.

Table 1 Microbial counts (Log CFU gdw^{-1}) in soil as function of time and trials

Trials[A]	Days				
	0	7	14	21	28
PCA[B]					
AS	3.53 ± 0.37i	7.61 ± 0.66eh	9.58 ± 0.24a	9.16 ± 0.11ab	8.98 ± 0.28ac
ASS	8.21 ± 0.25cf	8.06 ± 0.39dg	8.96 ± 0.30ac	9.03 ± 0.07ac	9.10 ± 0.37ab
ASSwS	8.21 ± 0.25cf	9.11 ± 0.15ab	8.74 ± 0.20ad	9.20 ± 0.31ab	8.85 ± 0.34ad
ASwR	3.53 ± 0.37i	7.29 ± 0.60gh	9.57 ± 0.21a	9.17 ± 0.11ab	9.06 ± 0.22ab
nAS	6.88 ± 0.20h	7.24 ± 0.08gh	7.11 ± 0.15h	7.41 ± 0.35fh	7.65 ± 0.15eh
nASS	8.45 ± 0.34be	7.26 ± 0.10gh	7.52 ± 0.08fh	7.35 ± 0.17gh	7.52 ± 0.18fh
nASSwS	8.45 ± 0.34be	7.30 ± 0.11gh	7.50 ± 0.15fh	7.48 ± 0.08fh	7.57 ± 0.13fh
nASwR	6.88 ± 0.20h	7.04 ± 0.02h	7.09 ± 0.09h	7.25 ± 0.13gh	7.34 ± 0.18gh
VIA[B]					
AS	2.12 ± 0.10p	7.77 ± 0.09ci	8.67 ± 0.04ac	6.91 ± 0.18io	6.61 ± 0.7lo
ASS	8.28 ± 0.02ae	7.19 ± 0.48gm	8.15 ± 0.25af	7.67 ± 0.15dk	8.01 ± 0.44ag
ASSwS	8.28 ± 0.20ae	7.72 ± 0.21ci	8.81 ± 0.08ab	7.18 ± 0.15gn	7.88 ± 0.32bh
ASwR	2.12 ± 0.35p	7.03 ± 0.56ho	8.85 ± 0.30a	7.25 ± 0.29fl	6.71 ± 0.14ko
nAS	6.85 ± 0.10io	7.17 ± 0.08go	6.90 ± 0.04io	6.23 ± 0.26no	6.90 ± 0.36io
nASS	8.35 ± 0.02ad	7.76 ± 0.24ci	7.49 ± 0.23dl	6.22 ± 0.29o	6.68 ± 0.09lo
nASSwS	8.35 ± 0.20ad	7.49 ± 0.11dl	7.35 ± 0.42el	6.97 ± 0.27ho	7.05 ± 0.12ho
nASwR	6.85 ± 0.35io	7.47 ± 0.11dl	7.07 ± 0.32go	6.60 ± 0.36lo	6.27 ± 0.08mo

Data represent the mean of four replicates of two independent experiments. For each media, data followed by the same letter are not significantly different according to Tukey–Kramer's multiple range test at $P < 0.05$

[A] Trials: *ASS* autoclaved soil inoculated with *S. maltophilia* LMG 6606, *AS* autoclaved soil, *nASS* non-autoclaved soil inoculated with *S. maltophilia* LMG 6606, *nAS* non-autoclaved soil, *ASSwS* autoclaved soil weekly inoculated with *S. maltophilia* LMG 6606, *ASwR* autoclaved soil weekly added with Ringer's solution, *nASSwS* non-autoclaved soil weekly inoculated with *S. maltophilia* LMG 6606, *nASwR* non-autoclaved soil weekly added with Ringer's solution

[B] Media: *PCA* plate count agar, *VIA* vancomycin–imipenem–amphotericin B agar

The trials not inoculated with *S. maltophilia* LMG 6606 (AS and nAS) showed levels of count on VIA particularly high. At harvest, the hypocotyls were analysed for TMC and stenotrophomonad concentrations and no statistical significant differences were registered for the microbial concentrations of the radishes cultivated in the eight different conditions (results not shown). TMC were in the range 7.07–7.63 Log CFU g^{-1}, while the counts detected on VIA were at least 1 Log cycle lower for each trial followed.

Monitoring of *S. maltophilia* LMG 6606

All ten presumptive stenotrophomonad isolates collected from VIA medium at the highest dilutions of soil samples were characterized at strain level by RAPD-PCR analysis. DNA from the pure culture of *S. maltophilia* LMG 6606 was used for strain recognition. The comparison of the polymorphic profiles is shown in Fig. 1. *S. maltophilia* LMG 6606 dominated the microbial community found on VIA medium for the trials ASS, nASS, ASSwS and nASSwS until the 28th day of experimentation. Despite the high counts detected on VIA for the trials AS, nAS, ASwR and nASwR, no colony showed a RAPD profile superimposable to that of *S. maltophilia* LMG 6606 excluding, at least at the highest dilutions of soil samples, a cross-contamination among inoculated and un-inoculated trials. One main RAPD pattern was recognised for the stenotrophomonad isolates from the un-inoculated autoclaved soil (AS and ASwR) trials and another main profile for the isolates from the un-inoculated non-autoclaved soil (nAS and nASwR) trials.

The same procedure was applied on the stenotrophomonad isolates collected from VIA medium at the highest dilutions of radishes and it produced the same results (not shown) registered for soil: *S. maltophilia* LMG 6606 dominated the microbial community of radishes for the trials ASS, nASS, ASSwS and nASSwS, whereas no colonies collected from the trials AS, nAS, ASwR and nASwR shared the same *S. maltophilia* LMG 6606 RAPD profile.

The last data could not indicate whether the bacterium added to soil was adherent to the radish surface or in the inner part. Thus, at harvest, the radishes were also specifically investigated for the internal presence of *S. maltophilia* LMG 6606. The enrichment cultures obtained in VIA broth after incubation of the homogenized surface sterilized hypocotyls were streaked onto the corresponding agar medium. No colonies developed for the trials AS, nASS, nAS, ASwR and nASwR (results not shown) indicating that no stenotrophomonads internalized in these conditions. On the contrary, a development was observed for the trials ASS, ASSwS and nASSwS. These colonies, characterized by the same appearance (colour, morphology, edge, surface and elevation), were randomly collected and subjected to the RAPD analysis as reported above; all cultures shared the same profile of *S. maltophilia* LMG 6606 (results not shown) demonstrating its internalization. Interestingly, the only inoculated trial that did not show the internal presence of *S. maltophilia* LMG 6606 in radishes, at harvest, was nASS. To confirm the last data, the enrichment cultures from nASS were also streaked onto Nutrient Agar. Seventy-nine colonies were isolated and analysed by RAPD-PCR (results not shown), but none of them shared the same profile of *S. maltophilia* LMG 6606.

Plant growth

Seed germination and seedling emergence occurred after 3 days from sowing in non-autoclaved soil and 1 day later in autoclaved soil. Inoculation of soil with *S. maltophilia* LMG 6606 had no effect on seed germination.

During the cultivation period, average maximum and minimum temperatures inside the greenhouse ranged between 29.5 and 18.4 °C, respectively. Soil temperature did not greatly differ from air temperature (29.2 and 17.8 °C, respectively).

Plant growth was significantly influenced by soil treatment (Table 2). After 28 days from sowing, some differences were registered for the development of plants. The above-ground part showed only little differences in the number of leaves plant^{-1} among the eight trials, but leaf fresh weight of nASS and nASSwS was higher (14.4 and 17.6 g plant^{-1}, respectively) than that found for the other trials (10.5 g plant^{-1} on average). No differences were found in dry matter percentage as function of treatments, while leaf dimensions were influenced by soil autoclaving as well as by inoculation with *S. maltophilia* LMG 6606. The lowest leaf area was recorded for the plants of trials AS and ASwR (141.5 cm^2 on average) that differed significantly from that of the plants of trials ASS, ASSwS, nAS and nASwR (193.3 cm^2 on average). The non-autoclaved

(See figure on next page.)

Fig. 1 Monitoring of *S. maltophilia* LMG 6606 during the growth cycle of radish performed by RAPD-PCR profile comparison. Trials: *ASS* autoclaved soil inoculated with *S. maltophilia* LMG 6606, *ASSwS* autoclaved soil weekly inoculated with *S. maltophilia* LMG 6606, *nASS* non-autoclaved soil inoculated with *S. maltophilia* LMG 6606, *nASSwS* non-autoclaved soil weekly inoculated with *S. maltophilia* LMG 6606, *AS* autoclaved soil, *nAS* non-autoclaved soil, *ASwR* autoclaved soil weekly added with Ringer's solution, *nASwR* non-autoclaved soil weekly added with Ringer's solution. *Lanes*: *M* marker (GeneRuler 100 bp Plus DNA ladder, M·Medical Srl, Milan, Italy); *1*, *S. maltophilia* LMG 6606; *2–11*, colonies randomly collected from the highest dilutions of soil samples on VIA agar from each trial

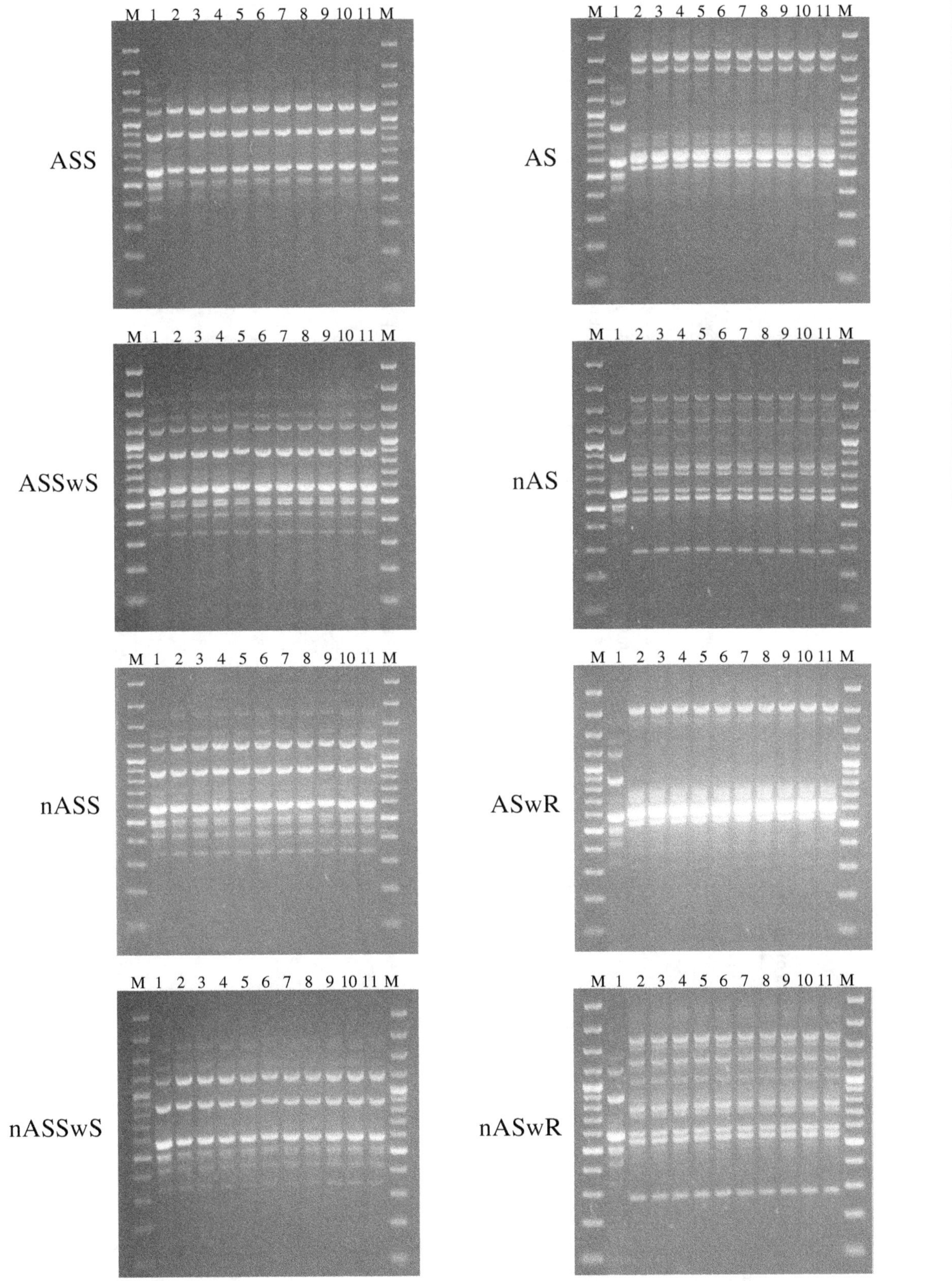
M 1 2 3 4 5 6 7 8 9 10 11 M
ASS
M 1 2 3 4 5 6 7 8 9 10 11 M
AS
M 1 2 3 4 5 6 7 8 9 10 11 M
ASSwS
M 1 2 3 4 5 6 7 8 9 10 11 M
nAS
M 1 2 3 4 5 6 7 8 9 10 11 M
nASS
M 1 2 3 4 5 6 7 8 9 10 11 M
ASwR
M 1 2 3 4 5 6 7 8 9 10 11 M
nASSwS
M 1 2 3 4 5 6 7 8 9 10 11 M
nASwR

Table 2 Agronomic and quality parameters of radish plants at harvest

Parameters	Trials							
	ASS	ASSwS	AS	ASwR	nASS	nASSwS	nAS	nASwR
Leaves								
Number (n. plant^{-1})	5.0 ± 0.3a	4.8 ± 0.2a	4.8 ± 0.2a	4.7 ± 0.2a	4.6 ± 0.2ab	4.6 ± 0.2ab	4.5 ± 0.3ab	4.1 ± 0.2b
Fresh weight (g)	10.5 ± 0.6c	10.7 ± 0.5c	10.3 ± 0.4c	10.8 ± 0.5c	14.4 ± 0.8b	17.6 ± 0.9a	10.3 ± 0.5c	10.3 ± 0.6c
Dry matter (%)	6.5 ± 0.3a	7.1 ± 0.4a	7.1 ± 0.5a	7.1 ± 0.3a	7.3 ± 0.4a	7.2 ± 0.4a	6.8 ± 0.3a	6.7 ± 0.4a
Leaf area (cm^2)	186.4 ± 8.7b	196.4 ± 9.3b	138.3 ± 6.9c	144.8 ± 7.2c	297.0 ± 16.9a	311.5 ± 14.6a	198.9 ± 9.7b	191.6 ± 9.8b
Roots								
Length (cm)	31.5 ± 1.6ab	33.6 ± 1.8a	28.1 ± 1.4ac	29.1 ± 1.5ac	26.3 ± 1.5bc	23.9 ± 1.1c	27.6 ± 1.4ac	27.8 ± 1.6ac
Fresh weight (g)	1.06 ± 0.05a	1.04 ± 0.04a	0.91 ± 0.05a	0.8 ± 0.04a	0.49 ± 0.02b	0.50 ± 0.03b	0.53 ± 0.04b	0.40 ± 0.03b
Dry matter (%)	6.2 ± 0.3b	6.2 ± 0.4b	7.2 ± 0.5b	7.3 ± 0.4b	7.7 ± 0.4b	8.1 ± 0.5b	12.8 ± 0.6a	11.7 ± 0.5a
Radish								
Fresh weight (g)	12.2 ± 0.7b	10.0 ± 0.6b	10.1 ± 0.5b	12.0 ± 0.6b	25.0 ± 1.4a	23.9 ± 1.2a	15.3 ± 0.8b	14.8 ± 0.9b
Dry matter (%)	3.9 ± 0.2a	3.9 ± 0.3a	3.9 ± 0.4a	3.8 ± 0.2a	4.1 ± 0.3a	4.3 ± 0.5a	4.3 ± 0.2a	4.5 ± 0.4a
Diameter (mm)	24.9 ± 1.2b	23.4 ± 1.2b	23.1 ± 1.3b	24.6 ± 1.2b	34.0 ± 1.7a	31.8 ± 1.6a	25.5 ± 1.3b	26.6 ± 1.5b
Firmness (N)	39.1 ± 2.1a	38.6 ± .0a	36.2 ± 1.8a	37.1 ± 1.9a	30.3 ± 1.7b	29.9 ± 1.5b	31.2 ± 1.6b	30.2 ± 1.5b
*L**	32.9 ± 1.7c	34.0 ± 1.6c	35.2 ± 1.8ac	34.2 ± 1.7bc	36.8 ± 1.9ac	38.7 ± 1.8ab	39.6 ± 2.0a	39.0 ± 2.0a
*a**	45.8 ± 2.3ab	46.0 ± 2.4ab	42.5 ± 2.1b	43.5 ± 2.2b	47.9 ± 2.6a	50.3 ± 2.5a	49.4 ± 2.9a	49.4 ± 2.5a
*b**	20.6 ± 1.0ab	20.7 ± 1.2ab	19.2 ± 1.0b	18.8 ± 0.9b	21.0 ± 1.1ab	23.2 ± 1.2ab	25.1 ± 1.3a	25.0 ± 1.4a
SSC (Brix)	2.2 ± 0.1b	2.2 ± 0.1b	2.8 ± 0.1a	2.6 ± 0.2ab	2.3 ± 0.1b	2.3 ± 0.2b	2.3 ± 0.1b	2.3 ± 0.1b
N–NO_3^- (mg kg^{-1} FW)	644.6 ± 33.2a	580.3 ± 29.8a	583.7 ± 29.2a	563.7 ± 28.2a	522.7 ± 25.1a	543.4 ± 27.8a	567.0 ± 28.4a	599.2 ± 30.2a
Ascorbic acid (mg kg^{-1} FW)	205.0 ± 10.3ab	197.0 ± 9.9ab	220.0 ± 11.0a	206.0 ± 10.3ab	182.0 ± 9.1ab	166.0 ± 8.3ab	132.0 ± 6.6b	192.0 ± 9.6ab
Plant WUE (g DM kg^{-1} H_2O)	2.03 ± 0.11b	2.03 ± 0.10b	2.0 ± 0.11b	2.14 ± 0.12b	3.51 ± 0.18a	3.88 ± 0.16a	2.39 ± 0.12b	2.33 ± 0.13b

Data within a row followed by the same letter are not significantly different according to Tukey–Kramer's multiple range test at $P < 0.05$

Percentages were subjected to angular transformation prior to perform statistical analysis soils inoculated with *S. maltophilia* LMG 6606, subjected or not to the weekly additions, determined the greatest total leaf area which was about 300 cm^2.

The roots of radish plants had a lower fresh weight in the trials with non-autoclaved soil than autoclaved soil trials, but dry matter content was higher only for nAS and nASwR. Root elongation was influenced by inoculation with *S. maltophilia* LMG 6606. However, this inoculums determined a decrease of root length in non-autoclaved soil trials, while an opposite behaviour was observed for the autoclaved soil trials.

After 28 day from sowing, radish hypocotyls reached commercial maturity. The highest average fresh weight was recorded in nASS and nASSwS (25.0 and 23.9 g, respectively), and was almost twice those reached by the hypocotyls of the other trials. A positive correlation was found between radish fresh weight and size. The biggest hypocotyls developed in non-autoclaved inoculated soils which overcame the diameter of 30 mm. No significant difference was found for dry matter percentage among the eight trials.

Colour modifications of radish were evaluated in terms of L^*, a^* and b^* values. Radishes grown in non-autoclaved soil had a darker colour especially for ASS and ASSwS trials ($L^* = 33.5$ on average), while trials nAS and nASwR determined the greatest values of L^* (39.6 and 39.0, respectively). The inoculation with *S. maltophilia* LMG 6606 did not influence neither the redness (a^*) nor the yellowness (b^*) of radish colour. The trials AS and ASwR were characterised by the lowest redness and differed significantly from non autoclaved trials. The only significant differences for yellowness were noticed between radishes from trials AS and ASwR (19.0 on average) and trials nAS and nASwR (25.0 on average).

Soil autoclaving determined an increase of radish firmness both in inoculated and un-inoculated trials (37.7 N on average), while a lower value was registered for those grown in non-autoclaved soils (30.4 N on average).

Soluble solid contents were almost comparable among the different trials, with the exception of AS that was significantly higher. Also nitrate and ascorbic acid content showed similar values among trials with no significant differences between inoculated and un-inoculated trials.

Stenotrophomonas maltophilia LMG 6606 increased significantly WUE for producing dry matter (DM) when inoculated in non-autoclaved soil trials: nASSwS and nASS ranged from 3.51 to 3.88 g DM kg^{-1} H_2O, respectively, against 2.15 g DM kg^{-1} H_2O on average for the other trials.

Multivariate data analysis

The results of the principal components analysis showed four principal components (PCs) with eigenvalues higher than 1.00 (Table 3), accounting for 56.5, 15.4, 7.7 and 6.34 % of the total variance, respectively. This indicated that the initial 18 variables could be expressed as a linear combination of four PCs explaining 85.9 % of the total variance. PC1 was mainly related to leaf fresh weight, leaf dry matter, leaf area, root length, root fresh weight, root dry matter, radish fresh weight, radish dry matter, diameter, firmness, ascorbic acid, plant WUE and color components (L^*, a^*, b^*); PC2 was related to leaf number and b^*; PC3 was related to SSC, and finally PC4 to nitrate content (Table 3). The projection of the original variables on the plane of the two first PCs could clearly illustrate such relationship as shown in the plot of loadings (Fig. 2a). The discrimination of the various trials can be visualized in the plot of scores (Fig. 2b) where three clusters could be clearly distinguished. The trials with autoclaved soil were close each other and located mainly in the negative part of *F1* axis; they were clearly separated from the trials with autoclaved soil that were located in the positive part of *F1* axis. Among these trials nASSwS and nASwR were clearly separated from nAS and nASS. The trial nASS also showed the greatest score for Factor 1.

Combining the information from the plot of loadings and scores, it can be inferred that nAS and especially nASS influenced positively fresh and dry matter of leaves and radishes, colour and firmness of hypocotyls and plant water use efficiency.

Table 3 Correlation of variables to the factors of the PCA analysis based on factor loadings

Variable	Factor 1	Factor 2	Factor 3	Factor 4
Leaf number	−0.236	*0.716*	0.242	−0.210
Leaf fresh weight	*0.809*	0.427	0.037	−0.065
Leaf dry matter	*0.781*	0.422	−0.070	0.082
Leaf area	*0.870*	0.310	0.072	0.182
Root length	*−0.773*	0.174	0.210	0.457
Root fresh weight	*−0.772*	0.452	0.206	0.017
Root dry matter	*−0.836*	−0.115	0.259	−0.137
Radish fresh weight	*0.940*	0.173	−0.075	−0.006
Radish dry matter	*0.956*	0.107	−0.063	−0.050
Radish diameter	*−0.841*	0.430	0.087	0.080
Firmness	*0.858*	0.242	−0.002	0.073
L^*	*0.794*	−0.518	0.088	−0.128
a^*	*0.771*	−0.398	0.328	0.027
b^*	0.506	*−0.726*	0.397	0.004
SSC	−0.282	−0.072	*−0.842*	−0.338
Nitrate	−0.020	−0.229	−0.392	*0.817*
Ascorbic acid	*−0.721*	−0.411	−0.079	−0.110
Plant WUE	*0.939*	0.283	−0.066	0.011

Values in italic within the same factor indicate the variable with the largest correlation

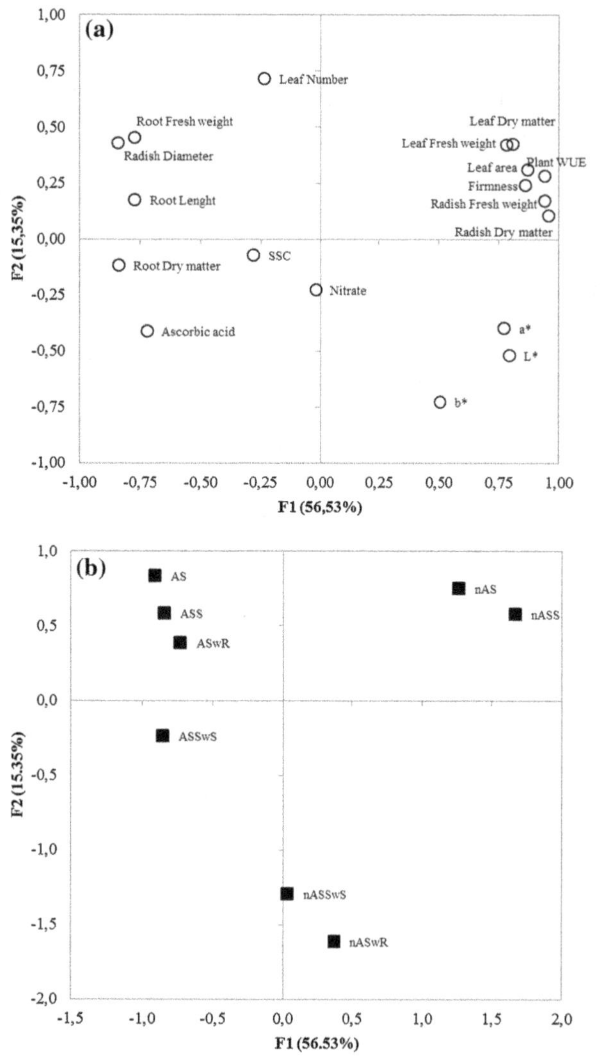

Fig. 2 Plot of **a** loadings (agronomic and quality parameters of radish plants at harvest) and **b** scores (trials) formed by the first two principal components from the PCA analysis. *ASS* autoclaved soil inoculated with *S. maltophilia* LMG 6606, *AS* autoclaved soil, *nASS* non-autoclaved soil inoculated with *S. maltophilia* LMG 6606, *nAS* non-autoclaved soil, *ASSwS* autoclaved soil weekly inoculated with *S. maltophilia* LMG 6606, *ASwR* autoclaved soil weekly added with Ringer's solution, *nASSwS* non-autoclaved soil weekly inoculated with *S. maltophilia* LMG 6606, *nASwR* non-autoclaved soil weekly added with Ringer's solution

Discussion

The positive effects of *S. maltophilia* on the plant growth has been reported for several species [36–39]. However, *S. maltophilia* is also reported to be a human pathogen. For this reason, its association with vegetables, especially when eaten raw, is detrimental for the final quality of these products. The aim of this work was to modify the microbial composition of soil with the addition of *S. maltophilia* LMG 6606 in order to simulate a massive contamination and to assess its effects on the growth of *R. sativus*.

The experimentation was carried out with soil amended with inorganic fertilizer; organic amendments were not included to avoid the transfer of consistent concentrations of microorganisms [33] commonly present in compost or manure. The trials were carried out in a greenhouse during the spring season, with an average air temperature of 24.0 and 23.5 ℃ reached in the soil. In this conditions, *S. maltophilia* LMG 6606 found a temperature range compatible with its growth.

Stenotrophomonas maltophilia may increase the plant growth not only by direct production of growth promoting substances but also indirectly by interacting with the native root microflora [40]. Kwok et al. [41] reported that the biocontrol ability of *S. maltophilia* might be enhanced in presence/combination with other soil microorganisms. For these reasons, our experimental design included autoclaved and non-autoclaved soils, in order to exclude or include, respectively, the interaction between *S. maltophilia* LMG 6606 and the resident soil microbiota. Plate counts specific for the added bacterium were performed on VIA medium that, to our knowledge, is still considered to be highly selective for the isolation of *S. maltophilia* [35]. The non-autoclaved un-inoculated soils showed an initial load of 6.92 Log CFU gdw^{-1} on VIA. Thus, for the trials carried out with non-autoclaved un-inoculated soil (trials nAS and nASwR), these populations were referred to as stenotrophomonads. This bacterial group was detected also in the soil soon after autoclaving, although at very low levels (2.13 CFU gdw^{-1}). Thus, the common protocol applied to sterilize soil (two autoclaving cycles at 121 ℃ for 70 min at 24 h interval) was not enough, in this study, to destroy completely the bacterial component and, as a consequence, the microbial interaction could not be totally excluded in the autoclaved soil trials, at T_0.

The trials artificially contaminated with more than 10^8 CFU gdw^{-1} of *S. maltophilia* LMG 6606 showed a decrease of these inocula during the first week of radish growth, but their number increased thereafter. De Boer et al. [42] registered an opposite trend during the first week of observation of sands inoculated with stenotrophomonads, but the stimulating effect was imputed to the development of fungi. However, the different behaviour of this bacterial population might be due to the different conditions tested in the two works; the soil amended with inorganic fertilizer used in our work and the sand used by De Boer et al. [42] represented two distinct ecosystems, characterized by differences in pH, water activity and oxidation–reduction potential. Furthermore, the differences can also depend on the strain-specific characteristics. A continuous decrease of *S.*

maltophilia inoculated at 10^7 CFU gdw^{-1} in different soil types with different management regimes at 28 °C was reported by Messiha et al. [43] who studied the antagonistic effects of *S. maltophilia* against *Ralstonia solanacearum*. Those authors stated that the reasons for the differences in survival of *S. maltophilia* in the various soils were not clear, but supposed a direct effect of nitrate and ammonium contents and pH. The increase in concentration of *S. maltophilia* LMG 6606, observed in this study from the second week, might be due to the production of methionine by *R. sativus* roots. *S. maltophilia* requires methionine [44] and colonizes mainly the rhizosphere of cruciferous plants which produce high concentrations of sulphur-containing compounds [45].

The persistence of *S. maltophilia* LMG 6606 was monitored by plate count and polymorphic profile by RAPD analysis, after isolation. The isolates collected from VIA agar at the highest dilutions of soils were analyzed and compared with the pure strain. The direct comparison of RAPD patterns allowed to confirm that *S. maltophilia* LMG 6606 dominated the stenotrophomonad group of soil in all the inoculated trials during the entire growth cycle of radishes. This approach was previously successfully applied on the recognition and monitoring of other food/spoilage bacteria tested in similar experiments [31, 33].

In our study, the strain *S. maltophilia* LMG 6606 determined a significant increase of the values of several parameters (leaf area, fresh and dry matter yield, radish size and WUE) indicating the growth-promotion of radish, even though the plants grown in presence of *S. maltophilia* LMG 6606 behaved differently in autoclaved or non-autoclaved soil. Plants exhibited a significantly greater growth in the latter condition. This might be explained with the fact that soil treatments, especially steaming or autoclaving, can result in soil toxicity [46].

Dry matter yield of control trials was similar to those obtained by other authors [47]. The increase in yield was significant only for non-autoclaved soil and ranged from 47 to 67 % for nASS and nASSwS, respectively. The growth increase may vary greatly depending on crop and PGP rhizobacteria strains used [48]. Antoun et al. [47] reported an increase of 15 % in the dry matter yield of radish using the strain *Bradyrhizobium japonicum* Tal 629.

The results of the principal components analysis showed four principal components (PCs). Combining the information from the plot of loadings and scores, it can be concluded that soil autoclaving negatively influenced growth and quality parameters, thus their use need further investigation. Furthermore, principal components analysis was able to differentiate the trials with respect to agronomic and quality trait of radish. Hence, soil autoclaving should be avoided, while the inoculation of non-autoclaved soil with *S. maltophilia* LMG 6606 only once before plant establishment was positively related with some growth parameter.

Radishes were microbiologically investigated at harvest. TMC and stenotrophomonad concentrations were not statistically different for all trials. Thus, inoculated trials produced radishes characterized by the same levels of microbial contamination of those obtained from control trials. The levels of TMC were in the range 10^6–10^7 CFU g^{-1} and were superimposable to those reported for radishes grown in hydroponic systems [31].

The isolates collected from the highest dilutions of hypocotyls confirmed that *S. maltophilia* LMG 6606 dominated the microbial community of radishes for all inoculated trials as assessed by RAPD analysis. However, to retrieve the exact location of this bacterium on the radishes, inside or outside the hypocotyls, they were surface sterilized and subjected to an enrichment in VIA broth. This procedure, developed specifically for radishes in a previous work [31], indicated that *S. maltophilia* LMG 6606 internalized in all inoculated trials except that carried out in non-autoclaved soil not subjected to the weekly addition (nASS).

Conclusions

The addition of *S. maltophilia* LMG 6606 to the non-autoclaved soil amended with inorganic fertilizer enhanced the productivity of radishes grown in greenhouse. *S. maltophilia* LMG 6606 colonized the external surface of the hypocotyls, but it did not internalize. Thus, a massive contamination of soil with *S. maltophilia* determines a growth promotion of *R. sativus*, but due to the hygienic implication of this bacterium, a sanitizing bath of hypocotyls before eating is mandatory. Works are being prepared to better evaluate the biocontrol activity of *S. maltophilia* in presence of other soil microorganisms in specific combinations.

Authors' contributions

AMi provided financial support of the experiment and conducted greenhouse experiment and agronomic and quality determinations, general conception and coordination of the experiments, interpretation of results and manuscript writing. AMa carried out the microbiological and molecular genetic analyses. GM provided general conception and interpretation of results and manuscript writing. LS provided financial support of the experiment, offered expert advice during the conduction of the assays, interpretation of results and manuscript writing. All authors read and approved the final manuscript.

Competing interests

The authors declare that they have no competing interests.

References

1. Rengel Z, Marschner P. Nutrient availability and management in the rhizosphere: exploiting genotypic differences. New Phytol. 2005;168:305–12.
2. Smit B, Pilifosova O, Burton I, Challenger B, Huq S, Klein RJT, Yohe G. Adaptation to climate change in the context of sustainable development and equity. In: McCarthy JJ, Canziani OF, Leary NA, Dokken DJ, White KS, editors. Climate change 2001: impacts, adaptation and vulnerability. Contribution of working group ii to the third assessment report of the intergovernmental panel on climate change. Cambridge: Cambridge University Press; 2001. p. 877–912.
3. Richardson AE, Barea JM, McNeil AM, Prigent-Combaret C. Acquisition of phosphorus and nitrogen in the rhizosphere and plant growth promotion by microorganisms. Plant Soil. 2009;321:305–39.
4. Biswas JC, Ladha JK, Dazzo FB. Rhizobia inoculation improves nutrient uptake and growth of lowland rice. Soil Sci Soc Am J. 2000;64:1644–50.
5. Dobbelaere S, Vanderleyden J, Okon Y. Plant growth-promoting effects of diazotrophs in the rhizosphere. Crit Rev Plant Sci. 2003;22:107–49.
6. Glick BR. The enhancement of plant growth by free-living bacteria. Can J Microbiol. 1995;41:109–17.
7. Glick BR, Liu C, Ghosh S, Dumbroff EB. Early development of canola seedling in the presence of the plant growth-promoting rhizobacterium *Pseudomonas putida* GR12-2. Soil Biol Biochem. 1997;29:1233–9.
8. Richardson AE. Prospects for using soil microorganisms to improve the acquisition of phosphorus by plants. Aust J Plant Physiol. 2001;28:897–906.
9. van Loon LC, Glick GR. Increased plant fitness by rhizobacteria. In: Sandermann H, editor. Molecular Ecotoxicology of Plants, vol. 170. Berlin: Springer-Verlag; 2004. p. 177–205.
10. Kloepper JW, Lifshitz R, Zablotowicz RM. Free-living bacterial inocula for enhancing crop productivity. Trends Biotechnol. 1989;7:39–43.
11. Kloepper JW. Plant growth-promoting rhizobacteria as biological control agents. In: Metting Jr FB, editor. Soil microbial ecology-applications in agricultural and environmental management. New York: Marcel Dekker; 1993. p. 255–74.
12. Lambert B, Joos H. Fundamental aspects of rhizobacterial plant growth promotion research. Trends Biotechnol. 1989;7:215–9.
13. Denton M, Kerr KG. Microbiological and clinical aspects of infection associated with *Stenotrophomonas maltophilia*. Clin Microbiol Rev. 1998;11:57–80.
14. Banerjee M, Yesmin L. Sulfur-oxidizing plant growth promoting rhizobacteria for enhanced canola performance. 2002. US Patent 07491535.
15. Berg G, Marten P, Ballin G. *Stenotrophomonas maltophilia* in the rhizosphere of oilseed rape—occurrence, characterization and interaction with phytopathogenic fungi. Microbiol Res. 1996;151:19–27.
16. Park M, Kim C, Yang J, Lee H, Shin W, Kim S, Sa T. Isolation and characterization of diazotrophic growth promoting bacteria from rhizosphere of agricultural crops of Korea. Microbiol Res. 2005;160:127–33.
17. Lockhart SR, Abramson MA, Beekmann SE, Gallagher G, Riedel S, Diekema DJ, Quinn JP, Doern GV. Antimicrobial resistance among Gram-negative bacilli causing infections in intensive care unit patients in the United States between 1993 and 2004. J Clin Microbiol. 2007;45:3352–9.
18. Brooke JS. *Stenotrophomonas maltophilia*: an emerging global opportunistic pathogen. Clin Microb Rev. 2012;25:2–41.
19. Salerno A, Pierandrei F, Rea E, Colla G, Rouphael Y, Saccardo F. Floating system cultivation of radish (*Raphanus sativus* L.): production and quality. Acta Hort. 2005;69:87–92.
20. Giusti MM, Wrostald RE. Characterization of red radish anthocyanins. J Food Sci. 1996;61:322–6.
21. Jing P, Zhao SJ, Ruan SY, Xie ZH, Dong Y, Yu LL. Anthocyanin and glucosinolate occurrences in the roots of Chinese red radish (*Raphanus sativus* L.), and their stability to heat and pH. Food Chem. 2012;133:1569–76.
22. Lu ZL, Liu LW, Li XY, Gong YQ, Hou XL, Zhu XW, Yang JL, Wang LZ. Analysis and evaluation of nutritional quality in Chinese radish (*Raphanus sativus* L.). Agr Sci China. 2008;7:823–30.
23. Curtis IS. The noble radish: past, present and future. Trends Plant Sci. 2003;8:305–7.
24. Watt JM, Breyer-Brandwijk MG. The medicinal and poisonous plants of southern and eastern Africa. Edinburgh and London: E. & S. Livingstone; 1962.
25. Sgherri C, Cosi E, Navarri-Izzo F. Phenols and antioxidative status of *Raphanus sativus* grown in excess copper. Physiolia Plantarum. 2003;118:21–8.
26. Takaya Y, Kondo Y, Furukawa T, Niwa M. Antioxidant constituent of radish sprout (kiware-daikon), *Raphanus sativus* L. J Agric Food Chem. 2003;51:8061–6.
27. Basu TK, Temple NJ, Garg ML. Antioxidants in human health and disease. Wallingford: CABI Publishing; 1999.
28. Ghiselli A, D'Amicis A, Giocosa A. The antioxidant potential of Mediterranean diet. Eur J Cancer Prev. 1997;6:15–9.
29. Meyer AS, Yi OS, Pearson DA, Waterhouse AI, Frankel EN. Inhibition of human low-density lipoprotein oxidation in relation to composition of phenolic antioxidants in grapes. J Agric Food Chem. 1997;45:1638–43.
30. Fahey JW, Zalcmann AT, Talalay T. The chemical diversity and distribution of glucosinolates and isothiocyanates among plants. Phytochemistry. 2001;56:5–51.
31. Settanni L, Miceli A, Francesca N, Cruciata M, Moschetti G. Microbiological investigation of *Raphanus sativus* L. grown hydroponically in nutrient solutions contaminated with spoilage and pathogenic bacteria. Int J Food Microbiol. 2013;160:344–52.
32. Marques APGC, Pires C, Moreira H, Rangel AOSS, Castro PML. Assessment of the plant growth promotion abilities of six bacterial isolates using *Zea mays* as indicator plant. Soil Biol Biochem. 2010;42:1229–35.
33. Settanni L, Miceli A, Francesca N, Moschetti G. Investigation of the hygienic safety of aromatic plants cultivated in soil contaminated with *Listeria monocytogenes*. Food Control. 2012;26:213–9.
34. Brandl MT, Haxo AF, Bates AH, Mandrell RE. Comparison of survival of *Campylobacter jejuni* in the phyllosphere with that in the rhizosphere of spinach and radish plants. Appl Environ Microbiol. 2004;70:1182–9.
35. Kerr KG, Denton M, Todd N, Corps CM, Kumari P, Hawkey PM. A new selective differential medium for isolation of *Stenotrophomonas maltophilia*. Eur J Clin Microbiol Infect Dis. 1996;15:607–10.
36. Idris A, Labus Chagne N, Korsten L. Efficacy of rhizobacteria for growth promotion in sorghum under greenhouse conditions and selected modes of action studies. J Agric Sci. 2009;147:17–30.
37. Sturz AV, Matheson BG, Arsenault W, Kimpniski J, Christie BR. Weeds as a source of plant growth promoting rhizobacteria in agricultural soils. Can J Microbiol. 2001;47:1013–24.
38. de Freitas JR, Banerjee MR, Germida JJ. Phosphate-solubilizing rhizobacteria enhance the growth and yield but not phosphorus uptake of canola (*Brassica napus* L.). Biol Fertil Soils. 1997;24:358–64.
39. Fages J, Arsac JF. Sunflower inoculation with *Azospirillum* and other plant growth promoting rhizobacteria. Plant Soil. 1991;137:87–90.
40. Kloepper JW, Schroth MN. Relationship of in vitro antibiosis of plant growth-promoting rhizobacteria to plant growth and the displacement of root microflora. Phytopathology. 1981;71:1020–4.
41. Kwok OCH, Fahy PC, Hoitink HAJ, Kuter GA. Interactions between bacteria and *Trichoderma hamatum* in suppression of *Rhizoctonia* damping-off in bark compost media. Phytopathology. 1987;77:1206–12.
42. De Boer W, Gunnewiek PJAK, Kowalchuk GA, Van Veen JA. Growth of chitinolytic dune soil beta-subclass *Proteobacteria* in response to invading fungal hyphae. Appl Environ Microbiol. 2001;67:3358–62.
43. Messiha NAS, van Diepeningen AD, Farag NS, Abdallah SA, Janse JD, van Bruggen AHC. *Stenotrophomonas maltophilia*: a new potential biocontrol agent of *Ralstonia solanacearum*, causal agent of potato brown rot. Eur J Plant Pathol. 2007;118:211–25.
44. Ikemoto S, Suzuki K, Kaneko T, Komagata K. Characterization of strains of *Pseudomonas maltophilia* which do not require methionine. Int J Syst Bacteriol. 1980;30:437–47.
45. Debette J, Blondeau R. Presence de *Pseudomonas maltophilia* dans la rhizosphere de quelque plantes cultivee. Can J Microbiol. 1980;26:460–3.
46. Warcup JH. Chemical and biological aspects of soil sterilization. Soils Fertil. 1957;20:1–5.
47. Antoun H, Beauchamp CJ, Goussard N, Chabot R, Lalande R. Potential of *Rhizobium* and *Bradyrhizobium* species as plant growth promoting rhizobacteria on non-legumes: effect on radishes (*Raphanus sativus* L.). Plant Soil. 1998;204:57–67.
48. Lucy M, Reed E, Glick BR. Applications of free living plant growth-promoting rhizobacteria. Antonie Van Leeuwenhoek. 2004;86:1–25.

PERMISSIONS

All chapters in this book were first published in CBTA, by Springer; hereby published with permission under the Creative Commons Attribution License or equivalent. Every chapter published in this book has been scrutinized by our experts. Their significance has been extensively debated. The topics covered herein carry significant findings which will fuel the growth of the discipline. They may even be implemented as practical applications or may be referred to as a beginning point for another development.

The contributors of this book come from diverse backgrounds, making this book a truly international effort. This book will bring forth new frontiers with its revolutionizing research information and detailed analysis of the nascent developments around the world.

We would like to thank all the contributing authors for lending their expertise to make the book truly unique. They have played a crucial role in the development of this book. Without their invaluable contributions this book wouldn't have been possible. They have made vital efforts to compile up to date information on the varied aspects of this subject to make this book a valuable addition to the collection of many professionals and students.

This book was conceptualized with the vision of imparting up-to-date information and advanced data in this field. To ensure the same, a matchless editorial board was set up. Every individual on the board went through rigorous rounds of assessment to prove their worth. After which they invested a large part of their time researching and compiling the most relevant data for our readers.

The editorial board has been involved in producing this book since its inception. They have spent rigorous hours researching and exploring the diverse topics which have resulted in the successful publishing of this book. They have passed on their knowledge of decades through this book. To expedite this challenging task, the publisher supported the team at every step. A small team of assistant editors was also appointed to further simplify the editing procedure and attain best results for the readers.

Apart from the editorial board, the designing team has also invested a significant amount of their time in understanding the subject and creating the most relevant covers. They scrutinized every image to scout for the most suitable representation of the subject and create an appropriate cover for the book.

The publishing team has been an ardent support to the editorial, designing and production team. Their endless efforts to recruit the best for this project, has resulted in the accomplishment of this book. They are a veteran in the field of academics and their pool of knowledge is as vast as their experience in printing. Their expertise and guidance has proved useful at every step. Their uncompromising quality standards have made this book an exceptional effort. Their encouragement from time to time has been an inspiration for everyone.

The publisher and the editorial board hope that this book will prove to be a valuable piece of knowledge for researchers, students, practitioners and scholars across the globe.

LIST OF CONTRIBUTORS

Michael James Van Oosten, Olimpia Pepe, Stefania De Pascale, Silvia Silletti and Albino Maggio
Department of Agricultural Science, University of Naples Federico II, Via Università 100, Portici, Italy

Barbara J. Cade-Menun
Agriculture & Agri-Food Canada, Swift Current Research and Development Centre, Gate 4 Airport Road, Swift Current, SK S9H 3X2, Canada

Zakaria Hazzoumi, Youssef Moustakime and Khalid Amrani Joutei
Laboratory of Bioactive Molecules, Faculty of Science and Technology, Sidi Mohamed Ben Abdellah University, Road of Imouzzer, B. P. 2202, Fez, Morocco

Johnson Masaka, Marshall Mutambu and Rangarirayi Mhindu
Department of Land and Water Resources Management, Midlands State University, Private Bag 9055, Gweru, Zimbabwe

Kudakwashe Muringaniza
Department of Geography and Environmental Studies, Midlands State University, Private Bag 9055, Gweru, Zimbabwe

Peteh Mehdi Nkebiwe and Torsten Müller
Fertilisation and Soil Matter Dynamics (340 i), University of Hohenheim, 70593 Stuttgart, Germany

Günter Neumann
Nutritional Crop Physiology (340 h), University of Hohenheim, 70593 Stuttgart, Germany

Inocencio Junior Oliveira, José Roberto Antoniol Fontes and Aleksander Westphal Muniz
Embrapa Amazônia Ocidental, AM 010 Road, km 29, Manaus 69010970, Brazil

Bruno Fernando Faria Pereira
Universidade Federal do Amazonas, Av. Gen. R. Otávio Jordão Ramos 3000 - Coroado I, FCA, DEAS, Manaus 69077-000, Brazil

F. Lelario, M. G. Bonomo, G. Salzano, S. Milan, L. Milella and S. A. Bufo
Department of Sciences, University of Basilicata, 85100 Potenza, Italy

L. Scrano and S. De Franchi
Department of European Cultures (DICEM), University of Basilicata, 85100 Potenza, Italy

Marios Drosos, Davide Savy, and Alessandro Piccolo
Centro Interdipartimentale di Ricerca sulla Risonanza Magnetica Nucleare per l'Ambiente, l'Agroalimentare e di Nuovi Materiali (CERMANU), Università di Napoli Federico II, Via Università 100, 80055 Portici, NA, Italy

Michael Spiteller
Institut für Umweltforschung (INFU) der Fakultät Chemie und Chemische Biologie, Technische Universität Dortmund, Otto Hahn Straße 6, 44221 Dortmund, Germany

Ouahid Elasri
Biochemistry and Biotechnology Laboratory, Mohamed First University, Oujda, Morocco.

Mahmoud Salem
Agricultural Engineering Research Institute, Agricultural Research Center, Nadi El Said St., Giza, Egypt

Mohamed Ramdani
Ecology, Water and Environment Laboratory, Mohamed First University, Oujda, Morocco

Oulya Zaraali
Applied Biology and Pathology Laboratory, Abdelmalek Essaâdi University, Tetouan, Morocco

Latrach Lahbib
Faculty of Sciences Semlalia, Cadi Ayyad University, Marrakech, Morocco

Assunta Nuzzo and Alessandro Piccolo
Centro Interdipartimentale di Ricerca sulla Risonanza, Magnetica Nucleare per l'Ambiente, l'Agroalimentare ed i Nuovi Materiali (CERMANU), Università di Napoli Federico II, Via Università, 100, 80055 Portici, NA, Italy

Antonio De Martino and Vincenzo Di Meo
Dipartimento di Agraria, Università di Napoli Federico II, Via Università, 100, 80055 Portici, NA, Italy

Matteo Giaccone, Chiara Cirillo, Pasquale Scognamiglio, Maurizio Teobaldelli, Alessandro Mataffo, Antonio Pannico and Boris Basile
Department of Agricultural Sciences, University of Naples Federico II, Portici, Naples, Italy

Adriano Stinca
Department of Environmental, Biological and Pharmaceutical Sciences and Technologies, University of Campania Luigi Vanvitelli, Caserta, Italy

Barbara Immirzi, Gabriella Santagata and Mario Malinconico
Institute of Polymers, Composites, and Biomaterials, CNR, Pozzuoli, Naples, Italy

Udomsak Kongmuang and Chaovayut Phornpimolthape
Faculty of Public Health, Bangkokthonburi University, Khet Taweewatana, Bangkok 10170, Thailand

Ikuharu Morioka
Graduate School of Health and Nursing Science, Wakayama Medical University, Mikazura 580, Wakayama 641-0011, Japan

Anjana J. Atapattu
Agromony Division, Coconut Research Institute, Bandirippuwa Estate, Lunuwila 61150, Sri Lanka
Key Laboratory of Tropical Forest Ecology, Xiahuangbanna Tropical Botanical Garden, Chinese Academy of Sciences, Mengla 666303, Yunnan, China
University of Chinese Academy of Science, Beijing 100094, China

B. D. Rohitha Prasantha
Department of Food Science & Technology, Faculty of Agriculture, University of Peradeniya, Peradeniya 20400, Sri Lanka

K. S. P. Amaratunga
Department of Agriculture Engineering, Faculty of Agriculture, University of Peradeniya, Peradeniya 20400, Sri Lanka

Buddhi Marambe
Department of Crop Science, Faculty of Agriculture, University of Peradeniya, Peradeniya 20400, Sri Lanka

Fatima Zahra Sadiki and Mostafa El Idrissi
Laboratory of Molecular Chemistry and Natural Substances, Department of Chemistry, Faculty of Sciences of Meknes, Moulay Ismail University, Bp 11201, Zitoune, Meknes, Morocco

Mohammed Sbiti
Laboratory of Microbiology, Military Hospital Moulay Ismail, Meknes, Morocco

Adrae Lemrhari
Laboratory of Plant Biotechnology and Molecular Biology, Faculty of Science, Moulay Ismail University, BP 11201, Zitoune Meknes, Morocco

Adriana Trifan and Oana Cioanca
Department of Pharmacognosy, Faculty of Pharmacy, University of Medicine and Pharmacy, "Gr. T. Popa", Iasi, Str. 16University Str., Iasi 700115, Romania

Paula Alexandra Postu and Lucian Hritcu
Department of Biology, Faculty of Biology, Alexandru Ioan Cuza University of Iasi, Bv Carol I, No 11, 700505 Iasi, Romania

Jesper T. N. Knijnenburg, Florentine M. Hilty and Michael B. Zimmermann
Department of Health Sciences and Technology, Human Nutrition Laboratory, Institute of Food, Nutrition and Health, ETH Zurich, Schmelzbergstrasse 7, 8092 Zurich, Switzerland
International College, Khon Kaen University, Khon Kaen 40002, Thailand

Riaan Buitendag and Janro Oelofse
Biopher (Pty) Ltd., 5 Ednau Smallholdings, Abrahamskraal Road, Mooiwater, Bloemfontein 9330, South Africa

Ismail Cakmak
Faculty of Engineering and Natural Sciences, Sabanci University, 34956 Istanbul, Turkey

Anne F. Grobler
DST/ NWU Preclinical Drug Development Platform, Faculty of Health Sciences, North-West University, Potchefstroom 2531, South Africa

Clovis B. Tanyi, Christopher Ngosong and Nelson N. Ntonifor
Department of Agronomic and Applied Molecular Sciences, Faculty of Agriculture and Veterinary Medicine, University of Buea, Buea, South West Region, Cameroon

Denisse Bender, Maximilian Schmatz, Senad Novalin, Foteini Chrysanthopoulou, Regine Schoenlechner and Stefano D'Amico
Department of Food Science and Technology, BOKU- University of Natural Resources and Life Sciences, Vienna 1180, Austria

Renata Nemeth, Sandor Tömösközi and Kitti Török
Department of Applied Biotechnology and Food Science, Budapest University of Technology and Economics, Budapest 1111, Hungary

Maria Amargianitaki and Apostolos Spyros
NMR Laboratory, Department of Chemistry, University of Crete, Voutes Campus, 710 03 Heraklion, Crete, Greece

Fabio Stagnari, Angelica Galieni and Michele Pisante
Faculty of Biosciences and Technologies for Agriculture Food and Environment, University of Teramo, Coste Sant'Agostino, Teramo, TE, Italy

Albino Maggio
Department of Agricultural Science, University of Naples Federico II, Via Università
100, Portici, Italy

V. Raut and I. Shaikh
Department of Biotechnology, Badrinarayan Barwale College, Jalna, MS, India
Dr. Babasaheb Ambedkar Marathwada University, Aurangabad, MS, India

B. Naphade
Department of Microbiology, Badrinarayan Barwale College, Jalna, MS, India
Dr. Babasaheb Ambedkar Marathwada University, Aurangabad, MS, India

K. Prashar
Department of Chemistry, Badrinarayan Barwale College, Jalna, MS, India
Dr. Babasaheb Ambedkar Marathwada University, Aurangabad, MS, India

N. Adhapure
Department of Biotechnology and Microbiology, Vivekanand Arts Sardar Dalipsingh Commerce and Science College, Aurangabad, MS, India
Dr. Babasaheb Ambedkar Marathwada University, Aurangabad, MS, India

Anteneh Argaw
College of Agriculture and Environmental Sciences, School of Natural Resources Management and Environmental Sciences, Haramaya University, Dire Dawa, Ethiopia

Angaw Tsigie
Ethiopian Institute of Agricultural Research, Holleta Agricultural Research Center, Holleta, Ethiopia

Alessandro Miceli, Alessandra Martorana, Giancarlo Moschetti and Luca Settanni
Department of Agricultural and Forest Science, University of Palermo, Viale delle Scienze 4, 90128 Palermo, Italy

Index

www.ingramcontent.com/pod-product-compliance
Lightning Source LLC
LaVergne TN
LVHW080918210926
848485LV00004B/675